Practical Drilling & Production Design

Dr. Douglas Bynum Jr.

PRACTICAL DRILLING & PRODUCTION DESIGN

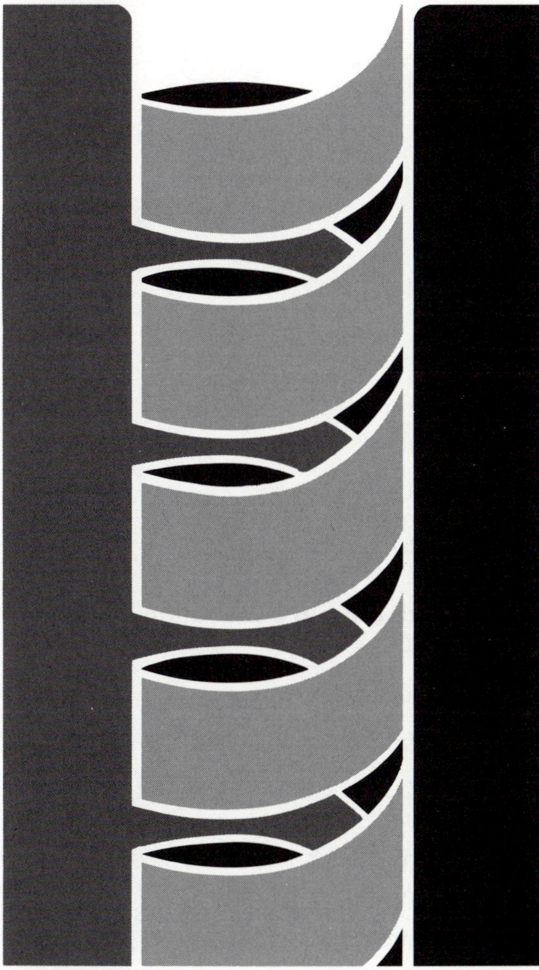

PennWell Books
PennWell Publishing Co.
Tulsa, Oklahoma 74101

ACKNOWLEDGMENTS

Special thanks are given to Kathryne E. Pile, the editor, without whose assistance and cooperation this book would not have been possible.

PennWell Publishing Company
1421 South Sheridan Road/P.O. Box 1260
Tulsa, Oklahoma 74101

Library of Congress Cataloging in Publication Data

Bynum, Douglas.
Practical drilling and production design.

1. Oil well drilling, Submarine. 2. Petroleum engineering. I. Title.
TN871.3.B96 622'.3382 81-15774
ISBN 0-87814-180-4 AACR2

Printed in the United States of America

1 2 3 4 5 86 85 84 83 82

CONTENTS

CHAPTER 6 **STATIC STRUCTURAL DESIGN** **421**

CHAPTER 7 **DYNAMIC STRUCTURAL ANALYSIS** **475**

CHAPTER

1 DRILLING DESIGN AND ANALYSIS

1.1 CAPACITY OF DAMAGED MASTS

Masts often get damaged during operation or during rig-down and transport for a variety of reasons. Damage is the amount of warp or bow in a section of leg in a panel between the horizontal girts. The bow is the maximum distance from the heel of the angle to a straight line between adjacent girts.

Bow during drilling or workover operations usually occurs when large hook loads are applied before jacking the substructure to correct for differential settlement of the foundation. Bow is sometimes caused during transport between drilling sites when a section of the mast is supported on a leg rather than at the girts.

The maximum static hook load applied to the mast must be reduced if any of the legs in any section have significant bow. Instead of going through a lot of complicated calculations, design charts are presented here which the toolpusher can use to determine allowable hook load as a function of bow in most of the existing drilling and workover masts. These charts are applicable for any mast that uses structural angles for the legs. This covers the smallest workover rigs up to about 1.3 million pound GNC drilling masts using 8 × 8 × 1⅛ angles for the legs. This is the largest standard angle available from U.S. steel mills. Masts with a larger GNC use beam legs.

It is important to know the allowable safe hook load on damaged masts since you may possibly continue drilling operations at a reduced hook load when the bow is not too great. The mast is usually more convenient to repair during rig-down and rig-up at the next site rather than by stopping drilling operations.

Code Requirements

Most drilling masts are designed to the API Codes given in Specs 4A and 4D for older masts. API Spec 4E has been used since March 1974. This later spec gives the four mast design conditions of: a) erection loads, b) maximum static hook load with no winds and no setback, c) storm winds from any direction with full setback and no hook load, and d) survival winds from any direction with no setback and no hook load.

The API 4E code requires the use of the AISC code for determining structural allowable stresses as a function of the type of materials used. The AISC code uses the material designations and mill tolerances specified by the ASTM codes.

The API code does not explicitly cover mast bow and its effect on hook capacity. However, the AISC code covers this implicitly, and the structural allowables are given in terms of unity checks. The first unity check is the stress due to the applied axial load divided by the stress allowable for the given slenderness ratio. The second unity check is the flexural stress caused by the bow divided by the flexural allowable stress. The sum of these two unity checks must be less than unity to satisfy the AISC code.

U.S. Mill Practice

The ASTM A6 code allows a tolerance of plus or minus 2.5% on the specified cross sectional area of a structural angle. It is, therefore, possible to have a range or variation in area of 5% and still be within acceptable mill practice.

This code allows a camber or bow of ⅛ in. times the total length of the section divided by five. Structural shapes are normally ordered in lengths of 20, 40, or 60 ft. A 40-ft length of structural angle can, therefore, have a bow of up to one in. as it leaves the steel mill and still be within accepted tolerance. The tolerance applies to either or both legs of the angle. This demonstrates why mast legs must be straightened and held on jigs and fixtures while the girts and braces are welded on.

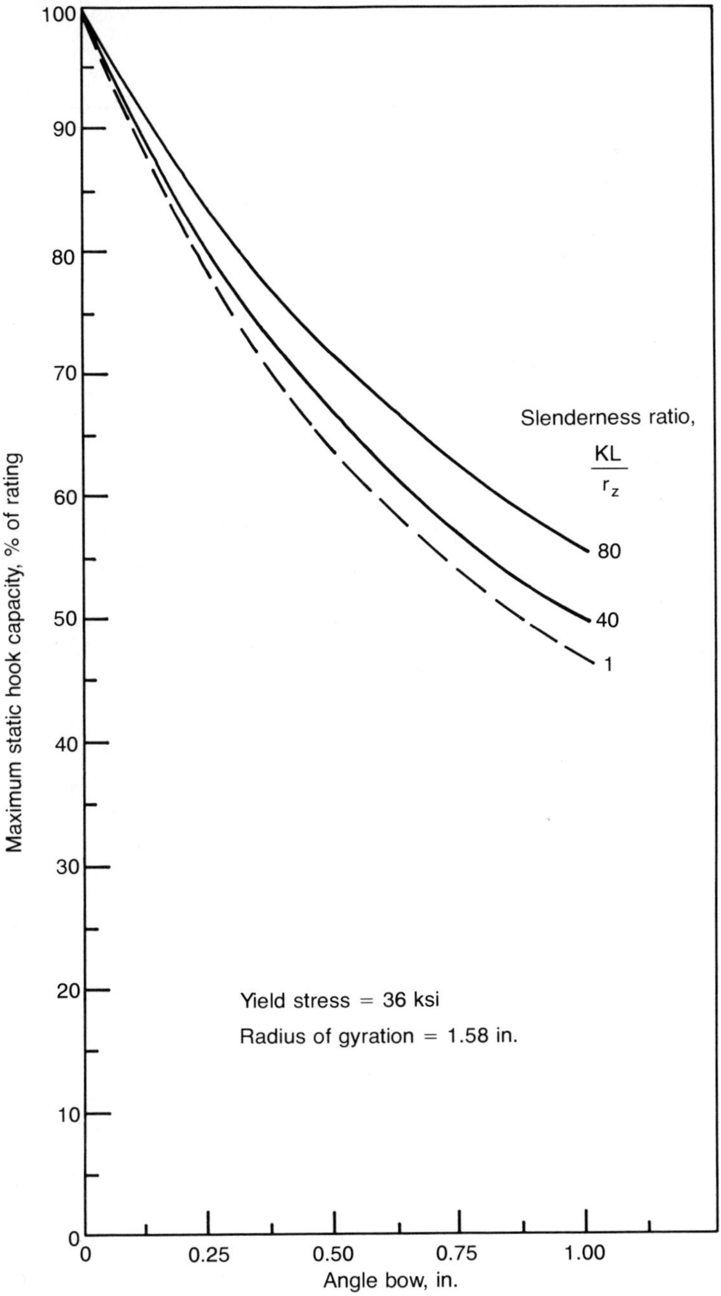

FIG. 1.1 Capacity of 8 × 8 angle

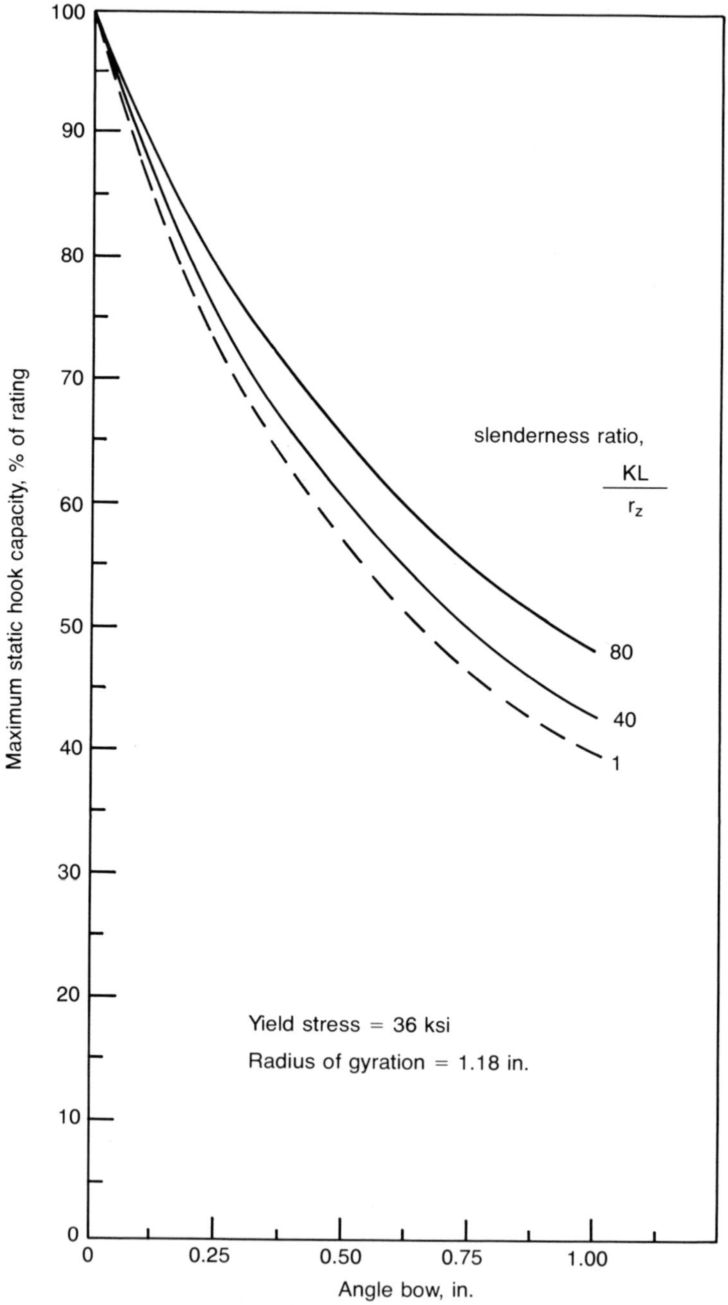

FIG. 1.2 Capacity of 6 × 6 angle

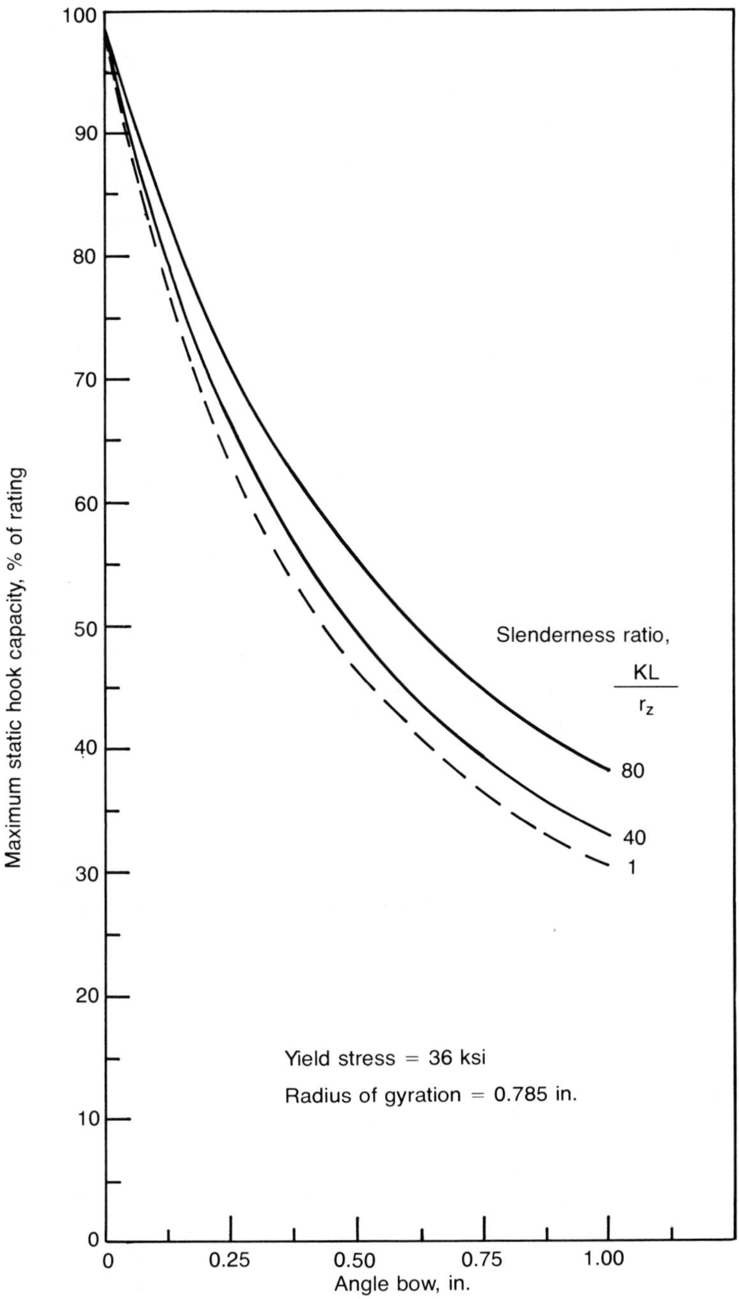

FIG. 1.3 Capacity of 4 × 4 angle

In addition, the A6 code allows the toe of an angle to be out of square, measured perpendicular to the toe, by plus or minus 1.5 degrees or 3 times the leg width divided by 128. For an 8-in. angle this gives plus or minus 3⁄16 of an in. For this reason, the design charts are for angle bow measured at the heel and not at the toe of the structural angle.

Procedures for Analysis

The properties of the structural angle, from strength of materials theory and the unity check equations per AISC, are given in table 1. Combining these relations gives the working equation for the ratio of allowable axial force with bow divided by the allowable force without bow. This ratio is the percent of maximum rated static hook load or the mast capacity with bow.

The capacity is a function of the minor radius of gyration, the bow, the leg length of the structural angle, and the ratio of axial allowable stress divided by flexural allowable stress. The allowable axial stress is determined by the slenderness ratio where the minor

TABLE 1
Equations and nomenclature

$I_z = r_z^2 A$ $\qquad c = 0.353\,B$ $\qquad UC_o = \dfrac{P_o}{AF_a}$

$UC_f = \dfrac{P_f}{AF_a} + \dfrac{P_f ec}{I_z F_b}$ $\qquad \dfrac{P_f}{P_o} = \dfrac{r_z^2}{r_z^2 + 0.353\,Be\,\dfrac{F_a}{F_b}}$

Slenderness ratio, $\dfrac{KL}{r_z}$:	1	40	80
Corresponding to A			
Stress ratio, $\dfrac{F_a}{F_b}$ at $F_y = 36$:	1	0.888	0.711
$F_y = 50$:	1	0.861	0.633

A	Cross-sectional area
B	Width of angle
c	Maximum distance from neutral axis
e	Eccentricity or bow of angle
F	Allowable stress
I	Area moment of inertia
K	Effective length factor
L	Length of panel
P	Axial force
r	Radius of gyration
UC	Unity check

Subscripts

a	Axial
b	Bending
f	Final
o	Initial
y	Yield
z	Minor axis

radius of gyration is used, since it gives a lower allowable than using the major radius.

The range in leg thickness varies between ¼ and ¾ in. for standard 4 × 4-in. angles and between ½ and 1⅛ in. for standard 8 × 8-in. angles. To simplify the design charts, the average minor radius of gyration for all leg thicknesses with a given leg width was used in the calculations since the variation is only about plus or minus one percent. This approximation changes the allowable axial stress to less than one-half of 1% and, therefore, insignificantly changes the computed capacity.

Discussion of Results

The design charts are given for 4, 6, and 8-in. angles, which cover most drilling and workover masts fabricated from angle sections. In each case, the hook capacity as a function of bow is given for three values of the slenderness ratio. The lower line is a limiting condition and is essentially for a block of steel. The slenderness ratio for masts rarely exceeds the upper bound of 80, and most masts have a slenderness ratio near the middle curve, which is for a slenderness ratio of 40.

The K factor varies with the type of mast design. From hand calculations as well as from computer analyses, we find that a K factor of 0.8 is appropriate for the back legs of most masts. For the front legs or ramp side, the K factors can be taken as 1.3 for masts having fan braces to the girts and 1.8 for front legs without fan braces.

At first review, it seems strange that the capacity increases with increasing slenderness ratio for a given bow. This is because the

TABLE 2
Stress ratio variation with slenderness ratio and yield strength

Slenderness ratio, KL/r_z	Yield strength, F_y = 36 ksi: Stress allowable, ksi: Axial, F_a	Yield strength, F_y = 36 ksi: Stress allowable, ksi: Bending, F_b	Yield strength, F_y = 36 ksi: Stress ratio, $\frac{F_a}{F_b}$	Yield strength, F_y = 50 ksi: Stress allowable, ksi: Axial, F_a	Yield strength, F_y = 50 ksi: Stress allowable, ksi: Bending, F_b	Yield strength, F_y = 50 ksi: Stress ratio, $\frac{F_a}{F_b}$
1	21.56	21.60	0.998	29.94	30.00	0.998
25	20.28	21.60	0.938	27.75	30.00	0.925
40	19.19	21.60	0.888	25.83	30.00	0.861
50	18.35	21.60	0.849	24.35	30.00	0.811
75	15.90	21.60	0.736	19.99	30.00	0.666
80	15.36	21.60	0.711	19.01	30.00	0.633

TABLE 3
Mast capacity in percent of maximum static rated hook load (for yield strength, F_y = 36 ksi)

Slenderness ratio, $\frac{KL}{r_z}$	Stress ratio, $\frac{F_a}{F_b}$	Angle	Average radius of gyration, r_z	Bow or eccentricity, e, in. 0.125	0.250	0.500	0.750	1.000
1	1	8 × 8	1.58	87.5	77.9	63.8	54.0	46.8
1	1	6 × 6	1.18	84.0	72.4	56.8	46.7	39.7
1	1	4 × 4	0.785	77.7	63.6	46.6	36.8	30.4
40	0.888	8 × 8	1.58	88.8	79.9	66.5	57.0	49.8
40	0.888	6 × 6	1.18	85.5	74.7	59.7	49.7	42.5
40	0.888	4 × 4	0.785	79.7	66.3	49.5	39.6	32.9
80	0.711	8 × 8	1.58	90.8	83.2	71.2	62.3	55.3
80	0.711	6 × 6	1.18	88.1	78.7	64.9	55.2	48.0
80	0.711	4 × 4	0.785	83.0	71.0	55.1	45.0	38.0

stress allowable ratio decreases with increasing slenderness ratio since the allowable flexural stress is not a function of the slenderness ratio.

The calculations used in preparing the charts had a yield strength of 36 Ksi or an ASTM A36 type of steel. The allowable stress ratio decreases more for higher strength steel than for A36 as the slenderness ratio increases. But this changes the capacity only slightly, and a grade 50 steel (with a L8 × 8, KL/r = 80, e = 1.0) would have, for example, only about 5% more capacity than an A36 steel. For all practical purposes, the given charts can be used for any grade of steel normally used in mast designs.

To illustrate the considerable reduction in mast capacity with an 8 × 8-in. angle leg for a slenderness ratio of 40, table 3 shows that with a ½-in. bow, the mast would only have about 67% of the rated hook load. With a one-in. bow the mast capacity drops to 50% of rated hook load. The reduction increases with decreasing angle size. With the same slenderness ratio of 40, the mast capacity of a 4-in. angle is 50% for ½-in. bow and 33% for a one-in. bow.

These charts show that the smaller sizes of structural angles are much more sensitive to bow than the larger angles. This explains why workover rig masts experience more failures than drilling rig masts.

1.2 MAST CAPACITY WITH WINDS AND SETBACK

How to determine the allowable mast hook load, acting simultaneously with various wind loads and percentages of full setback,

TABLE 4
Nomenclature

Symbol	Variable	Units
c	Percent setback constant	numeric
c_h	Wind height coefficient	numeric
c_s	Wind shape coefficient	numeric
d	Diagonal distance across racking board	ft
F_D	Setback dead load	kips
F_L	Setback live (wind) load	kips
H	Hook capacity	percent
P	Pressure	psf
V_k	Velocity	knots
V_m	Velocity	mph

has been a problem for a long time. This is because it is a complicated problem and the governing codes do not specify how it is to be done. However, the manner in which mast design specifications are given imply how it can be done. Equations for the reduced hook load with wind and setback are given below along with an operations type of chart to help determine the allowable hook load for all conditions (figure 1.4).

Code Specifications

The API Specification 4E gives four design conditions that the mast must satisfy. These are erection loads, maximum static rated hook load, storm winds from any direction with full setback and no hook load, and survival winds from any direction with no setback and no hook load. The dead weight of the mast and its equipment must be included in each case. Specified wind pressures are 22.5 psf for storm winds and 40 psf for survival winds. Using a shape coefficient of 1.25 and an average height coefficient of 1.1, these pressures correspond to velocities of 80 mph (70 Kt) storm winds and 107 mph (93 Kt) survival winds.

The setback load consists of two components, the dead weight of the setback and the wind load of the setback. The maximum setback lean is specified as 2.5 degrees. The force on the racking platform at 84 ft elevation is the dead weight of the setback times the 2.5-degree moment arm divided by the distance of 84 ft from the drill floor to the racking baord. The live load on the mast at the racking board is the sail area of the setback times the distance to the center of pressure divided by 84 ft. The maximum setback wind loads occur with the wind direction perpendicular to the mast diagonal, so the diagonal distance of the racking board is used to compute the sail area of the setback.

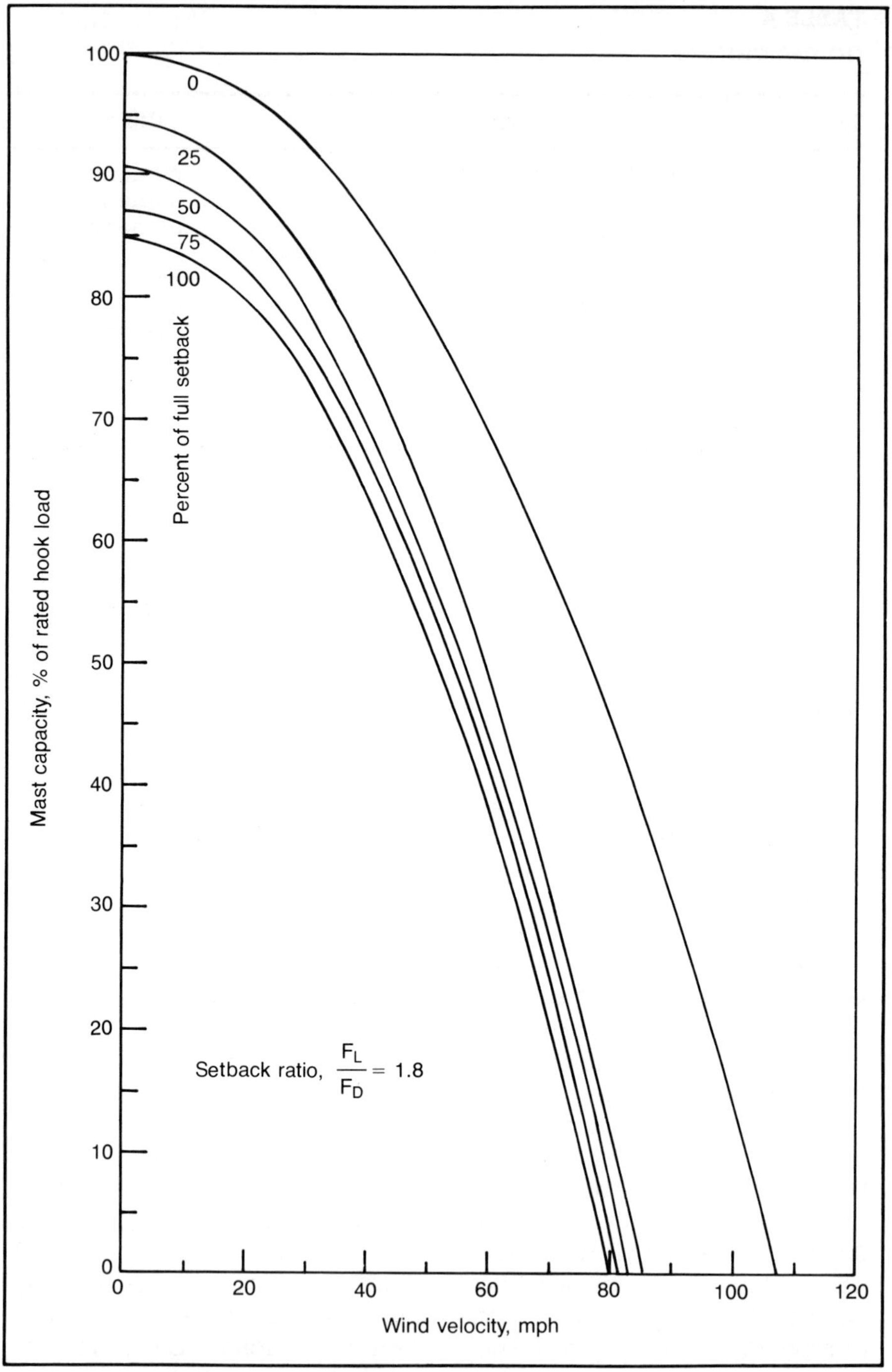

FIG. 1.4 Mast capacity with setback and winds

Procedures

The perfect mast would have each of its members stressed the same in each of the four design conditions. This is not practical, and the wind stresses are usually less than the hook load or erection stresses. In the following tables, we assumed that the member stresses are the same in all four loading conditions so we could obtain the reduced hook load with setback and winds. This gives conservative results when the wind load stresses in the mast are less than the rated hook load stresses; that is, the mast capacity with setback and wind will be slightly greater than calculated here.

The essence of these procedures is that stresses in the mast due to hook load or wind load are equivalent. The code prescribes a 40 psf survival wind and a 22.5-psf storm wind. The only difference in these two conditions, besides wind pressure, is that the mast has a full setback with storm winds. This implies that the total setback forces on the mast is the difference in pressures of 17.5 psf.

The procedure for determining the two setback component forces on the mast at the racking board are given above. The setback ratio is the setback wind force on the mast divided by the setback dead load force on the mast. Most masts have a value of about 1.8 for the setback ratio. This means that the dead weight component is 1 divided by 2.8 or 35.7% of the total setback forces.

The differential 17.5 psf times 35.7% gives an equivalent loading of 6.25 psf on the mast with setback dead load and of 11.3 psf equivalent on the mast with setback wind load. The 6.24 psf equivalent is 15.6 percent of the 40 psf survival wind. Therefore, the mast hook capacity must be reduced by 15.6% for full setback and no wind; that is, the mast capacity with full setback and no wind is 84.4% of the rated maximum static hook load with no setback and no wind load.

These concepts were put into the form of equations. The last equation in table 5 gives the relationship for determining the hook capacity with any wind velocity and with any percent of full setback. The influence of various percentages of setback weight is determined by the value of the setback weight substituted into the constant c in the last equation.

Discussion of Results

Using the developed equation for the above concepts, table 6 was constructed. This gives the percent mast capacity for hook with 0%, 25%, 50%, 75%, and 100% of the setback with any wind velocity. For example, in determining the allowable hook load on a 600-kip maximum rated static hook with 50% of the full setback and 30-mph winds, the table reads 79%. The allowable hook load is then 0.79 times 600 or 474 kips.

TABLE 5
Equations for reduced mast capacity

1 mph = 0.869 knots, 1 knot = 1.151 mph

$P = 0.00338\ V_k^2 c_h c_s = 0.00254\ V_k^2 c_h c_s$

For masts, $c_s = 1.25$, $c_h = 1.1$ average

$$P = 0.00460\ V_k^2 = 0.00350\ V_m^2$$

$$F_D = \frac{90\ W \sin 2.5}{2\ (84)} = 0.0233\ W \text{ kips}$$

$$F_L = \frac{22.5(45)(90)d}{84} = 1.084\ d \text{ kips}$$

$$\frac{F_L}{F_D} = \frac{1.084\ d}{0.0233\ W} = \frac{46.5\ d}{W} = 1.8 \text{ average}$$

$$F_D = \frac{F_T}{2.8} = 0.357\ F_T \text{ average} = 0.357(40 - 22.5) = 6.24$$

$$F_L = (40 - 22.5) - 6.24 = 11.3$$

$$P_L = a\ V_m^2 = 11.3 = a\ 80^2,\ P_L = 0.00177\ V_m^2$$

$$c = \frac{F_T}{1 + \dfrac{F_L}{F_D}} = \frac{17.5}{1 + \dfrac{46.5\ d}{W}},\ G = \frac{17.5 - F_D}{80^2}$$

$$H = 100\left[1 - \frac{0.00350}{40}V_m^2 - \frac{G}{40}V_m^2 - \frac{c}{40}\right]$$

The table was constructed using a setback ratio of 1.8, but this value applies to a large range of masts. To be sure, the setback ratio for any particular mast should be determined from the given equation. If the calculated ratio is more than 1.8, the actual mast capacity will be slightly greater than that given by table 6.

If the calculated safety factors for the API wind load conditions are much greater than the safety factor for the maximum static rated hook load condition, the mast capacity will be greater than the values given in table 6. For example, assume that the mast analysis gives safety factors on code of 1.0 for rated hook and 2.0 for survival winds and that the mast has a 1.8 setback ratio. With no setback and 53 mph winds, the hook capacity would be 88% of rated hook instead of the 76% given by the table. With full setback and 40 mph winds, the hook capacity with these same safety factors would be about 74% of rated hook instead of the 64% given by the table.

Table 6 could be used universally. This is with the understanding that the percent capacity of the rated hook for most mast designs will be slightly greater than those given by the table which are conservative.

TABLE 6
Operating data

Wind velocity, mph		Mast capacity, % of rated hook				
	Setback:	0	25%	50%	75%	100%
	c:	0	2.14	3.81	5.16	6.24
0		100	94.6	90.4	87.1	84.4
10		99.1	93.2	89.0	85.7	83.0
20		96.5	89.3	85.1	81.9	79.1
30		92.1	82.7	78.5	75.2	72.5
40		86.0	73.5	69.3	66.0	63.3
50		78.1	61.6	57.4	54.1	51.4
60		68.5	47.1	42.9	39.6	36.9
70		57.1	30.0	25.8	22.5	19.8
80		44.0	10.2	6.0	2.7	0
90		29.1				
100		12.5				
107		0				

1.3 HIGH-STRENGTH STEEL LIMITS FOR DRILLING STRUCTURES

What grade of steel should be used in drilling structures such as masts, derricks, ponys, and substructures? Here are some guidelines.

Grade means the specified minimum yield strength of a particular steel designation as given in the ASTM codes. It is of little consequence that the mill test reports give values that usually exceed the associated minimum strength, because the strength will vary with each lot. Normally, it is not practical to track a steel lot through a large fabricated structure where several lots would be involved.

The appropriate grade of steel to use depends upon cost, strength to weight ratio, usage or application, and maintenance or repair. To reduce lost revenues in downtime with drilling rigs, it is desirable to do as much repair as possible in the field. This immediately eliminates the applicability of the very high grades of structural steel because of welding problems.

The grade of steel to use depends upon structural purpose as dictated by the codes as well as practical considerations. The load conditions are dictated by API 4E, and the structural allowable stresses are governed by the AISC codes for masts and substructures. During rig-up and down, the mast sections are picked up and set on the flatbeds. With the dynamics of off-loading ponys and substructures, these structures are always much more robust than masts. Because of the transportation and handling conditions with substructures, it is not sensible to use high strength steels even if there were no cost premium.

Steel Cost

The cost of a given tonnage of steel of any grade varies weekly. It also varies tremendously with quantity per single purchase. With this understanding, average costs were used in figure 1.5 to show the variation of steel cost with yield strength. These values will be used in the cost indexes to determine economic limits of high strength steel.

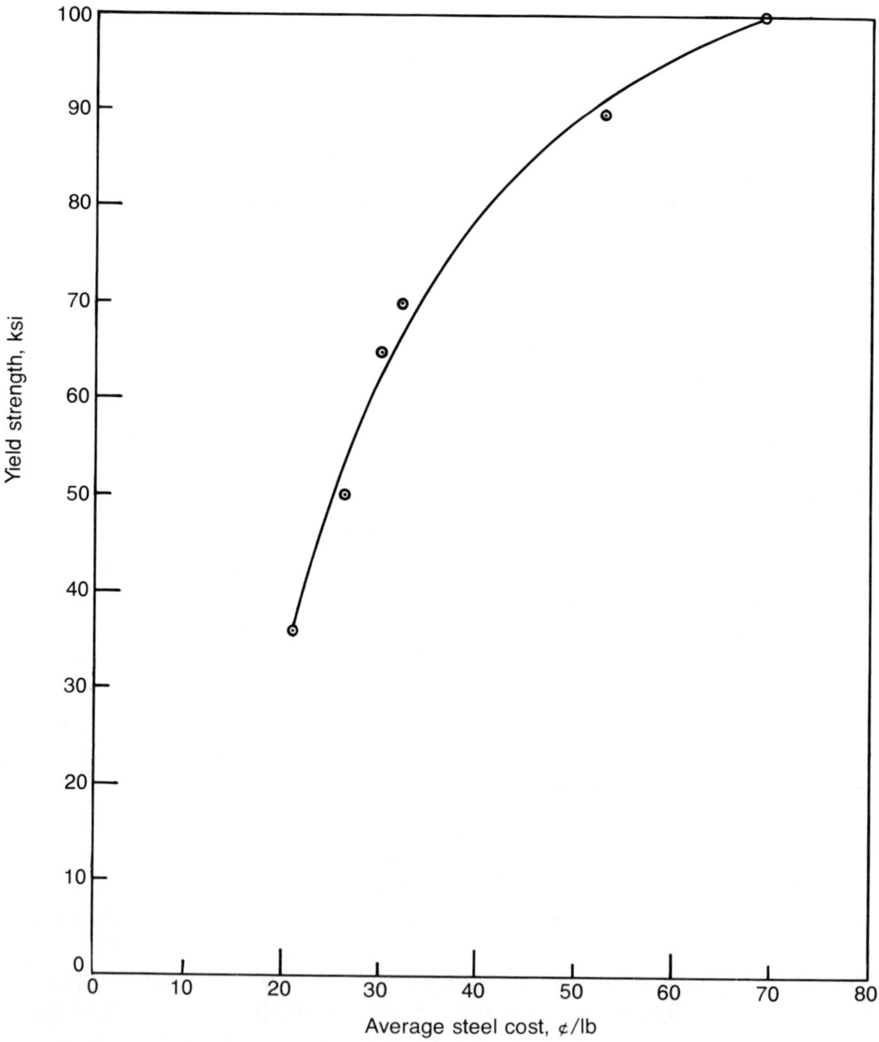

FIG. 1.5 Steel cost variation with strength

Strength to Weight Ratio

The strength to weight ratio normalized on ASTM A36 vs. the slenderness ratio is given in figure 1.6 for pure compression loads. The slenderness ratio is the effective length factor, K, times the member length, L, divided by the minor radius of gyration, R, of the cross section. Most of the members in a mast and in a substructure are subjected primarily to compression loading. However, lower members of a mast are also subjected to flexural stresses due to the secondary bending movements resulting from deflections.

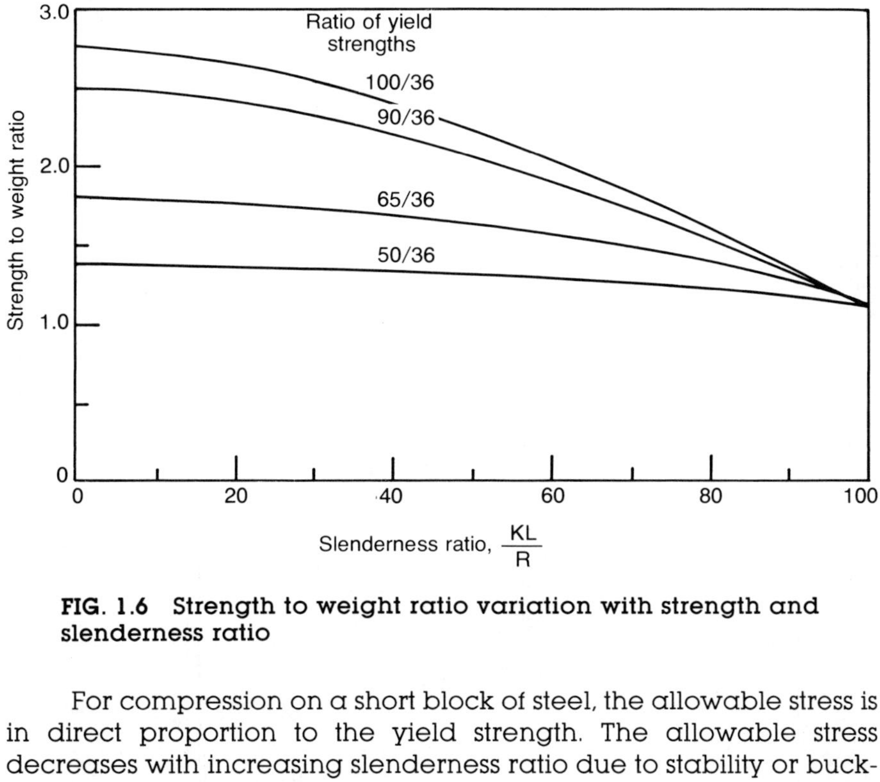

FIG. 1.6 Strength to weight ratio variation with strength and slenderness ratio

For compression on a short block of steel, the allowable stress is in direct proportion to the yield strength. The allowable stress decreases with increasing slenderness ratio due to stability or buckling considerations. As figure 1.6 shows, the curves converge to the Euler buckling criteria which is an AISC allowable of 10.3 Ksi at a KL/R of 120 for all grades of structural steel. The safety factor on yield built into the code varies from 1.67 to 1.92, and the slenderness ratio varies from unity to 120.

An intermediate type of failure can occur with structural shapes which have a large width to flange thickness ratio, and the shear center is significantly off the centroid of the cross sectional area. In this case, torsional buckling failure can occur with only axial loads applied. Design for torsional buckling will be discussed in more detail later.

Cost Index

The cost effectiveness index is determined by dividing the cost ratio from figure 1.5 by the strength to weight ratio from figure 1.6.

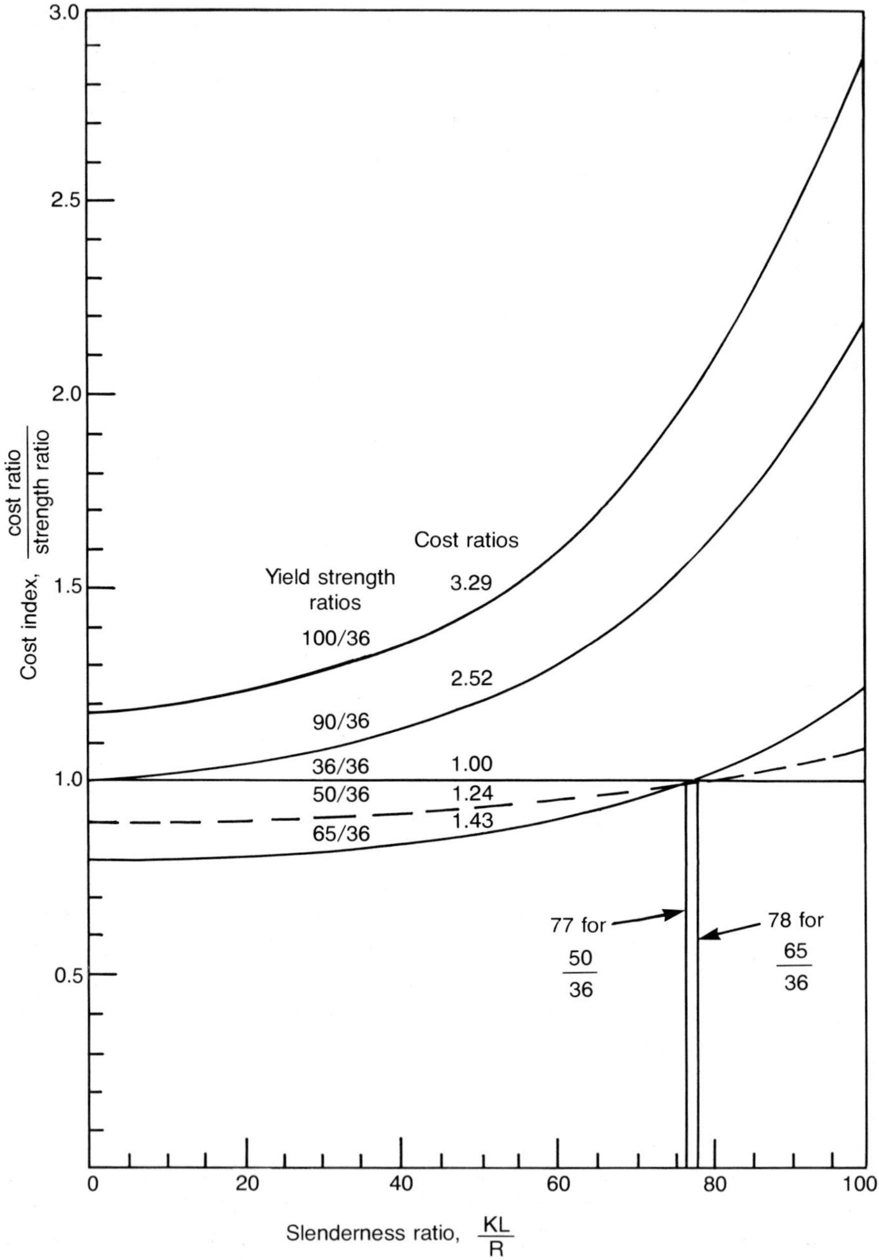

FIG. 1.7 Cost index variation with strength and slenderness ratio

The results are plotted in figure 1.7. An index of unity is for ASTM A36, and materials with an index less than unity indicate a beneficial cost effectiveness.

The calculations demonstrated in figure 1.7 show that only grades 50 and 65 are cost effective. The limits in terms of slenderness ratio were about the same for these two grades, 77 for grade 50 and 78 for grade 65. For typical mast panels having a slenderness ratio of around 50, the cost index is about 6% better for grade 50 and about 15% better for grade 65 relative to grade 36.

Conclusions

From figure 1.6 note that for a slenderness ratio of 50, grade 65 is about 64% better than grade 36 and grade 50 is 32% better. This is especially important in self-erecting masts where the legs constitute 40% to 50% of the stem weight and where it is important to minimize dead weight. In addition to welding considerations, these calculations show why a well designed mast will have steel legs of grade 50 and secondary members of grade 36.

With other uses of steel, such as pinned compression struts for wing deck extensions on offshore platforms, repairs are seldom required. Therefore, grade 65 steel would be cost and functionally effective to use.

1.4 LUG CONNECTIONS AND PIN KIT DESIGN

Pin kits are used primarily for two reasons: temporary connections or for ease in field construction. Temporary connections include the lug plates holding the drilling mast sections together, the pin kits to connect the spreader beams to a mast substructure, and the rope connections for lifting sling beams.

Sometimes additional production equipment, added to small offshore production platforms, encroaches on the space set aside for the workover rig. Components of modular type workover rigs can then be placed on add-on struted wing decks. Pin connected wing decks cost less and are much easier to install offshore than all welded construction.

The type of pin plate connection covered here is limited to loads acting longitudinally in tension or compression along the beam with single pins in the plates at the end of the beam. The procedures for analyses, including appropriate stress concentration factors, are discussed and design charts are presented. The given pin kit capacities are for treble or quarble plates with or without cheek plates.

TABLE 7
Nomenclature

Symbol	Variable	Typical Units
A	Area	in.^{2}
B	Bearing contact width	in.
C	Bearing contact fraction, b/d	numeric
D	Pin diameter	in.
d	Diameter of cheek plate	in.
E	Young's modulus of elasticity	psi
e	Strain	in./in., %
F	Allowable stress	ksi
f	Induced stress	ksi
G	Longitudional lug margin	in.
H	Lateral lug margin	in.
I	Area moment of inertia	in.^{4}
K	Stress or strain concentration factor	numeric
N	Number of pin shear planes	numeric
P	Force	kips
R	Lug radius	in.
S	Safety factor	numeric
t	Male lug thickness	in.
UC	Unity check	numeric
Y	Clearance	in., mils
Z	Deflection	in., mils

Subscripts

a	Axial	n	Net	N	Normal
B	Bearing	t	Total	G	Geometrical
C	Combined	e	Elastic	y	Yield
f	Flexure	p	Plastic	u	Ultimate
s	Shear	P	Pin	M	Male
t	Tension	L	Lug	F	Female

Code Restrictions on Design

The standards and guidelines for masts, derricks, and substructures are dictated by the API 4E code. This code, in turn, dictates the use of the AISC Building Code for structural allowable stresses. While the AISC is the best and most appropriate stress code to use in rig design, it was written by and for building and bridge designers and some of the design parameters are slightly different than those of drilling rigs.

For example, a treble mast is about the same height as a 13-story office buiding, and the same types of steel and welding codes are used in both classes of structures. But the portability requirement of a drilling rig changes the design procedures. We will try, therefore, to determine to what extent the AISC criteria is good for drilling rigs.

Because of the portability requirement of masts, the bearing points in the lugs are slightly different each time the structure is rigged up. This causes the pin holes to wallow. In contrast, the loads are essentially acting constantly on the same point and in the same direction in the multistory building. The stress allowables for buildings, then, should be more liberal than the requirements for drilling rigs.

The basic AISC stress allowables and restrictions on pin plates are given in figure 1.8 where the use of stress concentration factors are implicit in the reduction of stress allowables. By using the mid-range of the AISC edge margin limits and common sense, we have the design dimensions given in figure 1.9, listed in terms of the pin diameter.

Allowable stress function of yield stress:

ALLOWABLE STRESS FUNCTION OF YIELD STRESS:

$F_b = 0.9\ F_y$ (1.1)

$F_f = 0.6\ F_y$ (1.2)

$F_t = 0.45\ F_y$ (1.3)

$F_s = 0.4\ F_y$ (1.4)

ALLOWABLE GEOMETRY LIMITS:

$D \leqslant 1.5\ G$ (1.5)

$D \geqslant 1.25\ H$ (1.6)

$4t \geqslant H$ (1.7)

FOR CIRCULAR LUGS:

$G = H = R - \frac{D}{2}$ (1.8)

$1.30\ D \geqslant R \geqslant 1.17\ D$ (1.9)

FOR DESIGN PURPOSES, USE:

$R = 1.25\ D$ (1.10)

$G = H = 0.75\ D$ (1.11)

FOR TEAROUT FAILURE:

$P = 0.30\ A_sF_u + 0.50\ A_tF_u$ (1.12)

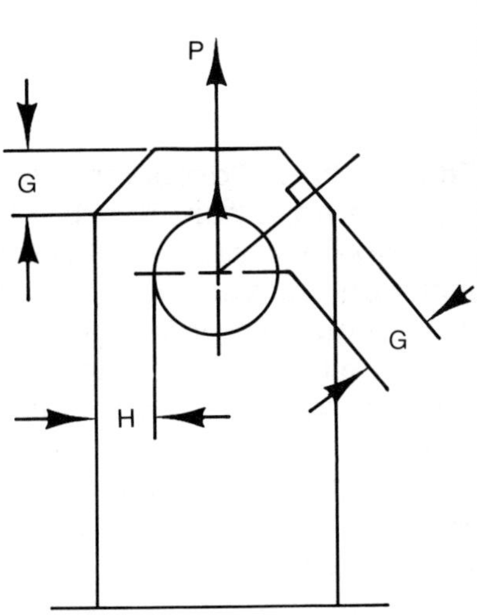

FIG. 1.8 AISC pin connection specs

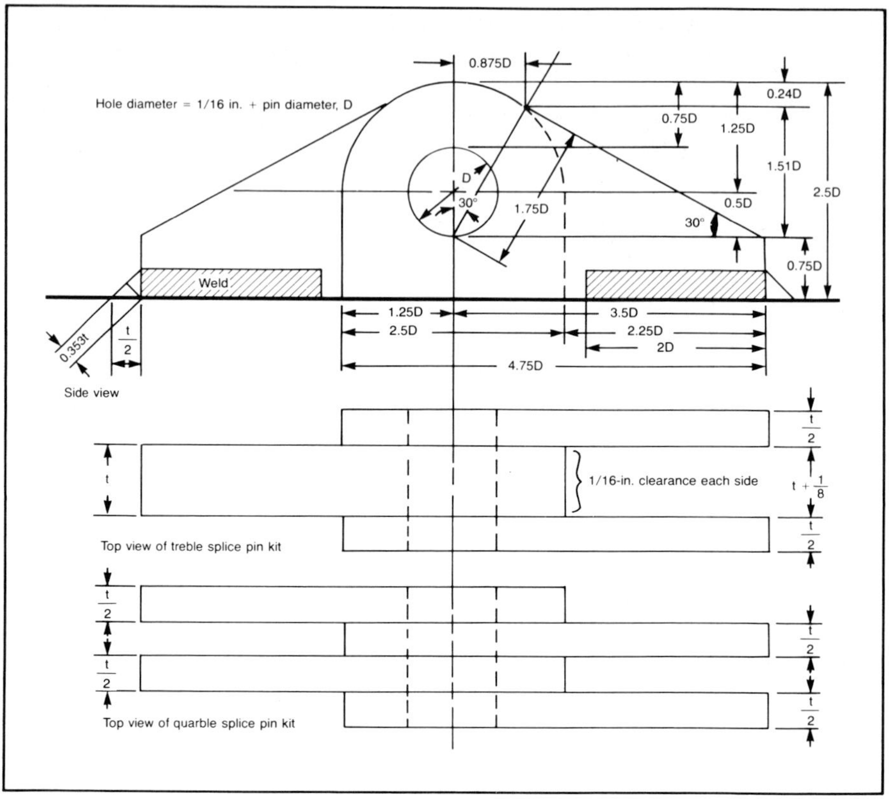

FIG. 1.9 Pin kit dimensions

Tensile Stress Factors in Lugs

We have considerable knowledge regarding stress factors at the edge of holes in plates where the hole is vacant or contains an inclusion. But very little data is available on stress factors with pins pulling on holes in lug plates.

The tensile stress factors based on net section, using Heywoods stress factors based on total section, are both given in figure 1.10. Note that the minimum total section stress factor and a mid-range AISC edge margin criteria both occur at a 40% ratio of hole diameter to lug diameter. In other words, lug radius equals 1.25 times the pin diameter.

This design data is for a tight fit pin. The given stress factors were specifically for a pin 0.0005 in. greater in diameter than the hole. There is meager evidence that the stress factor will be about 50% greater for clearance to a hole diameter ratio of as low as 0.2%. With a constant 1⁄16-in. oversize pin hole used in practice for drilling rig

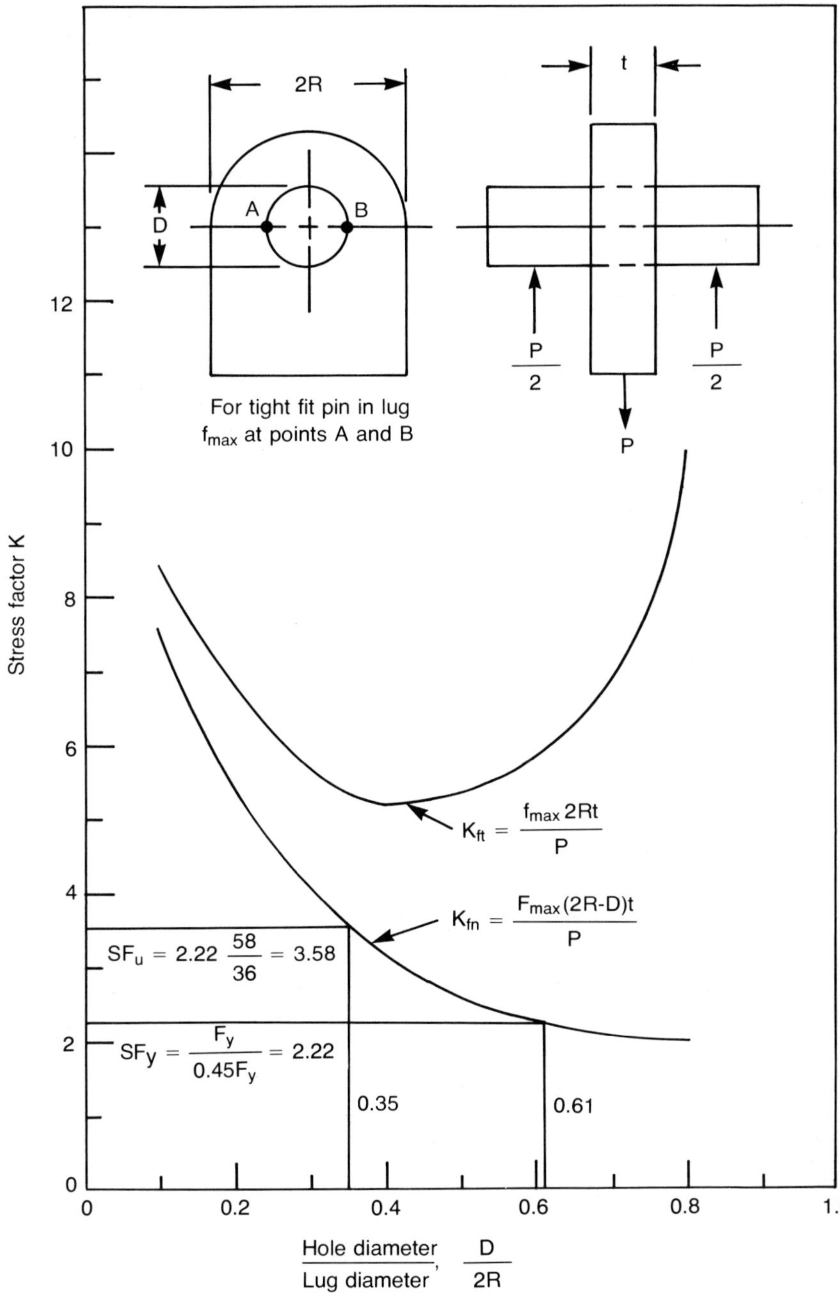

FIG. 1.10 Tensile stress factor in pin kits

design, diameter ratios of 1.2% on a 5-in. pin to 6.2% on a one-in. pin have clearance. So, the one-in. pin will have a much higher stress concentration factor than that given in the chart.

The chart also shows a net stress factor of 3.2 at a hole to lug diameter ratio of 40%. While this is greater than the 2.22 factor implied in AISC by the reduced tensile allowable of 45% of yield at the diameter ratio of 61%, the safety factor on ultimate tensile strength at a diameter ratio of 40% is about 12%. This means that the lug will yield locally a small amount, the forces will be redistributed, and the design will be safe for the case where the load is always applied in the same direction.

Elastic Stress Limits in Bearing

The smaller the pin diameter in relation to the lug hole diameter, the greater the elastic bearing stress factor. As the clearance becomes very large, the pin approaches line contact on the lug, giving extremely high stresses. The Roark equations from the theory of elasticity given in figure 1.11, were solved to determine the extent of the bearing stress factors.

Basic Roark Equations:

$$B = 2.15\sqrt{\frac{P\,D_H D_P}{tE(D_H - D_P)}} \qquad (1.13)$$

$$f_{b\,max} = 0.591\sqrt{\frac{PE(D_H - D_P)}{t\,D_H D_P}} \qquad (1.14)$$

$$Z_{max} = \frac{1.82\,P}{tE}\left(1 - 2\log_e \frac{B}{2}\right) \qquad (1.15)$$

$$\text{If } D_H - D_P = \frac{1}{16} = \text{constant and } E = 3^{\underline{7}} \qquad (1.16a, b)$$

$$P \approx t\left(\frac{f_b D}{809}\right)^2 \text{ and } B \approx \frac{1.27\,P}{f_b\,t} \qquad (1.17a, b)$$

$$\text{If } f_b = f_y = 36 \text{ then } \frac{P_{max}}{t} = 1.98\,D^2,\ B = 0.07\,D^2,\ K_{EB} = \frac{36\,Dt}{P_{max}} \qquad (1.18a, b, c, d)$$

Which gives the following results:

D, in.	P_{max}/t, kip/in.	B, in.	Z, mils	$\frac{P_{max}}{Dt}$	K_{fe}
1	1.97	0.07	0.924	1.97	18.2
2	7.89	0.279	1.42	3.94	9.13
3	17.8	0.629	2.32	5.93	6.07
4	31.4	1.118	3.03	7.85	4.58
5	49.2	1.747	3.39	9.84	3.65

FIG. 1.11 Equations of elasticity

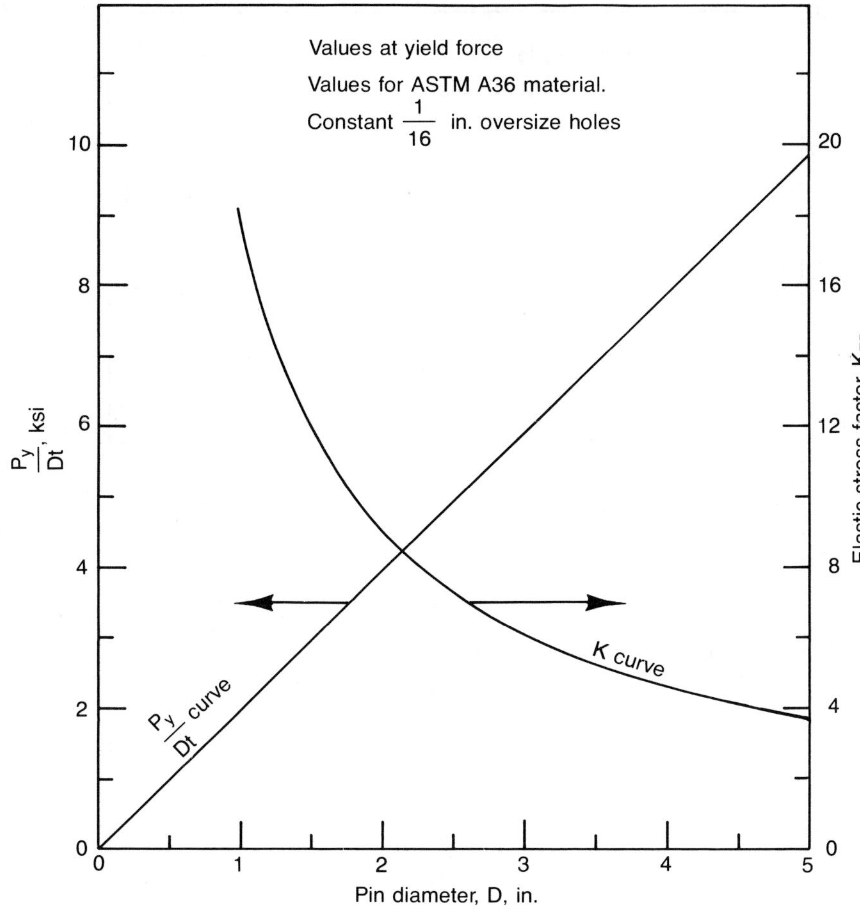

FIG. 1.12 Yield force and elastic stress factors in bearing

The results in figure 1.12 show that the stress factor varies from 3.6 for a 5-in. pin to 18.2 for a 1-in. pin where a 1/16-in. clearance is held constant. This is based on an ASTM A36 lug material where the allowed bearing is taken as 100% of the yield strength instead of 90% of yield as allowed by the AISC code.

For reasons of economy in fabrication and ease in rig-up, the given clearance must be used in all but very special applications in the design of drilling rigs. The extremely high bearing stress concentration factors mean that rig pin kits cannot be designed on an elastic basis. This also means that the AISC criteria for using projected area will not give suitable results in rig design.

The pin kits must be designed using the theory of plastic structural analyses. One way to do this is by equating the combined elas-

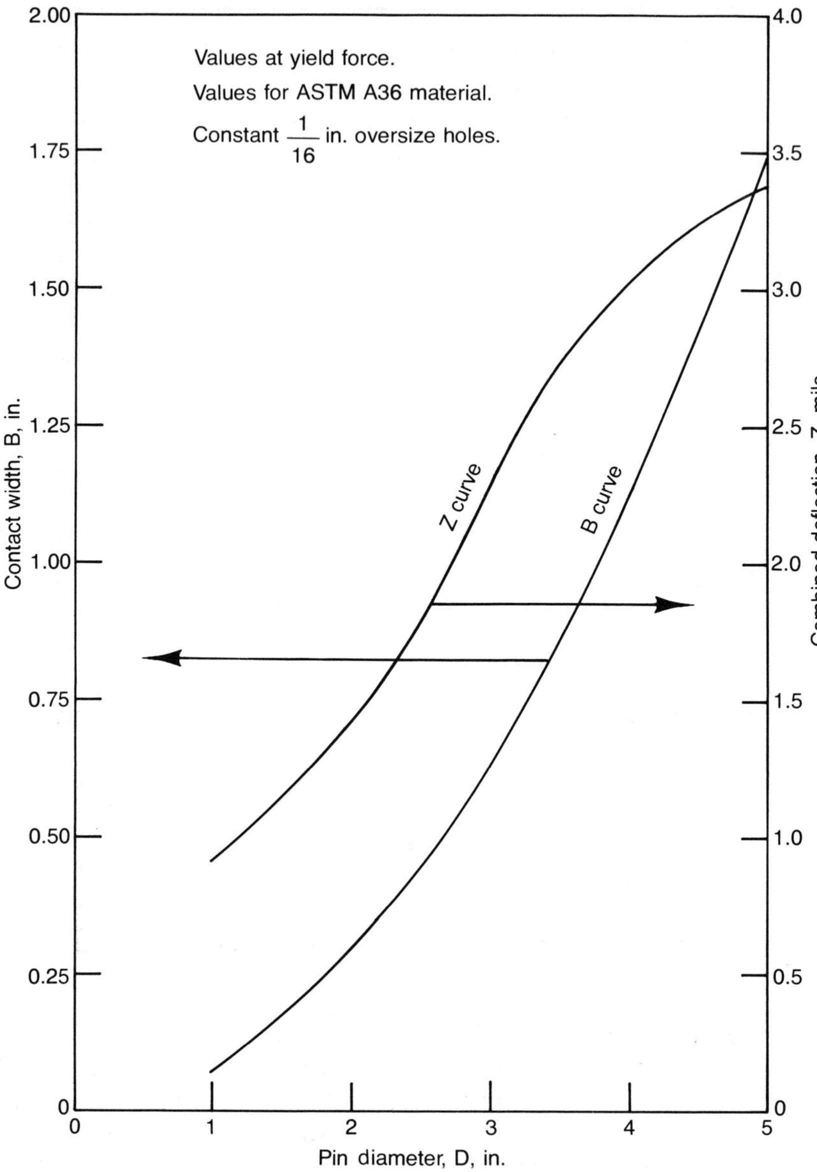

FIG. 1.13 Elastic pink kit bearing functions

tic deflection of the pin and the lug as given in figure 1.13 with the geometrical clearance ratio to get an acceptable bearing width.

Acceptable pin-kit bearing design will then depend on a specific application. The information given here on pin kit capacity is satisfactory for unidirectional lug loads. It cannot be used, however, for the foot pins in masts or connector plates in grasshopper type substructures where the load direction rotates during rig up. For these applications, you must reduce the results for pin kit capacity given here or consider the resulting excessive wear and frequent repair and maintenance.

Plastic Strain and Stress Factors in Bearing

The first step in doing the plastic bearing analysis is to work out the geometry. Equations and results to determine the projected bearing area as a function of clearance are given in figure 1.14.

The constructed plastic stress-strain relation for ASTM A36 material is given in figure 1.15. This was done by assuming that one-half of the minimum ultimate elongation of 20% at rupture will occur at the maximum tensile strength. It was then assumed that this point is connected to the yield point by a parabolic relationship.

To keep the plastic analyses managable, we selected a projected width fraction of 66%. As shown in figure 1.16, the geometrical clearance at this width fraction is not a strong function of pin diameter, so we used a constant clearance of 10 mils in the analysis.

Using the equations in figure 1.17, the results in figure 1.18 show that the plastic strain factor varies from 2.9 for a 5-in. pin to 10.8 for a 1-in. pin. But the plastic stress factor only varies from 1.13 with a 5-in. pin to 1.36 with a 1-in. pin. Since the ultimate to yield strength ratio of A36 is 1.61, a 66% bearing width ratio is satisfactory for unidirectional load applications. For design purposes, this means using a stress factor of 1.5 and a bearing allowable of 90% of yield.

Pin Kit Design Procedures

The AISC code gives relations for combined shear and tension on bolts but does not explicitly give general allowable stresses where shear is included. In the case of pin kits, the shear stress and the flexural stress in the pin should be included.

The lateral load distribution is assumed in several different ways with either concentrated load or uniform pressure (figure 1.19). The true load distribution is actually a nonlinear pressure across the lug. To simplify this analysis, the flexure on the pin was taken as a concentrated load. It acted as a cantilever with a moment arm equal to the female lug thickness, which is one-half the thickness of the male plate for a treble plate pin kit.

There is usually some warning when lug plates begin to yield. The wallow of the hole becomes more evident, and the rig begins to vibrate more. Failure of the pin in shear is abrupt so the pin criteria should be more stringent than the lug criteria. Some designers do a combined shear and flexure unity check. Mohr circle stresses (figure 1.20) are used in the analysis here since this gives vectorial stress addition rather than algebraic stress addition by the use of unity checks.

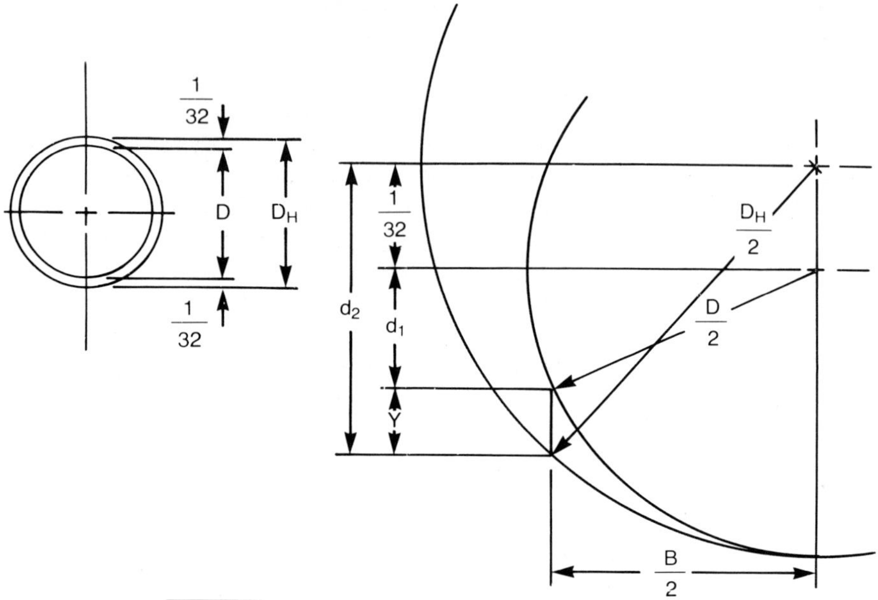

$$d_1 = \frac{\sqrt{D^2 - B^2}}{2} \qquad (1.19)$$

$$d_2 = \frac{\sqrt{D_H^2 - B^2}}{2} \qquad (1.20)$$

$$Y = d_2 - d_1 - \frac{1}{32} \qquad (1.21)$$

$$C \equiv \frac{B}{D}, \quad Q \equiv 1 - C^2 \qquad (1.22)\ (1.23)$$

Combining above equations gives:

$$Y = \sqrt{0.25\,QD^2 + 0.0312\,D + 0.000976} - 0.5\,D\sqrt{Q} - 0.0312 \qquad (1.24)$$

				Clearance Y, mils				
C	C^2	Q	D:	1	2	3	4	5
0.2	0.04	0.96		1.03	1.46	1.87	2.27	2.67
0.4	0.16	0.84		2.96	3.31	3.60	3.87	4.14
0.6	0.36	0.64		7.23	7.53	7.64	7.69	7.72
0.7	0.49	0.51		11.4	12.0	12.3	12.5	12.6
0.8	0.64	0.36		18.4	19.5	19.9	20.1	20.3

FIG. 1.14 Geometric pin-lug clearance

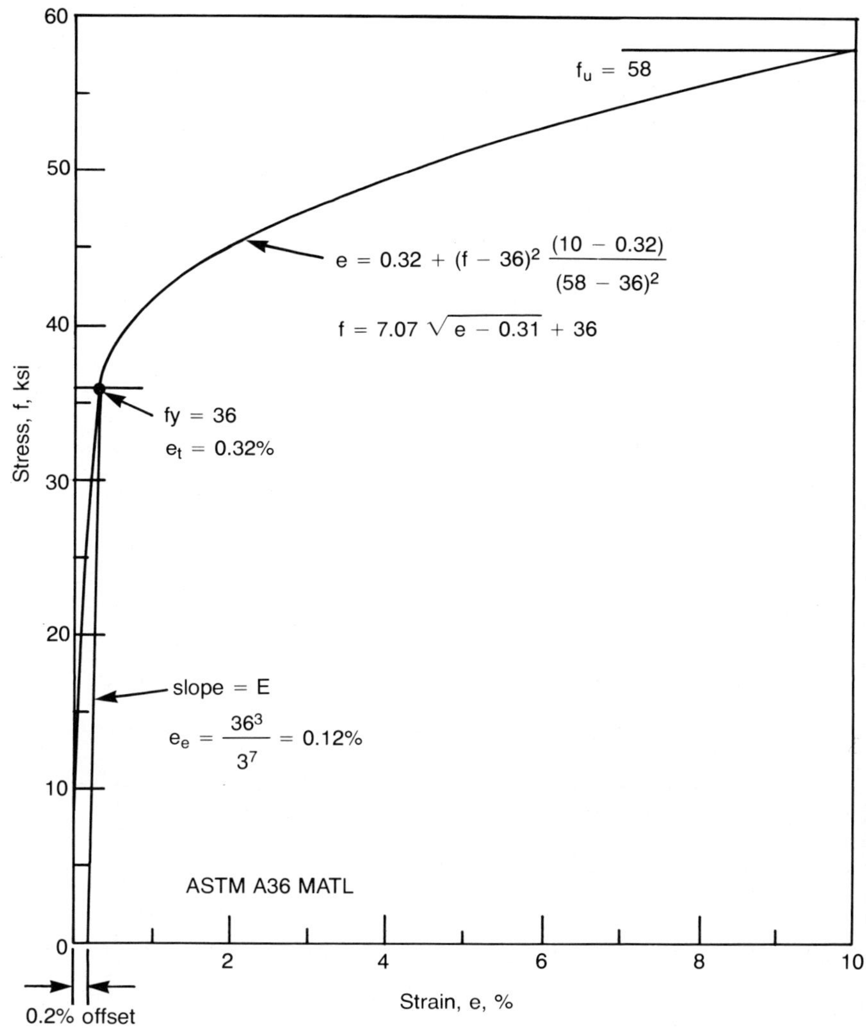

FIG. 1.15 Constructed plastic stress-strain relation

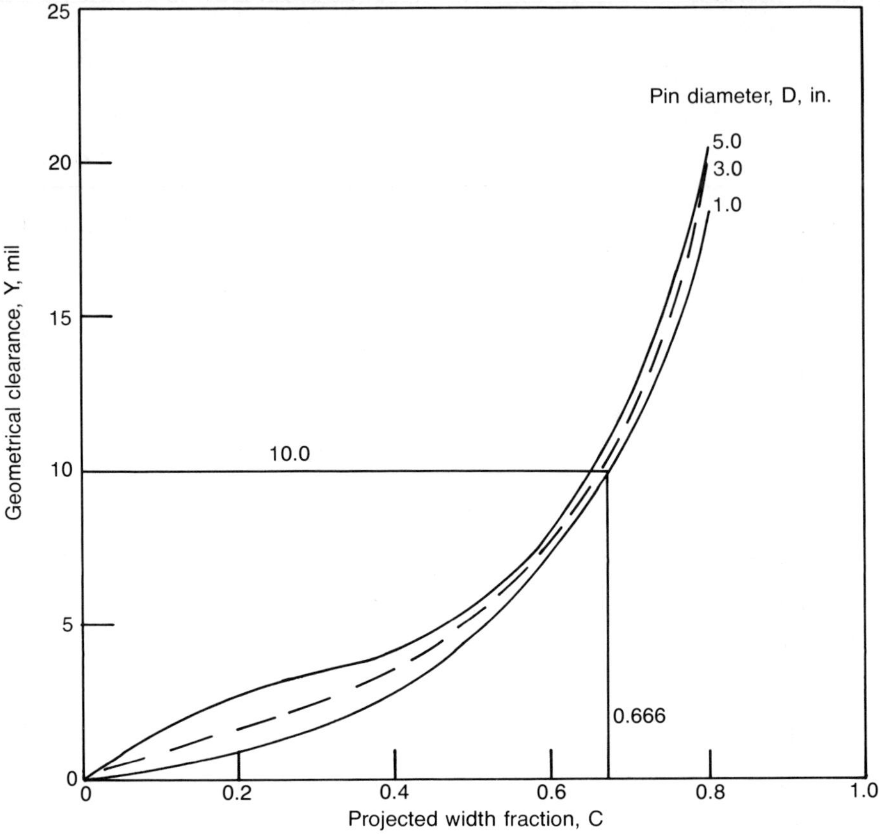

FIG. 1.16 Geometrical clearance and projected width

Pin dia, D, inch	1	2	3	4	5
Elastic deflection, Z_y, mil	0.924	1.42	2.32	3.03	3.39
Clearance-deflection ratio, Y_G/Z_y	10.8	7.04	4.31	3.30	2.94
Plastic strain, e_p, percent	3.45	2.25	1.37	1.05	0.940
Plastic stress, f_p, ksi	49	46	43	42	41
Plastic stress factor, k_{fp}	1.36	1.27	1.19	1.16	1.13
Normalized contact ratio, C_N	0.492	0.527	0.563	0.577	0.592

Where:

$$Y_G = 10.0 \text{ mil at } C_G = 0.67 \qquad (1.25)(1.26)$$

$$f_y = 36.0 \text{ ksi}, \; e_y = 0.32\%, \; e_p = 0.32 \frac{Y_G}{Z_y} \qquad (1.27)(1.28)(1.29)$$

$$K_{pf} = \frac{f_p}{f_y}, \; K_{pe} = \frac{Y_G}{Z_y}, \; C_N = \frac{Y_G}{K_{pf}} \qquad (1.30)(1.31)(1.32)$$

FIG. 1.17 Contact ratio normalized on plastic stress factor in beading

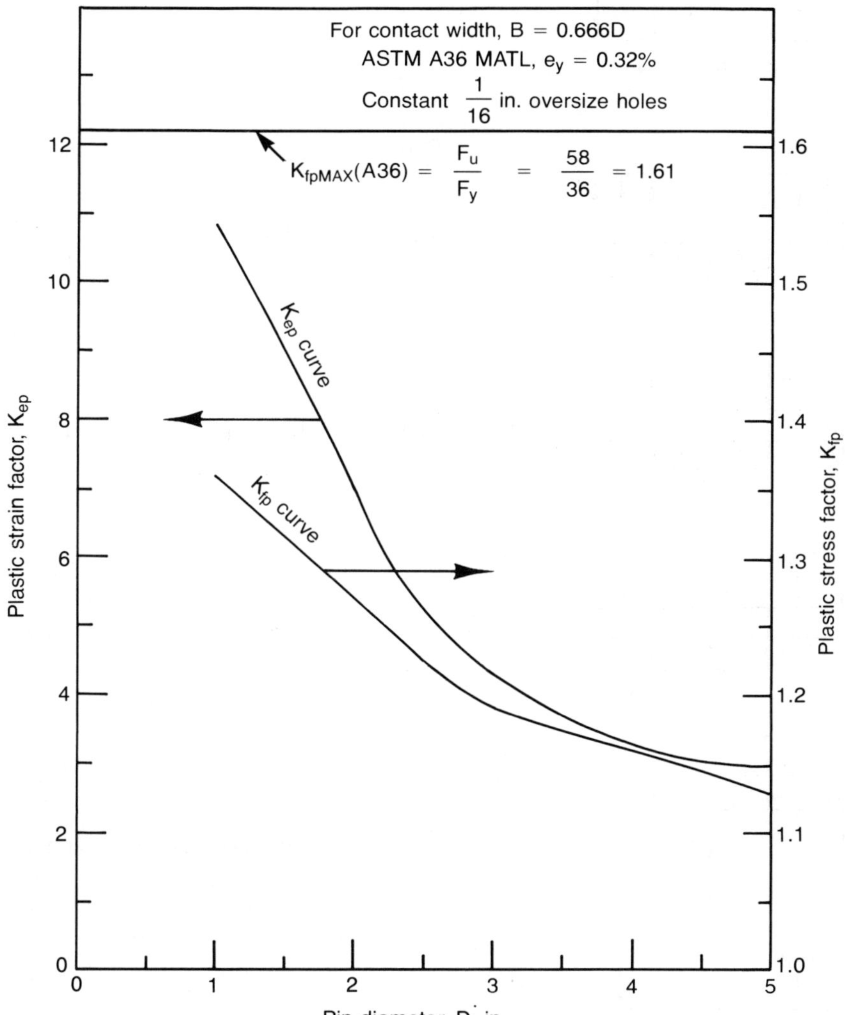

FIG. 1.18 Plastic stress and strain factors in bearing

Load distribution	Moment	Comment	
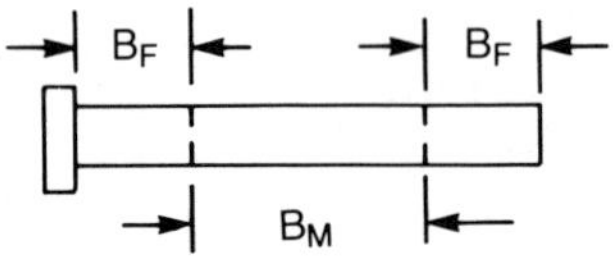			
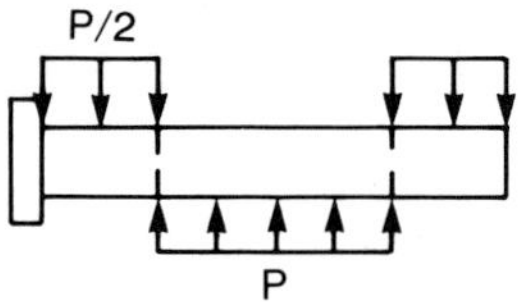	$\frac{PB_M}{12}$ or $\frac{PB_F}{2}$	Tight fit	(1.33) (1.34)
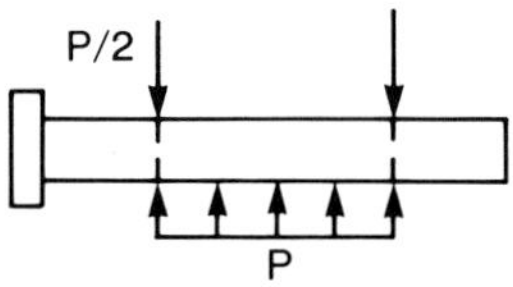	$\frac{PB_M}{8}$	Loose fit	(1.35)
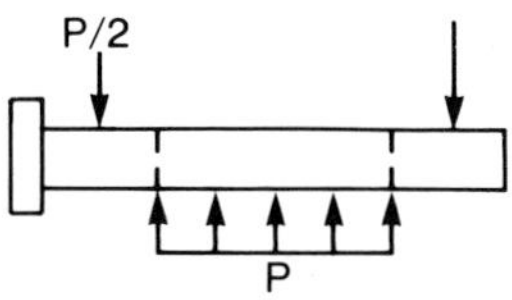	$\frac{P(B_M + B_F)}{8}$	Loose fit	(1.36)
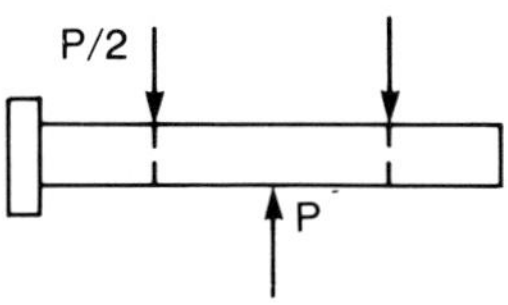	$\frac{PB_M}{4}$	Nonuniform load	(1.37)
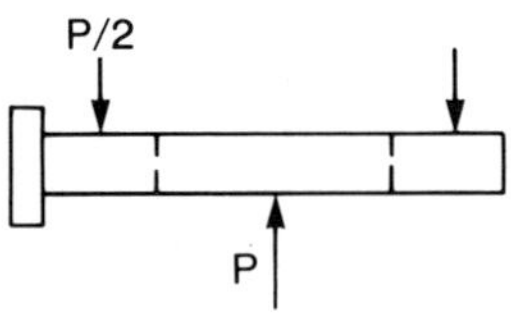	$\frac{P(B_M + B_F)}{4}$	Nonuniform load	(1.38)

FIG. 1.19 Maximum pin moment

For the pin:

$$f_s = \frac{P_s}{0.785\ D^2 N} = 0.4\ F_y \text{ or } \frac{P_s}{D^2 F_y} = 0.314\ N \qquad (1.39)(1.40)$$

$$f_f = \frac{M_c}{I} = \frac{P_f}{2}\frac{t}{4}\frac{D}{2}\frac{64}{\pi D^4} = 0.6\ F_y \text{ or } \frac{P_f}{D^2 F_y} = 0.472\ \frac{D}{t} \qquad (1.41)(1.42)$$

$$f_{s\,max} = 0.5\sqrt{f_f^2 + 4f_s^2} \text{ or } \frac{P_{s\,max}}{D^2 F_y} = \frac{0.63}{\sqrt{\frac{4}{N^2} - \left(\frac{t}{D}\right)^2}} \qquad (1.43)(1.44)$$

$$f_{n\,max} = 0.5\ f_f - f_{s\,max} \text{ or } \frac{P_{f\,max}}{D^2 F_y} = \frac{0.945}{\frac{t}{D} + \sqrt{\frac{4}{N^2} + \left(\frac{t}{D}\right)^2}} \qquad (1.45)(1.46)$$

For the lug where $2\ R = 2.5\ D,\ t_m = 2\ t_F$: (1.47)(1.48)

$$P_s = 0.4\ F_y\ (2\ Dt) = 0.8\ F_y Dt \qquad (1.49)$$

$$P_t = 0.45\ F_y\ (2)(0.75\ Dt) = 0.675\ F_y\ Dt \qquad (1.50)$$

$$P_B = 0.9\ F_y\ Dt/k = 0.6\ F_y\ Dt \qquad (1.51)$$

$$P_w = 0.4\ F_y\ (0.5\ t \cos 45)(4\ D + t) = 0.141\ F_y\ t(4\ D + t) \qquad (1.52)$$

$$P_{to} = 0.3\ (2\ Dt)\ F_u + 0.5\ (2t)(0.75\ D) = 1.35\ F_u\ Dt \qquad (1.53)$$

FIG. 1.20 Pin kit design equations

The comparison of treble and quarble pin kits are given in figure 1.21 where each of the four plates in a quarble have the same thickness. The results show that the quarble has only a slight increase in capacity over the treble. However, it is a useful type of pin kit when the male lug thickness of a treble pin kit becomes excessive and cheek plates are not used. In comparing the maximum combined stress from Mohr's circle, we find that the basis of pin failure will be the maximum normal stress (figure 1.22).

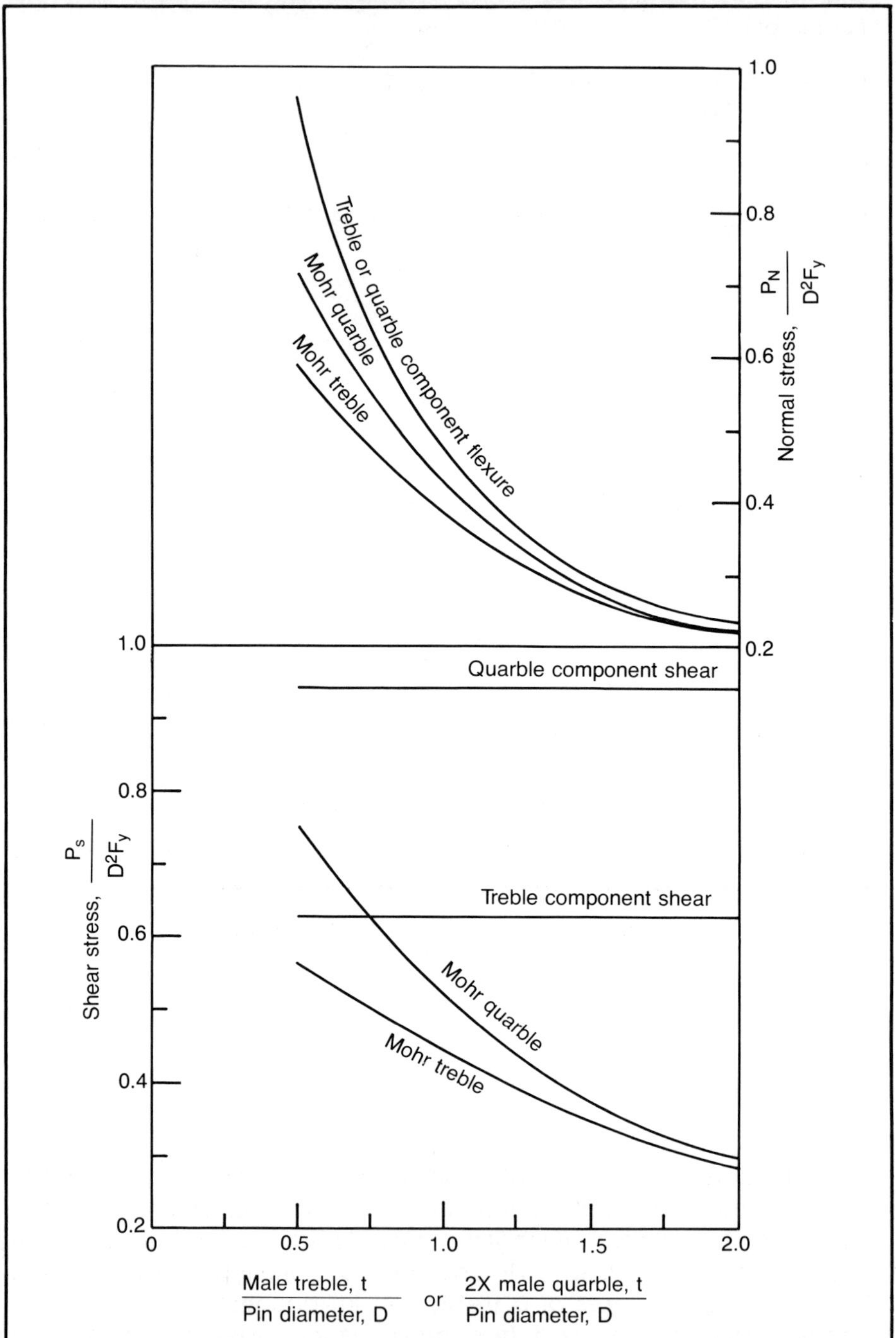

FIG. 1.21 Pin stress comparisons

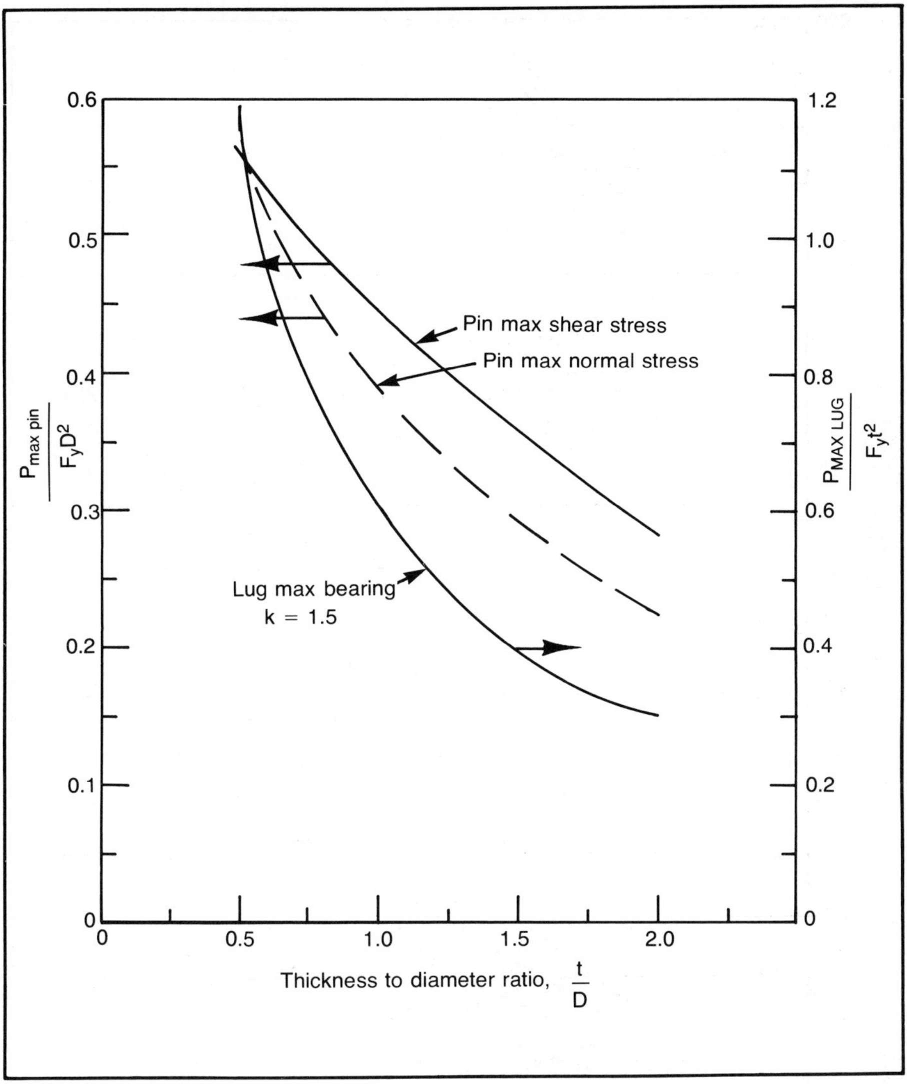

FIG. 1.22 Pin kit stress

Pin Kit Capacity

This discussion as well as the design charts were prepared using grade 50 lugs (figure 1.23) and 36 grade lugs (figure 1.24), with 100 ksi yield strength pins used in both cases. Other charts will have to be constructed for grades of steel different from those specified here. As expected, the lug bearing stress dictates until very thick lugs are used. Then the pin stress dictates.

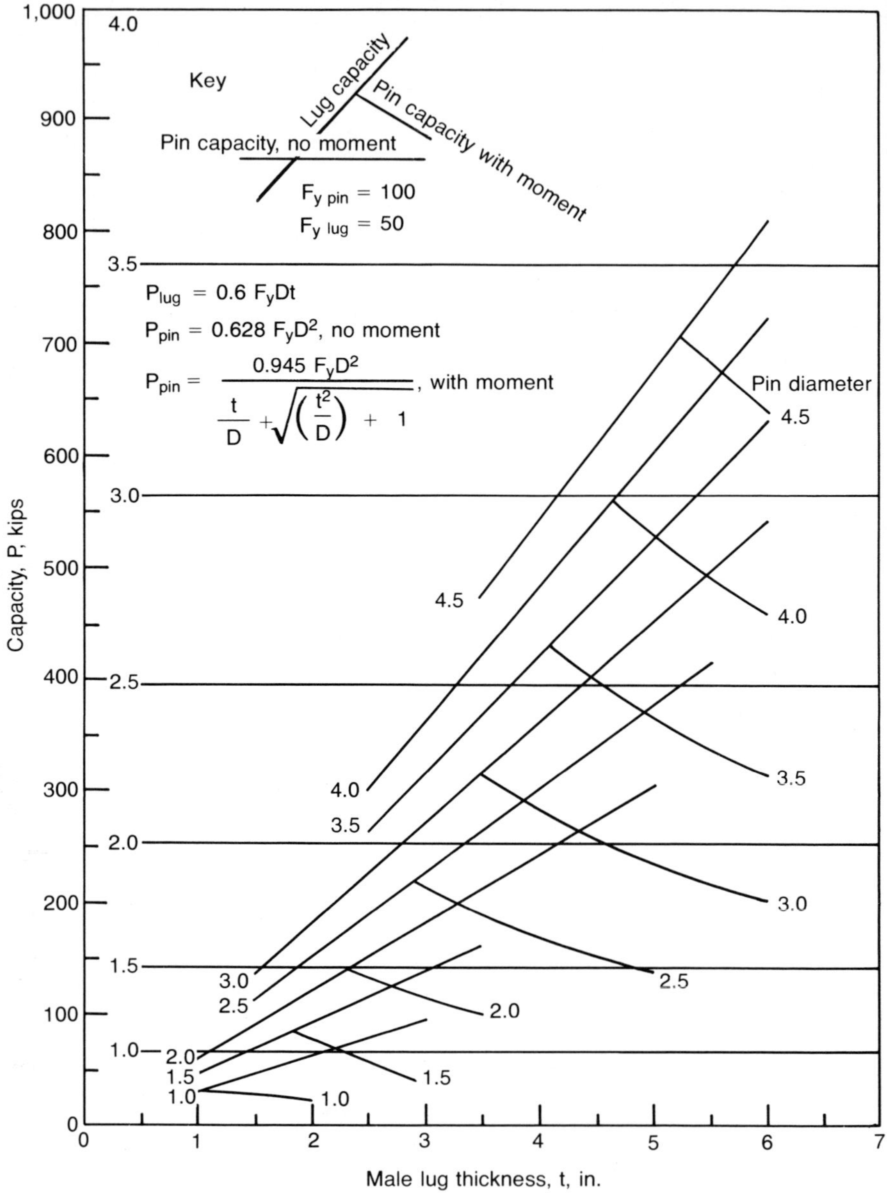

FIG. 1.23 Pin kit capacity for grade 50 lugs

Consider a 3-in. grade 100 pin in grade 36 lug plates. Testing shows that the pin-kit capacity will be 195 kips for a 3-in. male lug thickness, 260 kips for a 4-in. thickness, and 240 kips for a 5-in. thickness.

These charts are based on maximum bearing stress in the lug

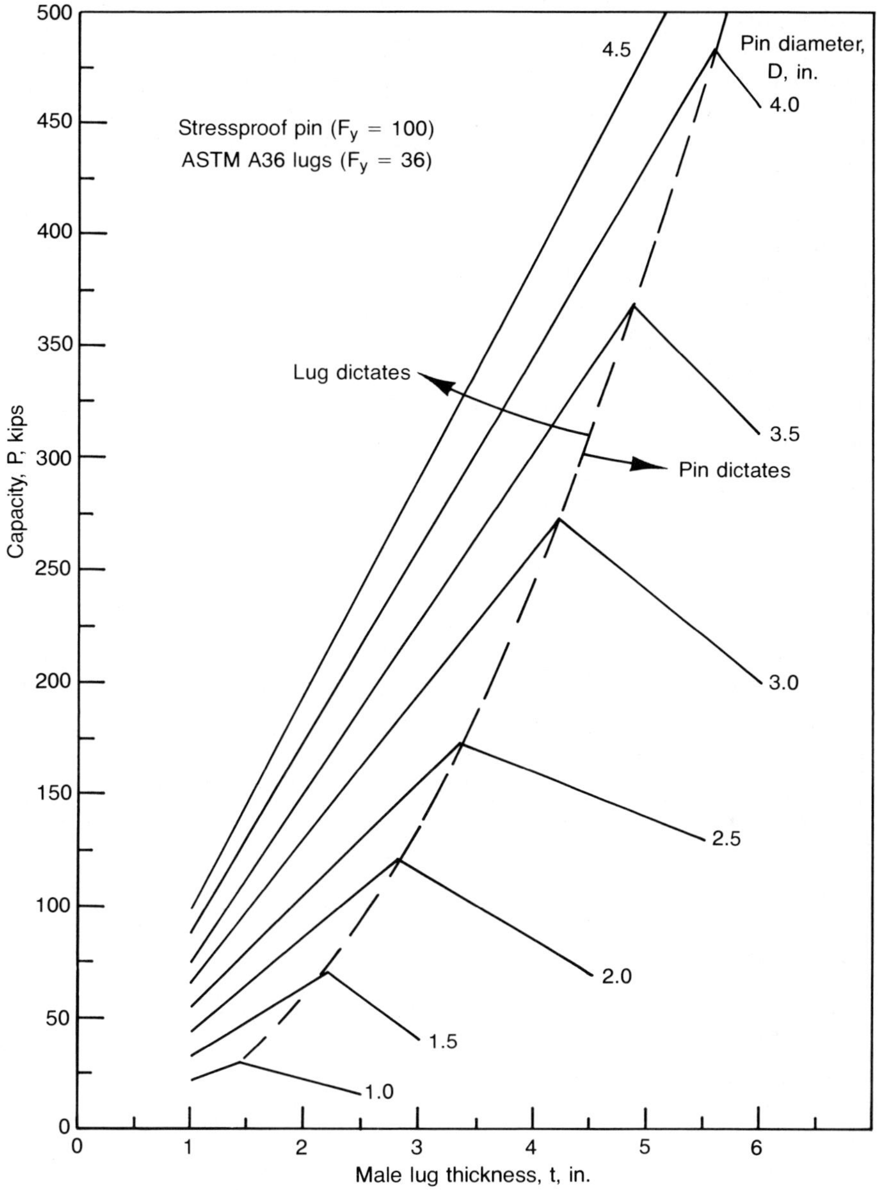

FIG. 1.24 Pin kit capacity for grade 36 lugs

so the pin kit capacity is the same for tension and for compression loads. As shown in the pin-kit design equations, bearing dictates with all of the given assumptions. This is because the allowable lug tension force is 12% greater and the lug shear force is 33% greater than the allowable lug bearing force.

Cheek Plate Design

Cheek plates are used when the lug plates begin to appear excessively thick. The first alternative is to use a cheek plate on each side of the male lug. The second alternative is to use cheek plates on the male and on the outside of the females (figure 1.25).

Each of the cheek plates on the male lug should have a thickness equal to or less than one-half of the male lug thickness. Cheek plates used on the female lug should have a thickness equal to or less than the female lug thickness. With either alternative design, the cheek plate should have a fillet weld all around with the weld leg distance equal to the cheek plate thickness. With these guidelines, the tensile, shear, and bearing capacity of the lug as well as the shear stress in the cheek plate weldment will be as good as the full thickness pin-kit lugs without cheek plates.

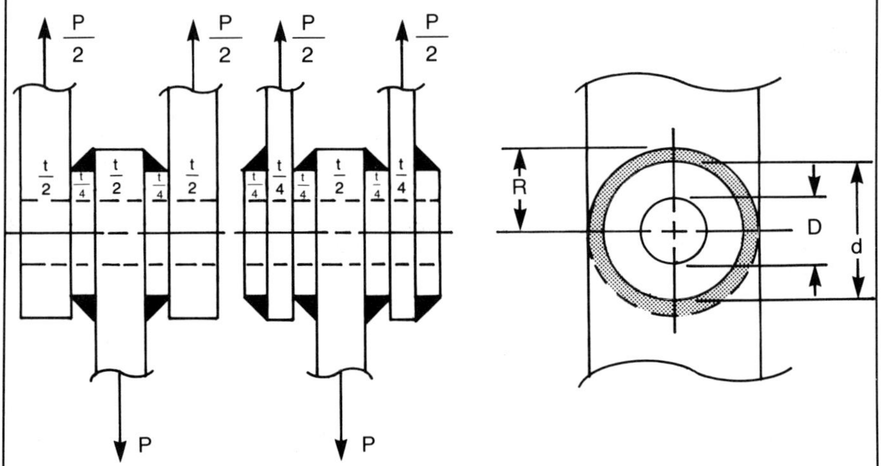

$$R = 1.25\ D \qquad (1.54)$$

$$d = 2\left(R - \frac{t}{4}\right) \qquad (1.55)$$

$$K_{\substack{net \\ tension}} = 3.2 \text{ at } \frac{D}{2R} = 0.4 \qquad (1.56,\ 1.57)$$

$$F_{us} = \frac{F_{ut}}{\sqrt{3}} = 0.577\ F_{ut} \qquad (1.58)$$

$$f_t = \frac{KP}{2A} = \frac{3.2\ P}{2t\ (0.75D)} \Rightarrow \frac{P}{Dt} = 0.468\ F_u \qquad (1.59)\ (1.60)$$

$$f_s = \frac{P}{2A_w} = \frac{P}{2\pi D \frac{t}{4} \cos 45} = \frac{F_u}{\sqrt{3}} \Rightarrow \frac{P}{Dt} = 0.640\ F_u \qquad (1.61)\ (1.62)$$

FIG. 1.25 Cheek plate arrangements

TABLE 8
Nomenclature

Symbol	Variable	Typical Units
A	area	sq. in
a	maximum distance from neutral axis	in.
B	web length	in.
b	shear width	in.
F	allowable stress	ksi
f	induced stress	ksi
G	distance parameter	numeric
H	mud tank height	ft.
I	moment of inertia	$in.^4$
L	length of flange plate	in.
l	span length	in.
M	mud weight	ppg.
m	moment	in.-lb.
N	first area moment of inertia	$in.^3$
P	pitch of stiffeners	in.
p	pressure	psi
R	inertia ratio	numeric
t	plate thickness	in.
W	weight	lb.
w	plate width	in.
y	centroidal distance	in.
Z	section modulus	$in.^3$

Subscripts

e	effective	q	per foot
f	flexure	s	shear
L	angle	V	vee crimp
p	plate	W	W crimp
		y	yield

1.5 DRILLING MUD PIT DESIGN

The basic design considerations and practical design data are given here for fabrication of the two types of drilling mud pits. While the discussion is directed primarily toward mud pit design, the data is equally suitable for box shape containers for other fluids such as water or fuel tanks and various liquid chemical storage.

The two basic types of mud pit construction are composite wall and crimp wall design. Composite wall design usually consists of a flat plate with equally spaced vertical rolled sections welded to it. Various types of rolled sections can be used, but angle sections oriented with an angle toe welded to the plate are the most common. Use of a T section is also cost effective. Design with a channel section is much less efficient than with L or T sections.

Crimp or corrugated wall design is flat sheet bent on equal pitch, called V crimp, or flat sheet bent with a small distance between the V shaped bends, which is the so called W crimp design. The crimp wall design will usually weigh more but the total cost will usually be less than composite design. This is because wall thickness should not be less than about 3⁄16 or 1⁄4 in. for handling and other practical considerations, regardless of theoretical calculations. Also, the cost disparity between the two types of design is due to the cost break down where labor cost in fabrication is presently about three times the material cost. Crimp wall design requires less welding than composite design. The disadvantage of crimp wall design is the requirement of large press facilities whereas composite design can be constructed in almost any fabrication shop.

Composite Wall Design

The basic design equations for composite wall design are given in figure 1.26. The moment equation has the necessary conversion factors already included so that the moment is given in terms of the height of the mud pit in feet and the mud weight is given in units of pounds per gallon.

There are no specific codes dictating the design of mud pits. For the composite type of design, the maximum allowable working stress was taken as the yield strength of the material. A safety factor is built in since a mud pit is never filled to the top. The wall was considered as a cantilever beam. This means that the structural reinforcement from the support beams for mud mixing and processing equipment was neglected. The support beam reinforcement increases the safety factor further.

Shear lag occurs in panels where the plate width is wide relative to the stiffener width. This means that the plate at the stiffener carrys full stress but the flange stress in the plate decreases to minimum halfway between the stiffeners for very wide thin plates. The extent of shear lag depends on the inertia ratio, span length to pitch ratio, type of beam supports, and load distribution.

This complicated theoretical business is simplified in figure 1.27. The effective width is the amount of the plate that takes full flange stress. Some typical composite beam properties for a three-ft pitch were calculated and are given in figure 1.28 where the effective width fraction was obtained from figure 1.27, Case 3.

While the above discussion is exciting background information, it does not get to specific pit design. Suitable plate thickness and stiffener sizes for pit fabrication based on the preceeding information are given in table 9 for mud pit heights of 6, 9, and 12 ft.

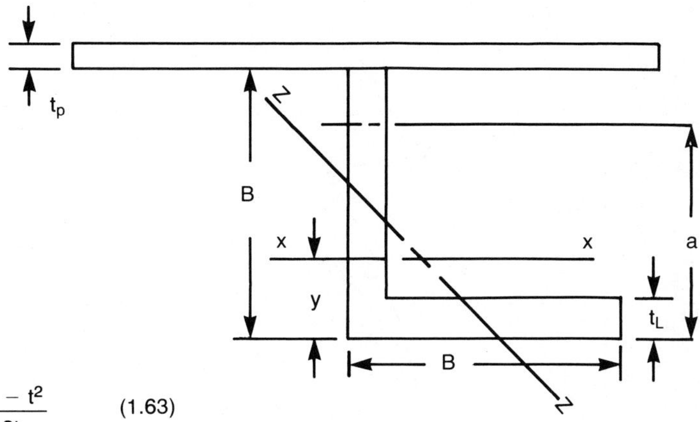

$$Y_L = \frac{B^2 + Bt - t^2}{4B - 2t} \qquad (1.63)$$

$$I_{XL} = \frac{1}{3}\left[t(B-y)^3 + By^3 - (B-t)(y-t)^3\right] \qquad (1.64)$$

$$I_{ZL} = I_X - \frac{tB^2(B-t)^2}{4(2B-t)} \qquad (1.65)$$

$$F_{ye} = (412\,\frac{t}{B})^2 \qquad (1.66)$$

$$p = \frac{62.4\,M\,H}{144\,(8.33)} = 0.0520\ MH\ psi \qquad (1.67)$$

$$w = \frac{p}{2}(144)H = 3.74\ MH^2 \quad lb/ft\ width \qquad (1.68)$$

$$m = \frac{MH}{3} = 15.0\ MH^3 \quad in.lb/ft\ width \qquad (1.69)$$

$$Z = \frac{I}{c} \qquad (1.70)$$

$$f_f = \frac{M}{Z} \qquad (1.71)$$

$$(a - y)\,A_L = (B + \frac{t_p}{2} - a)A_p \qquad (1.72)$$

$$I_c = I_p + I_L + (B + \frac{t_p}{2} - a)^2 A_p + (a - y)^2 A_L \qquad (1.73)$$

$$F_f = F_y \qquad (1.74)$$

FIG. 1.26 Composite design equations

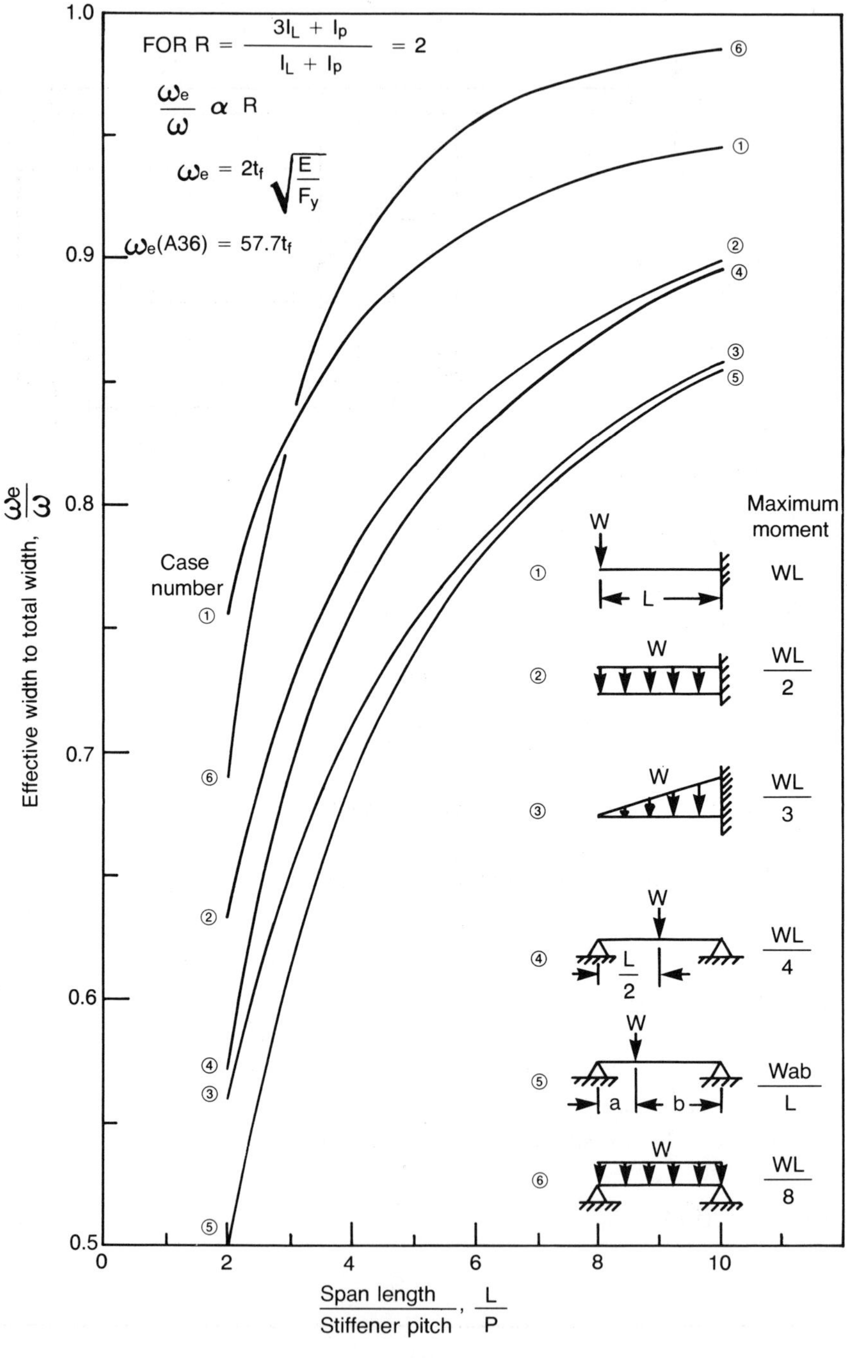

FIG. 1.27 Shear lag in built-up beams

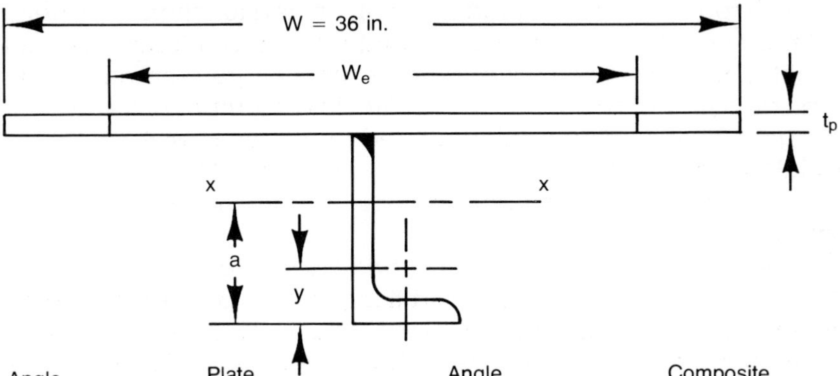

Angle Section	Plate			Angle			Composite		
	W_e	t_p	A_{pe}	A_L	I_L	y_L	a	I_{cx}	Z_{MIN}
3x2½x¼	20.1	0.250	5.02	1.31	1.17	0.911	2.65	6.23	2.35
4x3x¼	20.1	0.250	5.02	1.69	2.77	1.24	3.38	13.24	3.91
4x4x¼	20.1	0.250	5.02	1.94	3.04	1.09	3.26	15.8	4.87
4x3x5/16	23.4	0.312	7.30	2.09	3.38	1.26	3.50	16.9	4.83
5x5x5/16	23.4	0.312	7.30	3.03	7.42	1.37	4.03	37.9	9.40
6x6x5/16	23.4	0.312	7.30	3.65	13.0	1.62	4.63	62.8	13.5
6x6x⅜	23.4	0.312	7.30	4.36	15.4	1.64	4.47	71.6	16.0
6x6x½	23.4	0.312	7.30	5.75	19.9	1.68	4.22	86.9	20.5
8x4x½	25.5	0.375	9.56	5.75	38.5	2.86	6.17	140	22.6
8x8x½	25.5	0.375	9.56	7.75	48.6	2.19	5.53	205	37.2

FIG. 1.28 Composite beam properties

TABLE 9
Composite mud pit wall design

For ASTM A36 and $f = F_y$ where $z_{min} = \frac{MH^3}{800}$, P = 36 in.

Mud pit height, H, ft	6	9	12
Effective width fraction, w_e/w	0.56	0.65	0.71
Effective width, w_e, in.	20.1	23.4	25.5
Plate thickness, t_p, in.	0.250	0.312	0.375
Plate area, A_p, in.^{2}	5.02	7.30	9.56
Plate inertia, I_p, in.^{4}	0.0261	0.0592	0.0112
Z_{min} for 10 ppg	2.70	9.11	21.6
14 ppg	3.78	12.7	30.2
18 ppg	4.86	16.4	38.8
Design Z_{min}	4.86	16.4	38.8
Design angle section	4 × 4 × ¼	6 × 6 × ⅜	8 × 8 × ½
Composite design z	4.87	16.0	37.2

The required section modulus is given as a function of pit height and mud weight. The variation in total pit cost as a function of mud weight in negligible. To cover all eventualities we recommend that all pit designs be for 18 ppg mud. With this assumption, the calculations give the following recommended angle sections: 4 × 4 × ¼ on ¼-plate for 6-ft height, 6 × 6 × ⅜ on 5⁄16-plate for 9 ft, and 8 × 8 × ½ on ⅜-plate for 12-ft height, all for 3-ft pitch and mild steel materials.

The exact stress ratios for thin wide plates are given in figure 1.29 for simply supported end conditions. In mud pit design, the stress ratios will be about 65% to 75% of the values given in the chart. This information can be used to check the plate stress midway between stiffeners for pit designs with greater stiffener pitch than recommended above. This stress check should use ultimate strength and

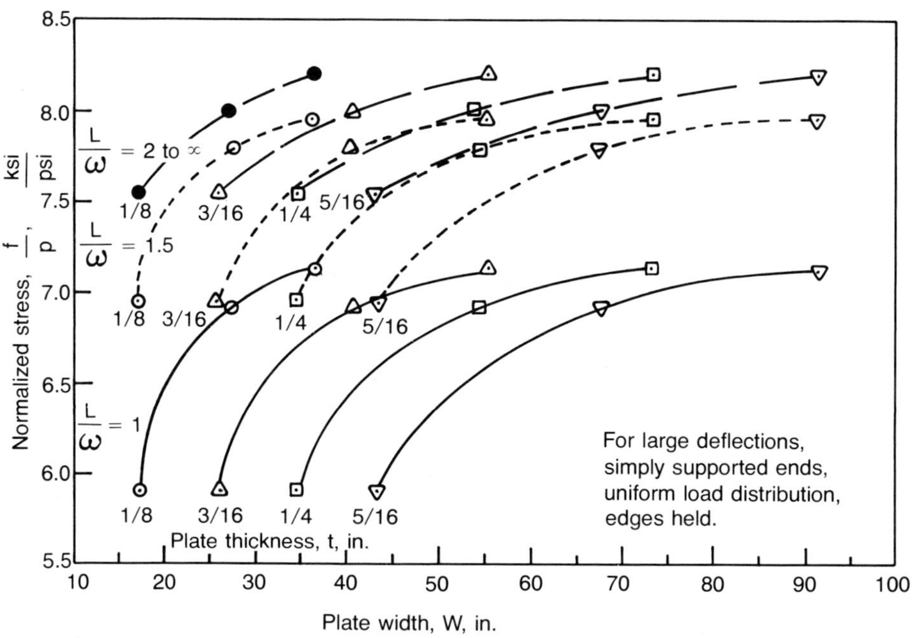

FIG. 1.29 Thin-plate, large deflection stress

not yield strength since the chart is for stringent calculations. If you use this procedure on stiffener pitch greater than three ft, the plate panels between stiffeners will have a slight bow with first maximum load application. The slight bow shape then allows greater load assumption by diaphragm stress rather than flexurally so no further yield or harmful effects results.

Crimp Wall Design

The basic design equations are given in figure 1.30 for both V and W crimp wall design. Design charts for the range of interest with mud pit design are given in figures 1.31 and 1.32. The leg length B and the plate width L are dimensions to the working points of the bent plate since the bend radius varies with the punch shape used in a particular press facility.

Because of the varying bend radius in practice, an allowable flexural stress of 80% of the yield strength was used in the calculations. The allowable shear stress was then taken as two-thirds of the allowable flexural stress.

$$I_V = \iint y^2 \, dx \, dy = \frac{t\,B^3}{12.1} \tag{1.75}$$

$$I_{VQ} = \frac{12\,I_V}{1.41\,B} = \frac{B^2\,t}{1.42} \tag{1.76}$$

$$Z_{VQ} = \frac{I_{VQ}}{c} = 1.98\,Bt \tag{1.77}$$

$$800\,F_y \geqq \frac{15.0\,MH^3}{1.98\,Bt} \tag{1.78}$$

$$\left[\frac{MH^3}{F_y}\right]_V \leq 106\,Bt \tag{1.79}$$

$$I_W = I_V + 2\,Lt\,(0.353\,B)^2 \tag{180}$$

$$I_{WQ} = \left(\frac{12}{2\,L + 1.41\,B}\right) I_W = \frac{0.497\,Bt\,(B + 3.03\,L)}{0.707 + \frac{L}{B}} \tag{1.81}$$

$$Z_{WQ} = \frac{I_{WQ}}{0.353\,B} = \frac{2.0\,Bt\left(1 + 3.03\,\frac{L}{B}\right)}{\left(1 + 1.41\,\frac{L}{B}\right)} \tag{1.82}$$

$$800\,F_y \geqq 7.49\,\frac{MH^3}{Bt}\,\frac{\left(1 + 1.41\,\frac{L}{B}\right)}{\left(1 + 3.03\,\frac{L}{B}\right)} \tag{1.83}$$

$$\left[\frac{MH^3}{F_y}\right]_W \leqq 106\,Bt\,\frac{\left(1 + 3.03\,\frac{L}{B}\right)}{\left(1 + 1.41\,\frac{L}{B}\right)} \tag{1.84}$$

Where $F_f = 0.8\,F_y$ and $F_s = 0.533\,F_y$ (1.85)

FIG. 1.30 Crimp wall design equations

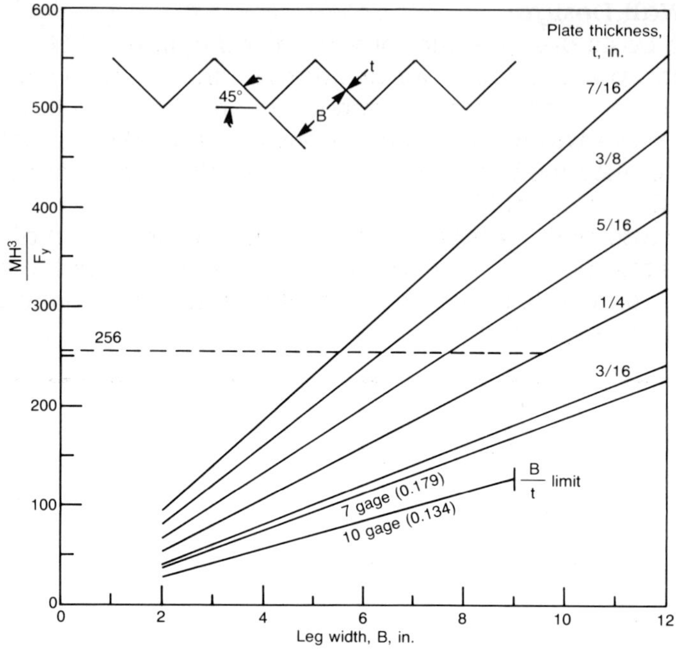

FIG. 1.31 V crimp moment design

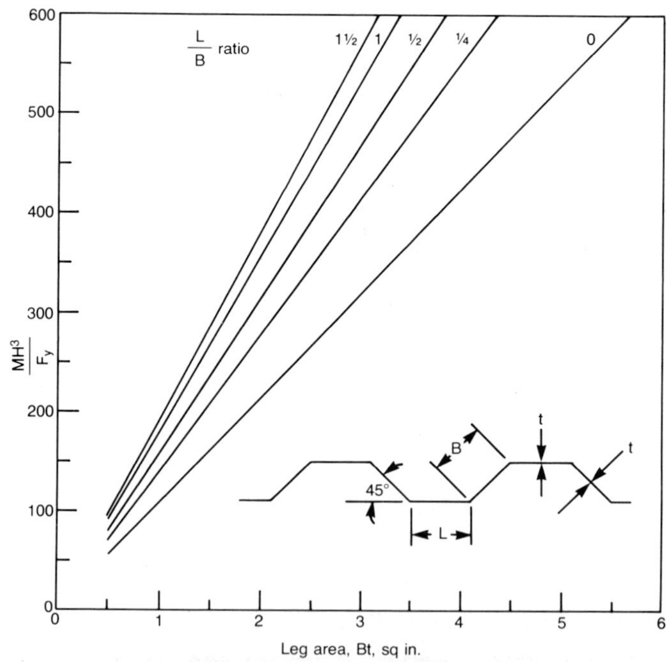

FIG. 1.32 W crimp moment design

As an example, consider determining the V crimp dimensions for an 8-ft-high mud pit with an 18-ppg mud weight. This gives an ordinate parameter of 256 in figure 1.31 for an ASTM A36 grade of steel (Fy=36) The graph shows that this can be done equally well with a ¼-in. plate and 9.5-in. leg or a 5⁄16 plate and 7.7-in. leg or a ⅜ plate and 6.4-in. leg.

The crimp wall beam shear equations were worked up as given in figure 1.33 and plotted in figure 1.34. This is the maximum shear stress at the center of the beam where the flexural stress is zero. Each of the three example V crimp beams had an area of about 2.4 sq in., and reading figure 1.34 shows that the beam shear is very low.

The base shear equations were developed and given in figure 1.35. This shows that the base shear stress is also very low for the example V crimp designs.

$$F_s = \frac{VN}{ID} \tag{1.86}$$

$$V_Q = 3.74\ MH^2 \tag{1.87}$$

$$D_Q = 1.41\ tG \tag{1.88}$$

$$G = \frac{12}{L + 0.707\ B} \tag{1.89}$$

$$I_Q = \frac{0.497\ Bt\ (B + 3.03\ L)}{0.707 + \frac{L}{B}} \tag{1.90}$$

$$N = \frac{2.11\ t\ (2\ L + B)}{0.707 + \frac{L}{B}} \tag{1.91}$$

$$533\ F_y \geqq 11.2\ \frac{MH^2\left(2\frac{L}{B} + 1\right)}{BtG\left(3.03\frac{L}{B} + 1\right)} \tag{1.92}$$

$$\frac{MH^2}{GF_y} \leqq 47.5\ Bt\ \frac{\left(3.03\frac{L}{B} + 1\right)}{\left(2\frac{L}{B} + 1\right)} \tag{1.93}$$

FIG. 1.33 Crimp wall beam shear equations

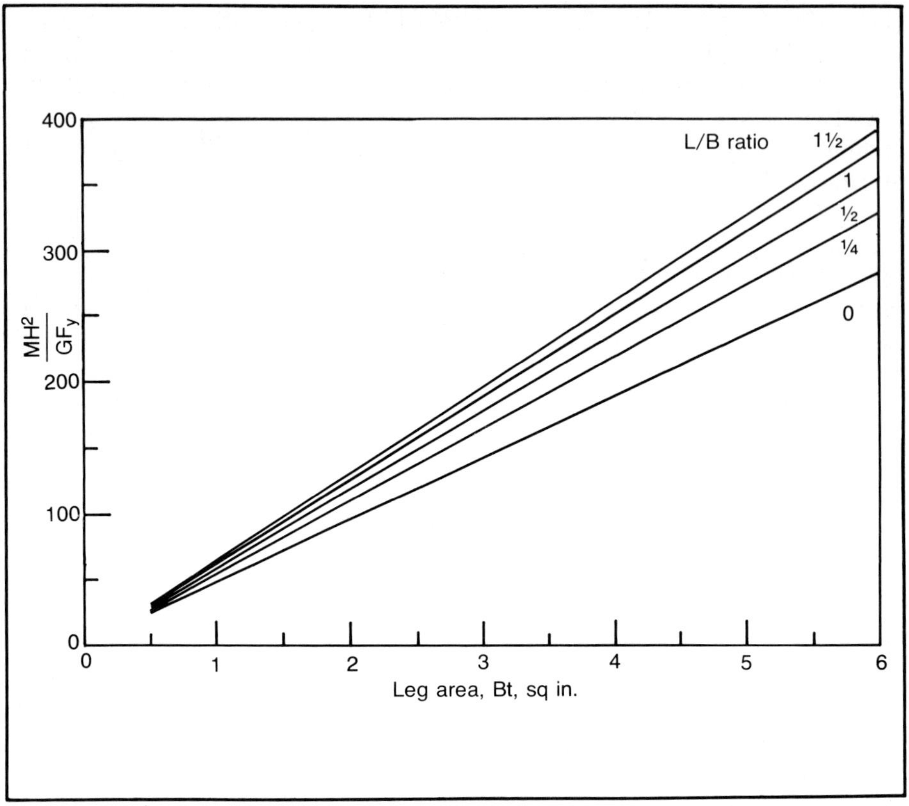

FIG. 1.34 Crimp wall beam shear design

$$W = 3.74\ MH^2 \tag{1.94}$$

$$A_{WQ} = 2t\ (L + B)\frac{(12 \cos 45)}{(L + B \cos 45)} = \frac{16.9\left(1 + \frac{L}{B}\right)}{\left(0.707 + \frac{L}{B}\right)} \tag{1.95}$$

$$533\ F_y \geqq \frac{W}{A_{WQ}} \tag{1.96}$$

$$\frac{MH^2}{F_y} \leqq 4840\ t\frac{\left(1 + \frac{L}{B}\right)}{\left(0.707 + \frac{L}{B}\right)} \tag{1.97}$$

FIG. 1.35 Crimp wall base shear equations

1.6 ELECTRIC RIG DESIGN

When cost effectiveness and performance of the three basic types of drilling rigs are compared, results show that electric rigs are competitive when the total rig cost is considered.

The three basic rig types are mechanical, DC-DC, and SCR AC-DC. The DC-DC electric rig is usually referred to simply as a DC rig, and the SCR AC-DC electric rig is simply called an SCR rig.

By mechanical rig we mean those using chains to connect the power from the engines to the drawworks, rotary, and slush pumps. On both DC and SCR rigs, DC motors are used to drive the mechanical equipment. In DC rigs, a DC generator furnishes power to a DC motor. With a SCR rig, AC generators are used, and the power is converted in an SCR set from AC to DC and transmitted to the motors.

There are two basic differences and advantages of electric rigs over mechanical rigs: the electricity can be piped around electric rigs which gives unlimited versatility in spotting the equipment, and electric rigs have less downtime. Basic differences between electric rigs are small. With a DC rig, the motors have to be assigned to particular generators. With an SCR rig the power is pulled off a common bus.

Factors to Consider in Rig Selection

The factors having the largest influence on the decisions regarding rig size and type are: operator preference, contractor practice and experience, and special job requirements. Special requirements include weight limitations (such as with helicopter rigs), logistics, and the extent of knowledge on the reservoir.

Our primary concern is drilling rigs for offshore fixed platforms. However, most of the information on power equipment will also apply to land rigs.

In several of the following example systems, we see that the cost of the power equipment on DC rigs varies from 20% to 44% more than mechanical and costs from 6% to 25% more for SCR than for DC. However, the differential cost of the power equipment as a function of the total rig cost only varys from 3% to 6% over mechanical for a DC rig and from 2% to 5% over a DC rig for an SCR rig.

We emphasize that the numbers given in this report should not be taken too seriously. Purpose and intent here is to give indications and approximate values.

A big factor in selecting rig type is the consequences due to downtime. This is a function of operator spread. For example, the total cost for making hole with an offshore exploratory rig is roughly

TABLE 10
Relative comparisons of rig types

Consideration	Mechanical	DC	SCR
Power equip., cost	Minimum	More than mech. rig	More than DC rig
Power cost/rig cost	—	Few % greater than mech.	Few % greater than DC
Brand reliability	—	Better than mech.	Better than DC
Rig down time	—	Less than mech.	Less than DC
Flexibility for auxiliary equip. power	Bad	Better than mech.	Best
Flexibility for alternate power use	Difficult	Better than mech.	Easy
Training for repair	Easiest	More involved	More sophisticated
Maintenance	Mech. only	Mech. and electrical	Mech. and electrical/ electronics
Replacement of primary driving unit	Time consuming	Less than mech.	Same as DC rig
Time requirement for rig-up	Heavy	Less than mech.	Same as DC rig
Distribution loading	Maximum	Less than mech.	Same as DC rig
Overall structural weight (& cost)	Maximum	Less than mech.	Same as DC rig
Rig-up time on platform	—	Usually less than mech.	More than DC
Mobility for modular rigs	Poor	Good	Excellent
Flexibility in rig layout	Poor	Good	Excellent
Drill floor work area	—	More than mech.	Same as DC rig
Driller's control	—	Better than mech.	Same as DC rig
Independent rotary drive	—	Easier than mech.	Same as DC rig
Crew safety	—	Better than mech.	Same as DC rig

two times the day rate of the rig. The spread is, of course, lower on land rigs. If an engine goes out on a mechanical rig driving the drawworks, operations often have to be shut down. Operations should never be shut down due to an engine loss on an SCR rig since power to the drive motors is drawn from a common bus bar receiving energy from several engines.

A number of other factors to consider are shown in a relative comparison of rig types in table 10. Crew safety is a very important consideration. In DC and SCR rigs, the engines can be 50-100 ft away from the drill floor versus about 10 ft for mechanical rigs. The risk of a fire or explosion due to engine spark or backfire, is, therefore, reduced with DC and SCR rigs.

When selecting a rig, the contingency for ancillary power should be considered. The cost per hp is about $20 for motors versus about $48 for engines. This means that the additional cost for AC or DC drive on extra centrifugal pumps, air compressors, hydraulic motors and other equipment is a lot less with an SCR rig than a mechanical rig.

Being able to pipe power around the rig becomes very important with the large offshore exploratory vessels. With an SCR rig, the power from the same engine package can be used during drilling operations and to power the propulsion units during rig moves. This is possible but difficult with DC rigs, and impractical with mechanical rigs.

Rig Design

A small collection of drilling rig design rules of thumb is given in table 11. The numbers are rough estimates. An advantage of these design rules is that parametric studies and preliminary designs can be performed quickly using these estimates. The disadvantage in using these estimates is that the numbers are for common or general designs and could be misleading for special situations.

Our estimates in table 11 show, for example, that the pounds weight per hp is about 7 for motors versus about 16 for engines, or a 2.3 ratio. The consequences are that the substructure with a mechanical rig would be much heavier than with a DC or SCR rig, if the same design procedures were used in both cases. This effect is amplified when the additional weights of the compound and rotary drive on mechanical rigs are taken into account.

An independent rotary drive with an electric rig is structurally simple and lightweight on DC or SCR rigs. Independent rotary gives the driller better and smoother control. Independent rotary also makes for a cleanlier unobstructed drill floor and is easier to maintain or repair.

TABLE 11
Drilling rig design rules of thumb

Engine weight	16 lb/hp
Motor weight	7 lb/hp
Engine—generator weight	29 lb/kw
Engine deck load	270 lb/sq ft
Motor deck load	440 lb/sq ft
Engine—generator deck load	260 lb/sq ft
Engine deck area	16 hp/sq ft
Motor deck area	60 hp/sq ft
Engine—generator deck area	10 kw/sq ft
Engine cost	$2.90/lb
Motor cost	$2.90/lb
Engine—generator cost	$2.90/lb
Engine cost	$48/hp
Motor cost	$20/hp
Engine—generator cost	$84/kw
Drawworks cost	$2.80/lb
Mud pumps	$4.70/lb
Mud tanks, skids	$0.60/lb
Mud tank weight	85 lb/bbl
Platform quarters	$65/sq ft
Platform quarters (OSHA)	50 sq ft/man
Platform galley	$80/sq ft
Platform galley	$675/man
Potable water requirements	65 gal/man-day
Drilling rig cost	$160/ft
Drilling rig power	10 ft/hp

For purposes of setting up an exercise on a comparison of rig costs, we chose the rig power distribution given in table 12. It gives the power distribution for the conditions of drilling, stuck pipe, and pipe tripping or running casing for drawworks horsepower varying from 750 to 2000. Note that this distribution is very arbitrary. For example, there are circumstances where 1600 hp on slush pumps would be about right, whereas use of only one pump at 800 hp would be adequate for other situations. The least reliable power requirements given in table 12 are the estimated energys for rig accessories and for housekeeping. The power for housekeeping (that is, the energy for quarters, gally, rig lighting, etc.) obviously depends on the level of manpower being used.

TABLE 12
Design summary of rig power distribution

Rig rating (Maximum drawworks hp)	750	1,000	1,500	2,000
Rig drilling class	Light	Inter-mediate	Medium	Heavy
Drilling condition				
Drawworks	0	0	0	0
Rotary	400	400	800	800
Slush pumps	1,600	1,600	1,600	2,600
Accessories	400	400	400	400
Housekeeping	100	100	100	100
Total	2,500	2,500	2,900	3,900
Stuck pipe				
Drawworks	750	1,000	1,500	2,000
Rotary	0	0	0	0
Slush pumps	800	800	800	1,300
Accessories	200	200	200	200
Housekeeping	100	100	100	100
Total	1,850	2,100	2,600	3,600
Pipe tripping or running casing				
Drawworks	750	1,000	1,500	2,000
Rotary	0	0	0	0
Slush pumps	0	0	0	0
Accessories	200	200	200	200
Housekeeping	100	100	100	100
Total	1,050	1,300	1,800	2,300
All maximum (Reference Only)				
Drawworks	750	1,000	1,500	2,000
Rotary	400	400	800	800
Slush pumps	1,600	1,600	1,600	2,600
Accessories	400	400	400	400
Housekeeping	100	100	100	100
Total	3,250	3,500	4,400	5,900

The assumed total power for each condition and each rig horsepower is plotted in figure 1.36. This shows that the maximum power to all components is considerably greater than ever needed and is shown for reference only. The drilling condition can usually be used to determine the maximum rig power requirement.

Specifications and Description of Equipment Components

The most commonly used drilling motors are probably GE, and a summary of the specs on the 752 and the 761 are given in table 13. Since the production of the 752 has been so much greater than with the 761, the price is somewhat better. The cost of 761 is roughly 20% more than the 752 in terms of $/hp.

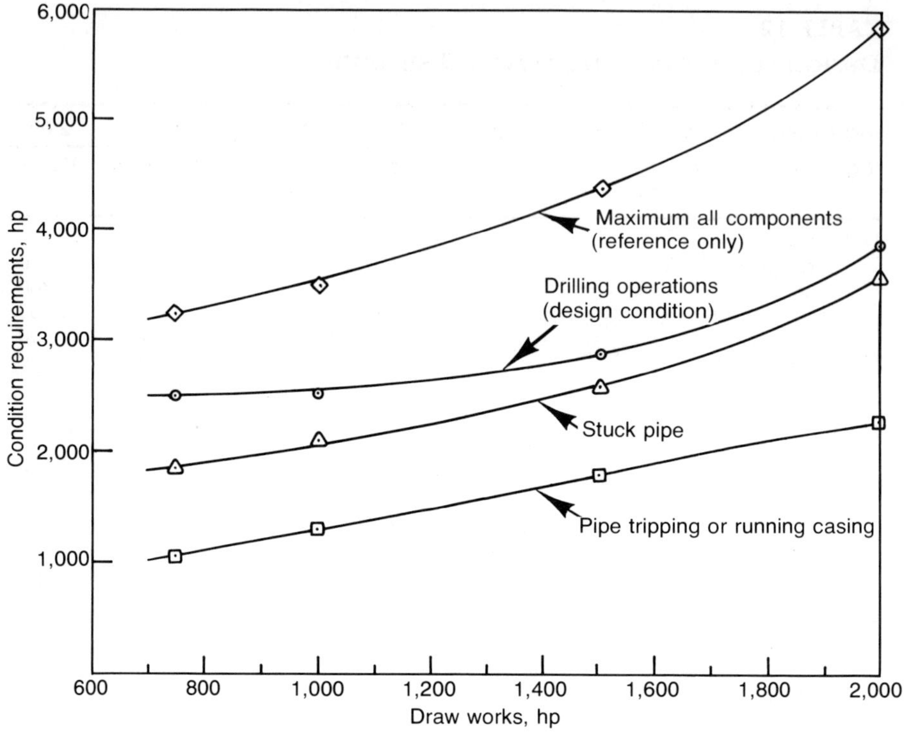

FIG. 1.36 Rig hp requirements as a function of operating condition

TABLE 13
Summary of design data on GE DC drilling motors

	Model 752	Model 761
Horsepower ratings		
Continuous	900	400
Intermittent	1,000	550
Weight, kips	7.2	3.9
Overall dimension, ft		
Length	4.9	3.3
Width	3.3	2.8
Height	4.8	4.5
Design functions		
Wt, lb/inter. hp	7.2	7.1
Deck area, sq ft	16.2	9.2
Deck load, psf	444.0	424.0
Inter. hp/sq ft	61.7	59.8

In selecting equipment from different supplies or various components from a particular vendor, required design capability and cost are not the only considerations. The ideal theoretical design might use a different size engine, motor, and generator for each function. But this is not practical when maintenances, repair, spare parts inventory, training, and other factors are considered.

For these reasons, all 752 motors will be used in our 750 hp DC and SCR rig examples. This gives over twice the capability for power to the rotary than is needed. Sometimes this would be the best decision to make; for other conditions a mix of 752 and 761 would be satisfactory. Similar reasoning applies to the selection of engines and generators.

Suppliers of the SCR units include the Baylor, GE, Hill-Hayes companies and others. Production of the SCR is very low compared to other components in the power system so the SCR is a custom job. Also, more than any other component, the SCR must be sized and built for particular rig requirements.

In putting the data together for sizing and costing of the various rigs in this exercise, we have assumed for simplification that cement pumps and power for these pumps would be leased or furnished by the operator. But an SCR should be specified to enable their use in power conversion to run the cement pumps whenever needed. For present purposes, the SCR is viewed only as a black box that accepts AC current and spits out DC current. All of the details that go into specification of the SCR will be ignored.

TABLE 14
Rig design data for Detroit Diesel Allison engine-generator sets

Engine drive	8V71	12V71	16V71	12V149	16V149
Kilowatt rating @ 1,800 rpm, 60 hz, standby condition					
Natural aspirated	200	300	400	540	725
Turbocharged	255	400	525	755	1,010
Turbocharged-intercooled	—	—	—	815	1,100
Overall dimensions, ft					
Length	8.3	9.4	12.8	12.1	13.6
Width	3.3	4.1	4.1	5.3	4.3
Height	5.0	5.2	5.2	—	—
Dry wt, kips	5.5	8.6	9.3	—	—

TABLE 15
Rig design data for caterpillar engines

Engine (all turbocharged–aftercooled)	399	398	379	353	349	348	346	343	
Cylinders	V16	V12	V8	6	V16	V12	V8	6	
Rpm, maximum and intermittent ratings	1,300				2,000				
Rpm, continuous rating	1,200				1,800				
SWAC hp									
Maximum	1,700	1,275	850	640	1,470	1,100	735	550	
Intermittent	1,425	1,075	715	535	1,220	920	610	460	
Continuous	1,125	850	565	425	970	725	480	365	
JWAC hp									
Maximum	1,550	1,170	780	585	1,360	1,020	680	510	
Intermittent	1,300	975	650	490	1,130	850	565	425	
Continuous	1,000	750	500	375	890	670	445	335	
Wt (includes radiator), kips	25.6	18.9	15.2	9.8	17.1	12.0	9.1	6.3	
Overall dimensions, ft									
Length	14.3	12.4	8.9	8.6	12.6	9.7	8.3	6.7	
Width	7.3	6.1	5.2	3.3	6.1	5.2	5.2	3.3	
Height (includes 8-in skid)	9.1	8.2	7.8	6.2	8.2	7.8	7.8	5.5	
Design functions									Avg.
Wt, lb/inter. SWAC hp	18.0	17.6	21.2	18.4	14.0	13.0	14.9	13.7	16.4
Deck area, sq ft	104.0	75.6	46.3	28.4	76.9	50.4	43.2	22.1	—
Deck load, psf	246.0	250.0	328.0	345.0	222.0	238.0	211.0	285.0	266.0
Inter. SWAC hp/sq ft	13.7	14.2	15.4	18.8	15.9	18.3	14.1	20.8	16.4

Note: Horsepower ratings are bhp without fan. For radiator cooling deduct 5.4% from SWAC continuous rating, for example.

Summaries of design data on Detroit Diesel engines and engine-generator sets for natural aspirated, turbochanged, and turbocharged-intercooled engines are given in table 14 for the engines commonly used on drilling rigs. Horsepower for both maximum and continuous ratings are shown.

Similar design data on Caterpillar engines and engine-generator sets for maximum, intermittent, and continuous ratings are given in tables 15 and 16. Horsepower ratings for both SWAC (separate water aftercooled) and JWAC (jacket water aftercooled) are shown.

For certain situations, the choice of other brands of engines might be better. Further, we emphasize that our *intent is not* to recommend nor to discourage the use of any particular brand of any of the components in the power systems discussed here. But without mention of any brand or model numbers, we end up with an uninteresting, sterile, academic piece.

In general, the facts seem to be that the Caterpillar is a heavy, sturdy, reliable, and standard machine. The Detroit is lighterweight than the Caterpillar in terms of lb/hp, and nearly every Detroit for drilling rigs is a custom design. To properly order or specify an engine requires a long list of requirements including size and type of muffler, silencer, spark arrestors, injectors, radiator, alternate cooling, type of paint, space allowables, weight allowables, and precise statement of intended usage.

Consequently, it is much easier to run parametric analyses and systems analyses, maintenance crew training is simplified, and spare parts inventory is reduced with Caterpillars as compared to Detroits. On the other hand, being able to specify a customized engine for special situations is sometimes advantageous.

Comparison of Electric Rigs

Using the assumed power distribution by function and by rig class given in table 12, we selected a list of components in the power system (table 17) for illustrative purposes. Power distribution circuits are given in figure 1.37 for the DC rigs and in figure 1.38 for the SCR rigs.

We tabulated the estimated cost of each component. The total cost of the power equipment for each rig type as a function of rig class is shown in figure 1.39. For the range of drawworks horsepower between 750 and 2000 hp, the cost of the power system varied \$340-590K for mechanical, \$500-810K for DC, and \$620-950K for SCR.

Ratios of the total power system cost as a function of rig class were calculated (figure 1.40). Results show that the incremental costs for mechanical to DC and from DC to SCR were minimum at about 1000 hp drawworks. Cost ratio of the power equipment at 1000 hp

TABLE 16

Rig design data for Caterpillar engine-generator sets

Engine drive (all TA)	399	398	379	353	349	348	346	343	
Rpm	1200				1800				
Kilowatt ratings @ 60 hz									
SWAC standby condition	900	730	485	365	810	615	405	305	
SWAC prime mover	875	665	440	330	720	550	360	275	
JWAC standby condition	900	675	450	335	750	565	375	285	
JWAC prime mover	800	600	400	300	670	500	335	250	
Weight (includes radiator), kips	28.6	20.6	17.7	10.0	19.2	15.1	13.4	7.8	
Overall dimensions, ft									
Length	19.0	12.2	12.0	10.4	12.3	10.8	8.8	10.7	
Width	7.3	6.1	5.2	3.3	6.1	5.2	5.2	3.3	
Height (includes 8-in. skid)	9.1	8.2	7.8	6.2	8.2	7.8	7.8	5.5	
Design functions									avg.
Wt, lb/standby SWAC kw	31.8	28.2	36.5	27.4	23.7	24.6	33.1	25.6	28.9
Deck area, sq ft	139	74.4	62.4	34.3	75.0	56.2	45.8	35.3	—
Deck load, psf	206	277	284	292	256	269	293	221	262
Standby SWAC kw/sq ft	6.5	9.8	7.8	10.6	10.8	10.9	8.8	8.6	9.2

Note: For radiator cooling, deduct 5.4% from SWAC continuous kw rating, for example.

TABLE 17

Summary of power equipment

Rig type	Mechanical				DC-DC				SCR			
Drawworks hp	**750**	**1000**	**1500**	**2000**	**750**	**1000**	**1500**	**2000**	**750**	**1000**	**1500**	**2000**
Driller's consoles	1	1	1	1	1	1	1	1	1	1	1	1
Drawworks compound	1	1	1	1	0	0	0	0	0	0	0	0
Rotary drivers	1	1	1	1	1	1	1	1	1	1	1	1
Pump drivers	2	2	2	2	2	2	2	2	2	2	2	2
DC switchgear set	0	0	0	0	1	1	1	1	1	1	1	1
DC controls	0	0	0	0	1	1	1	1	0	0	0	0
500 KVA transformers	0	0	0	0	0	0	0	0	2	2	2	2
SCR set	0	0	0	0	0	0	0	0	1	1	1	1
Engines												
D399 SWAC	0	0	0	4	0	3	3	4	0	0	3	4
D398 SWAC	3	4	4	0	2	0	0	0	3	3	0	0
Generators												
GE752	0	0	0	0	4	4	4	6	0	0	0	0
500 kw	0	0	0	0	0	2	2	2	0	0	0	0
700 kw	0	0	0	0	0	0	0	0	3	3	0	0
930 kw	0	0	0	0	0	0	0	0	0	0	3	4
Engine-generator sets												
Cat 346	2	2	2	2	2	0	0	0	0	0	0	0
Motors												
GE 752	0	0	0	0	4	5	5	7	4	5	5	7

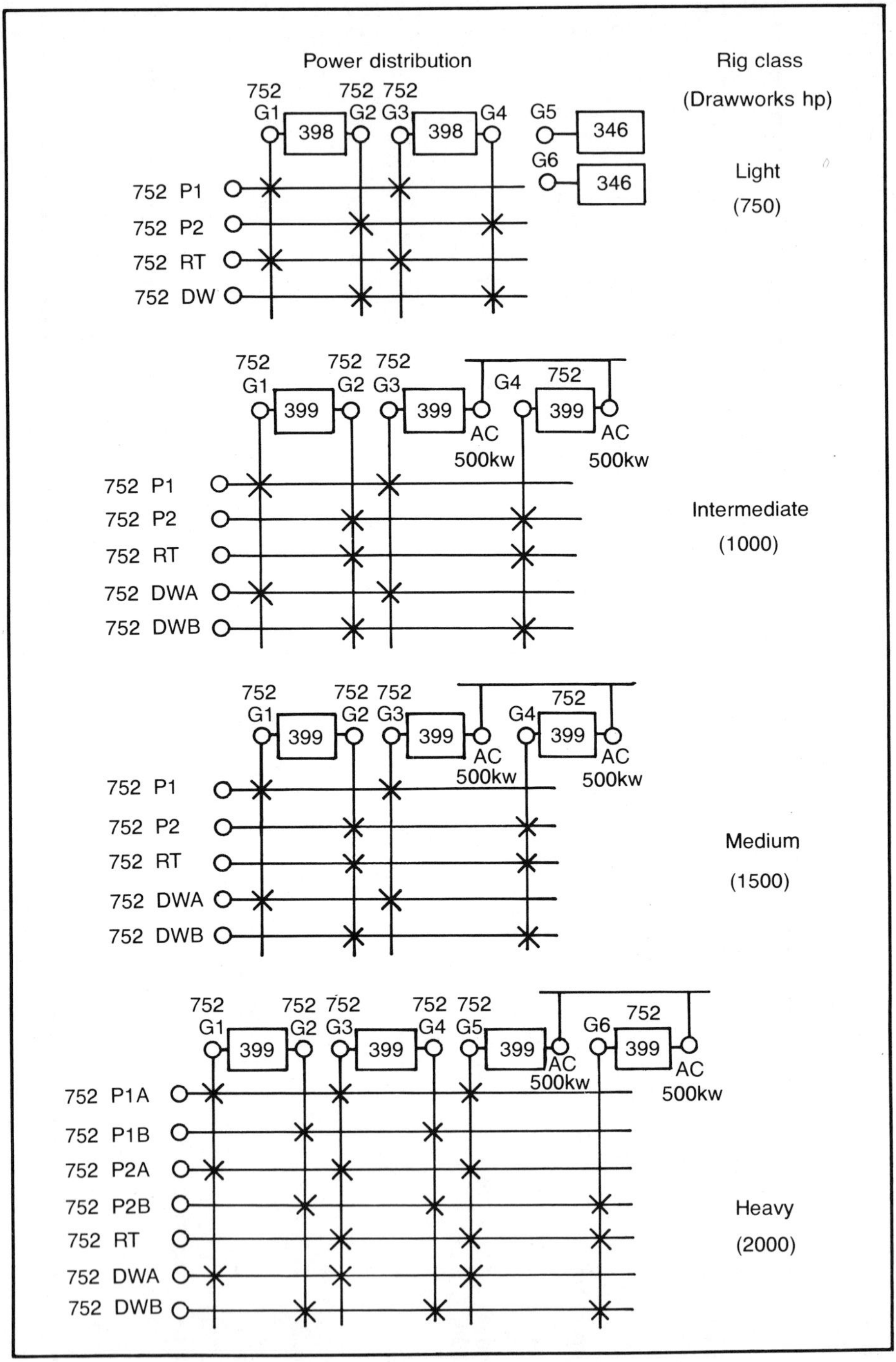

FIG. 1.37 DC-DC rig power distribution circuits

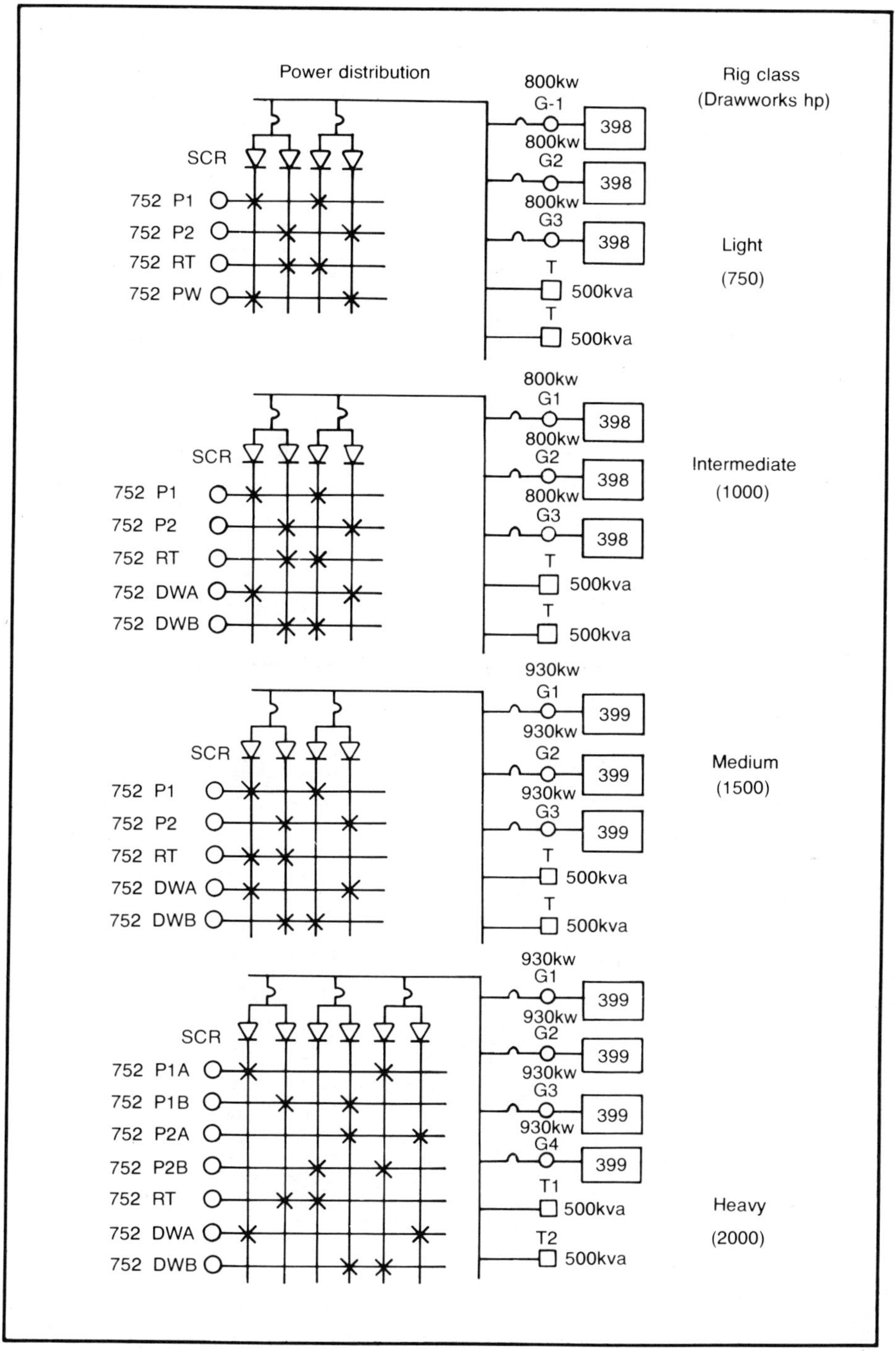

FIG. 1.38 SCR AC-DC power distribution circuits

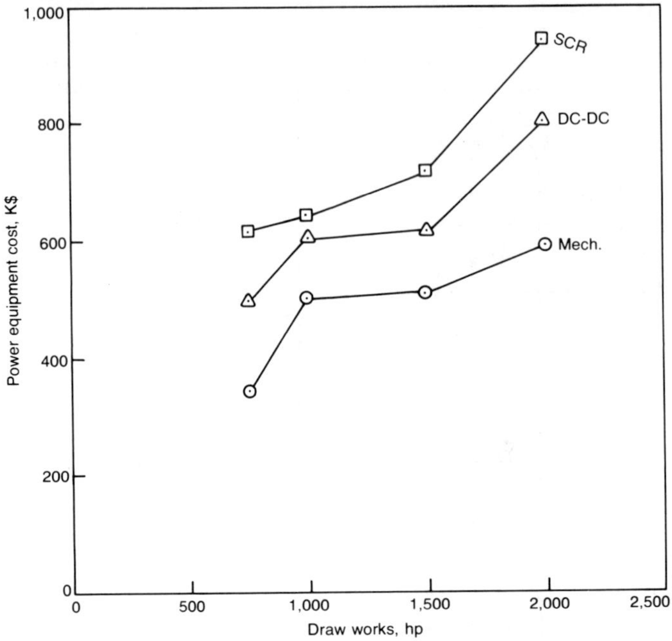

FIG. 1.39 Power equipment cost comparisons by rig type

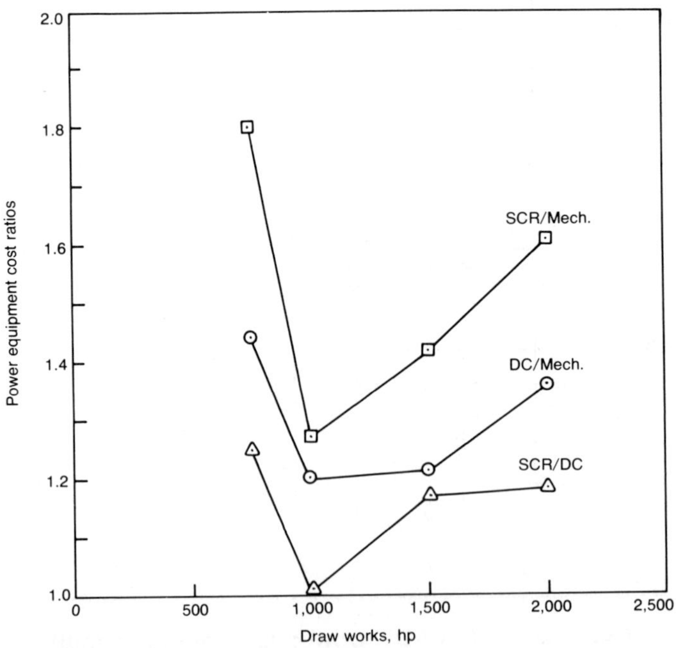

FIG. 1.40 Relative costs of power equipment by rig type

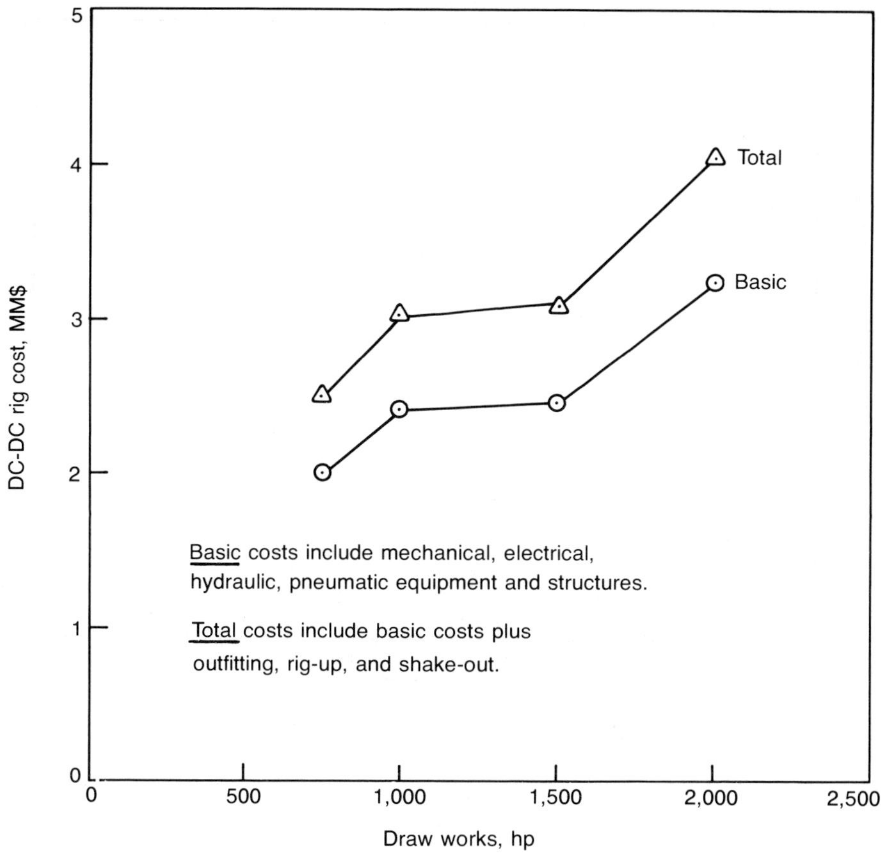

FIG. 1.41 Cost of DC-DC rigs by rig type

was about 1.20 for DC/mechanical; that is, the DC power system costs about 20% more than the mechanical system.

But these results could be misleading. Total cost and performance in making holes, not subsystem costs, should be reviewed. To do this properly, the influence of power equipment cost on the total rig cost must be determined. For reference purposes, ball-park type of estimates of the total cost of DC rigs as a function of rig class are given in figure 1.41. Normalized total rig cost in terms of thousand dollars per drawworks horsepower as a function of rig class is shown in figure 1.42.

Finally, combining the results in figures 1.39 and 1.41 gives the incremental costs of the power equipment as a function of both rig type and rig class (figure 1.43). This shows, for example, that with a 2000 hp drawworks rig, the DC power equipment makes the rig cost about 5% more than the mechanical rig; SCR makes the rig cost about 4% more than DC; and the SCR power equipment makes the rig cost about 9% more than a mechanical rig.

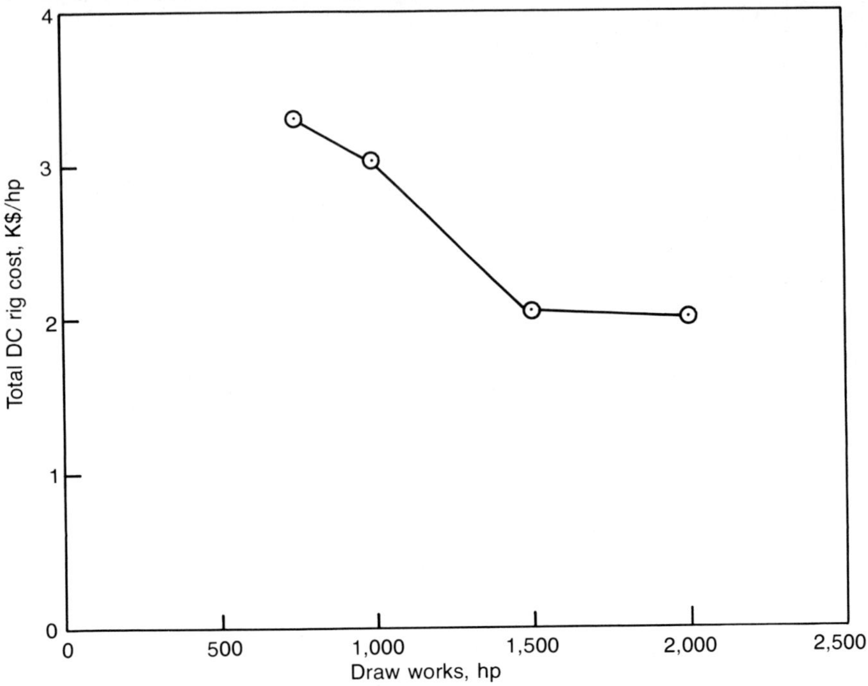

FIG. 1.42 Relative cost of DC rigs

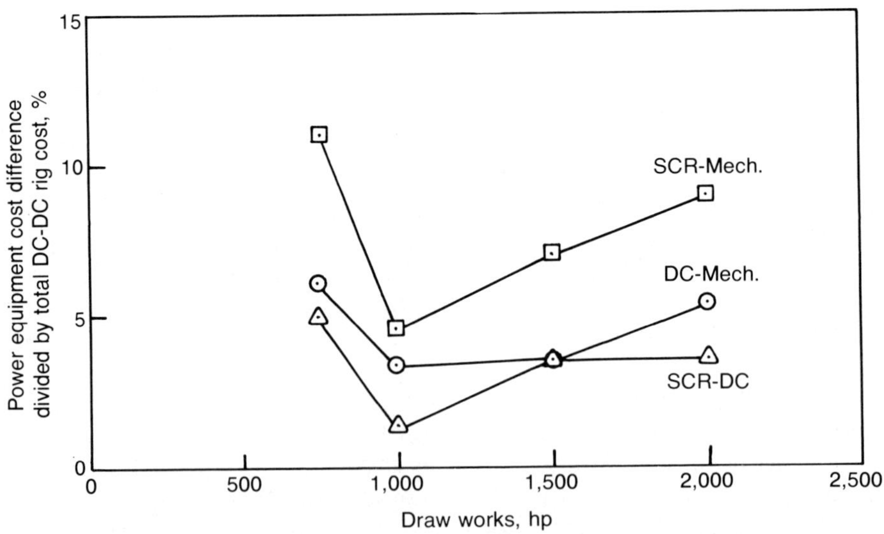

FIG. 1.43 Incremental costs of power equipment by rig type

These incremental cost percentages are somewhat underestimated as a consequence of using the total rig cost of the DC rig for comparison in each case. But consider that because there is more weight in the substructure of mechanical rigs, the cost of structural components is more for a mechanical rig than an electric rig, which partly justifies the given method of comparison.

The incremental cost percentages are the numbers that the contractor must plug into the equations to arrive at a profitable day rate. The operator, however, must consider the overall costs. Assuming a cost spread to the operator for drilling, transportation, supply and other factors of about two times the drilling day rate, the incremental cost percentages to the operator are roughly one-half of the incremental increase in the day rate.

All of this means that if the day rate for a 750-hp SCR rig is 12% more than the day rate for a mechanical rig, it would be more profitable for the operator to use the SCR rig if it can make 6% more hole in a given time period than a mechanical rig. On a short term basis, there should be no difference in drilling rates for the different types of rigs. On a longer term basis, when downtime due to equipment repair or replacement is taken into account, the slight increase in cost of using an SCR rig instead of a mechanical rig can result in net cost savings.

REFERENCES

1. *Drilling and Well Servicing Structures,* API Spec 4E, 2nd ed., March 1974.
2. *Manual of Steel Construction,* 8th ed., 1980, AISC.
3. *ASTM Standards,* part 4, structural steel, 1979.
4. Bynum, D., Jr., L. U. Rastrelli, and R. C. Dehart, "Photoelastic Observations Using Tolyene Di-Isocyanate Polyurethane," *Exp. Mech.,* July 1964.
5. Bynum, D., Jr., "On the Accuracy of Fringe Multiplication with Birefringent Coatings," *Exp. Mech.,* July 1966.
6. Bynum, D., Jr. and R. C. Dehart, "Fillet and Groove Stress Concentrations," *Exp. Mech.,* June 1964.
7. Bynum, D., Jr. and M. M. Lemcoe, "Birefringent Coating Analysis of Laterally Loaded Perforated Plates," *Exp. Mech.,* June 1963.
8. Bynum, D., Jr., "Birefringent Coating Analysis of Tensile Plate Containing a Heterogeneous Inclusion using Fringe Multiplication," *M. S. Thesis,* May 1961, Univ. Wash. Library.

9. Agarwal, R. B., and D. Bynum Jr., "Photoelastic Strain Factors in Flexible Concrete," *Matls. and Structures*, V5, N29, 1972.
10. Heywood, R. B., *Designing by Photoelasticity*, 1952, Chapman Hall Publ.
11. Frocht, M. M., and H. N. Hill, "Stress Concentration Factors around a Central Circular Hole in Plate Loaded through a Pin in the Hole," *Trans. ASME*, V 62, 1940.
12. Levy, G., and F. C. Smith, "Stress Distribution near Circular Hole Loaded by Pin," *7th Intl, Cong, Appl. Mech.*, 1948.
13. Howland, R. C. J., "On the Stresses in the Neighborhood of a Circular Hole in a Strip under Tension," *Phil. Trans. Roy. Soc.*, V229, 1930.
14. Roark, R. J., *Formula for Stress and Strain*, 1954, McGraw-Hill.
15. Peterson, R. E., *Stress Concentration Design Factors*, 1953, John Wiley and Son.

CHAPTER

2 OFFSHORE MOBILE RIGS

2.1 CALCULATION AND EVALUATION OF RIG FATIGUE

The preliminary procedures for performing fatigue analyses on offshore mobile rigs usually give conservative results and enable an approximate evaluation to be made very quickly. More involved and refined procedures must be used in final design and evaluations.

One hundred year storm data is given for six different world areas. The data is used in examples to show fatigue damage to a rig subjected to weather conditions greater than the environmental criteria used to design the rig.

General Approach

To keep this section brief, it is necessary to make a number of general assumptions. For example, in going through the fatigue analyses, it is assumed that adequate structural integrity is maintained.

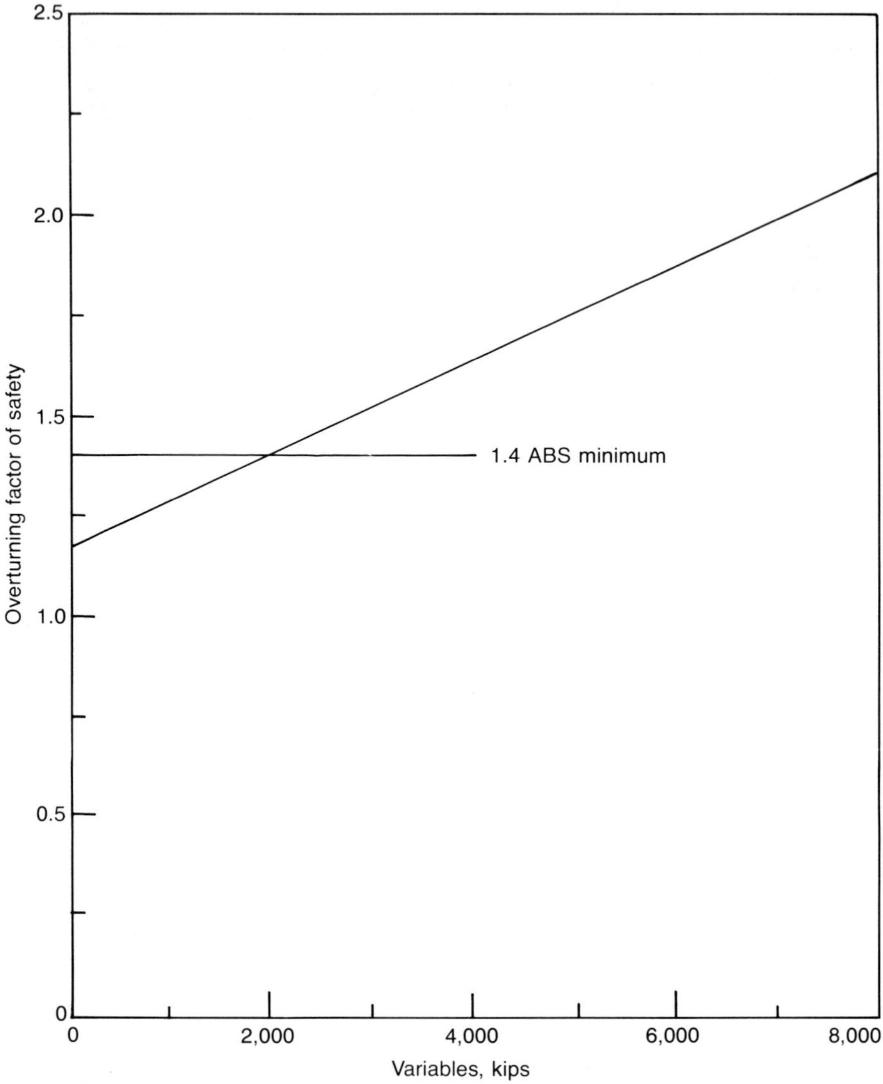

FIG. 2.1 Overturning factor of safety for typical jackup rig

The overturning factor of safety due to environmental loads is a function of the rig variables, i.e. consumables as well as ballast, for both floaters and jackups. Additionally, overturning factor of safety on a jackup is a function of water depth and air gap. It is assumed here that adequate overturning safety is sufficient and that the effect of sufficient variables is reflected in overall stress analyses.

The maximum allowable wave height for a given mobile rig is a nonlinear function with maximum allowable wind velocity. The combination of maximum wave and maximum wind is also a function of water depth for jackups.

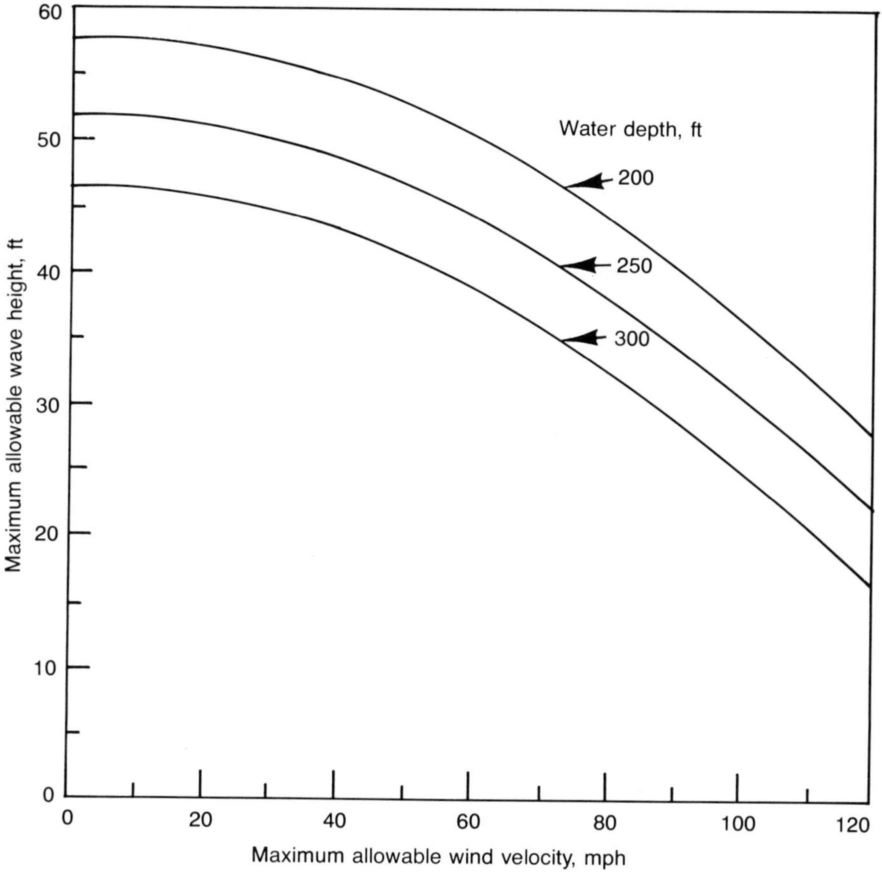

FIG. 2.2 Variation of design criteria for typical Gulf of Mexico jackup rig

To simplify our analyses, it is assumed here that the components of oscillatory stress due to waves and wind are equal in our most severe storm criteria considered. Also, the current causes a non-oscillating load, and it is assumed that the component of stress attributable to current can be taken as 10% of the total stress.

To get perspective, engineers have been concerned with cyclic failure of machinery for over 60 years. But it has only been about 30 years since Milton Miner published his important work in 1945 that for the first time put fatigue analyses on a practical basis. Fatigue analyses of offshore rigs first began to be considered only about nine years ago.

100-Year Storm Criteria

The usual accepted values for maximum winds and waves in a 100-year storm for 6 areas of the world are given in tables 18–21.

TABLE 18

Design criteria for 100-year storm

Area	Storm Case	1 Malaysia	2 Persian Gulf	3 Gulf of Mexico	4 Southern North Sea	5 Middle North Sea	6 Northern North Sea
Maximum wave height, ft.		33	35	39	50	75	100
Maximum one minute wind gust, knot		62	53	74	100	100	100
Design severity compared to:							
Malaysia		1	0.92	1.15	1.56	1.83	2.08
Persian Gulf		1.09	1	1.25	1.70	2.00	2.27
Gulf of Mexico		0.87	0.80	1	1.36	1.60	1.82
Southern North Sea		0.64	0.59	0.73	1	1.17	1.33
Middle North Sea		0.54	0.50	0.63	0.86	1	1.14
Northern North Sea		0.48	0.44	0.55	0.75	0.88	1

TABLE 19

Annualized 100-year storm wave height distribution

Number of Waves	Wave height, ft					
	Malaysia	Persian Gulf	Gulf of Mexico	Southern North Sea	Middle North Sea	Northern North Sea
10^7	4	4	4	4	4	4
10^6	8	9	9	10	14	18
10^5	12	13	14	17	24	31
10^4	16	17	19	22	34	45
10^3	21	22	24	30	44	59
10^2	25	26	29	37	55	73
10^1	29	30	34	44	65	86
10^0	33	35	39	50	75	100

Note: Assumes one 30 minute maximum storm at a single location in 100 years.

TABLE 20
Annualized 100-year storm stress distribution

Number of Cycles	Stress, psi					
	Malaysia	Persian Gulf	Gulf of Mexico	Southern North Sea	Middle North Sea	Northern North Sea
10^7	3.6	3.4	3.1	2.4	1.6	1.2
10^6	7.3	7.7	6.9	6.0	5.6	5.4
10^5	10.9	11.1	10.8	10.2	9.6	9.3
10^4	14.5	14.6	14.6	13.2	13.6	13.5
10^3	19.1	18.9	18.5	18.0	17.6	17.7
10^2	22.7	22.3	22.3	22.2	22.0	21.9
10^1	26.4	25.7	26.1	26.4	26.0	25.8
10^0	30	30	30	30	30	30

Note: Assumes 1.2 safety factor on 36 Psi yield strength.

TABLE 21
Relative fatigue severity

Number of cycles	Case 3 fatigue fraction	Case 3 rig in case 4 weather: Case 3 stress × 1.36	Case 3 rig in case 4 weather: Fatigue fraction	Case 3 rig in case 5 weather: Case 3 stress × 1.60	Case 3 rig in case 5 weather: Fatigue fraction
10^7	0	4.2	0	5.0	0
10^6	0	9.4	0	11.0	0
10^5	0	14.7	0	17.3	1540×10^{-4}
10^4	0	19.9	270×10^{-4}	23.4	556×10^{-4}
10^3	22.2×10^{-4}	25.2	71.4×10^{-4}	29.6	182×10^{-4}
10^2	4.34×10^{-4}	30.3	21.3×10^{-4}	35.7	58.8×10^{-4}
10^1	1.0×10^{-4}	35.5	5.88×10^{-4}	41.8	20.8×10^{-4}
10^0	0.2×10^{-4}	40.8	1.66×10^{-4}	48	6.7×10^{-4}
Sum	0.00278	—	0.0370	—	0.236
Reciprocal	360	—	27	—	4.2
FS on 20 years	18	—	1.35	—	X
Relative fatigue Severity	1	—	13.3	—	85.7
Relative stress Severity	1	—	1.36	—	1.60

These maximum storm conditions are usually taken concurrently with a two or three-knot current.

The 100 year maximum one minute wind gust for all of the North Sea is 100 knots. However, the 100 year maximum wave height is 50 ft at 52° latitude. The 100 year wave height increases by about 5 ft per degree to a storm criteria of 100 ft waves at 62° latitude.

To get a relative comparison of the design severity for different combinations of wind and waves, the assumption of equivalent stress is used here. For example, comparing the southern North Sea with the northern North Sea, the same wind but 50% reduction in wave height gives a 25% reduction in total stress. In other words, a Case 4 storm is 75% of the intensity of the Case 6 storm.

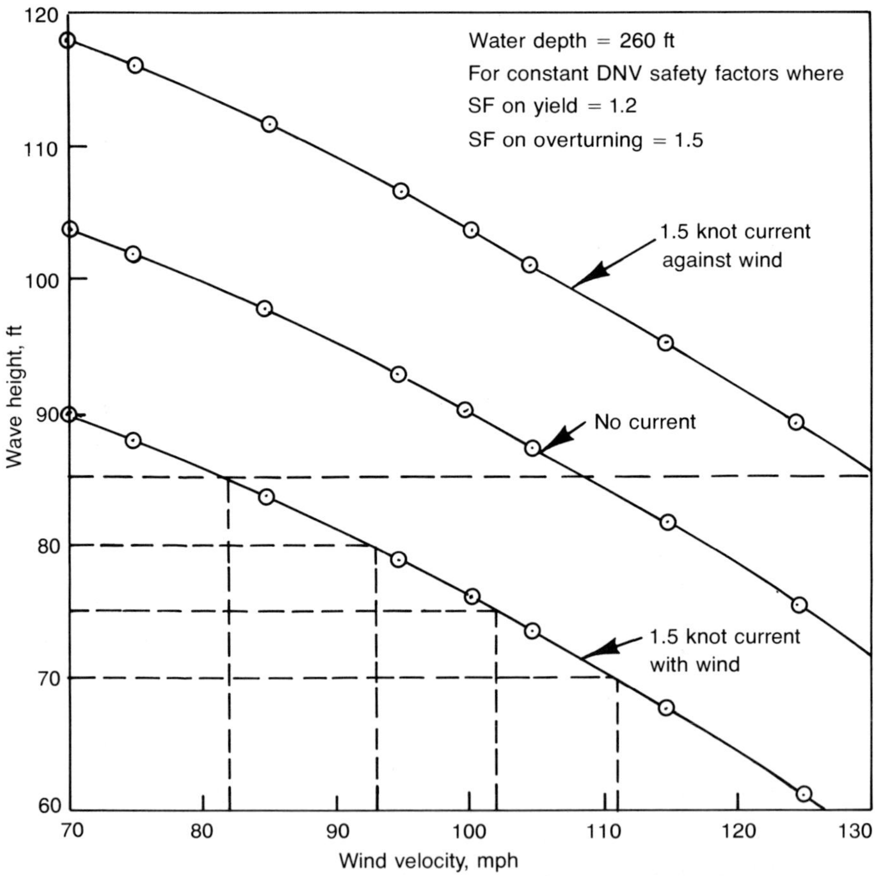

FIG. 2.3 Effect of current on allowable wind-wave conditions for typical North Sea jackup rig

This normalization scheme is then carried through for each storm condition considered. For example, if a rig designed for Malaysia, storm Case 1, was subjected to the environment of storm Case 6, the stresses would be increased by a factor of 2.08.

The actual severity indexes are a function of the given geometry and pay load for any particular vessel. Data such as given by the example of allowable wave heights vs. allowable wind velocity or water depth would have to be used to calculate the true severity indexes for any particular vessel.

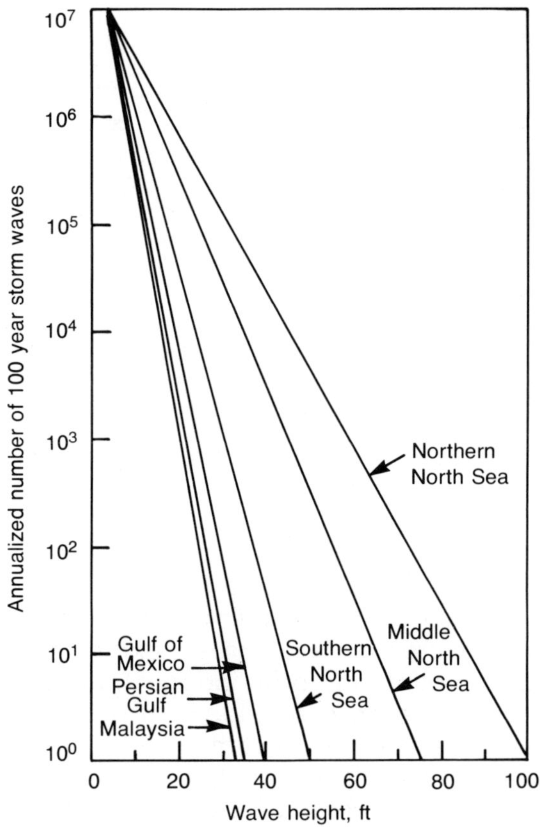

FIG. 2.4 Annualized 100-year storm waves

In the 100-year storm criteria, it is assumed that the storm lasts for about 30 minutes at any one location. With 17-20 second wave periods, this amounts to about 100 of the maximum waves. For purposes of calculation, these statistics are then annualized with the results given in the attached table. The number of waves at each wave height for each storm case are then tabulated.

Approximation of Fatigue Strength

Sufficient fatigue data on any particular material is hard to come by. This is because fatigue strength is dependent upon thickness in mill runs, surface finish, corrosion environment such as underwater or in salt water spray, stress levels, oscillatory stress component in total stress, and other factors. Even with all of the usage conditions well defined, the basic test data for fatigue are widely scattered and

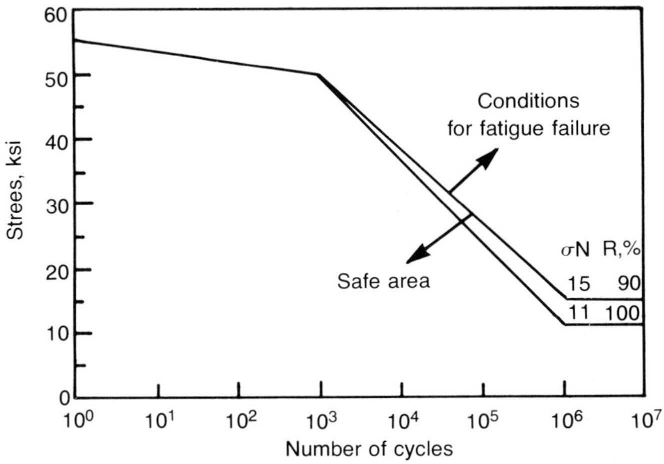

FIG. 2.5 Fatigue endurance strength

the actual fatigue strength must be deduced. With this in mind, the calculated fatigue life of a structure should then have a factor of safety of not less than two and preferably three or four.

When fatigue strength is not given explicitly, it is possible to make an approximation based on the mechanical properties. The fatigue strength drawn on a semilog plot runs from the ultimate strength at one cycle to 90% of ultimate at 1,000 cycles, and from 10 to 30% of ultimate strength at a million cycles, depending on the corrosion environment. Twenty percent of ultimate strength at 10^6 cycles is appropriate for surfaces protected from corrosion by painting or by use of sacrificial anodes.

Endurance life is taken as a million cycles. The endurance stress depends upon the oscillatory stress component. With the assumed steady state stress due to current equal to 10% of the total stress, the endurance strength is increased by 10% of the difference between the fatigue strengths at a thousand and at a million cycles.

The general case of cyclic stresses about non-zero mean stress is usually handled by modified Goodman diagrams. Construction of the diagram gives the decrease in the allowable range between maximum and minimum stress in a cycle as the mean stress level increases and is shown more clearly in figure 2.6.

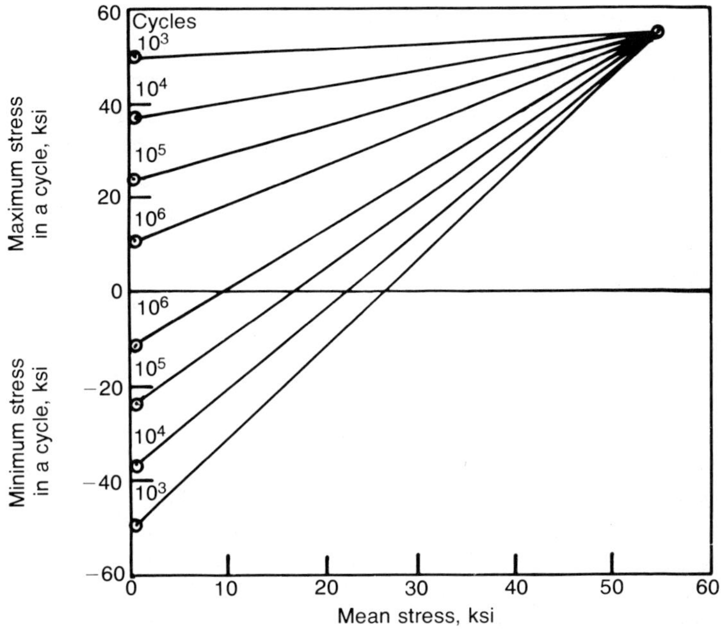

FIG. 2.6 Modified Goodman diagram for a low-strength steel in marine environments

Calculated Fatigue Life

Given the storm wave height distributions, the yearly stress distributions are determined on the basis of the 100-year storm criteria. It is assumed here that the stress is linear with wave height.

With the stress distribution and the fatigue strength given, the fatigue life can be calculated using Miners law. This states that the accumulative damage, AD, is the sum of the cycle ratios and must be less than one. In equation form, AD = sum (n/N), where n is the number of cycles at some stress level and N is the number of allowable cycles at that stress level. The factor of safety, FS, on fatigue is then the reciprocal of the accumulative damage, or FS = 1/AD. Going through the numbers with storm Case 3, for example, shows that the fatigue factor of safety is very high with a value of 18 on 20-year life.

What would be the fatigue life of a rig designed statically and dynamically for a 100-year Case 3 storm but subjected to 100-year

Case 4 weather? The summary of the procedure shows that the calculated life would be 27 years—or very marginal since the fatigue factor of safety is only 1.35 and a factor of two to four is desired.

What would be the fatigue life of a rig designed for Case 3 storms but subjected to Case 5 weather? The results show a yearly AD of 0.236 or a life of only 4.2 years.

The comparative results in regard to relative severity show that stress severity is not an indicator of fatigue severity. For example, compared to Case 3, stress severity is 36% greater for Case 4 and 60% greater for Case 5. But fatigue severity compared to Case 3 is greater by factors of 13 for Case 4 and 85 for Case 5.

Perhaps most importantly, the above procedures can be used to determine accumulative damage when the rig is caught in storms that exceed design conditions. For example, assume that a rig was designed for Case 3 storms but is exposed to two Case 5 storms in the first year of operation. The Case 5 storm causes some points to yield, but structural integrity could still be maintained, that is, no cracks. After these two Case 5 storms, how many years could the rig operate in areas having 100-year Case 4 storms?

Case 5 has an accumulative fatigue damage of 0.236 per storm or 0.472 for two Case 5 storms. Case 4 storms cause an accumulative damage of 0.0370 per year. Allowable years of operation left in Case 4 storm areas is then (1—0.472)/0.0370 = 14.2 years, or 7.1 years using a factor of safety of two.

It is emphasized that some of the assumptions used for the purposes of this article are not generally applicable. The fatigue life and accumulative damage would be calculated using results of structural analyses for the particular rig. The important point is that examples show that fatigue life of offshore rigs can be calculated in a straightforward manner.

It is not always possible to deploy an offshore rig in the area having weather no more severe than the design environment. It is necessary for management to maximize rig profits by extending the work areas of mobile rigs as much as possible. Management decisions regarding rig deployment must consider the risks in severe environments by taking quantitative assessment of accumulative fatigue damage.

2.2 SAFETY FACTORS FOR OFFSHORE RIGS

With offshore drilling rigs moving into deeper waters and rougher areas of the world, it is reasonable to expect the rigs to encounter more severe environmental conditions. The methods of formulating the environmental design criteria for today's rigs, however, are questionable. Traditionally, the rig design criteria are determined by calculating the statistical extremes of wind, wave, and current forces independently. These loads are then combined and used in the design of the unit without regard to their occurrence in time, direction, or space.

The element of arbitrariness in selection and application of design criteria and its effect on capital cost should thus be examined. It is unlikely that the proposed extreme wave, peak wind, and maximum combined current will occur simultaneously in the same space and direction. Building a rig to survive these combined criteria or to brave a higher factor of safety also greatly increases the cost of the rig. Thus, it may be economically advisable to accept a lower safety factor in the design of the rig but to spend more on the safety of the crew.

Safety factors for offshore mobile drilling units are determined by many sources, including governmental agencies, insurance companies as well as insurance history and past practice. The resulting "rules" are then administered by several classification societies and governmental regulatory agencies.

The rules have established general design requirements for safety and survival of existing mobile drilling rigs. Typically the criteria might involve 115 mph winds simultaneous with 60 ft waves. Depending on the type of rig, this condition could be equivalent to 80 mph winds with 75 ft waves. However, the cost of such a rig could possibly be reduced by $10 million if greater risk on its loss were accepted, up to a level of risk equal to past recorded losses.

In order to maintain the level of safety for rig personnel implied by the rules, an expenditure of perhaps $1 million in emergency life support systems would still mean a capital saving of about $9 million. Therefore, in establishing design guidelines for offshore rigs it would be helpful to accurately determine the probability of combined wind, wave, and current as well as other combined environmental conditions for particular situations.

The design of today's rigs thus requires a rational alternative to more accurately reflect the typical characteristics of the environment. This article outlines some probable circumstances in which combined environmental loads produce unduly extreme conditions. The risk of operating an offshore mobile rig or a fixed platform can be determined through procedures described herein. With such risk criteria, the operator of the unit can then evaluate whether or not the risk in the given environment is acceptable.

Risk Concept of Design

The risk concept of design (1) is an alternative to the traditional use of safety factors which represents the arbitrary selection of

TABLE 23
Nomenclature

H = Wave height, ft
L = Wave length, ft
N_a = Number of wind occurrences, numeric
N_i = Number of time intervals, numeric
N_w = Number of wave occurrences, numeric
P = Probability, numeric
P_a = Wind encounter probability, numeric
P_c = Tidal current encounter probability, numeric
P_j = Joint wind, wave, current probability, numeric
P_L = Probability of coincident location
P_p = Wind encounter probability in T_p, numeric
P_s = Storm encounter probability, numeric
P_t = Wave period encounter probability, numeric
P_w = Wave encounter probability, numeric
T = Wave period, sec.
T_a = Individual wind duration, sec.
T_c = Peak wind occurrence interval, hour
T_i = Wave platform intersection time interval, sec.
T_p = Duration of peak storm period, hour
T_r = Return period, year

design criteria recurrence periods. The newer and more extreme risk criteria must be considered in light of existing conditions. Balanced risk enables a particular design to be evaluated relative to other normally existing or normally accepted risks.

Using readily available literature, we find that the average size of the offshore mobile drilling rig fleet has been about 150 rigs over the past 24 years. There have also been 37 total rig losses out of 92 rig accidents (2) during this time interval. Using these figures a rig casualty probability of $\frac{37}{(24)(150)}$ or $\frac{1}{97}$ means there is one chance in 97 that any particular rig will be lost in a year. This represents about three times the normally accepted shipping losses for the same period (3, 4). It should, however, be emphasized that this is the probability of loss of the rig, not loss of human life. The probability of loss of human life has historically been much smaller.

In 10 of the 92 rig accidents cited above there were a total of 77 crew fatalities. Assuming a crew of 40 men per rig per tour and two tours per month, this gives a crew fatality probability of $\frac{17}{(24)(150)(40)(2)}$ or 3,740, i.e., one chance in 3,740 that a crew fatal-

TABLE 24

Probability	Annual Risk Comparisons
$\frac{1}{1,950}$	Accidental work fatalities
$\frac{1}{3,750}$	Accidental auto fatalities
$\frac{1}{6,700}$	Accidental home fatalities
$\frac{1}{10,000}$	Accidental public fatalities
$\frac{1}{970}$	Cumulative accident risk
$\frac{1}{3,740}$	24-year average crew fatalities (offshore drilling rigs)
$\frac{1}{333}$	1971 U.S. flagship losses
$\frac{1}{97}$	24-year average mobile drilling rig casualties

ity will occur on a particular rig in a one-year time interval. While it is agreed that rig safety for personnel should be improved, the fatalities are presently about half as much as in other working environments which have a fatality probability of $\frac{1}{1,950}$ (table 24). This means that working on an offshore rig is about as safe as working on the average onshore job and is half as safe as living in one's own home. On this basis, working offshore represents about the same danger as driving down a freeway.

Joint Occurrence

For this joint occurrence study and to illustrate the principles behind risk design, a storm model is assumed. In the model, waves exceeding 60 ft (maximum waves) will occur at random during a four hour period, T_p. Peak winds are assumed to occur at random in a 20 hour period, T_c, which includes the four hour period in which the maximum waves occur. Discrete time intervals, T_i, are taken as the period for one storm wave to pass through and in some way affect a part of the rig or platform. A 60 ft wave having a steepness of 10 (wave length divided by wave height) passes through a 200-ft-long platform in about 20 seconds. Thus, the number of intervals, N_i, of 20 second duration in the 4-hour storm period (3,600 seconds) are:

$$N_i = \frac{T_i}{T_i} = \frac{4\ (3{,}600)}{20} = 720 \tag{2.1}$$

A particular wave may or may not pass through the structure in the four hour period so the possible events are $2N_i - 1$. If a wave does pass through the structure in the four hour period, it can lead, lag behind, or be coincident with the basic time interval, T_i. Thus, the probability of encounter of a particular wave, P_w, in a particular time interval is:

$$P_w = \frac{3}{2N_i\text{-}1} = \frac{3}{2(720)\text{-}1} = \frac{1}{480} \tag{2.2}$$

The peak storm wind period, T_a, is usually taken as one minute. The probability that a particular storm wind will occur, P_a, during a

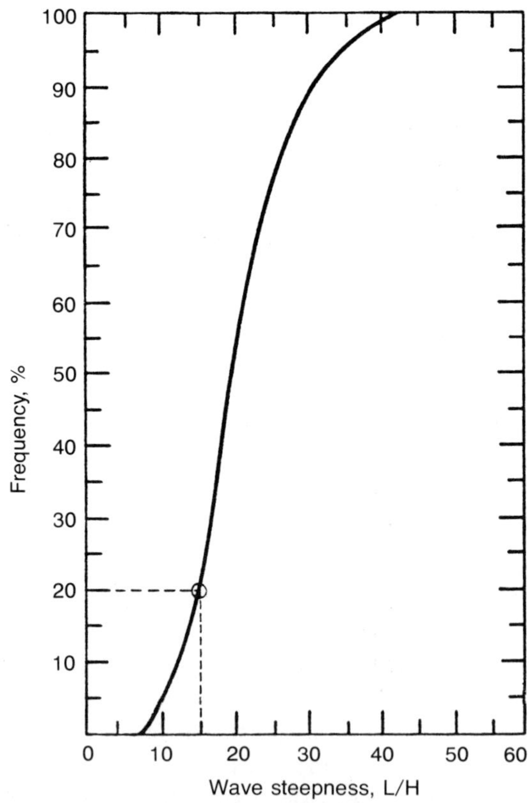

FIG. 2.7 Cumulative frequency distribution of wave steepness for North Atlantic

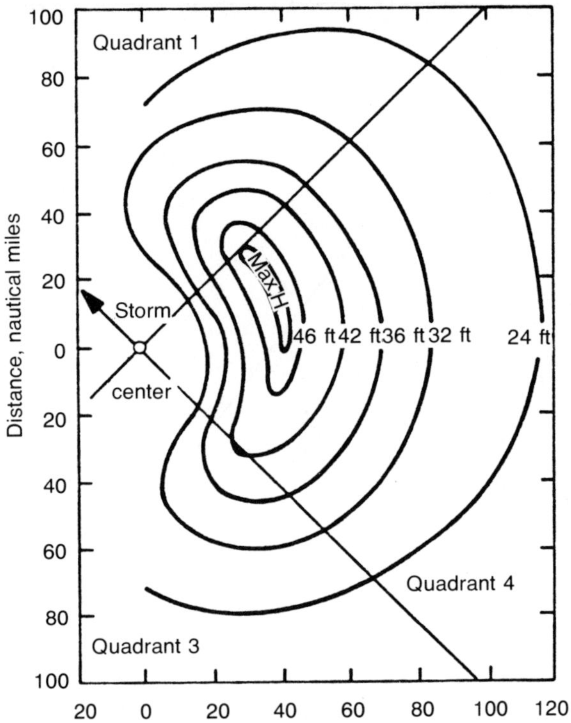

FIG. 2.8 Typical tropical storm model showing significant wave height distribution

particular time interval, T_i, in the four hour period, T_p, is:

$$P_a = \frac{(T_p)}{(T_c)}\frac{(T_a)}{(T_iN_i)} = \frac{(T_p)(T_a)}{(T_c)(T_p)} = \frac{T_a}{T_c} = \frac{1}{20(60)} = \frac{1}{1{,}200} \tag{2.3}$$

The useful rig life for an offshore mobile exploratory rig is normally assumed to be 20 years. The encounter probability, P_s, of a 100-year storm in the 20 year period is then:

$$P_s = \frac{20}{100} = 0.2 \tag{2.4}$$

Let P_t be the probability that a particular wave will have a particular period. This can be determined from Figure 2.7 since the wave length, L, can be computed from $L = 5.12\ \tau^2$ where τ is the wave period. The steepness, s, is defined as $s = \frac{L}{H}$ where H is the wave height. Thus, waves of high steepness (15 or greater) occur 80% of the total time as shown in figure 2.7

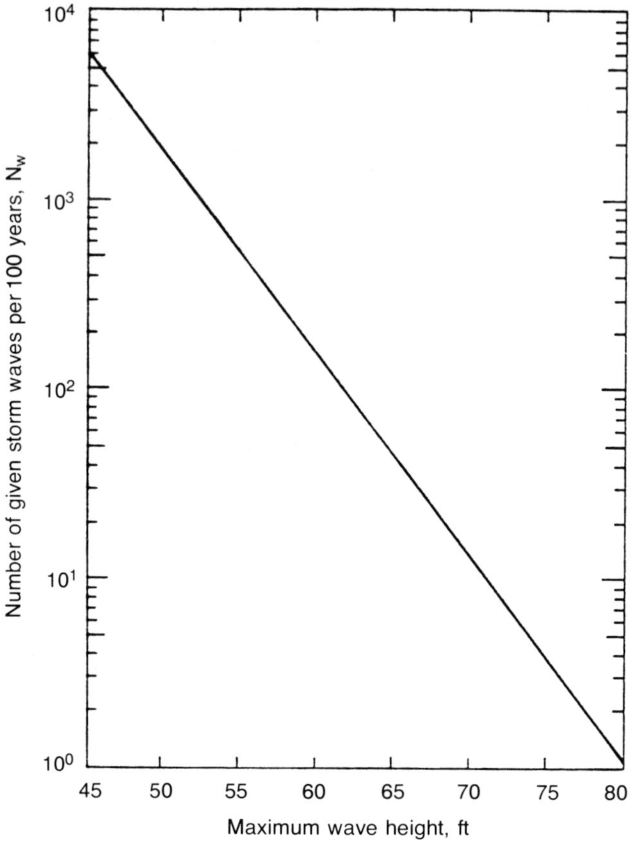

FIG. 2.9 Maximum wave height occurrence in middle North Sea

The tidal current could also be at any angle with the wind direction. P_C is the probability that the current will be coincident with the wind direction. For conservative purposes, it can be assumed that the current is either with or against the wind so that a value of 0.5 can be used for P_C.

There is a very remote possibility that the peak one-minute wind and the maximum storm wave would occur exactly at the same location and at the same time on a mobile rig or fixed platform. This is particularly questionable when considering that a storm can cover an area of several hundred miles in width. No information seems to be available in the literature on the subject, so for preliminary purposes we assume a conservative value of 0.25 for the probability of coincident location, P_L. This is a good estimate since storm models (5) show that the maximum wave heights usually occur in quadrant 4 of a storm (figure 2.8).

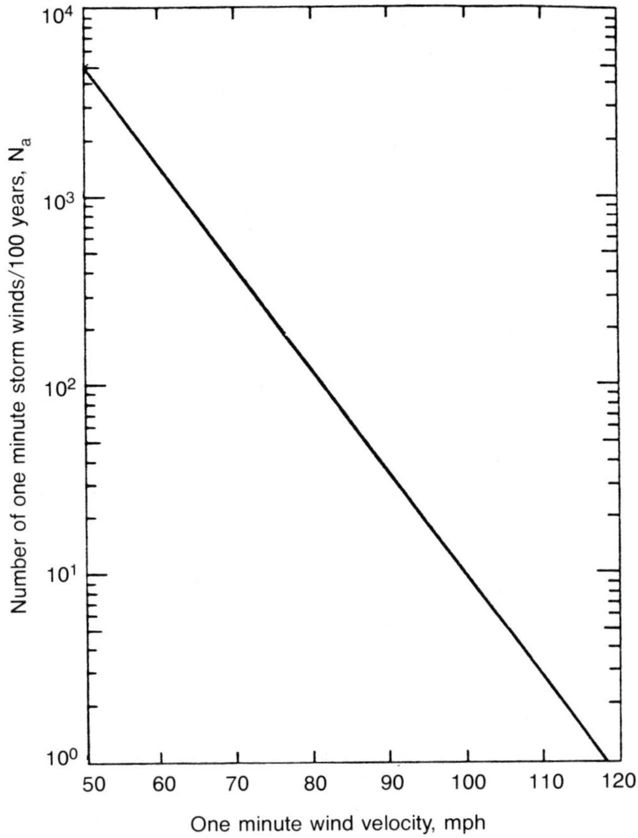

FIG. 2.10 Severe wind occurrence in middle North Sea

The joint probability, P_j, of simultaneous wind, wave, and tidal current coinciding cumulatively in a 20-year period is then:

$$P_j = (P_L P_c P_t P_s P_w N_w)\ (P_\alpha N_\alpha),$$

where N_w and N_α are the possible occurrences of particular waves and winds, respectively. These numbers can be obtained for the North Sea from figures 2.9 and 2.10.

Using the above assumptions, the joint probability is then reduced to $P_j = (3.47 \times 10^{-8})\ (N_w\ N_\alpha)$, for wave steepness greater than 15.

General Application

Carrying through with the mathematics indicated above, table 25 gives the criteria for 60- to 75-ft wave heights and 80 to 120 mph wind velocities in the middle North Sea (about 57° latitude). The

TABLE 25

Probability of 20 year combined environment

Steepness Greater Than 15				
Wind Speed, mph:	80–90	90–100	100–110	110–120
Wave Height, ft				
60–65	1.72×10^{-4}	5.00×10^{-5}	1.38×10^{-5}	2.78×10^{-6}
65–70	4.73×10^{-5}	1.37×10^{-5}	3.82×10^{-6}	7.63×10^{-7}
70–75	1.51×10^{-5}	4.37×10^{-6}	1.21×10^{-6}	2.43×10^{-7}
75+	4.30×10^{-6}	1.25×10^{-6}	3.47×10^{-7}	3.94×10^{-8}
Accumulative	2.38×10^{-4}	6.93×10^{-5}	1.92×10^{-5}	3.83×10^{-6}
Subtotal	3.31×10^{-4} for steepness greater than 15			
Subtotal	8.27×10^{-5} for steepness less than 15			
Total	4.15×10^{-4} for all steepnesses and one tidal current direction			
Grand Total	8.30×10^{-4} for all steepnesses and both tidal current directions			

TABLE 26

Design joint probability of combined environment over rig life

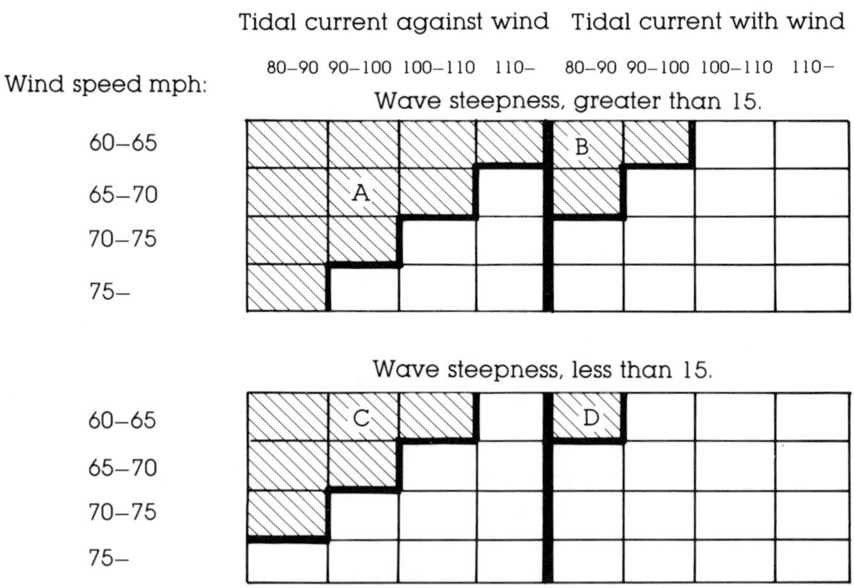

Area:	A	B	C	D	Total
Probability	3.27×10^{-4}	2.69×10^{-4}	7.80×10^{-5}	4.30×10^{-5}	7.06×10^{-4}

Note: This means that if design conditions represented by shaded area A above are satisfied, then with the risk analysis theory, the conditions indicated by shaded areas B, C, and D might also be satisfied.

results show that the chance is only 3.94×10^{-8} or about 1 in 25 million that 75-ft waves (with steepness greater than 15) coinciding with 120 mph winds and tidal current will occur in 20 years. The table also shows, for example, that the probability of joint occurrence of 70-ft waves and 80 mph winds is about the same as the probability of simultaneous occurrence of 60-ft waves and 100 mph winds. However, these two load combinations might produce the same or considerably different safety factors in design, depending on exposed wind areas, air gap, etc.

A schematic of the data in table 25 is given in table 26 to illustrate risk design. If the conditions of shaded Area A in table 26 are met by the given design, then additional conditions such as those indicated by shaded areas B, C, and D might also be met as a consequence. The probability of severe environments occurring which are not met by the design criteria is then the total probability of 8.30 $\times 10^{-4}$ less the probability of design criteria indicated by areas A through D, 7.06×10^{-4}, which results in 1.24×10^{-4}. This means that in a 20-year peiod, the given design criteria of A through D would be insufficient to withstand expected environmental loads only one chance out of 8,000 or 0.012% of the time.

Finite Risks

As a case example, we will examine the risks involved for a jackup design. The ETA Europe Class jackup is a deep water, high variables capacity rig specially designed for severe environmental conditions. Two such units were delivered in June and November 1976 for operation in the North Sea in up to 350 ft of water.

The 507 ft-long legs were designed for maximum conditions of 75-ft waves and 125 mph winds at 260-ft water depths and for slightly less severe criteria in 350-ft water depths. These conditions are substantially more severe than ever expected by a jackup.

The rigs were designed to meet the requirements specified by Det norske Veritas (DNV) in the latest "Rules for the Construction and Classification of Offshore Mobile Drilling Units," 1975, and were the first jackups to be built to DNV classification.

From the above finite risk approach discussion, it is clear that various combinations of environmental loads can be encountered, and the rig should, therefore, be designed accordingly. Design envelopes of environmental loads on station have been calculated for the ETA Europe Class jackup. Figure 2.11 depicts the allowable wave height versus allowable wind velocity for zero current, for a 1.5 knot current running against the wind and for a 1.5 knot current running with the wind.

The dotted lines in figure 2.11 show that for equal risk the rig can withstand any combination of waves up to 85-ft heights with any

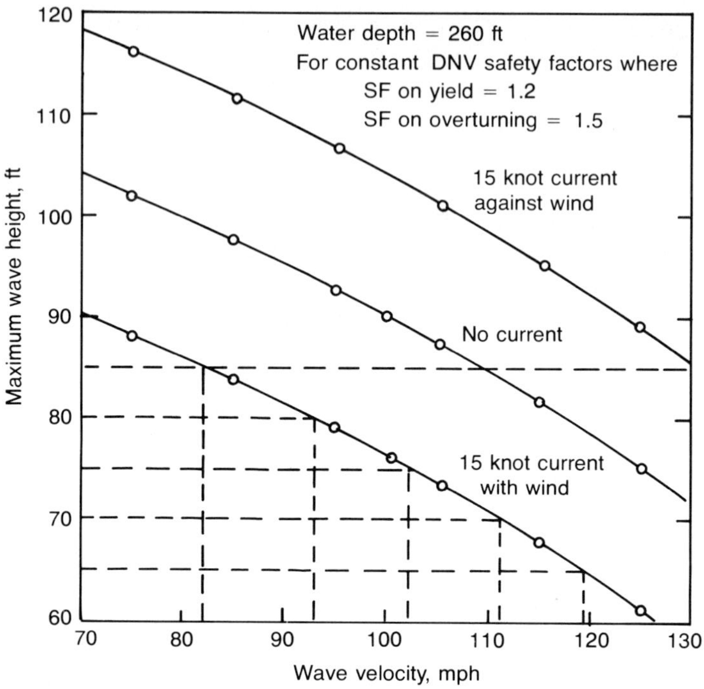

FIG. 2.11 Environmental criteria for ETA Europe class jackup

wind velocity up to 130 mph when the current is in the opposite direction of the wind. When the wave heights are 60, 65, 70, and 75 ft, the allowable wind velocities are 126, 119, 111, and 102 mph, respectively, with the wind and current taken in the same direction.

Keeping the DNV safety factors constant, the only combined environmental load excluded with this rig then is area E as shown in table 27. The shaded area represents acceptable design criteria. Area E, however, has a risk probability of 6.29×10^{-7} or one out of about 1½ million. Assuming that the rig is adequately designed for wave steepness greater or less than 15, area F in table 27 is then 1.57×10^{-7} or one out of about 6.4 million. The total probability of design exceedance for areas E and F is therefore 7.86×10^{-7}, compared to a total probability of 8.30×10^{-4} for combined environmental conditions in 20 years as shown in table 25.

Finally, these numbers can be expressed in a more readily understandable way. The probability of 8.30×10^{-4} or a chance of about one in 1,200 means that in a 20-year period, various combinations of environmental criteria given in tables 26 and 27 will occur at specific times. The combined loads will average a total of about 6.02 days in the 20-year period. Thus, the probability that design limits will be exceeded is only 7.86×10^{-7} (areas E and F) or one chance in 1.2 million. This probability represents a total of about 8 minutes during the 20-year period.

These critical 8 minutes, however, should be examined more closely. The rules require that the maximum stress shall not exceed 84% of the yield strength of the materials, which is the design standard of the ETA Europe Class jackup. If the most extreme combined environmental loads of 75-ft waves and 110 mph winds are acting concurrently in the same direction as the 1.5 knot currents, then the ETA Europe Class jackup will experience combined stresses that are on the order of 10–5% greater than design stress. This condition, however, will still be below the yield strength. In fact, the combined stresses would be 94.9% of yield.

TABLE 27
Design joint probability of combined environment over rig life

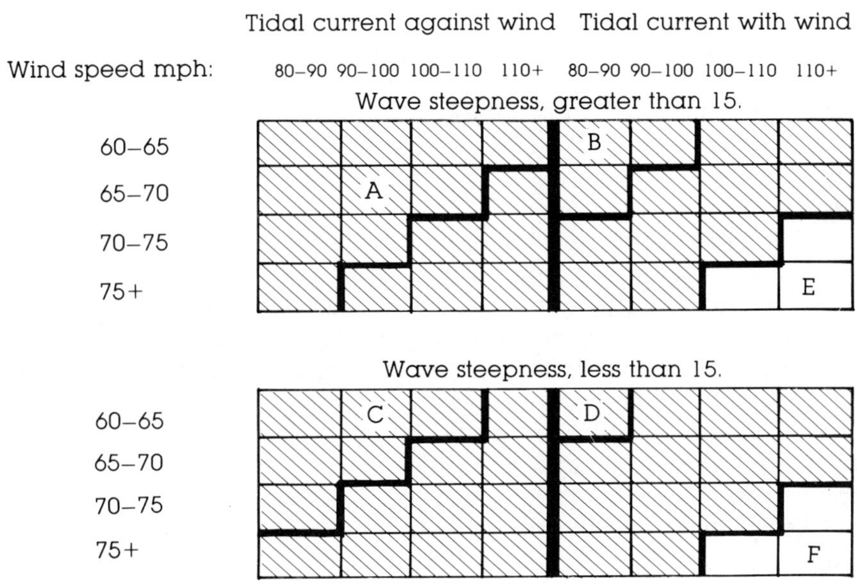

Area:	A	B	C	D	Total
Probability	3.27×10^{-4}	2.69×10^{-4}	7.80×10^{-5}	4.30×10^{-5}	7.06×10^{-4}

Note: Shaded areas represent criteria for which the ETA Europe Class jack-up design is sufficient.

As another example, if combined loads of 75-ft waves and 125 mph winds are acting with 1.5 knot currents, the rig experiences combined stresses that are about 20% greater than design criteria or only slightly greater than the yield strength. From other calculations, the factor of safety against loss of the rig from overturning is then reduced from 1.50 to 1.33.

Comparative Risk

Taking another approach, the adoption of current rules will imply that the safety of the jackup, i.e., the probability against loss, will be substantially greater than that corresponding with the past record of offshore drilling units. This point is explained in the following discussion.

As shown previously, the calculated 20-year joint probability for the middle North Sea is $3.47 \times 10^{-8} N_w N_a$ for all steepnesses. With the annual rig risk calculated as 1/97, the 20-year joint probability of loss from all causes is 20/97. All causes include loss due to environmental criteria, blowouts, ship collisions, sabotage, acts of God, revolution, etc. As a matter of interest, let us consider the effect of ignoring all risks except severe storm conditions and determine the savings possible as a result of this.

Calculating the above assumption so that $20/97 = 4.33 \times 10^{-8} N_w N_a$ and using figures 2.9 and 2.10, any number of weather combinations can be determined, such as 50-ft waves with 55 mph wind and no current in 260 ft water depth. The ETA Europe Class jackup has a 3.02 factor of safety against overturning with these environmental criteria.

If one next considers the relative costs of meeting a design code based on safety factors (such as the rules) vs. taking a risk equal to previous overall performance, then the tradeoffs are as follows. The ETA Europe Class jackup, for example, is designed for 75-ft waves and 125 mph winds with an overturning factor of safety of 1.5 in 260-ft water depths. However, the 20-year risk or joint probability is one in 13 million that 75 mph waves and 125 mph winds will occur simultaneously (1.5 safety factor). The risk with 75-ft waves and say 180 mph winds (1.0 safety factor) is one in 12 billion.

Using the safety factor approach, we find that when there is a safety factor of 15 and a risk of rig loss of one in 13 million, the total rig cost is estimated at about $35 million. Using the risk analysis approach, when there is a safety factor of 1.0 and the environmental risk is taken as equal to the total previous risk of 20/97, then the total rig cost is estimated at about $22 million.

In order to compensate for the higher risk to human life that the risk analysis approach indicates, it would be financially advantageous and wiser to spend one or two million dollars on emergency life support systems (survival capsules, devices to preserve life in storms) and thus save $11 or $12 million overall. This then may translate into operating procedures.

Specifically, if extreme weather is expected, if a blowout occurs, or a ship collides with the rig, use of a "foolproof" method of removing personnel could be advantageous. The cost of such a system might prove to be more attractive than paying a large initial penalty by making the rig working environment safer overall when it is already safer than most other work environments. The consequence of building stronger and larger rigs to protect against risk to human life rather than the risk for the property alone is significant—an increase of $10 to $20 million per rig.

The rules for the classification and safety of offshore rigs are revised periodically. Many of the organizations are currently reviewing the design criteria requirements for offshore mobile drilling units for potential readjustments. A reassessment of the rules might possibly allow for significant initial investment savings through the use of the risk analysis approach.

Conclusions

Methods are now available to the operator and contractor which can more accurately evaluate the actual risk of operating a particular rig or platform in various environments, and thus fully utilize the unit's operational capabilities.

A lot of money is being spent on resisting environmental conditions which are inconsistent with other risks. This is costly protection.

Reduction of this cost should be attempted and consideration given to spending more on other systems for safety of life and safety against other risks such as blowouts, collisions, fires. This is believed to be more economical overall.

Large sums of money are being spent on constructing the rigs to counteract conditions which have been shown to produce the least losses. As these conditions become more severe, the cost becomes very much higher, much more so than in the past.

In view of the results presented above as well as from the results of other internal studies, it appears that more concentrated attention should be devoted to safety of rig personnel. With the considerable interest by operators, contractors, as well as insurance firms, it seems obvious that an industry study on improved safety could be very cost effective.

Unless drastic changes are made in the trends of the rules which are becoming more stringent or unless use of a risk analysis is made, the construction cost of offshore structures may escalate to such a degree that the growth of the industry may be severely inhibited.

2.3 ESTABLISHING MULTIOPERATIONAL CAPABILITIES FOR RIGS

Offshore mobile drilling rig design has become more complex with increasing operational requirements and spiralling construction costs. Two fundamental problems must be considered in planning today's rigs: the economics of designing and equipping a rig to work efficiently under varying environmental conditions, and matching dollar capital investment against risk of rig loss.

A rig is generally designed and equipped for service in one particular area. As the search for oil goes into deeper and rougher waters, it is important for the rig owner and the operator to know the measure of risk involved in operating the rig in other locations. Can a rig designed for Gulf of Mexico criteria operate safely and efficiently in the North Sea, and vice versa? A study of typical environmental criteria, determining joint probability of occurrence, can establish a rig's operational capabilities and risks in various areas.

Evaluation

Several companies are currently evaluating realistic design criteria based on typical environmental criteria and comparative risks to given rigs in certain areas. For example, the ETA America Class jackup is designed to operate in 380 ft of water in Gulf of Mexico hurricane conditions and can be adapted to over 400 ft in nonhurricane conditions. This adaptability allows the rig to operate in various drilling areas.

In these evaluations, recognized sources for oceanographic and environmental data are used. The application of these data to the structural analysis of the rig and its response to environmental conditions are significant. Additionally, effective estimates of downtime due to environmental conditions and hence quantitative comparisons of the rig's operational efficiency in different locations can be provided.

Environment

Environmental conditions can significantly affect the economics of the drilling operation. The owner can save a substantial amount of capital investment by not "overbuilding" or "overequipping" the rig for too severe criteria. Or he might choose to "underbuild" the unit and accept the calculated risk due to exposure.

Environment also influences equipment specifications. A North Sea drillship may require a three-racker pipe handling system because of severe vessel motions. The same rig operating in the calmer seas of southeast Asia may require only a two-racker system or none at all. Special equipment can affect required drill floor space, causing possible rig refitting if a location change is planned.

The operator also has a critical interest in the rig's operational efficiency and suitability in various locations since the day rate is directly dependent on rig cost. A rig's relative success in a particular area can be estimated by considering potential downtime due to environmental conditions. Knowledge of expected downtime due to weather is vital to an economic evaluation of the type of rig to use. A heavy-duty rig with high operating costs may be able to operate 12 months out of the year while a lighter rig subject to greater downtime can be contracted at a lesser day rate.

Oceanographic and environmental data are historical in nature. Expected weather trends, however, have been broken. In 1973, two 100-year storms occurred in one month (November) in the North Sea. Thus, careful study and application of these data are critical in determining expected area operating conditions. By integrating environmental data, vessel motion, and structural analysis, design criteria can be significantly upgraded.

To show how environmental conditions are incorporated into rig design criteria, we examined drilling activity for several areas (table 28). Gulf of Mexico drilling has gone on steadily for years, while North Sea drilling developed more recently but has been highly concentrated. Performance records in these areas can be used to predict downtime and to estimate the capability of various rig types to withstand actual environmental conditions.

Southeast Asia

Southeast Asia has been relatively unexplored even though it has about 10% of the total offshore worldwide ultimate recoverable reserves, or half as much as the Middle East and twice as much as the US Gulf of Mexico. The saturation of exploratory wells per 1,000 sq miles in the North Sea is already about 16 times greater than in Southeast Asia. While the North Sea has about 10 times the offshore acreage leased as the Gulf of Mexico, Southeast Asia has 100 times the acreage. Thus, in future drilling activity Southeast Asia must be considered a profitable area for exploitation.

Malaysia

Table 29 presents sample data on typical environmental conditions for one part of Southeast Asia, offshore Malaysia. The basic data were converted to exceedance values, or percent of time the condition exceeds a specific level (figure 2.12). Offshore Malaysia, for example, 25% of the time in the winter (severest conditions) winds

TABLE 28
Summary of offshore drilling statistics

	Bathymetry, 0–200 meter water depths		Ultimate recoverable reserves of crude oil		1974 drilling activity (first-class rigs)				Cumulative drilling activity (first-class rigs, 1958–1974)	
					First Class Rigs Operating		Exploratory Wells Drilled			
Area	1,000 sq. Naut. Mi.	Percent	Billion Barrels	Percent	Number	Percent	Number	Percent	Rig-Years	Exploratory Wells
Southeast Asia	2,000	32	380	10	33	14	132	14	176	704
U.S. Gulf of Mexico	200	3	185	5	47	20	188	20	441	1,764
North Sea	110	2	105	3	45	19	180	19	199	796
Mideast	125	2	800	20	33	14	132	14	208	832
Other	3,795	61	2,480	62	78	33	312	33	708	2,832
World Total	6,230	100	3,950	100	236	100	944	100	1,732	6,928

Notes:
1. Reserves from Geological Survey Paper 885.
2. Rig counts from *Offshore* magazine.
3. Wells based on 4 wells/rig-year.
4. First class rigs exclude workover, coring, and submersible vessels.
5. Southeast Asia runs from Pakistan to Japan, including the area North of Australia.

Petroleum Times July 25, 1975

TABLE 29
Environmental data for Malaysia

Season	Summer Aug. 1973	Fall Oct. 1973	Winter Jan. 1974	Spring Apr. 1974	Average
Wind, knots			Percent of time		
Calm	4	8	1	10	6
1–3	8	17	3	24	13
4–10	60	54	40	55	52
11–21	27	18	46	10	25
22–33	1	3	9	1	4
>33	0	0	1	0	0
Sea, ft					
Calm	14	10	7	29	15
1–3	59	80	54	64	64
3–8	25	10	33	6	19
>8	2	0	6	1	2
Swell, ft					
None	21	50	59	80	53
1–6	79	40	15	15	37
6–12	0	10	26	5	10
>12	0	0	0	0	0

exceed 18 knots and wave heights are more than 4.4 ft. The mildest winds and waves occur in the spring, with calmest swells in the summer.

Comparison

Since a mobile rig is likely to operate in various areas during its lifetime, it is important to compare regional environmental data. Thus, plots of log exceedance vs. wind speed and wave heights were prepared for offshore Malaysia, Gulf of Mexico, southern North Sea, and Persian Gulf. The percent of time exceedance was converted to the number of annual one minute winds (figure 2.13), assuming one year is statistically representative. For example, about 500 one-minute 80 knot winds are expected per 100 years offshore Malaysia.

The percent of exceedance for waves was converted to the number of waves using the relations $L = 5.12T^2$ and $s = L/H$. (L is wave length in feet, H is wave height in feet, T is wave period in seconds, and s is steepness of 10). The results plotted in figure 2.14 show that about 20 waves of 40 ft height are expected to occur per 100 years offshore Malaysia.

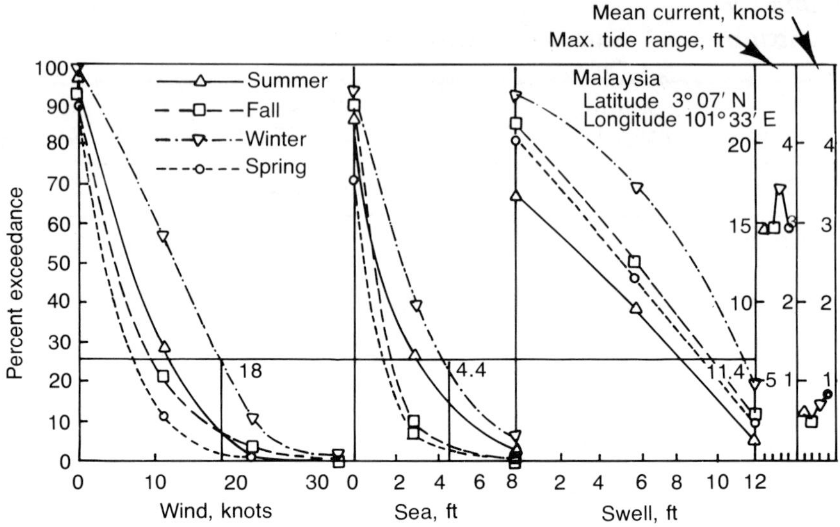

FIG. 2.12 Exceedance values for various conditions off Malaysia

Equal joint probability for combined environmental criteria were compared for several areas (table 30). As a rough rule of thumb the 100-year storm wave height in the North Sea is 50 ft plus about 5 ft per degree of latitude up to 62 degrees. The suitable 100-year storm wind condition for the North Sea is 100 knots.

The southern North Sea criteria of 100-knot, one-minute wind gusts are expected to occur about 5.5×10^3 times per 100 years (figure 2.13). In the Gulf of Mexico, 74 knot winds should occur the same number of times. About 700 waves of 39 ft maximum height will occur in the Gulf of Mexico in 100 years and 50 ft maximum waves will occur about 700 times per 100 years in the southern North Sea (figure 2.14).

Conditions

The joint probability of simultaneous environmental conditions is decisive in determining a rig's capability to operate in different areas with equal risk although actual wind, wave, and current conditions may be radically different.

Another study found that the joint probability of simultaneous wind, wave, and current forces coinciding cumulatively in the same location and from the same direction in a 20-year period is $Pj = 4.34 \times 10^{-8} N_w N_\alpha$. (The number of waves of a given height in 100 years is N_α: N_α is the number of one-minute wind gusts of a given velocity in 100 years.) Thus, the probability of 80 knot winds and 40 ft

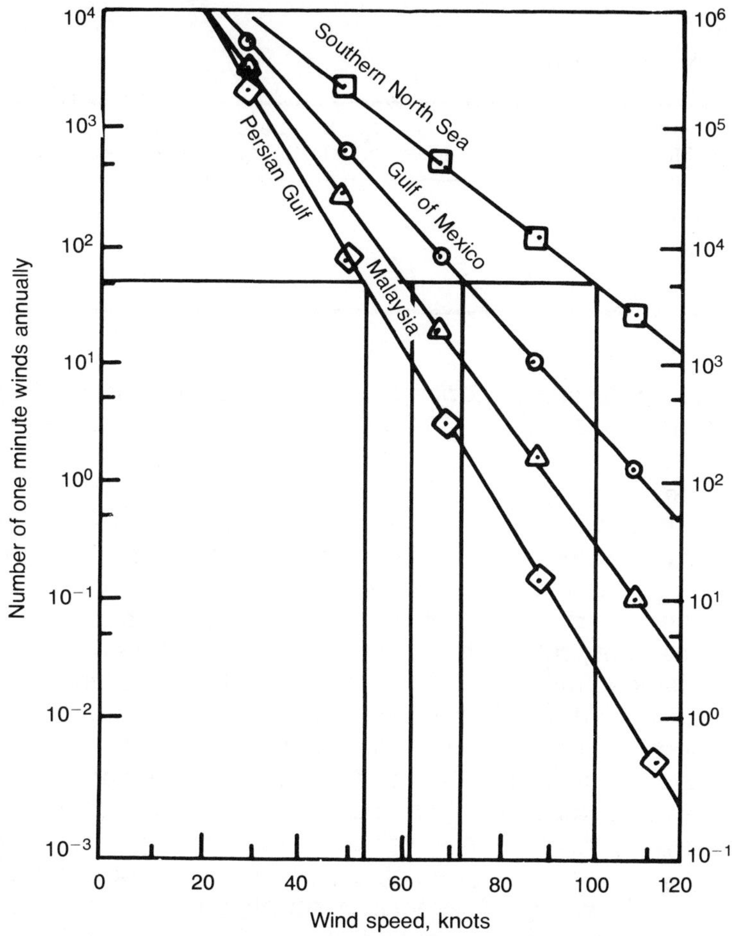

FIG. 2.13 Wind statistics comparison

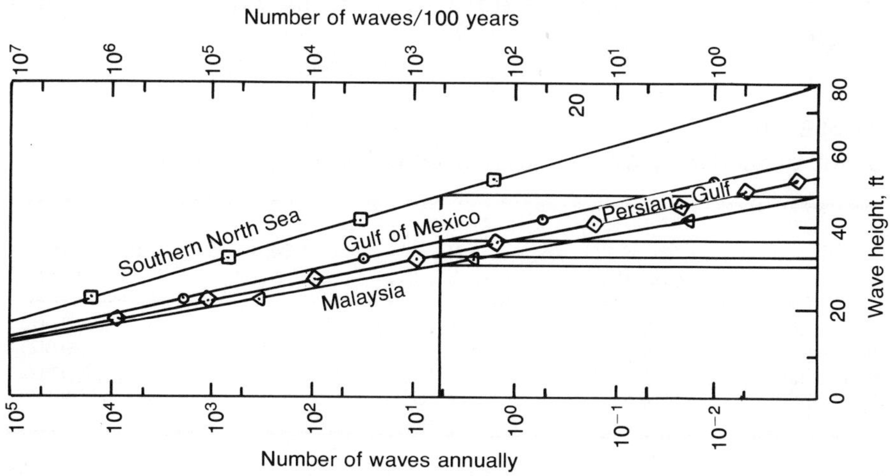

FIG. 2.14 Wave statistics comparison

TABLE 30
Equal joint probability

Area	Offshore Malaysia	Persian Gulf	Gulf of Mexico	Southern North Sea	Middle North Sea	Northern North Sea
Max. wave ht., feet	33	35	39	50	75	100
Max. wind gust, knots	62	53	74	100	100	100

waves occurring simultaneously offshore Malaysia is $Pj = 4{\cdot}34 \times 10^{-8}(20)(5000) = 4.34 \times 10^{-3} = 1/230$ or about 32 days in 20 years.

Probabilities

A comparison of probabilities of simultaneous 100-knot winds and 50 ft waves in the areas studied is given in table 31. The highest expectation for simultaneous occurrence of the selected conditions is in the southern North Sea, for a total of about 29 hours in 20 years.

With the joint probability of occurrence of combined winds and waves of specified magnitudes, rig downtime due to weather can be estimated. For total downtime, time spent waiting on supplies must be estimated. The limiting criterion for loading and off-loading supply boats is 6 to 8 ft seas. Offshore Malaysia, where criteria are mildest, waves are expected to exceed 6 ft only 20 days per year.

By studying supply schedules and other factors, downtime for supply operations can be predicted. For example, floating drilling rigs have longer periods of heave than supply boats. Rig and supply-boat motions are, therefore, out of phase. Wind speed alone can become a critical factor when drilling operations become unsafe for the crew, even though the rig itself is not in danger of structural failure.

TABLE 31
Joint probability of 50-ft waves and 100-knot winds

	N_w (no. waves)	N_α (no. winds)	$4.34 \times 10^{-8} N_w N_\alpha$	Joint probability in 20-year period
Malaysia	1.04×10^{-1}	4×10	1.80×10^{-9}	1: 50,000,000
Persian Gulf	6.5×10^{-1}	3.5×10^{0}	9.87×10^{-10}	1:100,000,000
Gulf of Mexico	7×10^{0}	3.6×10^{2}	1.09×10^{-8}	1: 10,000,000
S. North Sea	7×10^{2}	5.5×10^{3}	1.67×10^{-5}	1: 6,000

Safety Factor

With joint probability determined, an acceptable safety factor for rig design can be calculated which permits full rig utilization without "overbuilding" for expected environments. Rig cost could be significantly reduced if greater risk on its property loss were accepted, up to a level of risk equal to past recorded losses.

Consideration should perhaps be given to spending more on safety systems rather on unnecessarily enhancing structural design criteria. Large, heavy-duty rigs which are overbuilt for environmental criteria in most areas to avoid risk to life can involve construction cost increases of $10 to $20 million. Safety and life support systems can provide comparable protection for $1 to $2 million without sacrificing structural integrity.

Conclusions

Knowledge of expected area environmental conditions to be experienced by an offshore drilling rig gives an economic advantage to both the rig owner and operator. Structural analyses incorporating environmental conditions provide realistic design criteria to more precisely determine operational risks to the rig and to reduce construction costs by upgrading rig design. The optimum and most economical use of each rig is then assured.

Vital studies

Such studies may prove vital in the next few years. The present construction program for rigs is straining the limits of today's worldwide shipyard capacity. Vast areas such as Southeast Asia are available for exploration and have calm enough environmental conditions that existing rigs can safely and easily operate there. Adapting existing and new rigs for work in various drilling areas can thus provide faster, more economical mobility and flexibility of the offshore rig fleet in the race for oil.

2.4 ENVIRONMENTAL DESIGN CRITERIA FOR OFFSHORE MOBILE RIGS

The move of oil exploration and production into deeper and more severe waters is leading to the general use of more extreme environmental events for the design of offshore drilling units. The traditional approach in utilizing these environmental extremes has been to determine the design load by taking the extreme wave, the peak wind, and the maximum combined current and considering all to be acting simultaneously, in the same direction, and for the most vulnerable orientation of the drilling unit. Typically, a recurrence interval of 100 years is selected for the individual extremes. Thus, the recurrence interval is well over five times the lifespan of any mobile drilling unit. Now, the probability that these individual 100-year recurrent events will occur at the same point in time, if indeed they coincide at the same location at all, is clearly less than that of a 100-year recurrence. There is, then, an existing arbitrariness in the selection and application of design criteria. What is needed is an equally rational design alternative with respect to traditional standards of safety; yet, one reflecting more closely the physical and/or stochastic character of the applicable offshore environment.

In this section, we will attempt to formulate a rational alternative. Among other things, probable circumstances in which environmental events, such as extreme wave heights, winds, and currents might simultaneously combine to produce resultant extreme environmental load conditions are assessed. Possible risks associated with these combined events are estimated and compared with other normally existing risks (e.g., ship collisions).

As the research developed, it became apparent that certain physical aspects of storm conditions which bear critically on design criteria might also be in need of reevaluation. This applied particularly to the effect of maximum waves and wind-generated currents on marine structures. For this reason, a brief survey of the probable physical character of ocean storms and their components is included.

The Evolution of Contemporary Design Criteria

Existing mobile drilling units, with the exception of the newer generation of semisubmersibles, generally conform to the design guidelines established by the American Bureau of Shipping. This means survival in 100-knot winds with 60-ft waves. Depending on the type of unit, this condition can be equivalent to 70-knot winds with 75-ft waves. As an example, in terms of mid-North Sea operation, the former condition constitutes approximately a 100-year return period on 1-minute winds and a 1-year return period on maximum waves; whereas the latter condition constitutes approximately a 1-year return period on winds and a 100-year return period on maximum waves. The inconsistency which now arises in different return periods for particular design condition components is believed to reside not so much in the condition itself as in the use of the environmental statistics.

One difficulty with the formulation of a design condition is that the statistical extremes for wind, wave, and current are generated independently but later combined without regard for their occurrence in space, direction, or time. For example, in the October 1970 North Sea storm, the jackup rig Orion recorded 60-knot winds with significant seas of 35 ft. Yet, two days earlier, in this same storm, 87-knot winds but with only 25-ft significant waves were recorded at the rig. This kind of wind-wave association tends to give some support to historical design criteria (i.e., 60-ft wave simultaneously with 100-knot wind) as not having been that unrealistic for North Sea conditions. Certainly, an extreme wind is a probable event by itself, and so is the extreme wave. But how probable is the simultaneous loading of the drilling unit by both extreme wind and wave?

Another, perhaps minor, inconsistency is that in seven years of North Sea drilling, no definite records of wave heights in excess of 66 ft appear to have been made at or below 60° north. Yet, based on statistical predictions being accepted in some quarters, 68-ft wave heights should have occurred once and lower wave heights in excess of 60 ft repeatedly. The apparent elusive occurrence of such maximum waves must indicate that they are at least of a very short crested type and, perhaps, the result of a momentary multidirectional interference process. If maximum waves are indeed of this type, then something other than regular orbital partical velocities should be expected, with the probable result that wave forces would be lower.

Where individually overestimated extreme values are used of necessity, a realistic set of design requirements could still emerge if a realistic interaction of the events is considered. Yet, the present trend in some segments of the industry and of its regulatory bodies is to cumulatively impose the new extreme storm components, just as was done in the past.

Storm-Dependent Design Parameters

This section is concerned primarily with an assessment of storm characteristics which bear critically upon the formulation of design criteria, and, consequently, on the resultant marine structures. Clearly, wave height by itself has little meaning unless specified along with wave type and period, etc. Similarly, wind-generated surface currents imply no particular current loading until the current profile is specified. Actually, in most instances, the physical nature and behavior of extreme storm components (i.e., the extreme components of the extreme storm) are not yet reliably known. Just what is the behavior of a wave whose height is 2.97 $\overline{H}$ and which might occur in the zone of the maximum H_s? Does it produce a wave loading comparable to the decaying wave of the same height and length? For the most part, answers to such questions are only now beginning to be considered of some importance.

Ocean Storms. The ocean storms, which are of primary interest for determining the maximum environmental loads on offshore structures, are of the cyclonic types. Cyclones and anticyclones consist of regions of abnormally low and high pressure, respectively, around which the storm winds circulate. The storm track is determined by the general upper level air currents of the latitude in which the storm is traveling as well as being influenced by local pressure distributions. The storm system may be several hundred to a thousand miles in diameter and generally drifts at speeds of from 10 to 15 knots. In the northern hemisphere, the storm winds circulate in a counterclockwise direction around areas of low pressure and in a clockwise direction around areas of high pressure. In the southern hemisphere, the directions of circulation are reversed.

Cyclones differ not only in the regions in which they originate but also in the nature of the meterological conditions which produce them. The two major types of cyclone storms are the tropical cyclone and the extra-tropical cyclone. Tropical cyclones are born during conditions of unstable showers and slightly lowered pressure in regions with warm ocean surface temperatures and light winds. They never form in regions of markedly anticyclonic circulation, over land, nor in areas where the Coriolos force is too small to establish cyclonic circulation. Extra-tropical cyclones are the typical storms of the middle and high latitude. They generally originate as wave disturbances on slow-moving polar-front systems. The evolution of the disturbance and the resulting occlusion process break down the initial frontal equilibrium and air-mass distributions. Major polar outbreaks of cold air are usually associated with a vast polar anticyclonic cell which moves toward the equator.

Although most major storms tend to be highly individualistic, certain basic features of cyclonic storms can be identified. In order to simplify the discussion of the basic storm structure, the arguments and presentation which follow will be developed for tropical storms in the northern hemisphere. It should be understood that appropriate reversals in circulation, etc., are necessary if the arguments are applied to anticyclones and to storms in the southern hemisphere.

The typical distribution of wind velocities in a tropical cyclone is shown in figure 2.15. Since the translational speed of the storm adds vectorially to the circulatory winds of the storm, the wind velocities in the right-hand part of the storm quadrants 1 and 4 are generally greater than those in the left-hand part. Two to three hundred miles in front of the storm center, the winds have a direction which is at right angles to the storm track. Winds at the rear of the storm diminish after the passage of the storm center and are generally in a direction

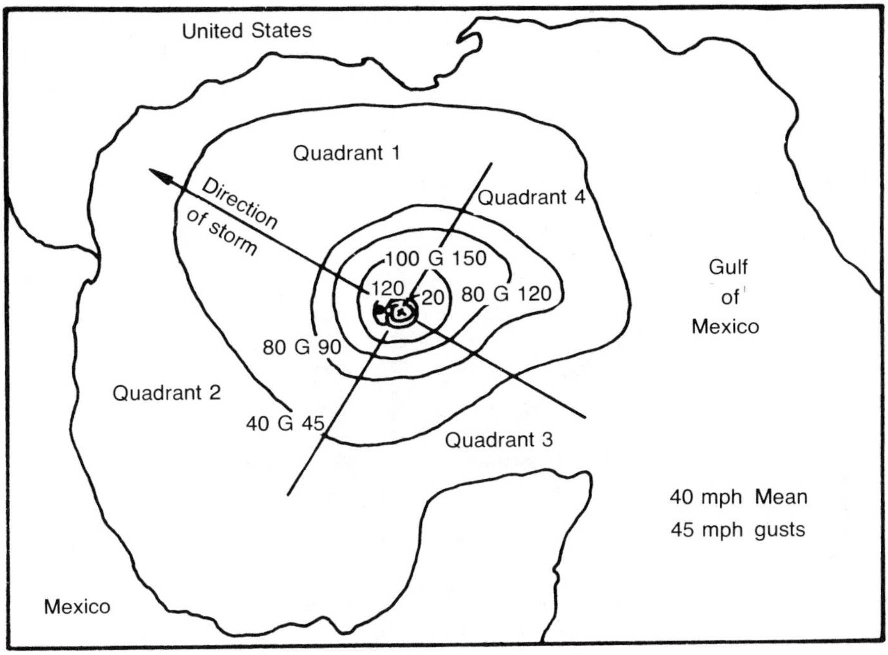

FIG. 2.15 Typical wind contours for a Gulf of Mexico tropical cyclone

opposite to those in the front of the storm. The greatest wind velocities are generally recorded in quadrant 4 some distance to the right of the line of advance of the storm center.

A model of the relative size and direction of the sea and swell developed by typical hurricane winds is shown in figures 2.16 and 2.17. The swells of the greatest length and magnitude, generated by the winds of the right-hand quadrants, precede the arrival of the main body of the storm. Sea and swell of moderate length and lesser magnitude move out to the left and right of the storm path. The longer waves generated in quadrant 4 move into quadrant 1 as the wave speed exceeds the translational speed of the storm. Thus, the probability of occurrence of the maximum storm wave is greatest in quadrant 1.

Although the simplified storm structure just presented will be typical of most tropical storms, some notable exceptions have been reported. In particular, the distribution of wave heights for the sea generated by the September 29, 1935, *Pacific Typhoon* (8) show that in this storm, the areas of maximum winds and seas tended to coincide in the right rear quadrant of the storm. This could be a result of the storm moving at a very fast rate at the time of the observations.

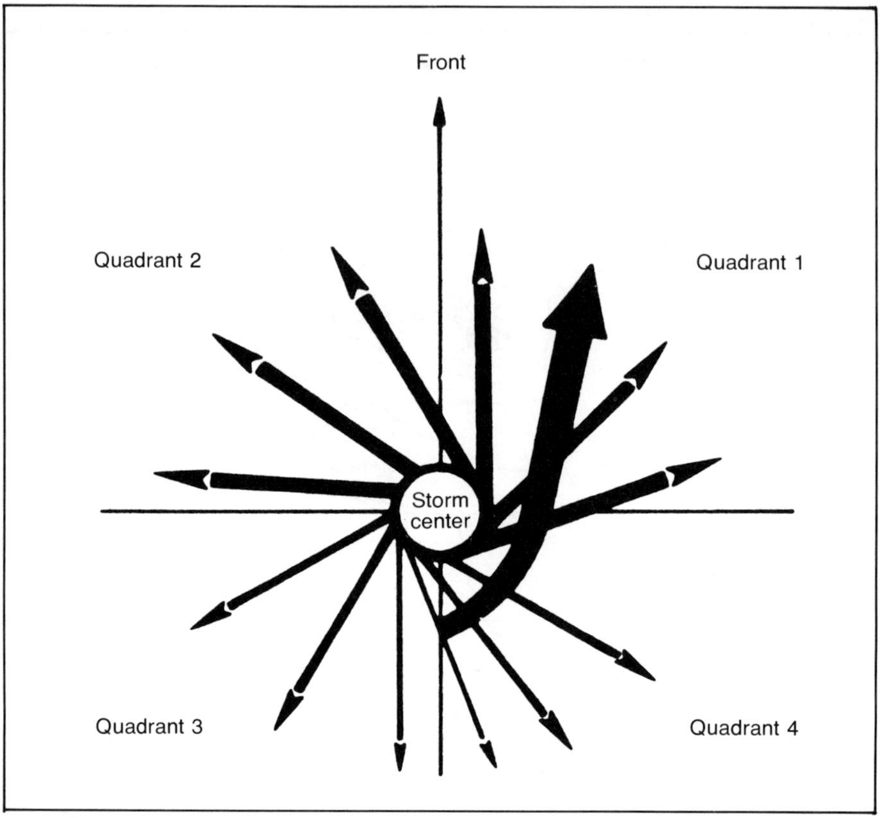

FIG. 2.16 Typical tropical storm model showing relative wave and swell heights and directions

The objective of the foregoing discussion on the basic structure of tropical storms has been to show that in most cases, the extreme winds and extreme waves in cyclonic storms do not tend to occur simultaneously at the same location in the storm. This tendency is supported by the fact that the longer waves propagate at faster speeds than the shorter waves and tend to move out of their area of generation. Thus, the probability of the simultaneous occurrence of the extreme wind and extreme wave at the same instant and location in a hurricane can be appreciably less than one.

Ocean Winds. Wind forces used for structural design purposes are commonly based on the extreme 1-minute average winds, appropriate for a selected return period. Winds used in the computations of wave heights and wind-currents, however, are based on mean wind speeds over longer time spans. The difficulty in selecting the extreme wind speed for design purposes generally centers on

TABLE 32

Nomenclature

F	=	Fetch length, naut. mi	P_j	=	Joint wind, wave, current probability, numeric
H	=	Wave height, ft	T	=	Wave period, sec
H	=	Average wave height, ft	T_i	=	Wave platform interaction time interval, sec
H_S	=	Significant wave height, ft	T_r	=	Return period, yr
H_m	=	Maximum wave height, ft	T_p	=	Duration of peak storm period, hr
L	=	Wave length, ft	T_a	=	Individual wind duration, sec
N_i	=	Number of time intervals, numeric	V_a	=	Wind velocity, mph
N_w	=	Number of wave occurrences, numeric	V_{wc}	=	Wind current velocity, knots
N_a	=	Number of wind occurrences, numeric	V_c	=	Velocity of current, knots
P	=	Probability, numeric	V'_{wc}, V'_c, V'_{wsc}	=	Wind direction component of V_{wc}, V_c, V_{wsc}, knots
P_w	=	Wave encounter probability, numeric	V_{wsc}	=	Wind-generated surface current, knots
P_p	=	Wind encounter probability in T_p, numeric	β	=	Probability parameter, numeric
P_a	=	Wind encounter probability, numeric	δ	=	Probability parameter, numeric
P_S	=	Storm encounter probability, numeric	μ	=	Eddy viscosity, lb-sec/sq ft
P_t	=	Wave period encounter probability numeric	θ	=	Angle between wind and wind current, deg
P_c	=	Tidal current encounter probability, numeric			

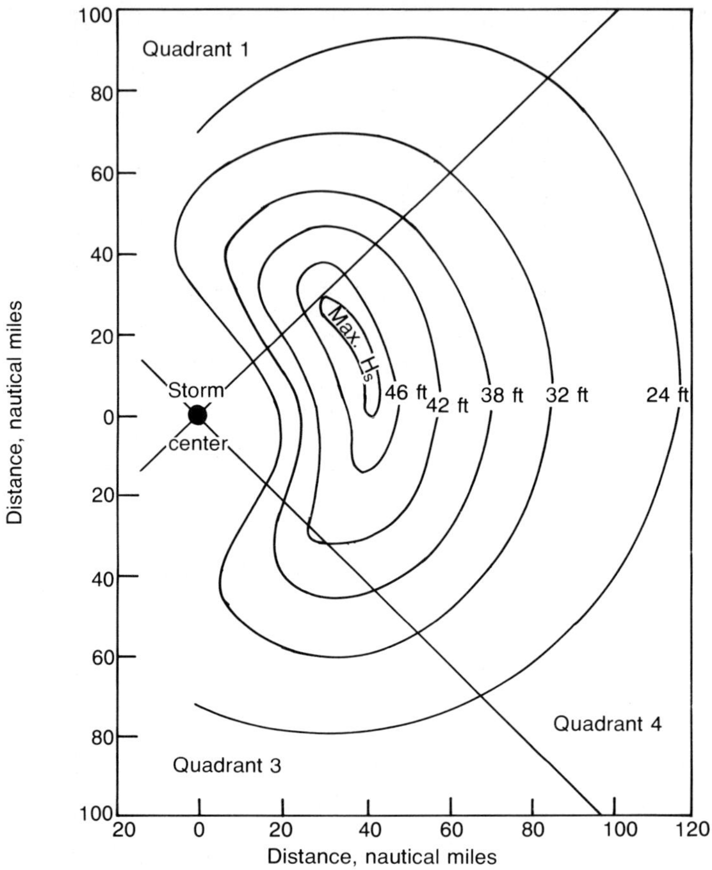

FIG. 2.17 Typical tropical storm model showing significant wave height distribution

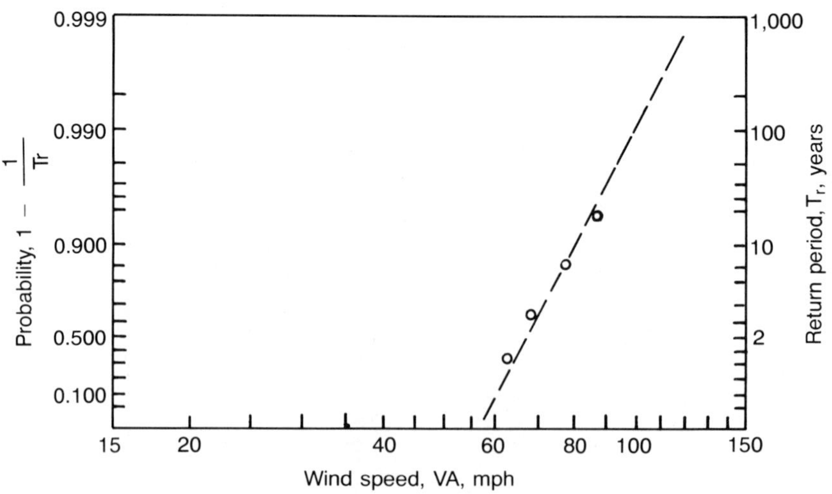

FIG. 2.18 Extreme wind speed vs. return period

the appropriate return period and the extreme winds to be applied for the selected return period.

It has been found (9) that return periods for extreme winds may be described by the relationship

$$T_r = [1 - e^{-}(V_a/\beta)^{-\delta}] - 1 \tag{2.5}$$

where V_a is the wind speed, and β and δ are parameters describing the probability distribution. This equation will plot data points as a straight line in the variables log $\ln(1 - 1/T_r)$ versus log V_a (figure 2.18). Thus, for a given operational location, wind speed data with known return periods can be used to estimate the general probability distribution for wind speed recurrence.

Ocean Waves. The significant wave height for a wind-generated sea can typically be determined from the generating wind speed, wind duration, and fetch length (figure 2.19). The recurrence of significant wave heights is often stated by probability distributions, such as that due to Weibull (10). Short-term wave recurrence data for a particular area of interest can be extrapolated to predict the significant wave heights for recurrence periods longer than the period covered by the available data. An example of the application of this distribution is given in figure 2.20. Once the significant wave height, H_s, is established, the frequency distribution of individual wave heights, H, may be described by the Rayleigh probability function,

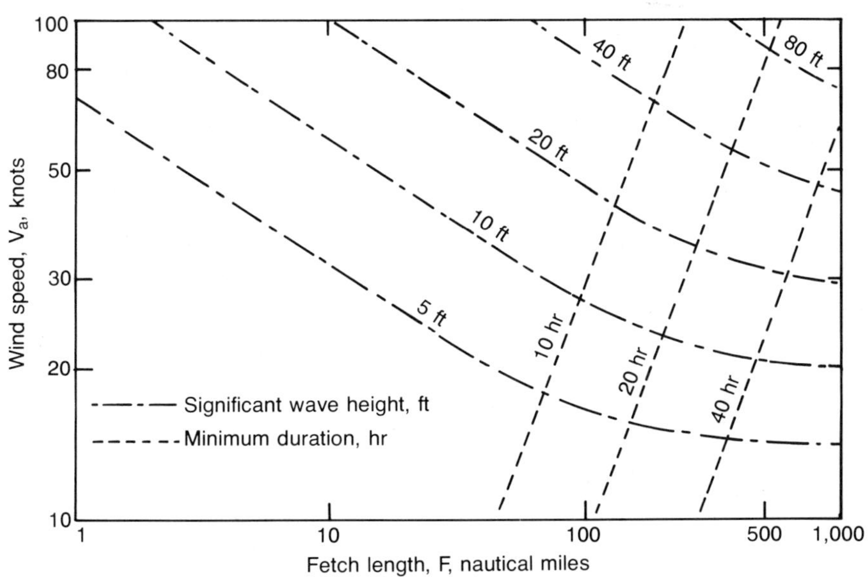

FIG. 2.19 General deep-water wave forecasting curves

$$P(H < x) = 1 - e - x^2/\bar{H}^2 \tag{2.6}$$

where $\bar{H} = 0.625\ H_s$ represents the average wave height.

Individual waves are known to occur with a distribution of wave periods limited only by the theoretical maximum steepness of H/L = 1/7 for first-order waves. For the purpose of the present work, a cumulative frequency distribution of inverse wave steepness was developed for a North Atlantic location (Marsden Squares 181-183), based on wave observations given by Hogben and Lumb (11). This distribution is shown in figure 2.21 and was constructed from the referenced data so it would exhibit the highest reasonable mean wave steepness for wave heights in the range of 19 to 50 ft. No definite correlation could be detected between wave height and wave steepness in the range of tabulated wave heights. Nevertheless, in view of the demonstrated distributive nature of wave steepness, the specification of an extreme wave steepness for the extreme design wave would not be reflective of its low probability of occurrence.

Ocean Currents. Currents which generally would need to be considered in designing offshore drilling units would include wave drift current, wind-generated current, and tidal current. Wave drift current is accounted for in the drag coefficient as it is presently estimated from wave force data and applied to wave force computations of the Morrison type. Currents should, of course, be considered since a 2-knot surface current can contribute substantially to the total wave force on a jacket-type structure, even under maximum wave conditions.

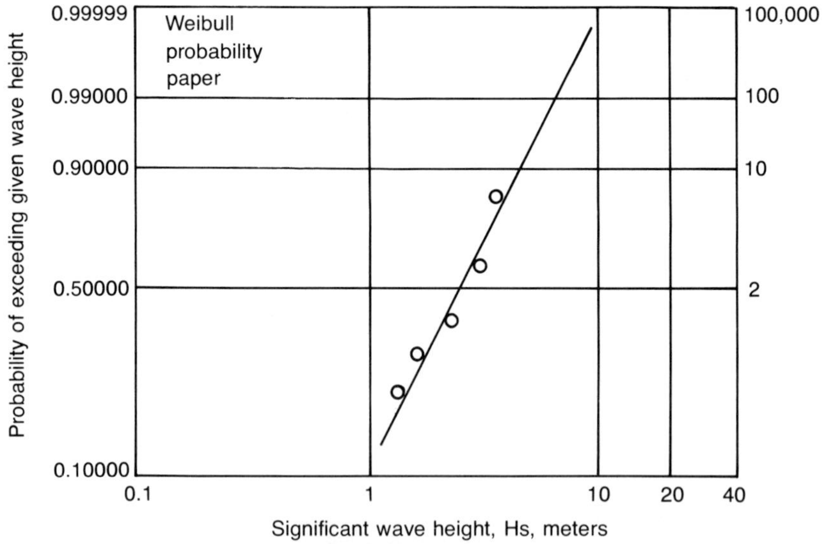

FIG. 2.20 Long-term North Sea wave height distributions

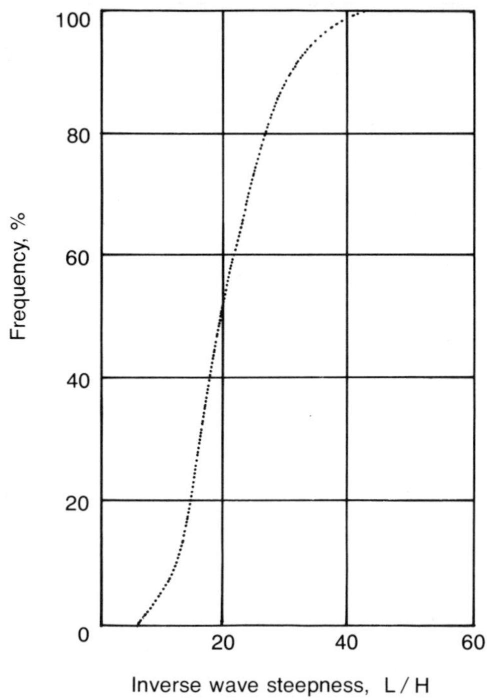

FIG. 2.21 Cumulative frequency distribution of wave steepness for wave heights from 23–32 ft for North Atlantic data

As a rule of thumb, wind-generated surface currents are often taken as 2% of the sustained wind. When using the surface current to establish a wind-generated current distribution with depth, the sustained wind should be interpreted as one no greater than that which generates the existing sea state. As an example, consider a sea state with 75-ft maximum waves, which could imply a significant wave height of $H_s = 75/1.86 = 40$ ft. Then from figure 2.19, it may be seen that generating winds for $H_s = 40$ ft are on the average about 60 mph (50 knots). Using this wind would then give a wind-generated surface current, Vwsc = 2% × 50 knots = 1.00 knot.

In view of the existence of Ekman-type currents (12), we have developed a wind-generated current profile which we feel is suitable for design purposes. This design current, which is shown in figure 2.22, acknowledges, in a conservative way, the angular deviation, with depth, of the wind-current from the wind direction. However, the characteristic Ekman current reversal at depth is not incorporated since it would actually reduce effective current loading. In

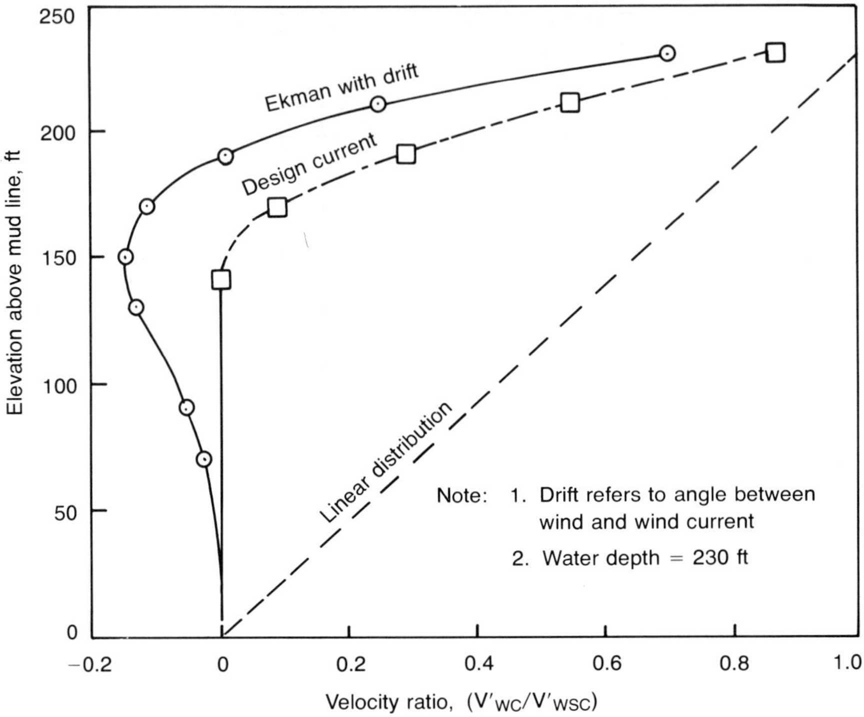

FIG. 2.22 Comparison of wind-direction current components

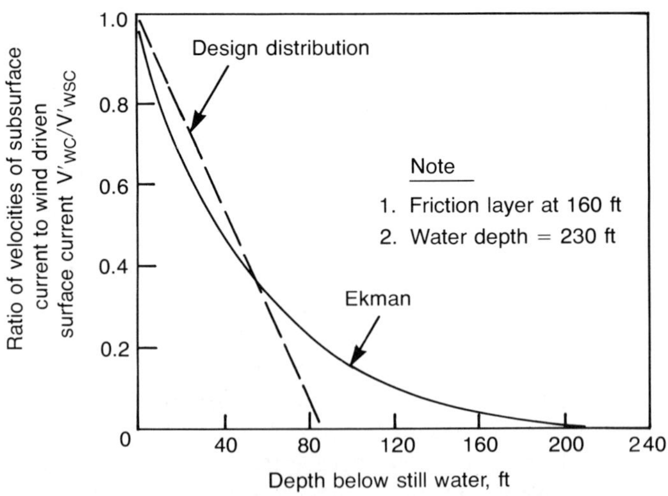

FIG. 2.23 Wind current magnitude distribution

figures 2.23 and 2.24, the proposed design current is compared with the constant eddy viscosity Ekman model. The Ekman distribution in these figures is based on an eddy viscosity of

$$\mu = 0.315 \text{ lb-sec/sq ft} \tag{2.7}$$

which would yield a friction layer depth of 160 ft at 55° north latitude. A comparison of the constant eddy viscosity Ekman model and a variable viscosity model, shown in figures 2.25 and 2.26, indicates that the design model developed in figure 2.22 should be considered for use in water depths equal to or greater than the friction layer depth. (*Note:* μ in figures 2.23 and 2.24 is not comparable to μ in figures 2.25 and 2.26.)

Tidal surface current in open ocean areas will vary inversely with water depth and rotate in direction at a roughly uniform rate in a diurnal cycle. The one particular feature of tidal current that could

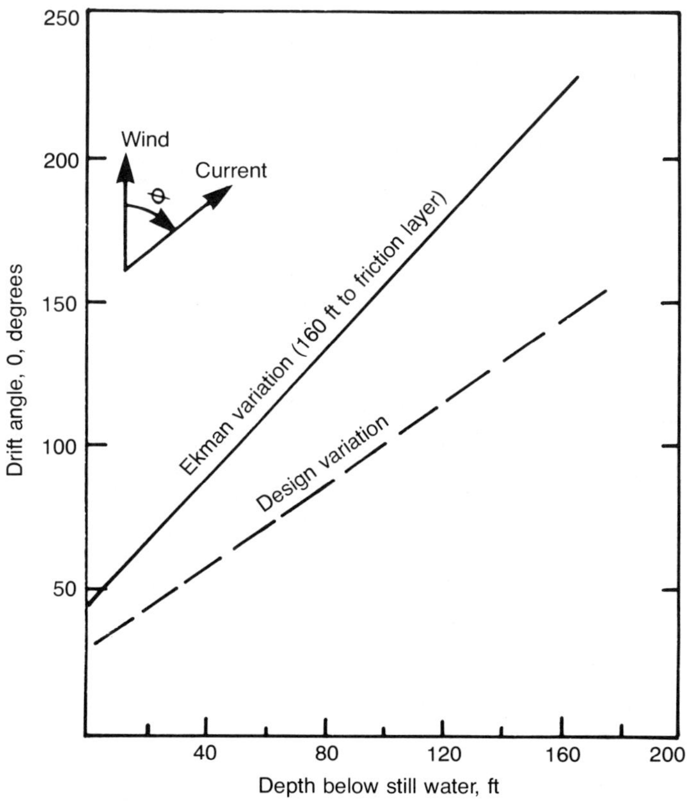

FIG. 2.24 **Variation of wind current direction with depth below mean water level**

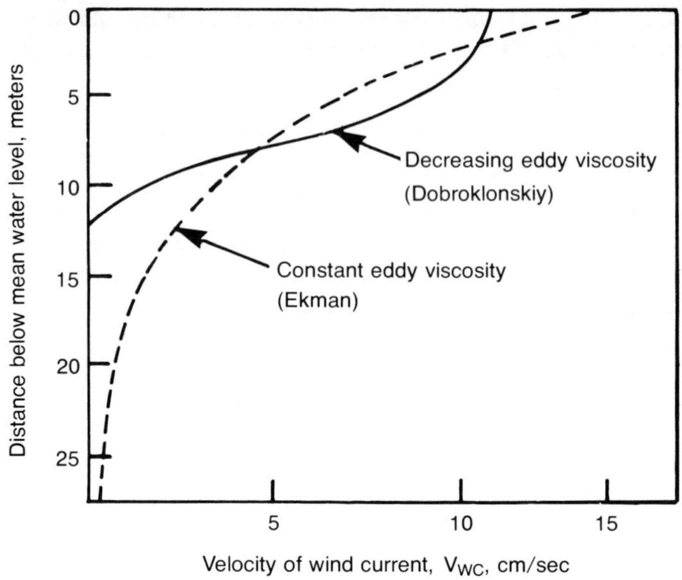

FIG. 2.25 **Comparison of wind current velocity profiles for two deep-water models (14)**

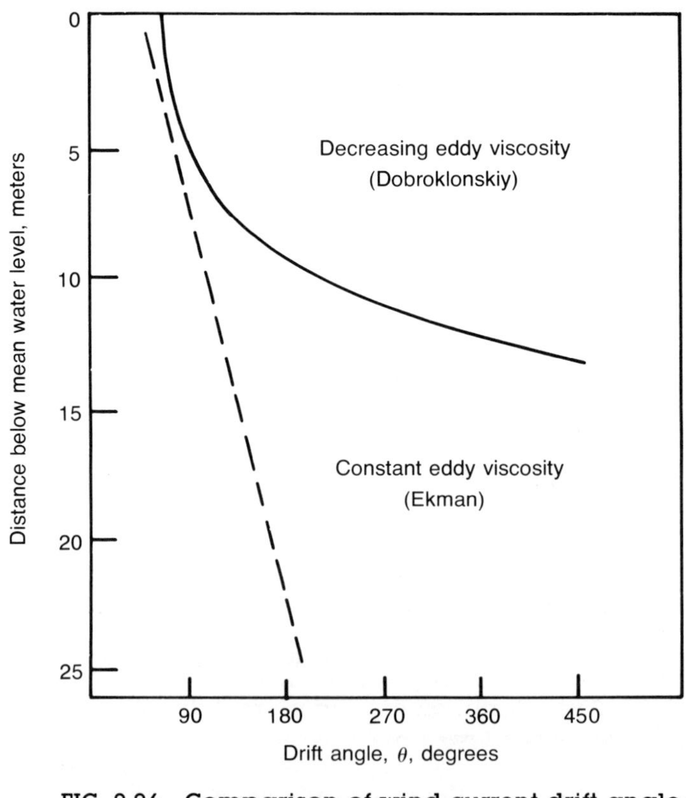

FIG. 2.26 **Comparison of wind current drift angle for two deep-water models (14)**

be brought to bear on formulating design criteria is the fact that it has a finite positive component in any given direction no more than 50% of the time.

The Risk Concept in Design

Risk acceptance has really always been a fact of life in the design process, but calling it by its proper name has usually been avoided. Some aspects of this circumstance were, and often still are, that certain risks were not quantifiable or that the risk implications would be misinterpreted. A flood control project might have been designed for a 100-year flood in the belief that this flood would not occur for another 100 years, commencing with the completion of construction. This kind of belief probably evolved more through misconceptions on the part of the public. But once the public attitude had formed, the criteria were reimposed on the engineer without regard to their real implications. As it is, in a period of 50 years, there is approximately a 40% chance that the 100-year flood will have occurred.

An alternative to the traditional factor of safety or the arbitrary choice of a design condition recurrence period would be to assess the design risk if at all possible, relative to other normally existing and normally accepted risks. Shipping loss rates could be cited as an example. United States flag losses for 1971 were at the rate of 0.3% per year (13). Also, table 33 lists some typically occurring risks in the loss of life category. However, the use of the loss of life basis in establishing design risk levels should be used with the greatest of care. Through individual actions, loss of life rates, such as the average of 1/670 home fatalities per 10 years, will be higher for some individuals but lower for others. The external imposition of a common risk level denies the influence of personal choice. It is thought that in just about every case, a sound decision could be based on financial loss versus cost of loss avoidance without disturbing the cumulative loss of life risk.

TABLE 33
Normal loss of life risk per ten years

Probability	Activity
1/195	Accidental work fatalities (9)
1/375	Accidental auto fatalities (9)
1/670	Accidental home fatalities (9)
1/1000	Accidental public fatalities (9)
1/97	Cumulative risk

About 1 in 350 ships are lost in collisions every year (14), and lives are lost as a consequence. But this does not imply directly that 1 in 175 losses would double fatalities, nor that 1 in 700 losses would cut them in half. In designing any marine system, one might, for example, design to a risk level that is comfortably lower than the ship loss risk but incorporate life-protection systems that ensure a life survival probability that is far higher.

The Joint Occurrence of Wind, Waves, and Current

Consider, for the purpose of demonstration, the environmental events which might occur at a particular location in a 100-year North Sea storm, and which thus might be applicable to a self-elevating drilling platform while on location. The demonstration is, of course, also to suggest the formulation of design criteria along similar lines.

The storm is characterized by certain wind intensities and durations, wave heights and periods, and currents. Since an on-station jackup is being considered, Rayleigh distributed individual wave heights should suffice for wave load prediction. A storm model is assumed in which the highest waves (exceeding 60 ft) will occur at random in a 3½-hr period. Peak winds are assumed to occur at random in a 24-hr period which contains the 3½-hr period. Individual events, other than wind-generated current, are assumed to be independent.

Since events in the storm model (e.g., peak wind duration) do not occur at a discrete point in time and in order to obtain a simple probability model for initial evaluation, proceed as follows. Divide the peak storm period, T_p (3½ hr) into suitably small time intervals, T_i. The shortest naturally available time scale is the interval during which one particular wave is passing through and affects in some way some part of the platform structure. Calculations indicated that a 60-ft wave (L/H = 10) would pass through a 200-ft-long structure in less than 20 seconds. A higher wave would pass through in a lesser time. Since a longer time interval will ultimately overestimate the encounter possibility, the basic interval is taken as

$$T_i = 20 \text{ sec}$$

One now asks, what is the probability of any one wave acting on the platform during any one interval of length T_i? First of all, there are

$$N_i = \frac{T_p}{T_i} = \frac{(3.5)\,(3600)}{20} = 630$$

intervals of 20-second duration in 3½ hours. Then, a particular wave occurrence (taken as 20 seconds in duration) need not coincide with any one of the intervals established in the foregoing. Thus, the total class of possible events in this discrete approach is $(2N_i - 1)$. Of this total, three events could possibly occur during a particular time interval (i.e., exactly coincident, overlapping from the preceding interval, and overlapping from the following time interval). Consequently, the probability of encountering a particular wave during a particular time interval would be

$$P_w = \frac{3}{2N_i - 1}$$

The encounter probability for a particular wind occurrence in the peak period, T_p, is approximately $P_p = 3.5/24$, since the wind has been assumed to occur with certainty in 24 hours though not necessarily in a particular 3½-hour interval. Generally, this will yield an overestimate on wind encounter, and it will be a particularly good estimate when the wind duration, T_a, is short relative to T_p (e.g., a 1-minute wind in T_p). Hence, the probability of a particular wind, T_a, occurring during a particular time interval, T_i, in T_p will be

$$P_a = \left(\frac{3.5}{24}\right)\left(\frac{T_a}{T_i} \Big/ N_i\right) = \left(\frac{3.5}{24}\right)\left(\frac{T_a}{T_p}\right) \tag{2.10}$$

When the joint probability of wind-wave coincidence in interval T_i is formed later, it will indicate only that both will occur during some part of T_i, not necessarily the same part, since T_i is extensive in time.

Again, an overestimate of final encounter probabilities will result when coincidence is defined in this way. One could avoid this problem by taking T_i to the infinitesimal limit and dealing with a probability density function. However, since the foregoing overestimates will turn out to be adequate for the present purpose, the latter was not pursued further.

Next, let P_s be the encounter probability for a 100-year storm in a 10-year period in order to establish a 10-year cumulative basis for the joint probabilities to be evaluated, that is,

$$P_s \cong 10/100 = 0.1 \tag{2.11}$$

Also, let P_t be the probability that a particular wave will have a particular wave period, and P_c the probability that tidal current will have a component concurrent with wind and wave direction at any particular time. The probability, P_t, will be estimated from figure 2.22, and P_c will be taken as one-half. The latter probability estimate is based on a uniform directional rotation of a diurnal tidal cycle. Consequently, the joint probability of simultaneous wind, wave, and tidal current coincidence in a 10-year period will be

$$P_j = (P_c P_w N_w)\, P_t\, (P_a N_a) \qquad (2.12)$$

The numbers N_w and N_a are the possible occurrence of a particular wave and wind category, respectively.

The number of occurrences of wave height groups are determined from probability distributions, such as given in figure 2.27 and listed in table 34. This table actually lists the total number of waves that are predicted to occur in a 100-year period, which is about 20% higher than the number predicted for a 100-year storm in the case of waves 60 to 65 ft high.

The number of occurrences of wind speed groups for 1-minute winds are determined from figure 2.28. This is an estimated probability distribution constructed with some readily available data and predictions. The distribution is intended to favor the highest wind speed predictions that have been made for mid-North Sea locations, irrespective of their occurrence in storms which could have generated wave heights in excess of 60 ft.

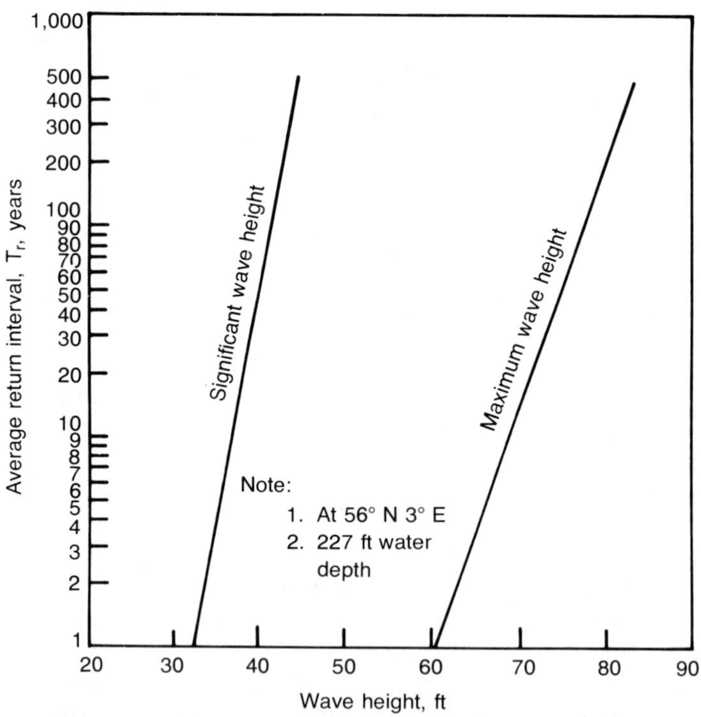

FIG. 2.27 North Sea return intervals for various wave heights (10)

TABLE 34
Wave height occurrences in North Sea

Wave height category, ft	Average number of waves per 100 years
60–64	80
65–69	22
70–74	6
75–79	2

56N 3E in 227 ft water depth (10).

TABLE 35
Wind recurrence

Wind velocity,		Number of occurrences in 100 years
knots	mph	
100	115	1
90–100	100–115	5
80–90	90–100	24
70–80	80–90	62

It is felt that some of the highest wind speeds did not occur in storms which generated the highest waves. Table 35 then lists the number of wind speed group occurrences as obtained from figure 2.28. The occurrence of 62 periods in 100 years in the 70 to 80-knot category will be assumed to indicate that in a 100-year storm, 62 one-minute periods occur with these winds. This is not quite realistic, obviously, but wind speed distributions of the form desired for joint probability estimates are not readily arrived at.

Further, probability estimates using peak predicted 1-hour, 3-hour, etc., winds appear to give results comparable to the foregoing assumption. Thus, it is felt that the foregoing assumption will at least reflect the situation in a 100-year storm and may not be too far from estimating the total occurrence in a 100-year period. A more extensive study of extreme wind frequency distributions, occurrence in extreme storms, etc., is obviously needed. In this regard, statistics, such as presented by Burridge (15), would be of definite value.

No reduced probability is presently included for wind and wave coincidence at the same point in space. This kind of coinci-

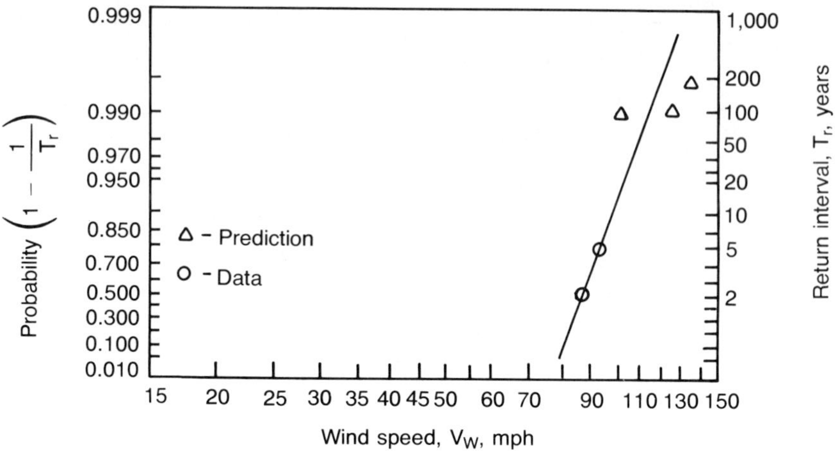

FIG. 2.28 Estimate of extreme 1-minute wind return periods for a mid-North Sea location

TABLE 36
Probability of combined events in ten years

Wind speed, kts:	100+	90–100	80–90	70–80
Wave ht., ft	**Inverse steepness greater than 15**			
75+	1.292×10^{-7}	6.458×10^{-7}	3.099×10^{-6}	8.008×10^{-6}
70–75	3.875×10^{-7}	1.937×10^{-6}	9.299×10^{-6}	2.402×10^{-5}
65–70	1.421×10^{-6}	7.104×10^{-6}	3.409×10^{-5}	8.808×10^{-5}
60–65	5.166×10^{-6}	2.583×10^{-5}	1.239×10^{-4}	3.203×10^{-4}
Cumulative	7.104×10^{-6}	3.552×10^{-5}	1.705×10^{-4}	4.404×10^{-4}
Subtotal	6.535×10^{-4}			
	Inverse steepness less than 15			
75+	3.644×10^{-8}	1.822×10^{-7}	8.746×10^{-7}	2.259×10^{-6}
70–75	1.093×10^{-7}	5.465×10^{-7}	2.623×10^{-6}	6.776×10^{-6}
65–70	4.008×10^{-7}	2.004×10^{-6}	9.619×10^{-6}	2.425×10^{-5}
60–65	1.497×10^{-6}	7.285×10^{-6}	3.496×10^{-5}	9.034×10^{-5}
Cumulative	2.003×10^{-6}	1.002×10^{-5}	4.808×10^{-5}	1.242×10^{-4}
Subtotal	1.843×10^{-4}			
Subtotal	8.378×10^{-4} for all values of steepness and for one tidal current direction.			
Total	16.756×10^{-4} for all values of steepness and for both tidal current directions.			

Note that probability assumed to be the same for either tidal current against wind current or tidal current with wind current.

dence is also in need of further investigation, since maximum seas are normally confined to a particular storm quadrant.

The estimated joint probabilities as discussed in the foregoing are tabulated in table 36. It indicates, for example, that the probability of encountering a wave in excess of 75 ft with an inverse steepness, L/H, less than 15 (wave period less than 14.8 sec), a wind speed greater than 100 knots and concurrent tidal current, in a 10-year period is 3.64×10^{-8} or 1 chance in about 27 million. Furthermore, the probability of encountering steep waves greater than 60 ft in height with winds exceeding 70 knots and with all currents is no more than 1 in 595 in a 10-year period.

Application of the Finite-Risk Concept

Consider now the selection of appropriate design criteria for a self-elevating drilling platform, based on the probabilities generated in table 36. If, for example, the combined conditions included within area A in table 37 are selected, without regard to other con-

TABLE 37
Design Probability of Combined Events in Ten years

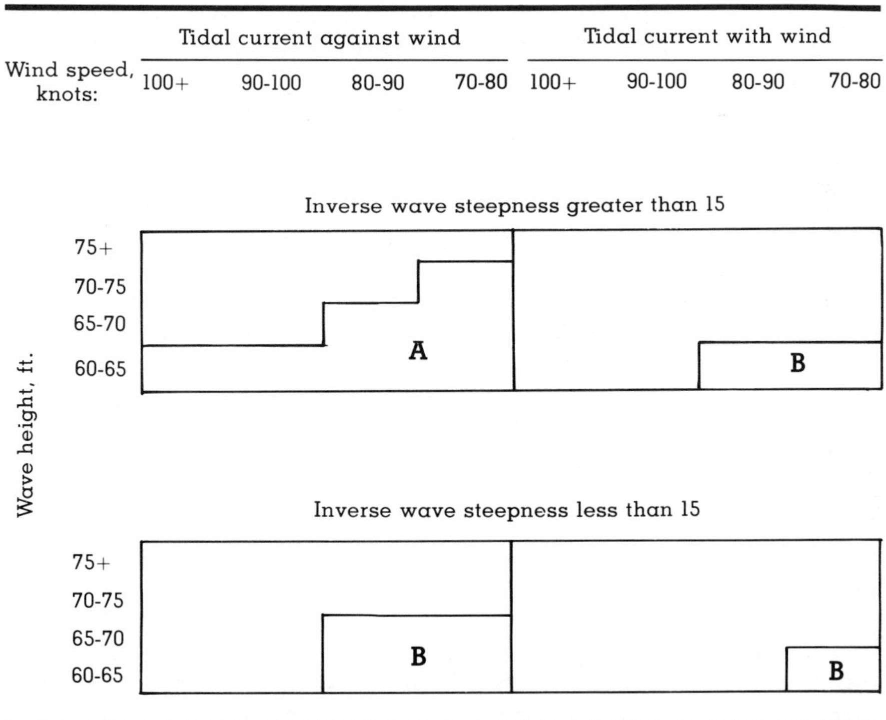

ditions which may as a consequence become satisfied, then the probability of encountering conditions not within area A (essentially the probability of exceedance of conditions within A) is composed of the sum of all probabilities not included in A. This sum would be 1/948 per 10 years or 0.11% per 10 years.

The conditions included within area A will, depending on the type of unit being designed, usually not be equally critical. Typically, 65-ft waves with winds greater than 100 knots may show the greatest factor of safety, while 75-ft waves with 80-knot winds may show the lowest. Also, additional combination of conditions may become automatically satisfied as a consequence of designing to those included with area A.

It is estimated that in a typical case those combined conditions contained within additional areas B would also be satisfied. Thus, by summing only probabilities not within both A and B, a probability of exceedance of 1/2780 per 10 years or 0.04% per 10 years is obtained.

In effect then, by designing to conditions of combined events as covered by area A, the probability of encountering a more critical condition in a 10-year period would be 1/2780. Restated, this may be interpreted to mean that if on the average, 2,780 similar units are operating each year for 10 years in the area covered by the analysis, then only one of them could expect to encounter a condition more critical than the maximum design condition during the 10-year period.

It is believed that the foregoing probabilities would indicate, in a conservative way, the inherent structural integrity of the bottom supported drilling unit design with respect to the chosen 100-year storm conditions. They would thus reflect that portion of the expectable loss rate not attributable to operational decisions. Typical operational decisions, which might be causative factors in higher actual on-location loss rates, would be operating with insufficient air gap or with inadequate footing penetration. It is further believed that the incorporation of appropriate total wind recurrence periods, together with the joint probability for spatial coincidence of particular winds and waves, will result in a cumulative design condition probability of exceedance not significantly greater than the 0.11% or possibly even the 0.04% per 10 years.

Conclusion

It has been demonstrated that the simple superposition of extreme environmental events produces a set of design criteria for mobile offshore drilling units whose actual probability of exceedance is far less than the recurrence of individual events or that of other normally accepted risks. We feel that the suggested approach could be the basis for the development of more realistic design criteria so that the cumulative risk of a particular design decision can be evaluated. The investigation also highlights the continuing need for improved hindcasting of storm conditions, together with a better understanding of extreme storm component behavior.

The approach presented here has been employed to investigate the on-station loads for a North Sea drilling operation. Out of this work has come the notion that an offshore facility should be evaluated with respect to a range of combined environmental conditions whose probability of exceedance entails some definable and acceptable risk level.

It was found that, at a risk level no greater than normal shipping losses (or even normal fatality rates), this range of conditions may be characterized by the simultaneous application of 100-year conditions of 80% of extreme H_{max} and extreme 1-minute wind on one end and the application of 100% of extreme H_{max} and extreme 1-hour wind on the other.

A sound stochastic approach to the assessment of environmental loads, in conjunction with an analysis of historical rig losses, would help to pinpoint the dominant causes of rig losses and, consequently, lead to the most effective approach for reducing such losses. Otherwise, mobile drilling units may be unnecessarily forced into a position which does not truly utilize their operational capabilities.

2.5 MARINE RISER DESIGN FOR EXPLORATORY RIGS

A riser can be simply described as a conduit from the platform to the ocean floor which transmits the drilling mud and serves as a guide for the drillstring. There are two broad classes of risers, those used for exploratory drilling operations and those used for production operations. In exploratory drilling, the riser tube running from the vessel down to the wellhead is normally called a riser while the first casing joint running from the wellhead slightly above the mudline down through the template to 30–100 feet below the mudline is called a conductor or surface pipe. In production operations, the riser tube running from the platform deck down to the wellhead is also sometimes called a conductor.

In exploratory drilling operations conducted with a jackup rig, the riser runs from the drilling deck down to the wellhead (figure 2.29). The drilling fluid is pumped down the drillstring. The mud returns in the annulus between the string and the riser in the water or between the string and the casing for that part of the drillstring below the subsea wellhead. Ordinarily, when oil is struck in exploratory

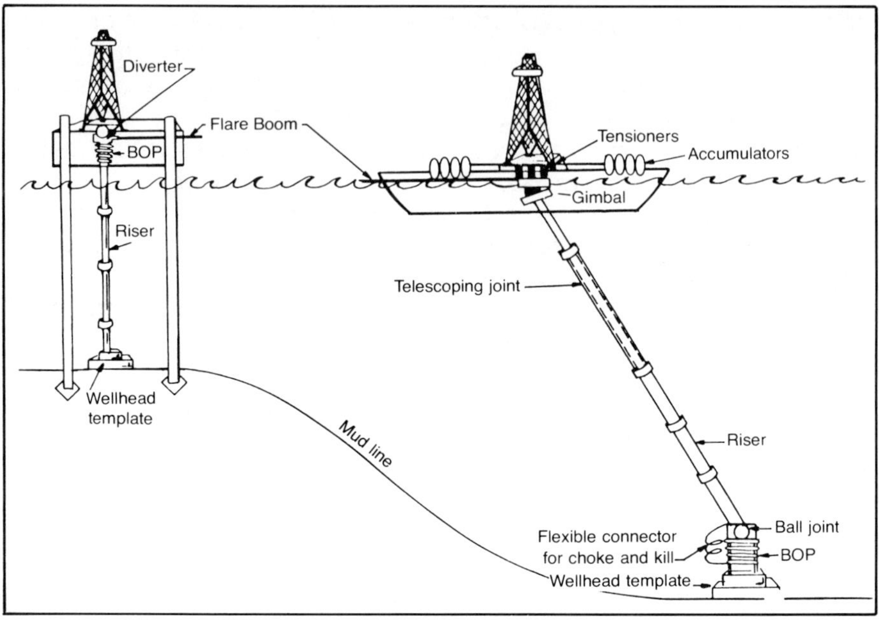

FIG. 2.29 Marine riser systems for exploratory jackup and floater drilling operations

drilling, the well is tested and capped and the rig moves on to another drilling site. In most areas, the regulations require that all equipment be removed from the ocean floor. The riser string can then be reused.

For floating drilling operations using a drillship or semisubmersible, more hardware is required to accommodate the vessel's heave, sway, and surge. Attention here is directed toward floating drilling operations since the riser-related problems are more complex.

Basic Components of an Exploratory Riser System

Figure 2.29 illustrates the arrangement of the riser system. It is important to note that the riser is not a simple piece of equipment but rather a complex system of component parts. A list of the sources for components of basic riser equipment and the manufacturers of each component has been prepared (table 38). We see that many manufacturers produce individual parts of the subsea riser system but only a very few supply the entire equipment package.

The riser string for a floating exploratory vessel is usually made up of 50-ft-long joints which can be stacked and stored on deck during transit from one drilling location to another. The ends of each joint have quick-disconnect couplings permanently attached to the

TABLE 38
Standard riser equipment*

Manufacturer	Headquarters	Joint coupling	Ball joint	Tele-scoping joint	Tensioners	Diverters	Motion compen-saters	Subsea BOP equip-ment	Subsea wellhead equip-ment	Flexible riser BOP connector
Aeroquip Corp. (Barco)	Jackson, Michigan		x							
A-Z Int'l Tool Co.	Houston, Texas			x			x			
Cameron Iron Works	Houston, Texas	x	x	x				x	x	
FMC Corp (Chicksan)	Houston, Texas								x	x
Gray Tool Co.	Houston, Texas	x							x	
Hydril	Los Angeles, Calif.							x		
Hydro Tech	Houston, Texas	x								
Johnston Div.	Houston, Texas						x			
Martin-Decker	Santa Ana, Calif.						x			
National Supply Co.	Houston, Texas	x	x	x					x	
Ocean Science & Eng.	Long Beach, Calif.	x								
Regan Forge & Eng.	San Pedro, Calif.	x	x	x		x		x	x	x
Rucker Shaffer	Houston, Texas	x	x	x	x	x	x	x	x	
Stewart & Stevenson	Houston, Texas							x		
Vetco	Ventura, Calif.	x	x	x	x	x	x	x	x	x
Western Gear Corp.	Houston, Texas				x		x			

*This table compiled from the 1974 *Composite Catalog* and represents only those manufacturers included within.

joint. The telescoping joint, which is at the upper end of the riser string, is usually designed for a maximum heave of between 15 and 30 ft.

A constant force tensioner system is attached to the top of the fixed outer barrel of the telescoping joint to provide enough axial force in the riser string to prevent buckling. The outer barrel and the riser string have lateral movement with vessel surge and sway but essentially no vertical movement with vessel heave. The vessel and the inner barrel of the telescoping joint move together vertically with vessel heave. The optimum tension is a function of water depth and operating conditions (mud weight, etc.), as shown in figure 2.30.

Ball joints on each end of the riser allow for rotation in any direction up to about 7–10 degrees. Actually, only a few operators insist on two ball joints, which offer more reliability than a single ball joint, because the use of two ball joints incurs greater costs and greater running time. The usual arrangement for floating drilling operations is a gimbal under the drilling deck and one ball joint attached to the top of the subsea BOP stack, which sits on the wellhead. The wellhead attaches to the base template that is set with the conductor pipe at the beginning of the operation. As drilling pro-

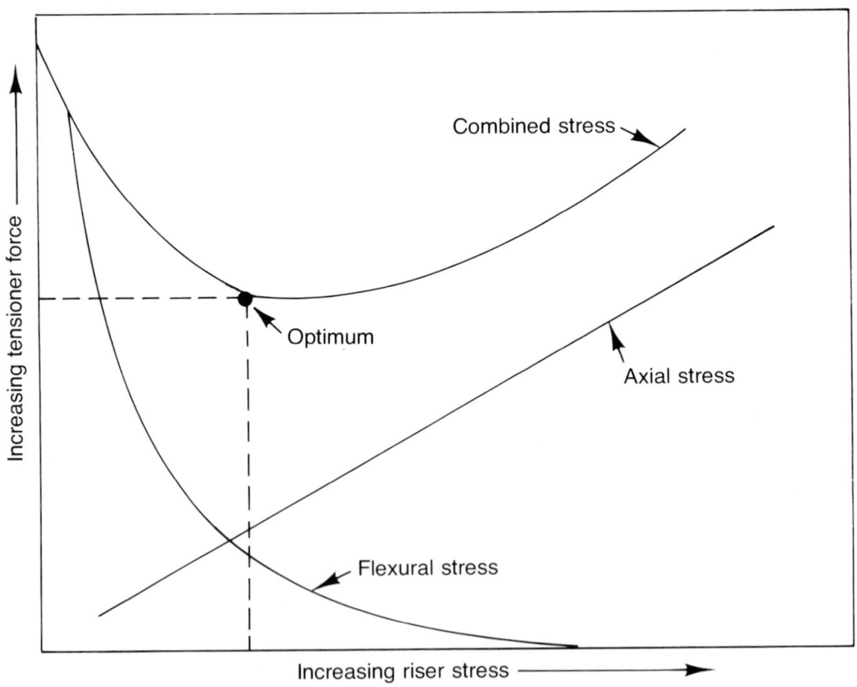

FIG. 2.30 Optimization of riser tensioners

gresses, casing supported by casing hangers in the wellhead is placed in the well bore. At intermediate depths, another casing string of smaller diameter is set inside the first casing string from other casing hangers in the wellhead. The depth and number of the various sizes of casing strings depend upon geological conditions.

Riser Coupling Joints. The first riser systems had the choke and kill lines strapped to the riser pipe. Pipe handling problems with this system proved to be extremely time-consuming. Most riser systems now use integral choke and kill lines, which are permanently attached to opposite sides of the riser and have their own connectors. When the riser joints are stabbed and quick-connected, the receptacles allow the choke and kill lines to be stabbed and automatically connected at the same time.

The BOP requirements have been the decisive factor in determining the diameter of the riser pipe and hence the riser wall thickness, required tensioner force, etc. In early drilling operations, a two stack, 20-in. BOP was used until the first few strings of casing were set. The 20-in. BOP was then changed out for a 13⅜-in. BOP for making the rest of the hole. Although this procedure is still used and preferred by some, the single stack is growing in usage. The first single stacks were 16¾-in. and required a 17½-in. underreamed hole for setting 13⅜-in. casing. Because of dissatisfaction with underreaming, a single 21¼-in. BOP was then used.

The heavy weight of these systems, which sometimes exceeded 200 tons, overstressed the system in rough weather. The proposed North Sea riser system is a single 18¾-in. stack which weighs half as much as the 21¼-in. system but does not require underreaming for the 17½-in. casing hole.

Choke and Kill Lines. The choke and kill lines run from the deck along the riser string down to the wellhead. At the lower riser ball joint, there are various schemes, such as looped pipes, to get the required flexibility in a jump line arrangement running from the bottom of the riser string (top of ball joint) around the ball joint to the BOP stack. The choke and kill lines control kicks in order to prevent them from developing into blowouts.

When a potential blowout is detected, mud is pumped down the kill line at the BOP stack to restore pressure balance in the hole. When excess gases occur, the bag and ram type BOPs are closed around the drillstring. The gas is relieved at the choke manifold on the BOP stack by running up the choke line on the riser string. As the gas expands, it proceeds up the choke, displacing more mud and traveling faster as the gas bubble or gas-entrapped mud approaches the mean water line. Without the choke, the gas would push out the annulus mud between the drillstring and the riser, and control from the weighted mud would be lost.

Telescoping Joint. There are two basic types of telescoping joints used with marine risers. The constant tension system (remote axial tensioning system) is most often used because maintenance is easier. This method uses a linkage system at the base of the drilling floor to maintain equal force on the several wire ropes attached to the outer barrel of the telescoping joint.

An alternative design of telescoping joint uses the direct axial tensioning method. This is a procedure where the seals and guide rings on the telescoping joint are designed to compensate for internal pressure so that the telescoping joint has the dual function of allowing vessel heave and acting as a direct tensioning piston.

A diverter is located at the top of the telescoping joint. Depending upon the magnitude of the kick, the gasified mud is valved either onto the shale shakers or to port.

Tensioners. The inner barrel of the telescoping joint is connected to the gimbal under the drilling floor of a floating vessel. Wire rope runs through pulley systems on the deck down to the top end of the lower (outer) barrel of the telescoping joint. In the past, when floating drilling operations were restricted to shallow water, dead weights were connected to the deck end of the wire rope to maintain riser tension and prevent buckling collapse of the riser string.

Constant tensioners are now used, having a total capacity ranging from 240–640 kips. The capacity of each individual tensioner is between 60 and 80 kips. The wire rope from the lower barrel of the telescoping joint is fed around pulleys on each end of a main hydraulic cylinder. Fluid in this cylinder is ducted to an air-over hydraulic bladder type of accumulator. The air end of the bladder accumulator is then ducted to a bank of air pressure vessels, and the relatively constant tension is maintained by constant pressure with large changes of air volume due to the compressibility of the gas. A typical tension force-stroke curve is given in figure 2.31.

Total mechanical and hydraulic friction in any tensioning system is about ± 15%. The variation from constant tension due to non-uniform pressure with varying volume depends upon the volume of the air pressure vessels used and is usually designed for 5–10%. The total variation from constant tension is then usually less than about 20–25%.

Riser tensioners are required for water depths greater than 250 ft. The critical buckling length, L_{cr} in ft, can be calculated by the relation:

$$L_{cr} = 2.75 \left(\frac{EI}{w}\right)^{1/3} \tag{2.13}$$

where E is the modulus of elasticity in psi, I is the moment of inertia in in.^{4}, and w is the weight of the riser in water in lb/ft.

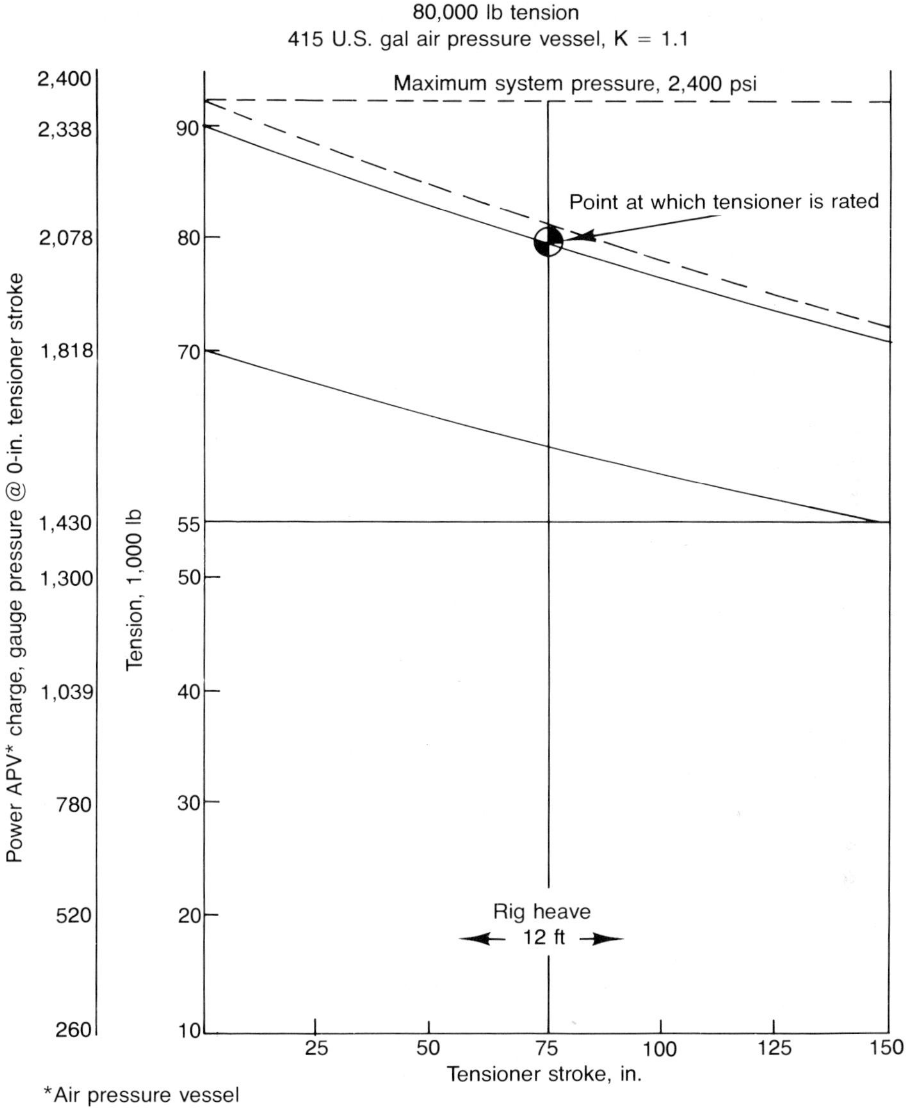

FIG. 2.31 Tension vs. stroke

Riser Tension Guideline. W. L. Clark of Hydril has developed a useful procedure for sizing the guidelines for riser tensioners. His procedure is summarized as follows:

The riser tension guideline is a rough order of magnitude calculation only. More complete and involved methods are required for complete accuracy. This method of calculation is based on the weight of the riser pipe, the suspended pipe, and the mud column.

Nominal tension equals weight of the riser system in water plus 20% so that positive tension will be maintained at all times during heave motions. For example:

20-in. OD riser × ½-in. wall in water = −33 lb/ft.
18 lb/gal mud = 264 lb/ft between ID of riser and OD of drillpipe
Nominal tension for 800 ft of water

$$T_{nom} = (800)\,(264 - 33)\,(1.20) = 222 \text{ kips} \tag{2.14}$$

For purposes of safety, the tension system must be operated at a level high enough to provide the minimum tension level (that required to maintain the bottom joint of the riser pipe in positive tension) after losing one tensioner unit. Thus

$$\text{Operating tension} = \frac{\text{minimum tension}}{\text{safety divisor}} = \text{OT}$$

Minimum tension riser system wt plus 5%

$$\text{Safety divisor} \; \frac{\text{no. of tension units less one}}{\text{no. of tension units}} \; \text{SD}$$

Therefore,

$$\text{S.D. for a four-tensioner system} = \frac{4-1}{4} = 0.75$$

$$\text{S.D. for a six-tensioner system} = \frac{6-1}{6} = 0.83$$

$$OT_4 = \frac{(800 \text{ ft})\,(231 \text{ lbs/ft})\,(1.05)}{0.75} = 259 \text{ kips} \tag{2.15}$$

$$OT_6 = \frac{(800 \text{ ft})\,(231 \text{ lbs/ft})\,(1.05)}{0.83} = 234 \text{ kips} \tag{2.16}$$

Production Risers

On fixed production platforms, the tubes running from the platform deck to each subsea wellhead are called connectors. After the oil runs up the connectors and through the processing equipment, it is then pumped from the deck down to the marine pipeline through the production riser. The three basic types of production risers are the J tube, the reverse J, and the bending shoe methods.

The production riser actually contains a bundle of pipes. These individual pipes include the production, solvent, and vent lines. The bundle may also include lines for gas lift, gas injection, water injection, air supply, and air return.

There are, of course, always exceptions to the conventional riser design mentioned above. One of these noteworthy exceptions is the phased operation in the Ekofisk field (29) in the North Sea. The initial oil discovery in the Ekofisk area was made in 1969 by the Phillips Norway Group with Phillips Petroleum as the operator. Development work began in November 1970, and production by sea loading began in the summer of 1971. This was the first development work in the middle North Sea, where the water depths go up to 230 ft. This Phase 1 test production was brought on stream by converting the Gulftide exploratory jackup into a production platform. In this case four wells were brought on stream with dual 4-in. pipelines running from each well to the Gulftide, connected to the caisson base, then up through a riser caisson located in the lattice jackup leg.

Design and Water Depth Limitations

Most floating exploratory vessels are now outfitted with a riser system for a maximum water depth of about 1,200 ft. Four or five new drillships in operation or under construction will have riser systems for the extreme water depth of 3,000 ft. A design analysis has been performed on one particular advanced riser system for 6,000 ft water depth. From that study, it appears that 6,000–8,000 ft of water is the practical limit for the conventional riser system.

There is a drawback with risers for over 6,000 ft water depths. The static head of heavy drilling fluid (e.g., 20-lb mud) could be about 7,500 psi at the mudline and proportionately greater with drilling depth compared to the normal water pressure of about 3,000 psi at the mudline. It is believed that this extraordinarily high pressure could cause severe fracturing of the formation. A possible solution might be to use gas injected mud in the drillstring and artificial gas lift for the mud column in the annulus between the drillstring and the riser with injection at the wellhead. A shortcoming of this approach is that mud monitoring and control for blowouts would become even more difficult trying to distinguish between the injected gas and a kick.

The major problem with deep-water riser systems for 2,000–6,000 ft water depths is finding a practical balanced design for variable water depths. For deep water, the riser string can become heavy enough from gravitational loading alone to severely stress the riser. The required tensioner sizes would then become enormous.

Augmented Riser Buoyancy. The maximum tension in the riser occurs at the top of the riser string and decreases with water depth. In very deep water, some type of added buoyancy is required to stay within the practical limits of the tensioner system. However, it must be remembered that the current drag force goes up as the square of the velocity and increases with riser diameter. This means that the outside diameter of the added buoyancy cylinders should be minimized. The optimum design riser has the smallest diameter and lowest possible wall thickness consistant with allowable stress levels and allowances for wear and abuse.

Plastic foam cylinders fastened to the riser have been used, and buoyant steel chambers have been suggested. The foam cylinders rapidly deteriorate, however, and require constant maintenance.

If a section of the riser system is made too buoyant, failure of a riser joint could hurl the riser up through the derrick floor like a missle. Drillers and roughnecks on the rigs look upon this possibility somewhat unfavorably. The requirement, then, for deep water riser systems is to make the riser neutrally or only slightly positively buoyant, with the buoyancy distributed over a long length of the riser.

Special Design Considerations. Sometimes it is desirable to make a first attempt at design with approximate methods, such as ignoring dynamic effects. When this is done, a safety factor of about 35% should be used on the riser design stress. This, of course, would not be satisfactory for large vessel motions in extreme environments which will be encountered in storms immediately before well abandonment. For example, in water depths of 500–1000 ft with typical riser designs, the dynamic stresses due to 36 ft waves would be 8–10 ksi, not counting the dynamic stresses due to vessel motions.

Normally, the outer surface of the riser should be smooth and with as few protuberances as possible. However, in exceptional cases special frames to preclude vortex shedding might be required. The water greatly dampens any vibrations due to resonant conditions, but special conditions where high currents could cause riser damage should always be investigated.

Analytical Review of Exploratory Riser Systems

Analyses for exploratory riser design must include a large number of variables. The combined stress in the riser is due to both axial and hoop or circumferential stress. Hoop stress is caused by internal pressure of the mud column and pressure with kicks. External forces on the riser result from hydrostatic pressure which is calculated for the given water depth.

The total axial stress is due to the tensioner load, riser buoyancy, riser weight, mud weight, current forces, and wave forces. With large surge or sway of the vessel, the dynamic forces on the riser string must also be considered. A schematic of these stresses is shown in figure 2.32.

Various procedures are used for analysis of the riser, depending upon the required objective. Simple catenary equations

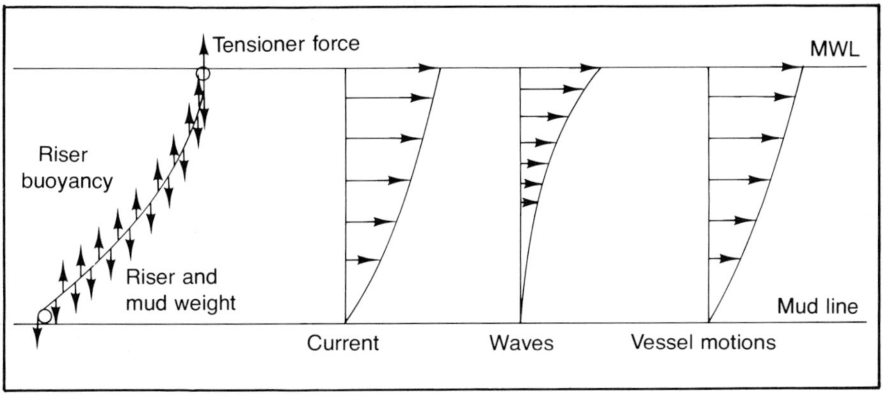

FIG. 2.32 Forces on the riser string

can be used to get a ballpark type of solution with results that may be off several hundred percent. This would be adequate for making rough comparisons of the effects of different water depths and diameters or wall thicknesses of the riser string. For more accurate correlations, finite difference procedures are used to solve differential equations. The finite difference approach gives excellent results for a very simple type of riser. However, a separate differential equation is required for each change in cross section, and the joints of different cross sections are tied together mathematically by compatibility conditions. The problem with the finite difference approach is that the procedure becomes totally unwieldy and impractical for analyzing a design with many variations in riser diameter, wall thickness, floatation material on the riser, or any other variables.

Conversely, the finite element approach for design and analysis finds considerable favor since this approach is virtually unencumbered by limitations or conditions. For example, the digital programming and computer usage is essentially the same for any number of variables, such as riser wall thickness, added buoyancy, etc. With the finite element approach, other variables can also be considered that cannot be done on a practical basis by any other method. This means that any current distribution can be considered (figure 2.33). Hydrodynamic drag effects of the riser couplings and the choke and kill lines can be accounted for. Loads due to any wave height can be considered. The influence of the telescoping joint can be fully analyzed, etc.

Fisher and Ludwig (24) have given generalized results that are very helpful for obtaining ballpark type answers. The value of their work is that the approximate response of any design can be quickly determined using their charts (figures 2.34 and 2.35). However, detailed and accurate analyses, such as with a nonlinear finite ele-

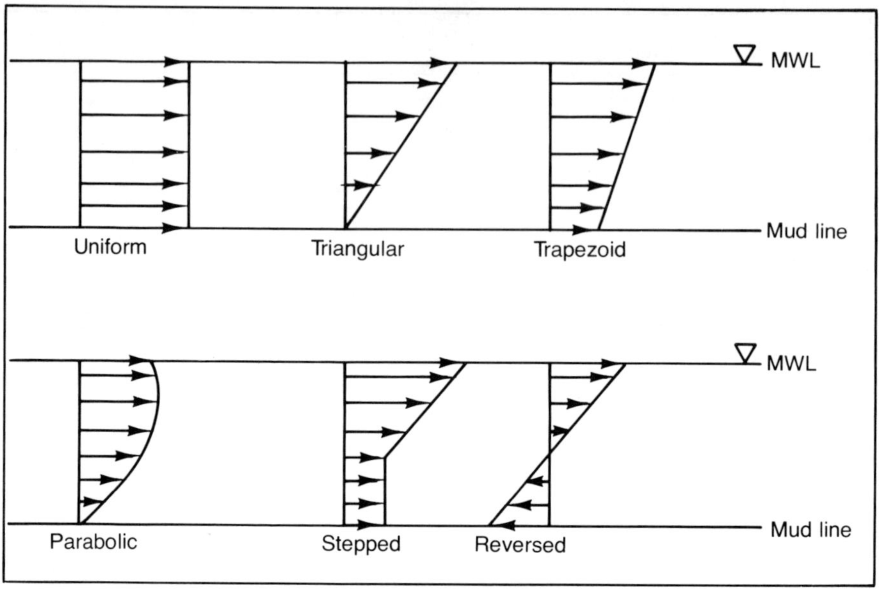

FIG. 2.33 Common design current profiles

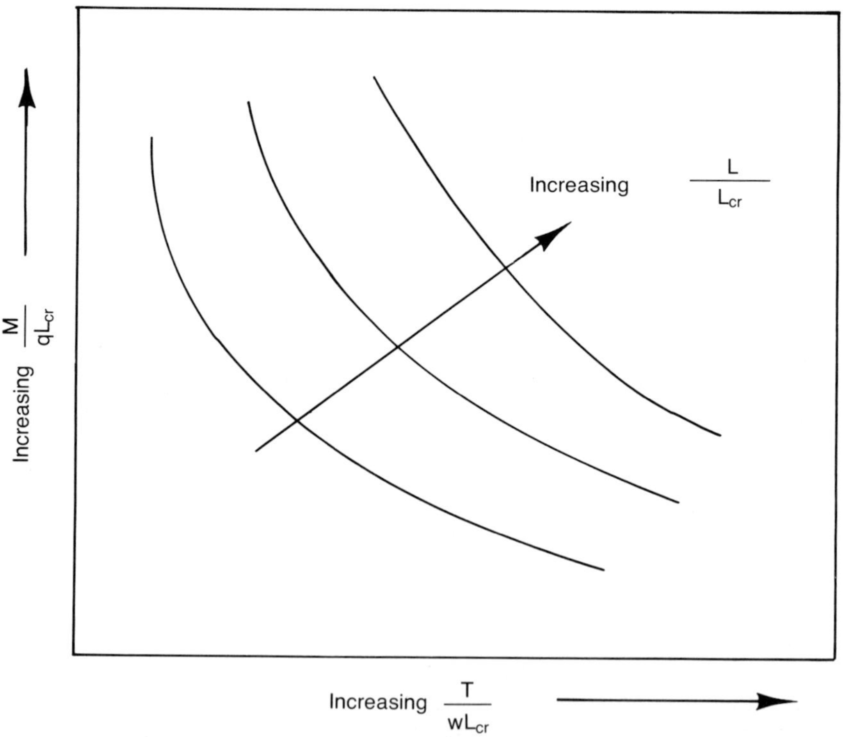

FIG. 2.34 Maximum moment for vessel offset or uniform current

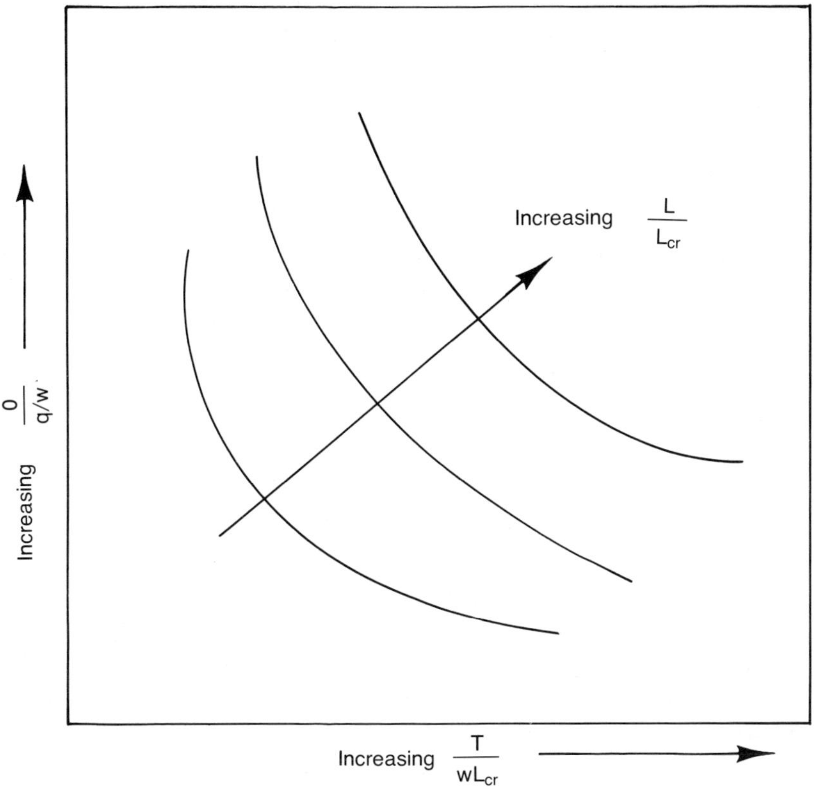

FIG. 2.35 Rotation of bottom ball joint for vessel offset or uniform current

ment routine, must be used for either deep water conditions or severe environmental conditions.

The digital computer riser program, ETA/FLEXRIS, is a nonlinear finite element program. Nonlinear means that the equations for displacement are second order rather than first. (Second order displacement functions are required because of the large displacements and relatively low stiffness of the riser cross section.) The nonlinear ETA/FLEXRIS program has all of the versatility of finite element programs. With ordinary standards of programming, the linear programs can be off by as much as 100%. The nonlinear programs give solutions with an accuracy of about 1%.

Conclusions

The following conclusions can be drawn:

1) The technology of riser design and operations for 200–800 ft water depths is well established. Special considerations are required

for water depths greater than 800–1,500 ft.

2) Innovative designs are needed for 1,500–3,000 ft water depths to reduce riser running time and to decrease the possibility of environmental damage due to extreme winds, waves, and vortex shedding.

3) Because of the possibility of fracture of the geological formation with excessive mud pressure, radically different designs and procedures may be required for 3,000–6,000 ft water depths.

4) Nonlinear finite element computer programs seem to be the best and most universal method for general design and analysis.

5) Computer simulated operations with various sets of wave and current forces and vessel motions can be used to set more realistic limits for drilling and hang-on conditions, thereby, decreasing weather related downtime.

2.6 JACKUP RIG DESIGN FOR THE NORTH SEA

In the rush to get as many offshore mobile drilling units as possible into the booming North Sea activity, the jackup, until recently the leader in the worldwide rig fleet, has been pushed into the background. Semisubmersibles clearly dominate current rig construction. Of the 135 offshore units constructed through 1977, over half (72) were semisubmersibles; 45 of these, for North Sea use. The semisubmersible's excellent performance and stability in the face of the severe North Sea criteria cannot be denied. The jackup's capabilities, however, meet many of the requirements for North Sea operation so well that renewed interest in this unit is growing as new jackups are designed specifically for the North Sea.

As a proponent of jackups for the North Sea, Engineering Technology Analysts, Inc. (a Houston consulting firm acquired by Baker Marine) integrates a fresh new philosophy with considerable experience in dealing with all types of offshore units into the design of its jackups. The ETA Europe Class jackup was designed specifically for use in the severe criteria of the North Sea in up to 350-foot water depths. Many of ETA's exclusive design techniques, including several innovations for which patent applications have been made, are used to overcome the limitations for existing jackups in the North Sea.

Depth Limitations

Part of the reason for the semisubmersibles' stronghold on the North Sea activity is the vast amount of area in water depths between 300–1,500 feet, too deep for many existing jackups (table 39). This area is four and one-half times greater than that for com-

TABLE 39
North Sea, Gulf of Mexico bathymetry comparison

	Area in square miles		
	0–300 ft	300–1,500 ft	0–1,500 ft
Offshore Louisiana	28,000	8000	36,000
Offshore Texas	22,000	5000	27,000
Gulf of Mexico (total)	182,000	60,000	242,000
United Kingdom	77,000	91,000	168,000
Norway	15,000	61,000	76,000
North Sea (total)	243,000	271,000	514,000

parable water depths in the Gulf of Mexico. What is often overlooked, however, is the fact that the North Sea offshore acreage extending from 0–300 foot water depths, the normal range for jackups, is 33% greater than that for the Gulf of Mexico. This area comprises 47% of the total offshore North Sea area; yet, jackups compose only 24% of the North Sea fleet. Although semisubmersibles are designed for 600+ foot water depths, 15 of the 29 semisubmersibles in the North Sea are operating in the 200–450 foot range, in many locations where jackups would be suitable.

Weather Restrictions and Stability

The most critical periods for a jackup occur when the unit is under tow and when it is going on and off location. The stability of the unit at these times is extremely sensitive to weather conditions, and these moves must be undertaken under favorable conditions. Customarily during long tows, the top of the legs are partially removed to ensure hull stability. If the leg is cut off, it must be rewelded when the unit has completed its tow. If mechanical disconnects are used for leg removal, there is the danger that the mechanical device, installed in one of the most highly stressed areas of the leg, will become faulty through long periods of submersion.

During short tows, often the legs are fully raised rather than cut off. This procedure requires favorable weather so that wind and wave forces will not upset the hull stability of the unit, which is now top-heavy. If the weather should become extreme during a short tow, there are two alternatives. The legs may be lowered to jack the hull up out of the water, providing of course that the water depth does not exceed the allowable leg length. In this case, the tow comes completely to a halt. Or, the unit may proceed with the legs partially lowered. This ensures excellent hull stability but greatly increases towing time and costs.

Favorable weather is again a requirement when the jackup is moving on and off location. The safety of the jackup during the transition from tow to getting on location generally becomes jeopardized in wave heights over 6 ft.

The influence of weather on the structural integrity of the semisubmersible is not the determining factor for the unit while it is in tow and going on and off location. In the transit mode, the limiting criteria is the ability of a unit of such extreme weight and mass to make headway in severe winds and waves. When going on and off location, the semisubmersible is restricted only by the ability of support vessels (in particular workboats) to deploy the mooring system. Except under very extreme weather conditions, the semisubmersible's ability to relocate is not hindered.

When abandoning the wellsite, the semisubmersible can quickly be underway after the mooring system has been released. The jackup must wait for the legs to be raised to a suitable towing position, which may entail a departure delay of several hours.

While in the drilling mode, the jackup is inherently a much safer rig than the semisubmersible, as it is the closest offshore equivalent to land drilling. The jackup's platform stability at this time is especially significant when considering the procedures for well testing. Another safety factor is that the blowout preventer is located on the platform for jackups (as opposed to beneath the surface for semisubmersibles) and is easily accessible for surveillance and repair.

Rig Costs

Initial investment costs of offshore mobile drilling units are rapidly increasing due to the rising cost of materials and labor. Inflation is worldwide, and today's larger and heavier units in particular require more and higher-priced steel.

The initial costs for jackup and semisubmersible rigs show an established trend toward escalation in cost, especially in the past years, with initial cost today approximately five times those in 1961. This is particularly important in planning future investments, as the day rates and operating costs are directly dependent on the initial investment cost.

In past years, the day rate was calculated to be approximately 0.10% of the original investment. Recently, however, that figure has changed to about 0.12%. The operator's cost on a daily basis are approximately double the day rate. It would not be unreasonable to expect total drilling costs of $50,000 to $60,000 per day for a North Sea drilling operation with a new, large semisubmersible in 400 feet of water.

Rig Efficiency

In addition to capital investment costs, rig efficiency plays an important role in offshore economics. Table 40 compares the effective time breakdown for both jackups and semisubmersibles: actual drilling time for jackups is quoted to represent 50% of the total time, compared to 45% for semisubmersibles. This effective time means that the jackup will make 11% more hole per year than the semisubmersible. Other authorities maintain that the jackup has a higher advantage.

Another check on rig efficiency is the number of wells per year that a unit drills. A review of the movements of 5 jackups and 5 semisubmersibles operating in the North Sea during a two-year period shows that the jackups, with 134 total rig-months service, drilled 47 wells at a rate of 1 well every 2.96 months, or 4.1 wells per year. The semisubmersibles, with 122 total rig-months service, drilled 37 wells at a rate of 1 well every 3.31 months, or 3.6 wells per year. This translates into 14% more hole per year for jackups.

TABLE 40
Effective time breakdown

Activity	% time	
	Jackups	Semisubmersibles
Drilling, tripping	50%	45%
Running casing	6%	5%
Conditioning hole	12%	10%
Logging, coring	12%	10%
Abandoning well	6%	5%
Surface BOP handling	1%	—
Subsurface BOP handing	—	5%
Platform elevation, lowering (No weather downtime)	1%	—
Anchor handling (No weather downtime)	—	5%
Weather downtime	6%	10%
Towing	6%	5%
Total time	100%	100%
Effective time	86%	75%

Source: Starink, A., West. F. G., "Various Types of Exploration Drilling Rigs for Non-Shallow Water Depths (50 ft.-600 ft.)"
R.I.N.A. Symposium on Offshore Drilling Rigs, 1970, London.

Overcoming the Limitations

ETA designed a new generation of offshore mobile drilling units for worldwide use in water depths up to 430 feet. The jackups are designed to meet the latest requirements of the American Bureau of Shipping.

Probably the most unique aspect of the ETA design procedure for jackup drilling units is the use of specialized, computer-based analytical techniques. ETA/MDUSAS (ETA Mobile Drilling Unit Structural Analysis System) evaluates the behavior of wind, operational, wave, and hydrodynamic effects on offshore structures and then checks the structure directly against the design criteria of A.B.S., Lloyd's Register of Shipping, det Norske Veritas, or A.I.S.C. ETA/NAME program (ETA Naval Architecture and Marine Engineering) analyzes the stability and motion response of unusual, nonshipshape hulls. Very close comparisons with model tests and full-scale performance results have been obtained.

The ETA Europe Class jackup can resist 75 ft waves and 125 mph winds and has a wave period up to 14 seconds. Particularly important for North Sea operations, ETA has designed a system enabling the Europe Class jackup to get on and off location in 10 to 15 foot seas, compared to the 5 to 7 foot limit on existing units. Optimum design of the hull by the use of comprehensive finite element routines provides the unit with a high factor of safety against overturning (figure 2.36). Det Norske Veritas has given the ETA Europe Class jackup design approval for Class 1A1 self-elevating platform.

The ETA Europe Class jackup is lighter in weight than comparable units with the same performance criteria. ETA's innovations in fabrication techniques allow easier, quicker, and more economical construction for this level of performance. Efficient structural framing provides the ETA North Sea jackups with lighter but stronger hull structures. Another significant weight and fabrication advantage leading to cost reductions is the design of the triangular space frame legs, allowing for maximum structural efficiency and simplicity. The legs are believed to be significantly stronger and more reliable than those on other jackups on a strength:weight ratio, due to the high structural efficiency of ETA's patented leg-joint designs.

The minimization of shipyard weight allows for a large consumables capacity of 9,500 kips for the ETA Europe Class jackup, almost twice that for any existing jackup. This provides operational convenience and economy in supply functions and allows the unit to be self-sufficient for much longer periods of time, which is an important factor in remote or hostile environments such as the North Sea. This unprecedented consumables capacity also makes the ETA Europe

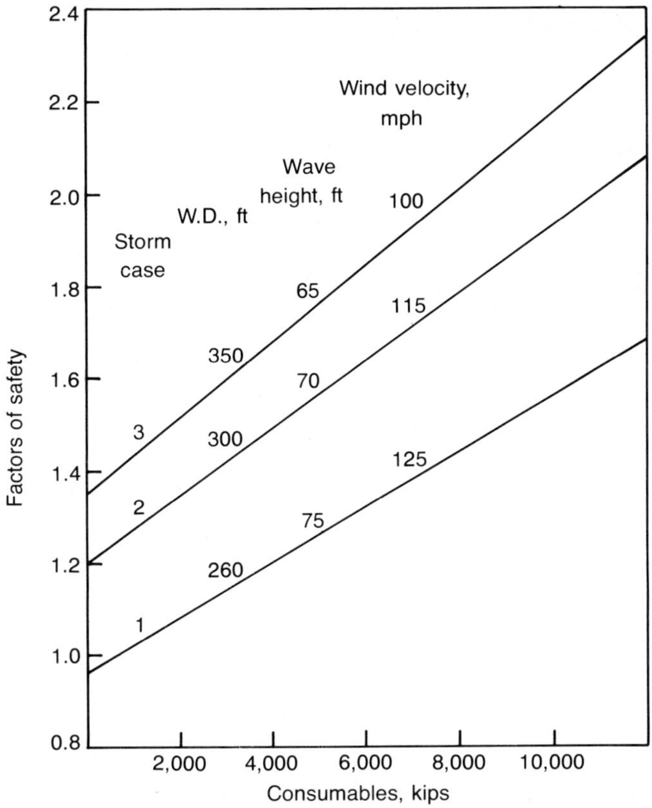

FIG. 2.36 Europe Class overturning safety factors

Class jackup an attractive candidate for a production platform. Considering that the skid-mounted type of production equipment weighs on the average about 1,000 tons per 250,000 bo/d production, the ETA jackups are capable of supporting the necessary production equipment for the most prolific fields.

Another, more innovative design for North Sea use is the ETA Mobile Monopod, a gravity-based, multiservice jackup for exploration, drilling, and production in 450 ft of water. A single center jackup leg connects the upper hull (with quarters and drilling equipment) and lower hull (with a capacity for crude oil storage of 220,000 bbl). The upper hull acts as an erection platform during leg construction and has sufficient buoyancy to support the entire flooded weight of the lower hull as an added safeguard against uncontrolled flooding. The Monopod is designed to accommodate almost any given North Sea soil condition and has a large consumables capacity of 8,500 kips.

The development of ETA's jackups for the North Sea is paralleled by the design of units for areas with less severe criteria. The ETA Asia Class jackup is designed for exploratory and developmental drilling to 25,000 feet in 250–300 ft water depths. It is suited for operation in relatively moderate environments such as Southeast Asia, offshore Brazil, West Africa, and the Middle East, and it has a consumables capacity of 6,700 kips. The ETA American Class jackup is a large unit which drills to 30,000 ft in up to 380 ft water depths in the U.S. Gulf Coast hurricane season criteria and to 430 ft water depths in nonhurricane season. This unit has an unusually large consumables capacity of 11,100 kips.

Innovations in the design of jackups, such as increasing leg length while maintaining a high consumables capacity extend the jackup's potential range of operation. Self-elevating units designed specifically for the North Sea will not replace the semisubmersible, but they may prove to be the most effective and efficient drilling mode for part of that area now being drilled by semisubmersible and drillships. Most particularly important to the rig owner, they require the lowest investment cost.

REFERENCES

1. Wiggins, J. H., "The Balanced Risk Approach to Earthquake Building Codes," Civil Engineering, Vol. 42, No. 8, pp. 55–59, August 1972.
2. Thobe, S., "Storms Are a Major Cause of Mobile Rig Mishaps Offshore" Offshore, pp. 23–28, June 5, 1974.
3. Lain, G., "Ship Collisions," Ocean Industry, p. 19, December 1971.
4. "377 Ships Lost at Sea Last Year," Houston Chronicle, Sec. 1, p. 23, November 15, 1972.
5. Besse, C. P. and LeBlanc, N. F., "An Application of Oceanographic Data in Offshore Structural Design," SPE Paper No. 1419, May 1966.
6. Hogben, H., and Lumb, F. E., Ocean Wave Statistics, Natl. Phys. Lab., H. M. Stationary Office, London, 1967.
7. Reusswig, G. H., Bozeman, J. D., and Bynum, D., Jr., "As Assessment of Environmental Design Criteria for Offshore Mobile Drilling Rigs," ASME Paper No. 73-PET-25, September 1973.
8. "Analysis of Winds, Wind Waves and Swells Over the Sea to the East of Japan During the Typhoon of Sep. 29, 1935," Mo. Weather Rev., Vol. No. 2, Feb. 1953, pp. 31–37.

9. Thom, H. C. S., "Distribution of Extreme Winds in the United States," Transactions of American Society of Civil Engineers, Vol. 126, 1961, p. 450.
10. Weibull, W., "A Statistical Distribution Function of Wide Applicability," ASME, Applied Mechanics Division, Annual Meeting, 1951.
11. Hogben, H., and Lumb, F. E., *Ocean Wave Statistics*, National Physics Laboratory, H.M. Stationary Office, London, 1967.
12. Neumann, G., *Ocean Currents*, Elsevier Publishing Co., Amsterdam, 1968, pp. 178–197.
13. "377 Ships Lost at Sea Last Year," *Houston Chronicle*, Section 1, Nov. 15, 1972, p. 23.
14. Lain, G., "Ship Collisions," *Ocean Industry*, Dec. 1971, p. 19.
15. Burridge, D. M., London Weather Centre Memorandum No. 22, May 1972.
16. Wiggins, J. H., "The Balanced Risk Approach to Earthquake Building Codes," Civil Engineering, Vol. 42, No. 8, Aug. 1972, pp. 55–59.
17. Glenn, A. H., "Storm Tide and Waves in the North Sea," Report to The Offshore Co., Aug. 29, 1972.
18. Glenn, A. H., "Meterological-Oceanograhic Conditions over the Texas-Louisiana Coastal Area During the Hurricane of 7–12 Sept. 1961," Report to The Offshore Co., Jan. 22, 1962.
19. Besse, C. P., and LeBlanc, N. F., "An Application of Oceanographic Data in Offshore Structural Design," SPE Paper No. 1419, Presented at AIME Symposium in New Orleans, La., May 23–24, 1966.
20. Meyers, J., ed., Handbook of Ocean and Underwater Engineering, McGraw-Hill, New York, 1969.
21. Dobroklonskiy, S. K., "Drift Currents in the Sea with an Exponentially Decaying Eddy Viscosity Coefficient" (Russian Academy of Science), Okeanologiya, Vol. 9, No. 1, 1969, pp. 19–25.
22. Kraav, V. K., "Computation of the Three-Dimensional Field of Tidal Currents in the North Sea," Okeanologiya (Russian Academy of Science), Vol. 10, No. 2, 1970, pp. 195–201.
23. Butler, H. L., Delfosse, C., Galef, A., and Thorn, B.J., "Numerical Analysis of a Beam Under Tension," *J. Struct. Div., ASCE*, Oct. 1967, pp. 165–174.
24. Fisher, W., and Ludwig, M., "Design of Floating Vessel Drilling Riser," *J. Petr. Tech.*, March 1966, pp. 272–280.
25. Harris, L. M., *Deepwater Floating Drilling Operations*, Tulsa: PennWell Publishing Co., 1972.
26. Kennedy, J. L., "High-Angle Holes, Supply Hurdles Complicate North Sea Drilling Work," *OGJ*, June 24, 1974, p. 132.

27. Morgan, G.W., "Riser Design Criteria and Considerations," *Petr. Engr.*, Nov. 1974, pp. 68–74.
28. Morgan, G.W. "General Aspects of Riser Design and Analysis Procedures," *Petr. Engr.*, Oct. 1974, pp. 36–48.
29. Wilson, R. O., and Martin, M.R., "Deepwater Pipelay for Central North Sea," OTC Paper 1855, 1973.

CHAPTER

3 OFFSHORE FIXED-PLATFORM DESIGN

3.1 GENERAL GUIDELINES FOR TUBULAR JOINT DESIGN

In the design of offshore truss structures, structural pipe is a more efficient material because of its high strength-to-weight ratio and uniform distribution of flexural loads when they are applied in any direction. More efficient and economical methods of tying these tubular sections together is of considerable interest in the offshore construction industry.

As an example of the weight savings that can be achieved with tubular structural members and joints, it is estimated that the Eiffel Tower in Paris could be built today with tubular steel members and weigh only about 200 tons. As constructed in 1889, it weighs about 7,000 tons.

Background

One of the problems in platform maintenance is adequate joint inspection for underwater joints. Bryan[(1)] discussed the results on a salvaged platform that was in service during 1955–58 in 61 ft of water and subjected to four tropical hurricanes.

While no failed points were evident before salvage, it was found after salvage that 65 out of about 230 welded joints were cracked. The cracks were attributed to fatigue, corrosion, tender impact, and the use of an inappropriate grade of steel.

Types of Tubular Joints

Some of the types of tubular joints previously used are shown in figure 3.1. The reinforcement shown in figure 3.1(g) is referred to as a

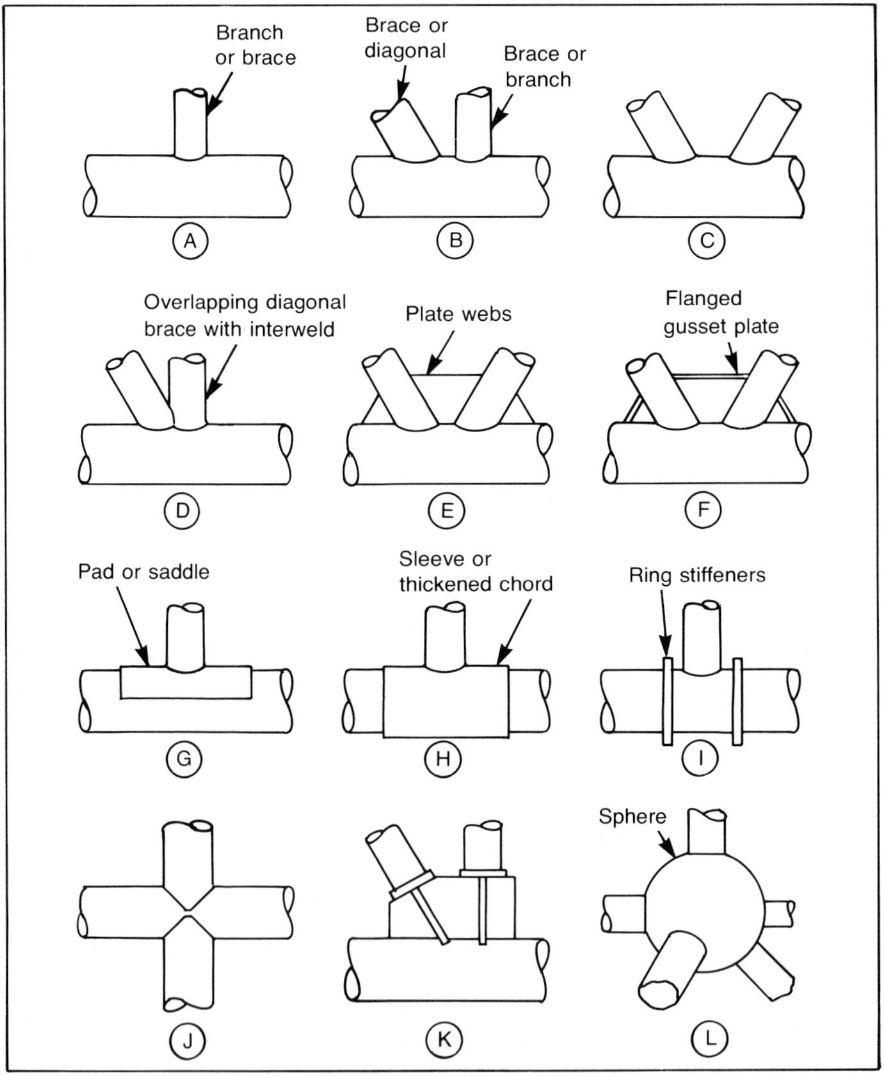

FIG. 3.1 Typical tubular connections

pad or saddle where the basic idea is to locally reinforce the chord.

Figure 3.1(h) could be either a shrink fit or a local increase in chord wall thickness with the latter obviously more economical.

Ring stiffeners shown in figure 3.1(i) have been proposed in a number of varieties. This is one of the more complicated joints to analyze, and the maximum stress concentration factor is often shifted from the chord brace intercept to the ring-cord brace intercept to the ring-cord intercept.

Plate webs in figure 3.1(e) are sometimes used for local reinforcement, but square cuts cause additional risers. Reinforcing plates should be scalloped to transfer the force flux and reduce or eliminate the local stress risers.

The stress risers developed by use of gussets as shown in figure 3.1(e), (f), and (k) become critical when considering the fluctuating loads that occur in real applications. They require special attention at the design stage.

Punch Shear

The most common procedure for designing tubular joints is the API punch shear method.[(2)] This consists of taking the ultimate brace force as the ultimate shear stress times the product of the brace circumference times the chord thickness.

Experimental work[(3)] has shown that the punch shear procedure can overpredict the ultimate allowable load by up to 40%. While this could mean an unsafe design by a wide margin, the punch shear method is still the best and quickest approximation for preliminary design work.

Stress Concentration Factors (SCF)

Barnoium and Thebault[(4)] have shown some typical results with T joints. In general, the stress concentration factors for 90° out-of-plane applied loads are roughly twice as much as at points in the plane of the chord and brace.

The SCF is normally based on either the maximum principal stress or the Von Mises stress, with the maximum principal strain being used less frequently as a criterion. The Von Mises SCF is usually about 15% less than the SCF derived by using maximum principal stress theory.

The calculated SCF obtained by finite element approaches is usually somewhat greater than the experimentally determined SCF for a given type of joint. Because of the computer cost of using smaller and smaller grids for the theoretical analyses, few reliable and fully developed programs are available.

Static Design Guidelines

The use of positive eccentricity tubular joints (figure 3.2) should be avoided whenever possible, since they have very high stress at the joint toe. The zero eccentricity joint is usually easier to fabricate than the negative eccentricity joint; however, the latter is more efficient because of the increased chord thickness normally associated with an interweld.

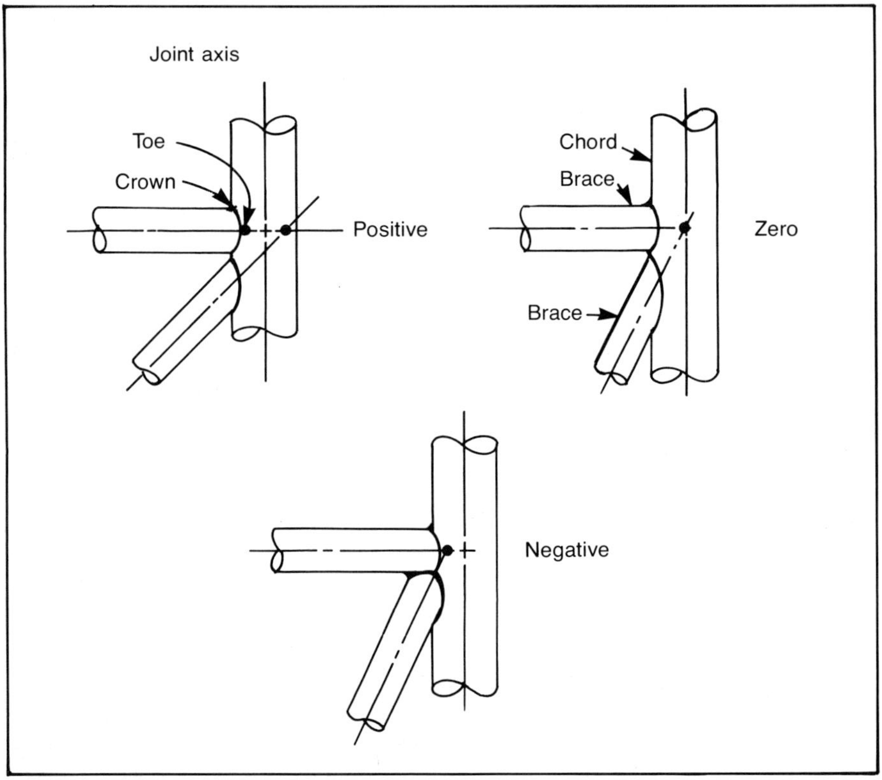

FIG. 3.2 Eccentricity definition

Most of the discussion here is primarily concerned with the maximum joint stress or the joint efficiency. The point of maximum stress actually varies considerably with the type of joint and the relative dimensions of the joint.

For example, with a chord diameter to thickness ratio near 40, the maximum stress occurs in the joint toe. When this ratio is decreased to approximately 20, the maximum stress is shifted from the chord to the brace near the crown.

Reserve Strength

The normal practice in tubular joint design is to use an allowable stress that is 60% to 80% of material yield strength. The reserve strength is then taken between design and yield plus an additional reserve between yield and ultimate strengths.

Tests by Noel and others[3] on a series of T joints indicated that first cracks could be initiated at load levels lower than 60% of material yield strength. They applied axial tensile loads on the brace of each joint and visually observed that cracks could occur at loads varying from 46% to 81% of ultimate strength. This indicates the necessity for determining and using the appropriate stress concentration factors in the joint design.

Data Extension for Design

For preliminary joint design and analyses, it is not always practical, nor economical, to perform expensive computer analyses or experimental tests. It is often useful to develop methods which normalize existing design data to approximate other dimensions that might be encountered.

This may be easily achieved by placing the existing data in nondimensional form, provided sufficient parametric data are available. To facilitate the use of these data, a concept has been developed to provide a joint coefficient C related to the joint efficiency E. This allows preliminary designs to be evaluated easily and quickly. When existing data are given in terms of joint strength relative to the brace strength, this method gives reasonable results.

Considering all the things that can go awry in the fabrication of tubular joints, the design should be such that the joint is always at least as strong as the ultimate strength of the most highly loaded brace.

There are four primary factors that must be considered for normalization of a given joint efficiency E in terms of a joint coefficient C and with a given eccentricity of Y or K type joints:

- Chord moment increases with decreasing brace diameter; therefore, the joint efficiency is divided by the brace diameter, d.
- The joint efficiency decreases as the cross sectional area of the brace increases; thus, the joint efficiency is multiplied by the product of the brace diameter and wall thickness.
- The chord stress decreases with increasing section modulus so the joint efficiency is divided by the product of the chord thickness times the square of the chord diameter.
- The brace should not punch/shear the chord. The punch/shear stress increases as the product of brace diameter times chord wall thickness decreases; thus, the joint efficiency should be multiplied by this product.

Equations and applications of this procedure are illustrated and discussed in the following sections.

T Joint Design

Tests conducted at the University of Texas on T joints[(3)] were performed with axial tensile loads, out-of-plane moment, and in-plane moment. In each case, the loads were applied on the brace up to the yield strength of the material. The joints were then tested to failure with axial load applied to the brace.

Normalizing the original data for the elastic loads gives the results shown in figures 3.3 and 3.4. This suggests that the maximum stress for the axial load are about 5 to 7 times greater than in-plane applied moment. The out-of-plane moment gave stresses that were two to three times greater than those produced by applying an in-plane moment to the brace.

The plastic ratio of a T joint is defined as the nominal brace stress at joint failure relative to the nominal brace stress when the hot spot stress in the joint is at the yield strength of the material. Normalizing the test data from the ultimate tests gives the plastic ratio functions shown in figures 3.5 and 3.6.

As an order of magnitude, the plastic ratio increased by a factor of two when the diameter to thickness ratio is doubled or when the brace to chord diameter ratio is doubled. Tests conducted at Southern Methodist Unviersity[(5)] on the axial loads in brace to chord configurations indicate that the diameter/thickness ratio has a significant influence on the ultimate brace force required to produce joint failure. These results are plotted in figure 3.7.

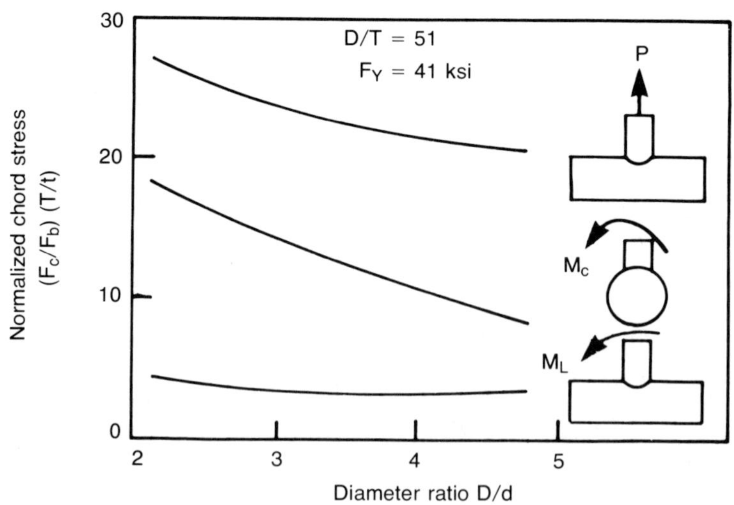

FIG. 3.3 Elastic stress function of diameter ratio

Some discretion must be exercised when using the attached graphs, since these curves were drawn as a best fit through the original data and are suggested for use only in preliminary design work.

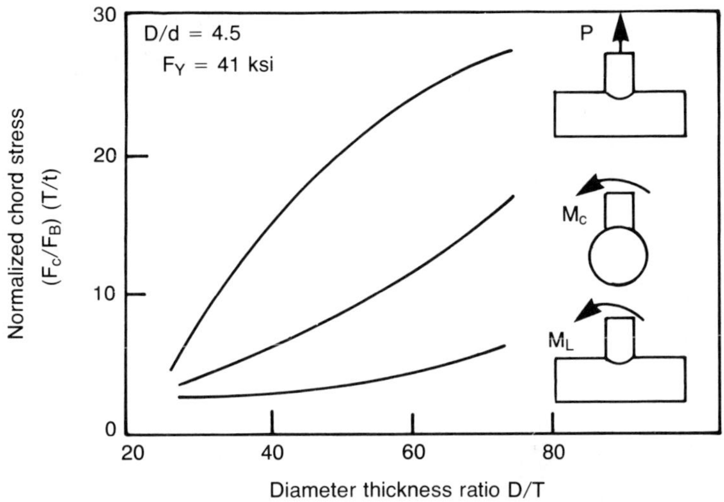

FIG. 3.4 Elastic stress function of diameter-thickness ratio

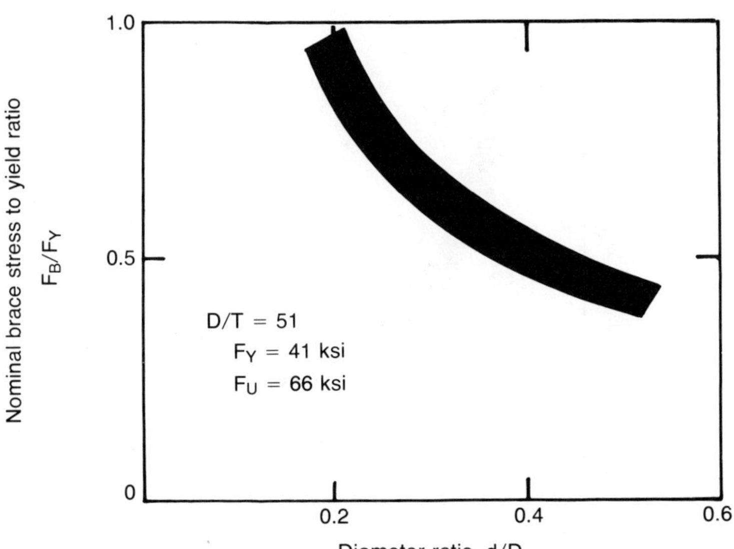

FIG. 3.5 Ultimate stress function of diameter ratio

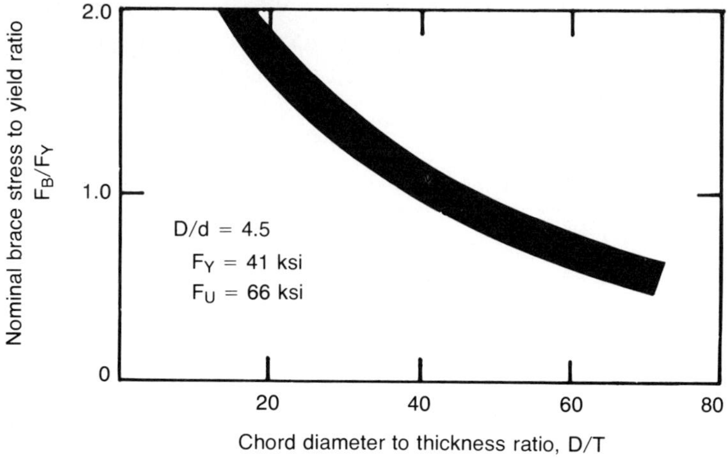

FIG. 3.6 Ultimate stress function of diameter-thickness ratio

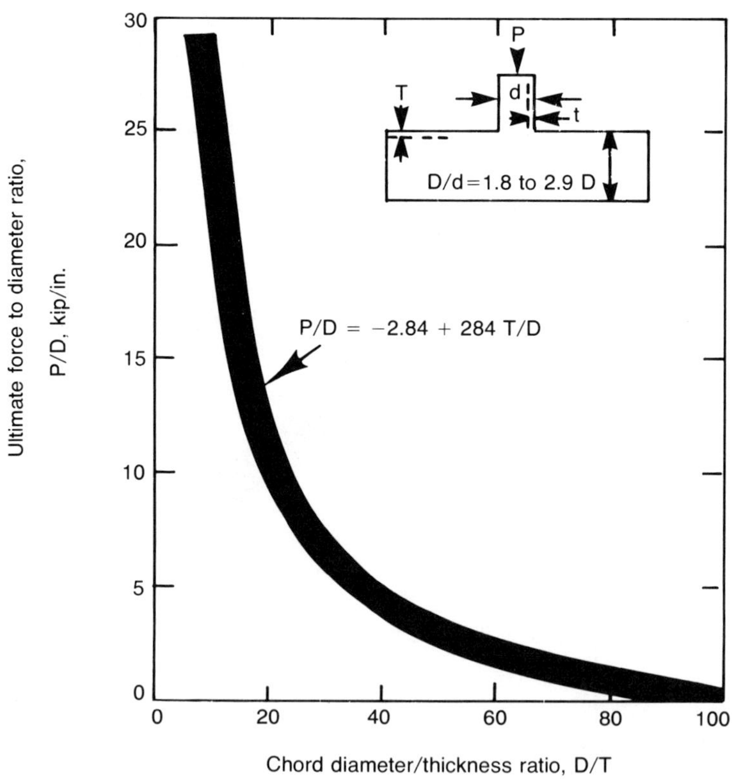

FIG. 3.7 Ultimate force for T. joints

T, Y, and TK Joint Design

A large number of tests were also conducted by Toprac and associates[6,7,8]. The results of their elastic tests were normalized by dividing the maximum chord stress by the product of the nominal brace stress and the brace diameter to thickness ratio. These results in figure 3.8 show that for any chord to brace diameter ratio, the T joint stress is about 3 times the Y joint stress or about 2¼ times the TK joint stress with 45° braces.

The Y joint stress is about ¾ of the TK joint stress for any diameter ratio where the chord diameter to thickness ratio is 51% and the thickness ratio near unity.

The dimensionless terms in figure 3.8 approximate a straight line relationship. Using this information with the T joint and 45° Y joint correlations, figure 3.9 was prepared for preliminary design and stress analyses with any brace angle.

The results of the elastic tests on TK joints are given in figure 3.10. These tests showed that the normalized stress is approximately linear with the diameter ratio for axial tensile loads applied to the T brace. Since all of the tests were with zero eccentricity joints, this work also shows the beneficial effects of brace interwelds.

TABLE 41
Nomenclature

Symbols

A	Cross-sectional area of brace, sq in.
C	Ratio of yield divided by ultimate strength of material
D, d	Diameter of chord and brace, respectively, in.
DF	Design factor
F	Stress, ksi
F_C	Stress in the chord, ksi
F_Y, F_U	Yield and ultimate strength, respectively, ksi
M_Y	Moment in brace at yield, in.-lb
P_Y	Axial brace force at yield, kip
P	Applied force on brace, kip
P_B	Allowable brace load, kip
SF	Safety factor
T, t	Thickness of chord and brace, respectively, in.
Z	Section modulus, cu in.

Subscripts

B	brace	P	plastic
C	chord	S	static
D	dynamic	T	"T" Joint
E	elastic	U	ultimate
K	"K" Joint	Y	yield
M	maximum	Y_1	"Y" Joint

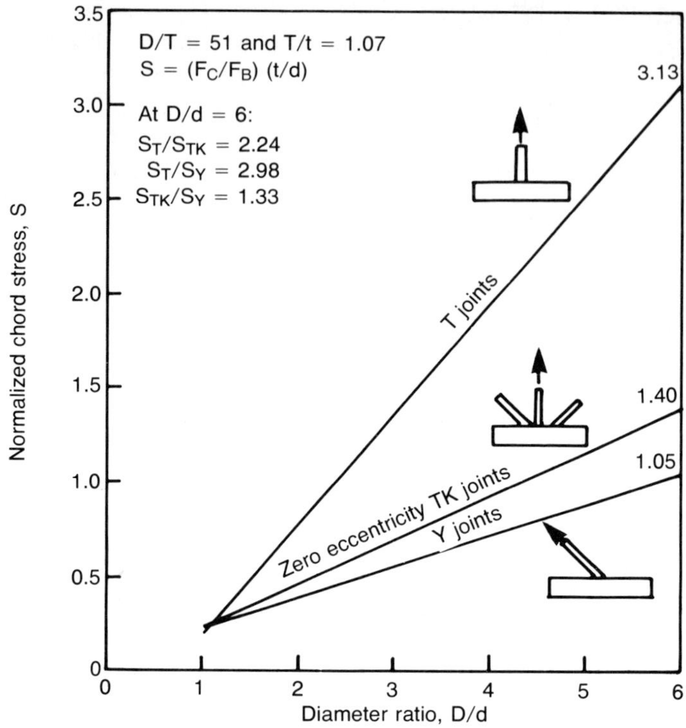

FIG. 3.8 Comparison of elastic joint factors

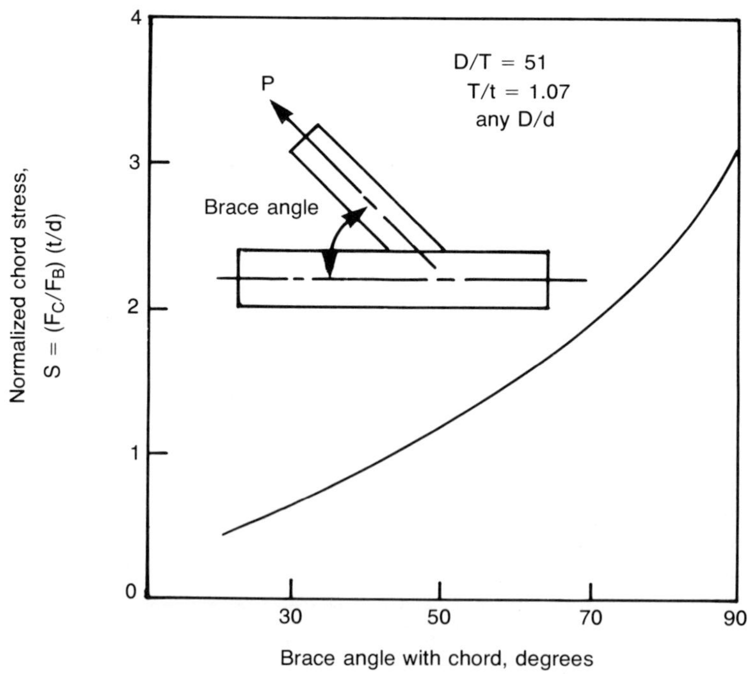

FIG. 3.9 Chord stress

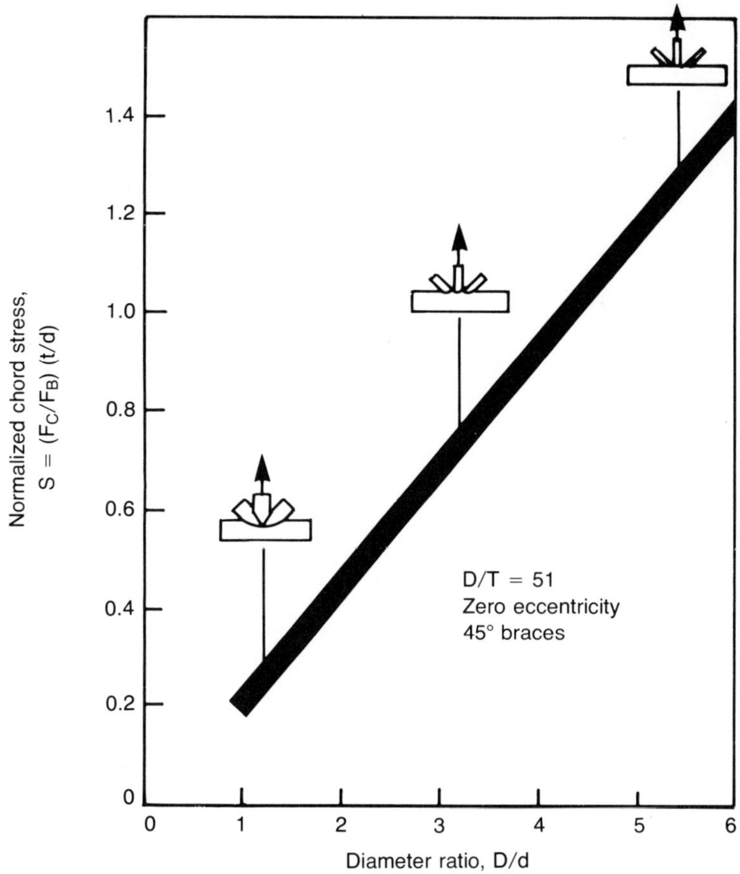

FIG. 3.10 Elastic stress factors in TK joints

Ultimate Strength of TY Joints

A thorough and comprehensive set of tests were performed at the University of California to determine the ultimate strength of 45–90° TY tubular joints.(9) The work is well done, but is cumbersome for the practicing design engineer.

To rectify this situation, the results for the complete set of tests were put in dimensionless form using methods of data extension. The final results are given in figure 3.11.

The joint efficiency E, in percent, is defined as the stress in the diagonal brace at failure of the joint divided by the ultimate stress in the brace based on a direct tensile force applied to the brace. Ultimate stress of the joint was defined at the point where the deformation increased with constant applied loads.

The original work also included data for a few joints having a negative eccentricity of 25%. In these joints, the centerlines of the T

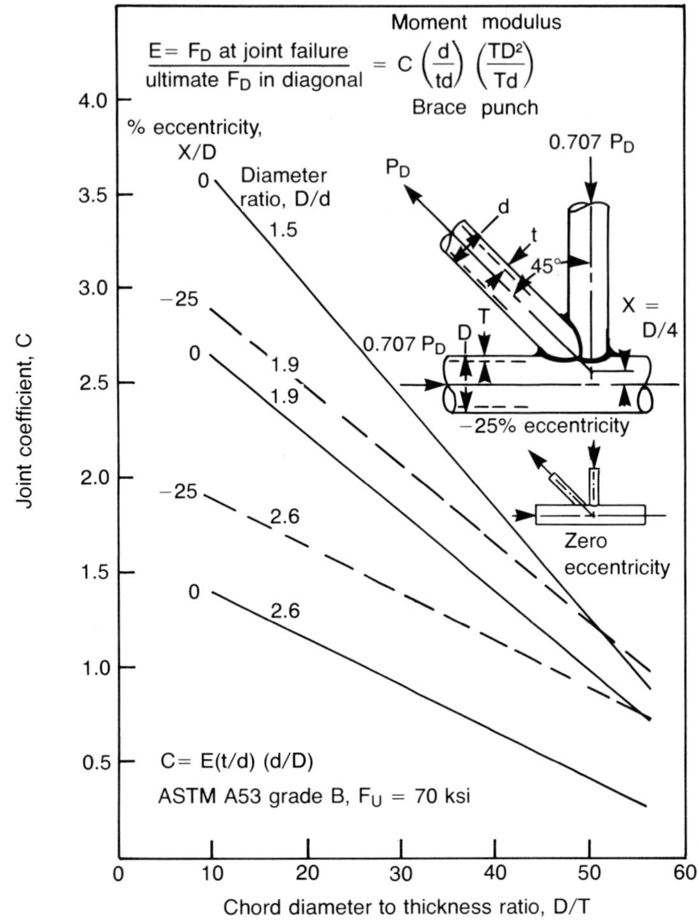

FIG. 3.11 TY joints ultimate strength coefficients

brace and the Y brace intersected at a point one fourth of the chord diameter between the chord centerline and the outside diameter of the chord at the brace intersection. Because of the limited data on the negative eccentric joints, they are often analyzed by extrapolation.

Using the slopes and intercepts from figure 3.11 the data was plotted and the equations representing the data were derived. The results are given in figure 3.12.

After detailed consideration of these data, a general equation for joint efficiency in terms of the ratios of the chord diameter to brace thickness, chord diameter to brace diameter, and chord diameter to chord thickness was also determined and is provided in figure 3.12.

To establish the accuracy of this equation, it was arbitrarily estimated that the experimental error was about 5%. The use of average diameter ratios could amount to another 5%, and plotting errors, 10%; thus, the final equation for ultimate TY tubular joint efficiency given in figure 3.12 could then reflect variations of up to 20%. This is considered acceptable for preliminary design.

Note that the test results summarized in figure 3.11 for TY joints were for a 70 ksi ulitmate strength steel. Similar normalization methods are discussed in the next section of the chapter on tubular joint design and could be used to develop a correction coeffiecient for ultimate capacity of TY joints as a function of yield strength.

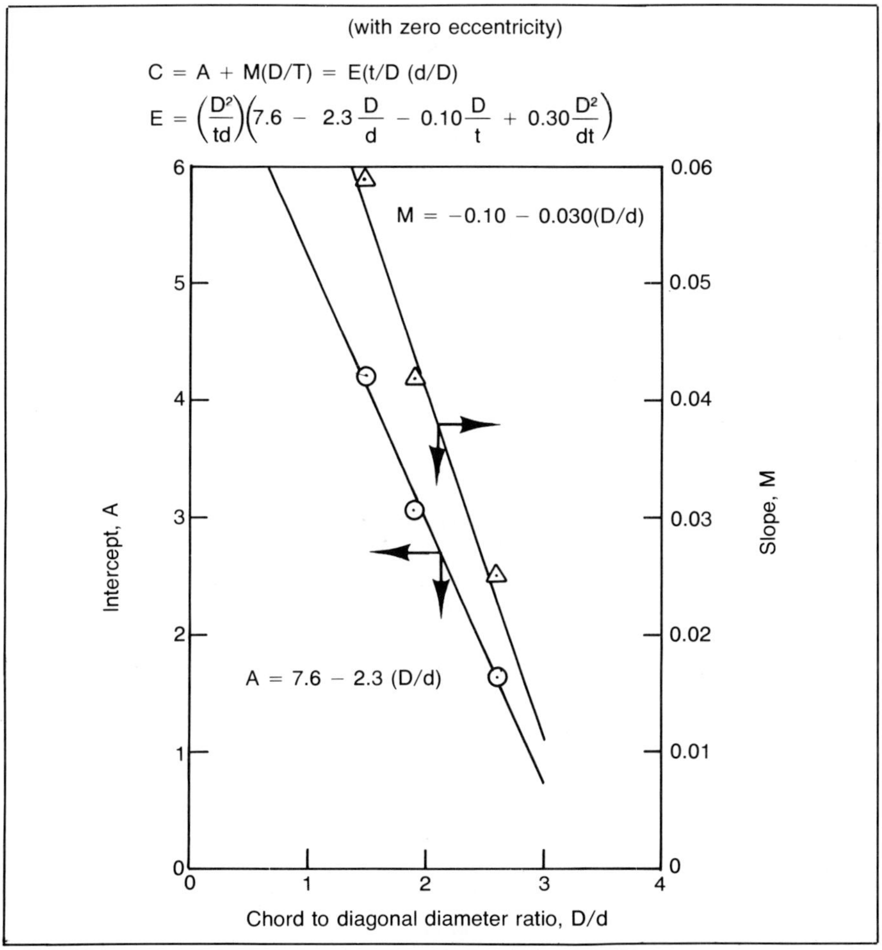

FIG. 3.12 TY joint efficiency

The literature review given here is not complete. The review includes most of the published work on tubular joint design that can be extended for general design purposes. Other general reviews on commonly used tubular joint designs[10,11] as well as special types of joints using gusset plates[12] or ring stiffeners are readily available in the literature.

3.2 PROCEDURES FOR ELASTIC AND PLASTIC JOINT DESIGN

For preliminary joint design, the initial diameter to thickness ratios can be selected from API criteria.[13] These values will preclude local buckling for loads up to the yield strength.

Diameter D		Wall thickness, T Minimum		Wall thickness, T Nominal	
in.	mm	in.	mm	in.	mm
24	610	0.49	12.45	1/2	13
30	762	0.55	13.97	9/16	14
36	914	0.61	15.49	5/8	16
42	1067	0.67	17.02	11/16	17
48	1219	0.73	18.54	3/4	19
60	1524	0.85	21.59	7/8	22

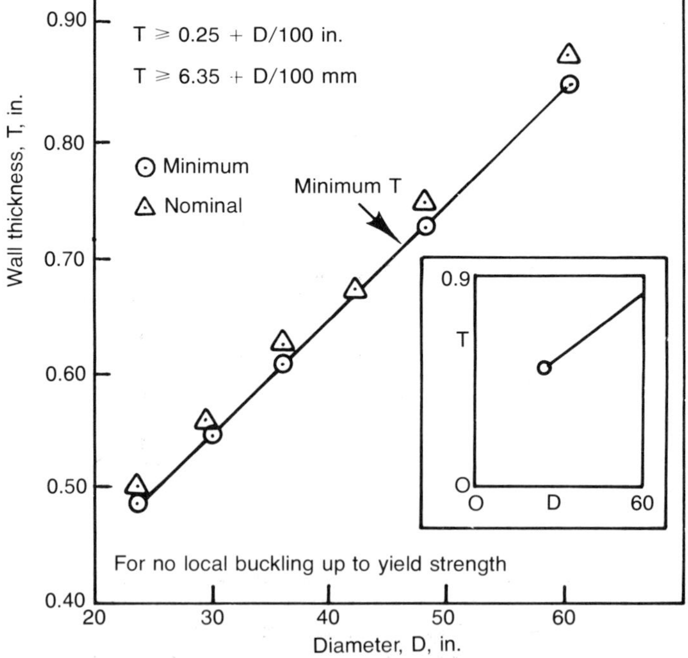

FIG. 3.13 Maximum D/T ratios

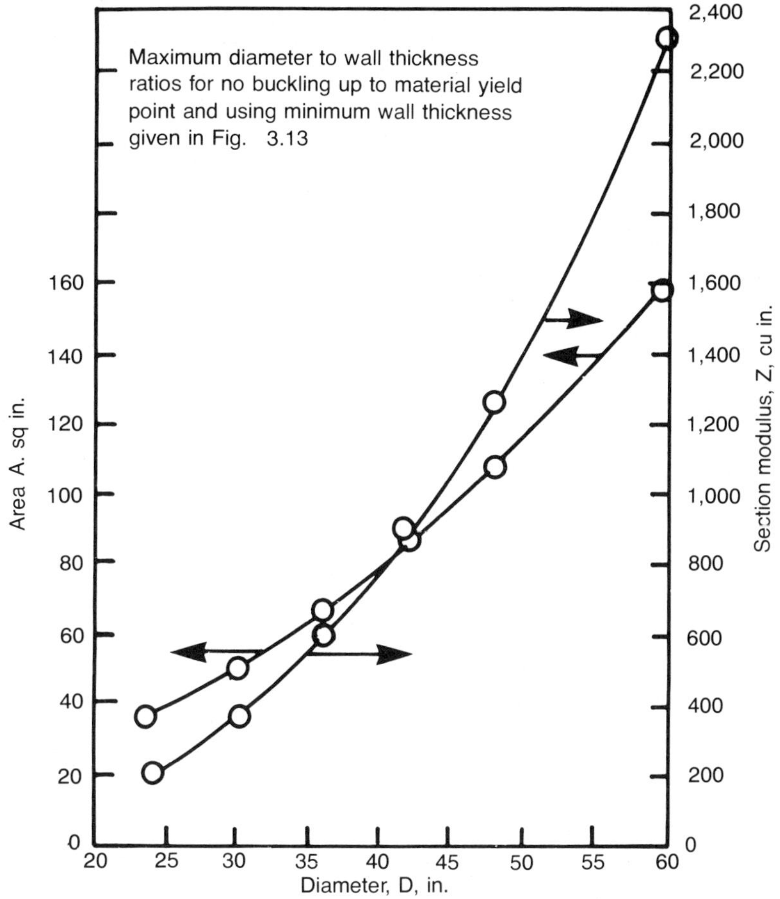

FIG. 3.14 Area and modulus for maximum D/T

For most tubular joints, this design guide is given in figure 3.13. The associated cross-sectional area and the section modules are given in figure 3.14. The allowable axial force and moment taken individually, and as a function of material yield strength, are given in figures 3.15 and 3.16.

Yield Fraction of Ultimate Strength

Details of a study on the mechanical properties of API grades 5L, 5LX, and 5LU pipe were given in a previous study.[14] The lowest yield strength is 25 ksi for grades 5L-A25 and the highest is 100 ksi for X 100.

Equations for the yield strength relative to ultimate strength and the yield percent of ultimate are provided in figure 3.17. A plot of

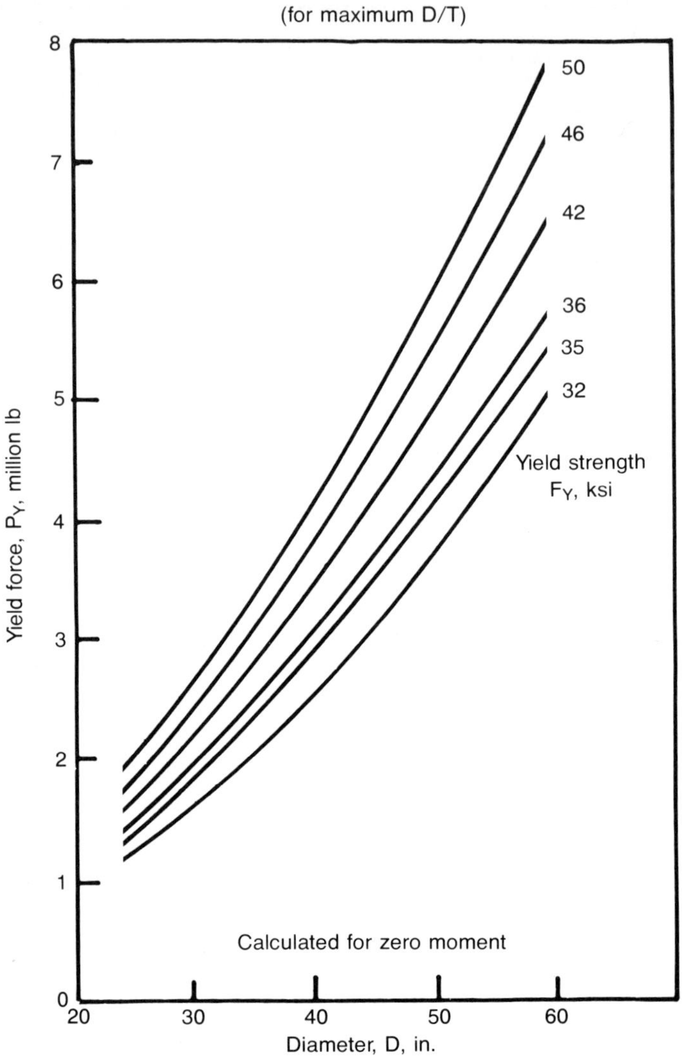

FIG. 3.15 Yield force vs. yield strength

these relationships indicates that the percentages vary from 56% for A25 pipe to 91% for X100 pipe.

Design and Safety Factors

The static design factor is the maximum elastic stress divided by the yield strength and is normally taken as 0.6 per the AISC code for elastic design. With dynamic loads, the elastic design factor is increased to 0.8 or 33% more than the value for elastic loads.

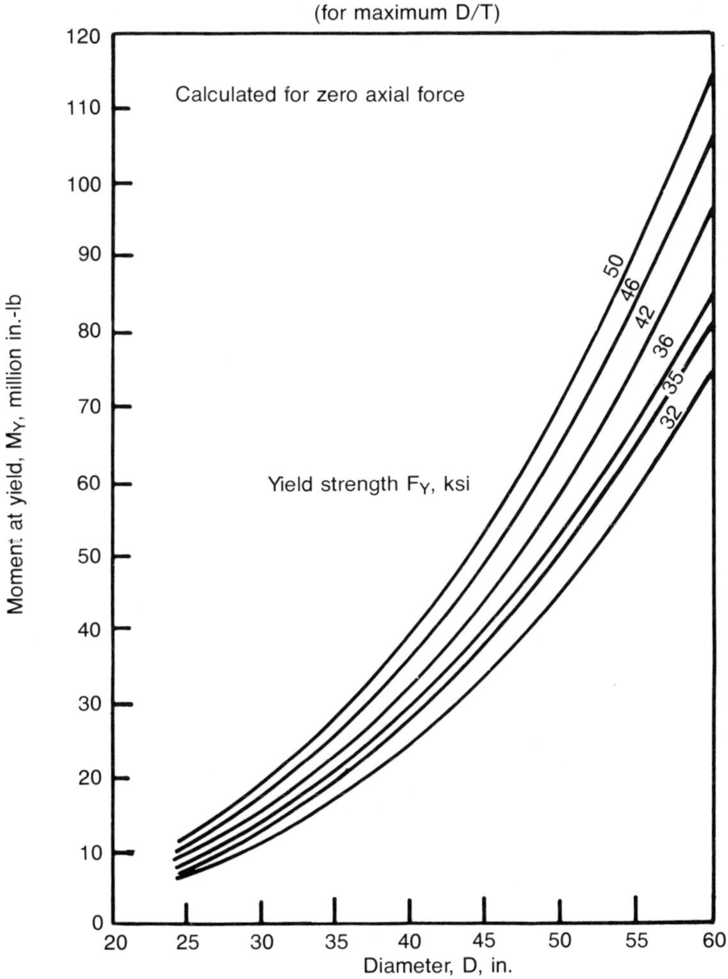

FIG. 3.16 Yield moment vs. yield strength

The safety factor in plastic analyses is defined as the ultimate joint stress where deformation increases with constant load divided by the yield strength of the material.

The safety factor is, by definition, the reciprocal of the design factor. For plastic analyses, the safety factor used normally varies from 1.5 to 2.0 and is often taken as 1.67. This is approximately the same value as the elastic static safety factor. The elastic and plastic safety and design factors are summarized in Table 42.

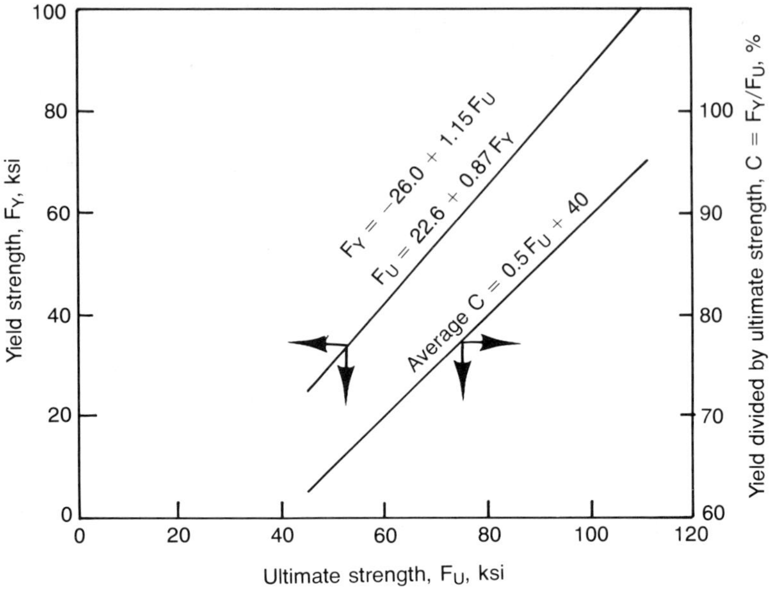

FIG. 3.17 Yield to ultimate strength

TABLE 42
Elastic and plastic safety and design factors

$DF_{SE} = \dfrac{F_{ME}}{F_Y} = 0.6$	(3.1)
$DF_{DE} = \dfrac{F_{ME}}{F_Y} = 0.8$	(3.2)
$DF_P = \dfrac{F_Y}{F_{MP}} = 0.6$	(3.3)
$SF_{SE} = \dfrac{1}{DF_S} = 1.67$	(3.4)
$SF_{DE} = \dfrac{1}{DF_E} = 1.25$	(3.5)
$SF_P = \dfrac{1}{DF_P} = 1.67$	(3.6)

Elastic Design Equations

Some of the most useful general design data on tubular joints is the work published by Reber in 1972.[15] His work was based on a general literature survey augmented by special tests and parametric computer analyses.

Results for the maximum elastic stresses in a T or Y joint are given in equations 7 and 8 of Table 43. It should be noted that equation 9 for the Y angle coefficient is a reasonable approximation. Equation 10 is a best-fit parabolic representation of the original data for K joints having a compressive load on one brace and a tensile load on the other brace.

The nomograph in figure 3.18 simplifies the calculation procedures for elastic analyses on the T, Y, and K joints. As an example, consider a 24-in. diameter chord having 0.8-in. wall thickness. By connecting the intersection point A on the scale with a 12-in. brace diameter, it is possible to determine that a chord stress of 0.60 ksi stress per kip is the applied T brace load. This is within about 10% of the value obtained from a calculation with equation 7 in Table 43.

The allowable brace load relative to the chord stress per equation 11 is then determined to be the design factor times the yield strength of the chord material divided by the hot spot stress factor provided by figure 3.18.

For instance, using an A36 material with a design factor of 0.8, the allowable stress would be 28.8 ksi. The allowable brace load would then be 28.8/0.60 or 48 kips.

Applying this assumption to equation 12, the required brace wall thickness would be only 0.044 in. Because of buckling considerations, as noted in figure 3.13, the minimum wall thickness required would be 0.375 in.

TABLE 43
Equations for elastic design of tubular joints

$$F_{MT} = \frac{F_C}{P} = \frac{0.62D.70}{T^{1.6}d^{1.1}}\frac{\text{ksi}}{\text{kip}} \tag{3.7}$$

$$F_{MY1} = F_{MT}(\sin^{1.5}\theta) \tag{3.8}$$

$$\sin^{1.5}\theta \approx 0.1 + \frac{\theta}{100} \tag{3.9}$$

$$\frac{F_{MEK}}{F_{MEY}} = 0.52 + 2.1\left\{\frac{d}{D} - 0.56\right\}^2 \tag{3.10}$$

$$P_B = \frac{(DF)(F_Y)}{\left(\frac{F_C}{P}\right)} \tag{3.11}$$

$$t_B = \frac{P_B}{(\pi d)(DF)(F_y)} \tag{3.12}$$

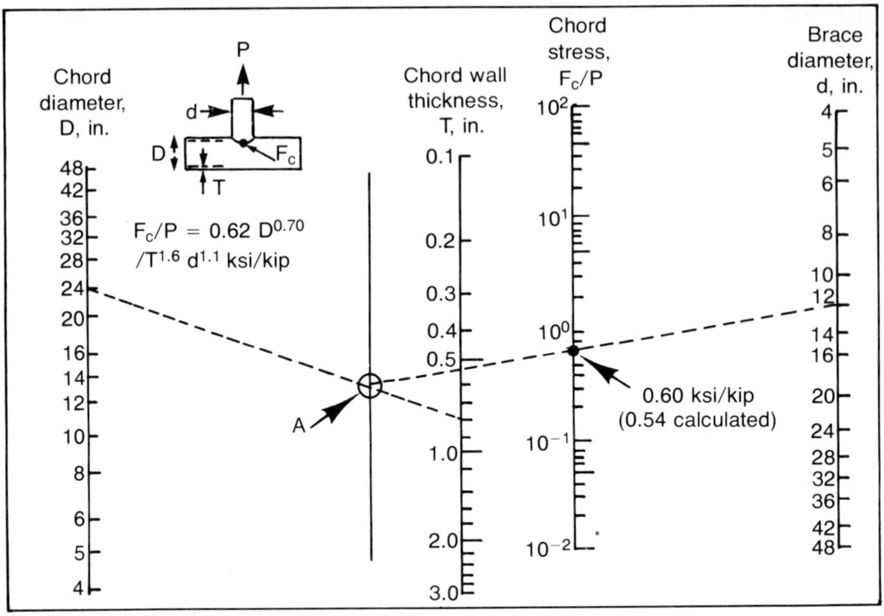

FIG. 3.18 Nomograph for elastic design of T joints

Plastic Design Equations

Reber reduced his accumulated data to a very brief and efficient procedure for plastic design by assuming that material with an average yield strength of 50 ksi would be used.(15) His results should be more than adequate for a first approximation.

For a second approximation, equations have been developed using modified data; i.e., fewer assumptions and results derived as a function of the actual material strength.

A third, and presumably more accurate, analysis would require special tests or additional computer work which generally lies outside the scope of this section.

The results of the second approximation are presented in Table 44. To further clarify these equations, it is helpful to look more closely at their derivations.

This analysis is based on the understanding that the plastic ratio is the load present in a given joint when the joints fail plastically, divided by the load in the brace when the hot spot stress in the joint equals the yield strength of the material.

Previous tests have suggested that the plastic ratio for K joints was linear with the diameter ratio for a given yield and with the yield strength for a given diameter ratio. The best fit of these data is given in equations 16 and 17 of Table 44. Combining these results, it

TABLE 44
Equations for plastic design of tubular joints

$\left(\frac{P_U}{P_Y}\right)_T = 8.3 - 5.0\,\frac{d}{D}$ at $F_Y = 38.5(36 - 41)$ksi	(3.13)
$C_K = 1.0 + \frac{38.5 - F_Y}{161}$	(3.14)
$\left(\frac{P_U}{P_Y}\right)_T \times C_K \left(\frac{P_U}{P_Y}\right)_T$	(3.15)
$\left(\frac{P_U}{P_Y}\right)_K = 4.0 - 2.2\,\frac{d}{D} \approx 4 - 2\,\frac{d}{D}$ at $f_Y = 71.7$ ksi	(3.16)
$\left(\frac{P_U}{P_Y}\right)_K = 4.0 - 0.019F_Y \approx 4 - 0.02F_Y$ at $\frac{d}{D} = 0.638$	(3.17)
$\left(\frac{P_U}{P_Y}\right)_K = 2.65 + \frac{68 - F_Y}{50} + \frac{0.68 - d/D}{0.5}$	(3.18)
$\left(\frac{P_U}{P_Y}\right)_{Y_1} = \left(\frac{P_U}{P_Y}\right)_K \left(\frac{F_{MEY}}{F_{MEK}}\right)$	(3.19)

is possible to derive equation 18 which eliminates the separate calculations in the two previous equations and provides results which have variations of about 4% in accuracy.

To further simplify the use of equation 18, a nomograph of this mathematical relationship for K joints is provided in figure 3.19. For example, a theoretical K joint made from 36 ksi yield material, and exhibiting a brace to chord diameter ratio of 0.6, would have a nomographically determined plastic ratio of 3.45.

The same mathematical techniques were also applied to T joints. Normalizing the data derived for equations 13 and 17 produces equation 14. This provides the necessary yield strength correction factor, which is then introduced into equation 15 and the plastic ratio for T joints is calculated.

By comparison, it appears that the plastic ratio for T joints is twice the value derived for K joints.

It is interesting that the T joint plastic ratio is also greater than the value derived for a Y joint. As indicated in equation 8, the elastic hot spot stress in a Y joint is less than that found in a T joint. The magnitude of decrease is directly related to the angle of the Y brace with respect to the chord.

This same general relationship exists with K joints. The reinforcing effect of the two braces on the chord attentuates the elastic hot spot stresses to a level which is roughly half that found in Y joints.

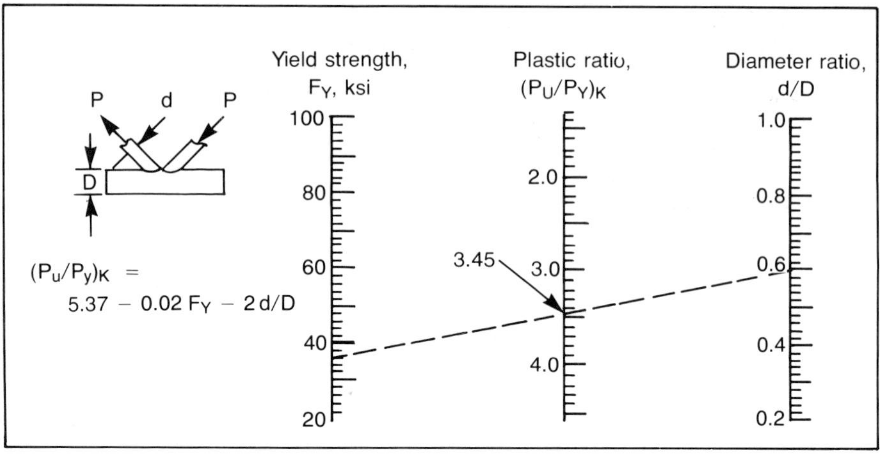

FIG. 3.19 Nomograph for plastic ratio in K joints

However, it should be noted that once the hot spot stress exceeds yield, the reserve strength between yield and failure is much less with a K joint than with a Y joint configuration.

In light of these conclusions, the plastic ratio for a Y joint can be taken as the K joint plastic ratio divided by the K to Y elastic ratio in equation 10. The final result for this relationship is given in equation 19 of Table 44. As might be expected, this analysis suggests that the plastic ratio for a Y joint is closer to that of a T joint and markedly different from the plastic ratio derived for a K joint.

3.3 HOW TO DESIGN FOR JOINT FATIGUE

The basic component for almost all fatigue analysis is an "SN" curve. This curve is usually developed from a plot of stress versus the number of cycles required to obtain a fatigue failure in a particular joint configuration. The extensive time and expenditures required to perform such tests often precluded the development of a full scatter, or plot, of basic fatigue data.

When comprehensive tests are conducted, such as the set of over 1,200 tests reported by Halford,[17] the scatter is so great that statistical analysis is required to effectively define the degree and magnitude of correlation between fatigue strength and number of cycles at a given stress level.

Once an SN curve is established, Miner's theory is applied to estimate the fatigue of life of a joint at veryjng cyclic stress levels.[18] This simple but profound theory states that fatigue damage is accumulative and proportional to the sum of the ratios of the number of cycles at various stress levels divided by the fatigue stress life cycles at that particular stress level.

Examples of this type of fatigue analysis, as applied to offshore rigs, are readily available from many references, including those listed.[19,20]

Scope of Fatigue Design

The scope of this discussion is applicable up to 10^7 or 10^8 cycles. For a great number of cycles, other procedures should be used, such as those discussed by Marshall.[21] He suggests the use of various standards including AWS-X, Lloyd's and British for the range of 10^7 and 10^{11} fatigue cycles.

The fatigue damage in the 10^7 to 10^{11} range of stress cycles is extremely long term and corresponds to the effects of 1 to 3-ft wave heights. Although it is not always evident, fatigue analysis has shown that a structure which is satisfactory in the offshore environment in the 10^2 to 10^7 range of stress cycles will usually have sufficient fatigue life for the low stress levels with greater than 10^7 cycles.

With the exception of laboratory environments, it is extremely difficult to suitably define the limits of all the variables that must be considered for most practical fatigue problems. With all of the limitations usually incurred, a good fatigue analysis should give results that are incorrect by a factor of two or three. This is particularly true in dealing with the ever-changing and unpredictable offshore environments.

Construction of SN Curves

It would be desirable to always have sufficient fatigue data readily available for application to offshore design. This is rarely the case, especially considering the severely corrosive environment of salt spray on the structure.

The following discussion explains a procedure to construct the SN curves (fatigue stress versus number of cycles to failure at that stress) from limited data.

This procedure should only be used as a first approximation analysis. It gives conservative design guidelines (underestimates the fatigue life), but the resultant structure should be satisfactory for the calculated fatigue life provided the parametric input is both sufficient and accurate.

Generally, the safe zone for design is anywhere in the area under the SN curve. The first fixed point in any situation is the ultimate stress, which corresponds to one cycle. For purposes of comparison, laboratory approximations are given first.

A second fixed point is established where the stress will be approximately 90% of ultimate at 5×10^4 cycles. A straight line can then be drawn from ultimate stress at one cycle to this point. Endurance life for the lab tests then usually falls to about 50% of ultimate strength of 10^7 and greater cycles.

A straight-line approximation can then be drawn from the second point to the third point of endurance life. The SN curve is then horizontal to the endurance life for any number of cycles.

One shortcoming of test data developed in the laboratory stems from the research environment. Testing personnel tend to take too much care in preparation and testing of specimens so that the test results can be duplicated. This includes good surface finish or, in some cases, may involve the use of polished specimens and an environment usually void of corrosive influences.

Results derived from this type of test would be appropriate for turbine blades or parts in internal combustion engines, but hardly suitable for offshore design.

For offshore design, a better second point approximation is the use of yield stress at 10^3 cycles. This is an important consideration because the yield percent of ultimate strength increases dramatically with increasing ultimate strength. The arbitrary use of the 90% ultimate rule for the second point is usually considered satisfactory for super high strength steels, but there is a matter of interpretation which could lead to incorrect use of this rule of thumb.

Low to super strength steels in offshore designs are normally defined as 25 to 100 ksi yield of strength or 45 to 100 ksi ultimate strength. In other industries where the amount of steel used is extremely small relative to offshore structures, such as the design of rocket motor cases, the use of 300 ksi ultimate strength steel is not uncommon.

The 90% rule should be applied only for steels having an ultimate strength of 100 ksi or greater.

The endurance life taken at 10^7 is an excellent approximation for lab tests. However, when surface imperfections are considered, 10^6 cycles for the initiation of the endurance life is a better approximation.

The percent of ultimate to be taken at endurance is a rough approximation at best. Consequently, reasonable values for use in offshore design can be taken as 40% of ultimate for inside offshore, 30% for outside offshore, and 20% for the splash zone.

Inside offshore is defined as well protected areas having no salt spray, such as the inside of platform modules. These are occasionally covered with fire wall insulation or at least well painted, but exposed to air with a high moisture content. Outside offshore is defined as the area between the platform deck level and the splash zone.

The splash zone is the area on fixed platforms or jackup legs between the wave crest and trough. Since the endurance strength must be reduced so much in the splash zone, tubular joints in platforms should be designed for placement as far away from the mean water level as possible. This type of placement also decreases wave loads in these areas.

These guidelines are applicable for situations without stress reversals. This implies that the stress oscillates anywhere between zero and some positive or negative value. For situations with stresses varying between positive and negative values, the reversal factors must be applied to further reduce the endurance limit.

The reversal factor can be obtained for the Goodman diagrams discussed later. These reversal factors are 55%, 60%, and 63% of endurance limits for no stress reversals taken up 20%, 30%, and 40% of ultimate respectively.

All of the above procedures have been applied to the lowest (A25) and the highest (X100) grades of API designated steels. The results of these procedures are given in figure 3.20.

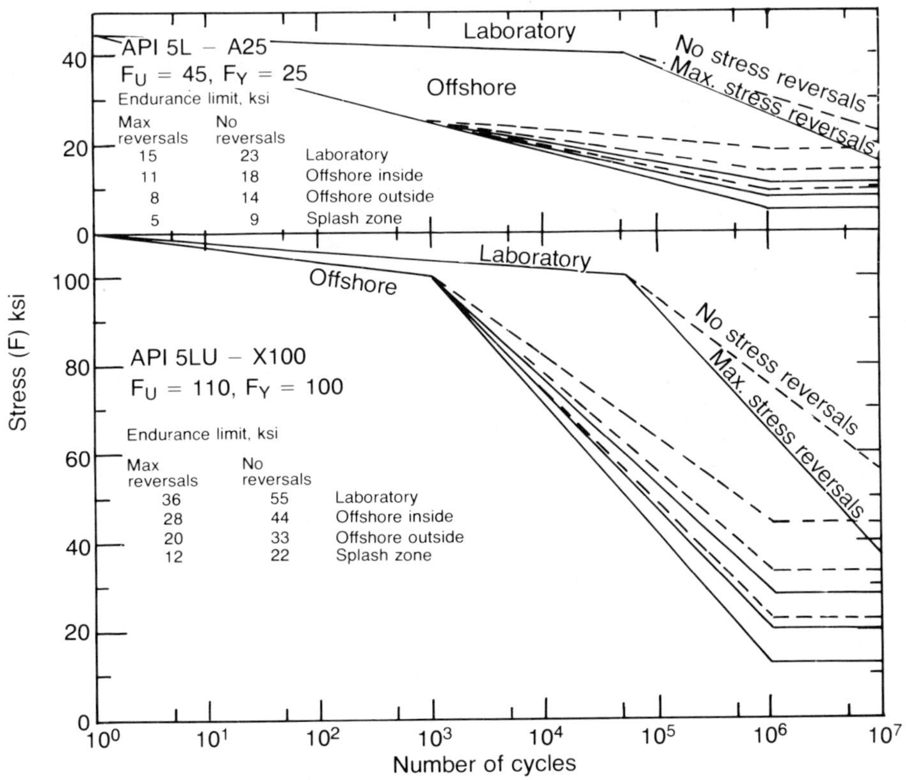

FIG. 3.20 SN curves for A25 and X100

Construction of Goodman Diagram

Because of the expensive nature of fatigue testing, it was long ago recognized that complete parametric fatigue tests for cyclic stresses at various stress levels, including stress reversals, would be cost prohibitive in most instances. These costs were virtually eliminated when sufficient test data allowed the Goodman hypothesis for stress reversals to be developed.

Goodman postulates that the endurance limit for maximum positive to maximum negative stress will be approximately between the positive limits of one-third of the ultimate stress.[22]

Lines which form the maximum to minimum stress reversal envelopes for infinite life are given in figure 3.21. This shows that a 33% maximum stress, with stress reversals, corresponds to a maximum of 50% of ultimate stress with no stress sign reversals.

To this graph, we have added our own approximations of 20%, 30%, and 40% of ultimate for the three most commonly encountered types of offshore environment. When the percent ultimate is 20, 30, 40, and 50, the graph shows corresponding stress reversal limits of 11%, 18%, 25%, and 33% of ultimate.

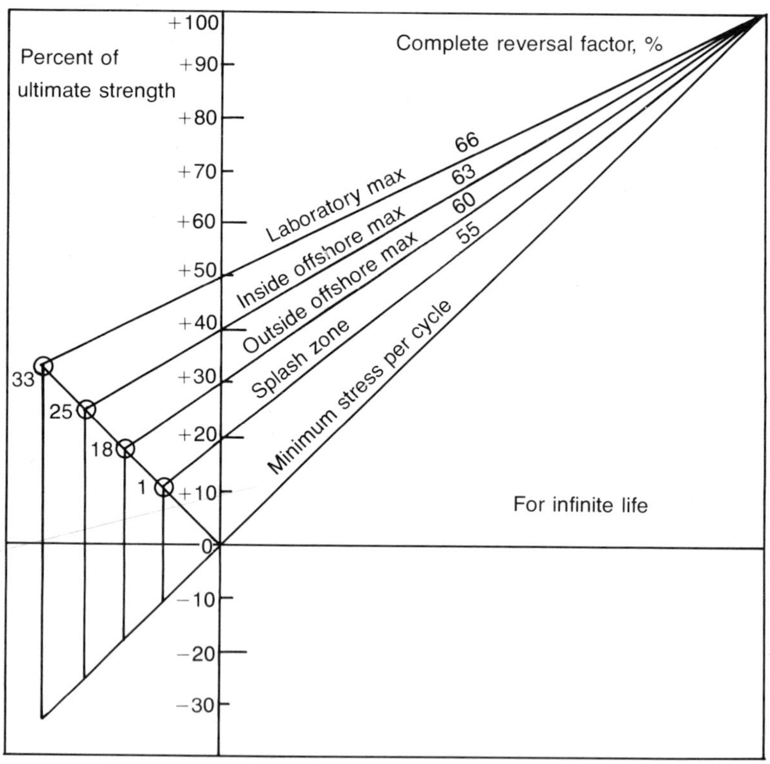

FIG. 3.21 Goodman diagram

Dividing these percentages, the percent of infinite life with maximum stress sign reversals, relative to maximum stress with no stress sign reversals, is shown to be 55, 60, 63, and 66.

The original type of Goodman diagram for fatigue stress sign reversals is still somewhat lacking since it fails to illustrate permissive oscillatory limits for infinite life. For this reason, the modified Goodman diagram came into favor. Results obtained from the modified Goodman diagram are shown in figure 3.22 for the arbitrarily selected Grade X100 steel. This graph is only for offshore outside conditions. Similar graphs can easily be constructed for other conditions or grades of steel.

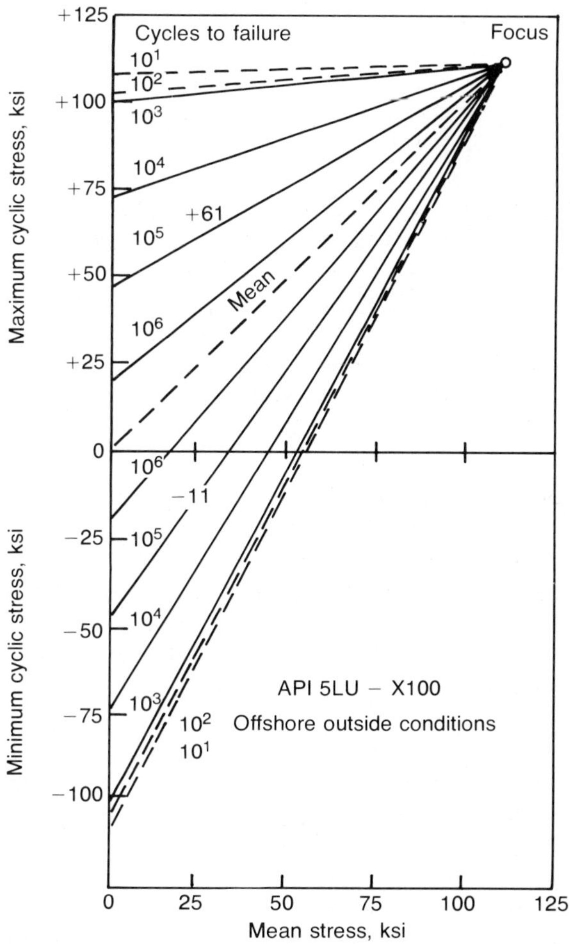

FIG. 3.22 Modified Goodman diagram

TABLE 45
Nomenclature

Symbol	
F	Stress, ksi
F_Y, F_U	Yield and ultimate strength, respectively, ksi
H	Maximum wave height, ft
N	Number of cycles
N_c	Cycles to crack initiation
N_f	Cycles to ultimate failure

The graph was constructed by making use of the reversal percentage of 60 as given in figure 3.21. Thus, the endurance limit at 10^6 cycles is the 100 ksi ultimate strength times 60% reversal factor, times the 30% factor for no stress sign reversals in the offshore outside environment or 20 ksi. The data points from figure 3.20 were also used at other fatigue cycle levels.

The use of the modified Goodman diagrams can be illustrated by the following example. If, for instance, it has been determined that the number of stress cycles will be a maximum of 10^5 and the maximum positive stress will be 61 ksi, then to what extent can negative cyclic stress be imposed?

Applying these parametric guidelines to figure 3.22, it is possible to determine quite readily that the minimum stress is a negative 11 ksi and the mean stress is 25 ksi.

Fatigue Tests on Tubular Joints

Bouwkamp fatigue tested several 35–90° TY joints having thin and thick walled chords in a 3.5% NaCl solution simulating sea water.(23) With the nominal stress in the Y brace oscillating about plus and minus 80% of the yield strength, the thin walled chord failed at about 100 cycles. Decreasing the chord diameter to thickness ratio 40 to 20 increased the fatigue life to nearly 10^4 cycles with the same loading.

Since the section modulus of tubular members increases proportional to the wall thickness and by the square of the diameter, it would be expected that doubling the wall thickness would decrease by a factor of two.

From the curve in figure 3.20 for A25 material, a slightly lower grade of steel than the A36 Bouwkamp used, it would appear that halving the stress would increase the fatigue life by a factor of roughly 100.

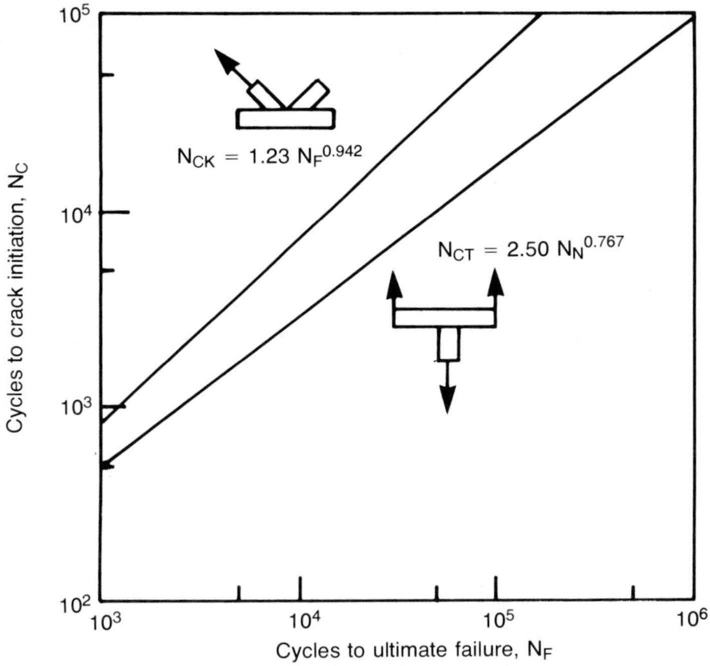

FIG. 3.23 Cycles at initial crack

Relative Fatigue Life

Mathematical relations have been given in the literature(24) for the fatigue life of several life types of tubular joints. New correlations were developed for this study by factoring out the load parameters and equating the results.

The relationship between the initiation of the first crack and the ultimate failure was determined from a series of fatigue tests on K joints as shown in figure 3.23. While the cyclic lifes are approximately the same as 10^4 cycles, the number of cycles for crack initiation for the same loads are about one-half of the number of cycles to ultimate failure when the cyclic loads are small enough for the endurance limit (one million cycles) to be reached.

For comparison a cursory fit of more recent data(25) on T joint fatigue was determined with results also given in figure 3.23. This suggests that when an ultimate failure occurs at a million cycles, fatigue crack initiation begins at about 5×10^5 cycles for the K joint and at about 10^5 cycles for the T joint.

Consequently, cycles to crack initiation with ultimate failure at the endurance limit is roughly a factor of five for K joints relative to T joints.

To be more definitive, the relations for ultimate failure to crack initiation shown in figure 3.23 were divided to give the equations in figure 3.24. The graph of these new equations shows how the number of cycles for crack initiation for a K joint, relative to a T joint, varies from approximately 1.6 at 10^3 failure cycles to 5.5 at 10^6 failure cycles.

A similar normalization procedure was used to relate the cycles to failure with K, T, and cross joints as shown in figure 3.25. The T joint was used as the base reference since more fatigue data is available for this type of joint. The results show that the fatigue life of the K joint is about 10 times better than the T joint for T joint failures at 10^3 cycles.

These data also indicate that the K joint is about 20 times better than the T joint for failure at the endurance life. The cross joint is shown to be about 12 times better than the T for fatigue failure at 10^3 cycles but decreases to about 8 times better than the T for fatigue failures at the endurance limit.

Effect of Integral Joints

Test results have illustrated the advantage of using integral joints in reducing the stress concentration in tubular connections.(26) During these tests, it was determined that the ultimate fatigue life was 2.4×10^4 cycles for these integral joints or approximately 3 times greater than the fatigue life of the fabricated joints.

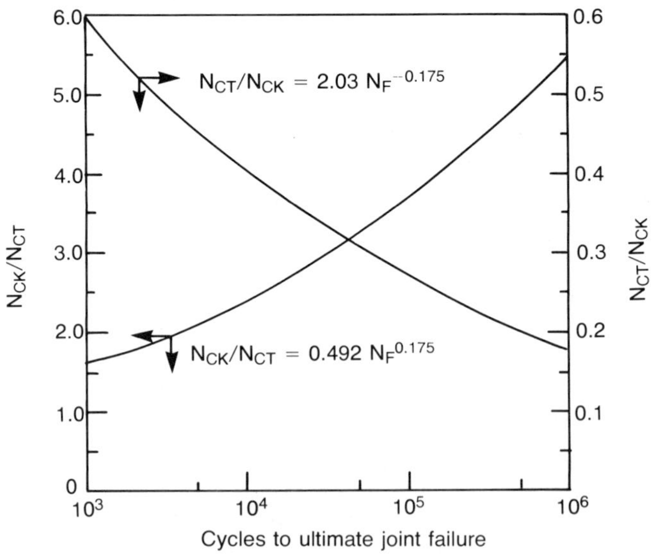

FIG. 3.24 Crack initiation ratios for T and K joints

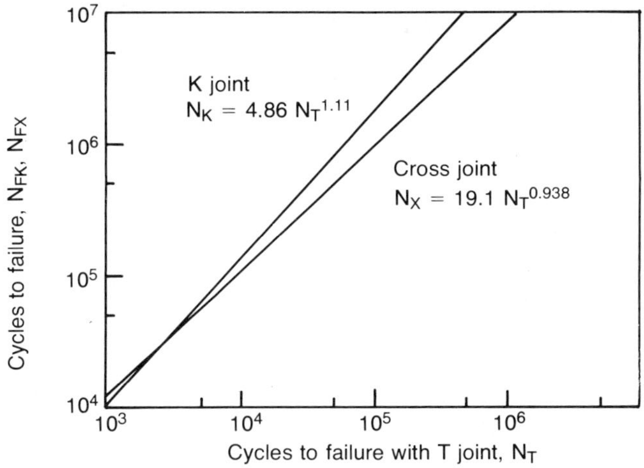

FIG. 3.25 Fatigue life relative to T joint life

More importantly, it was determined that the cycles to crack initiation were 9.2×10^4 cycles for the integral joints, which was nearly 15 times better than their fabricated counterparts.

It should be noted that the integral joint, either forged or cast, may cost considerably more than the field joint unless extremely large quantities of a particular joint are required. This suggests the possible use of integral joints for other locations.

Equal performance may also be expected from shop fabricated joints in which the integrity of the weld has been established and surfaces blended to eliminate stress risers.

For tubular joints in offshore platforms having very large loads, the alternatives are: first, consider the use of traditional types of fabricated tubular joints which often have high strength-to-weight ratios; second, consider forged or cast joints which have improved fatigue life but may be more expensive in small quantities even though they are cost effective in large quantities; and third, new types of fabricated or forged joints having improved strength to weight ratios.

Weather Data for Fatigue Design

In a previous exercise(27) to determine the suitability of given rig designs for work in different offshore environments, it was necessary to consider a large amount of statistical data on wind and wave loads from various parts of the world.

A brief summary of some of that previous data on wave and wind loads for six environmentally different areas of the world is given in figures 3.26 and 3.27. Use of this type of weather data in fatigue analysis is a widely accepted practice.(20)

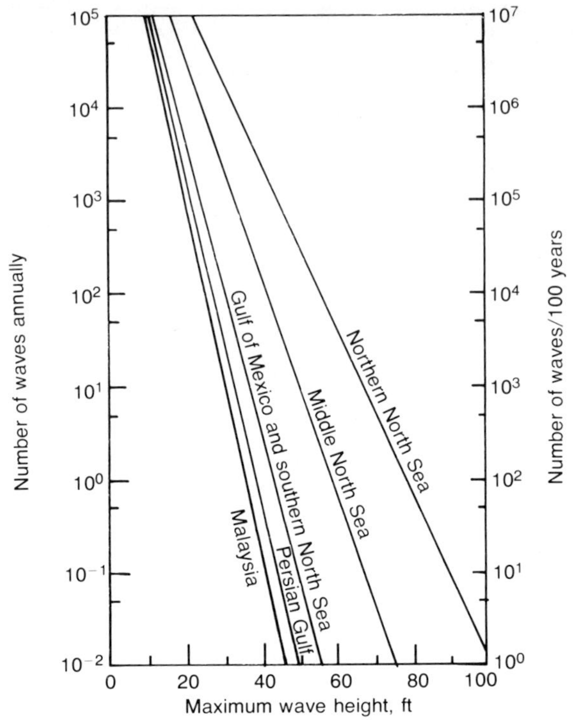

FIG. 3.26 Frequency of storm waves

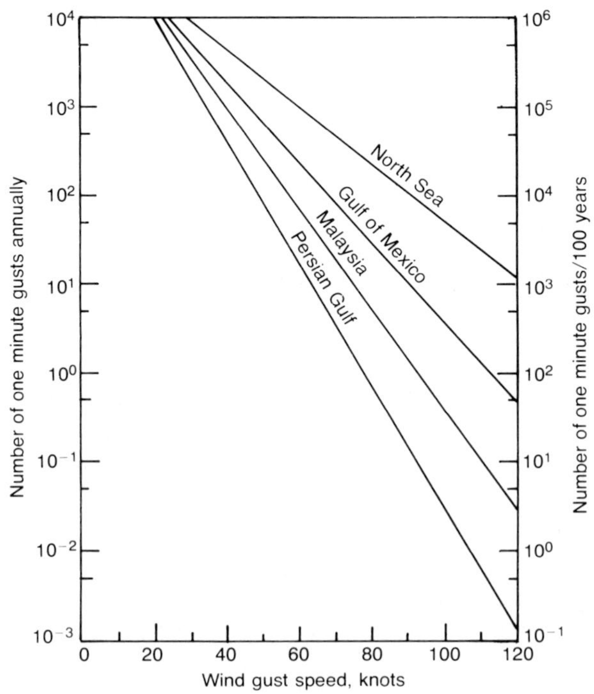

FIG. 3.27 Frequency of one-minute wind gusts

However, this weather data is suitable only for preliminary and comparative analyses. Final designs should use more specific data applicable to a particular and exact location.

3.4 ANALYSES OF ELBOW JOINT DESIGNS

Significant improvement in the strength to weight ratio of tubular joints could be achieved if the stress concentration factors normally found with fabricated joints were decreased. Do such structural fittings provide a substantial degree of increased joint efficiency when compared to joints fabricated from pipe?

These fittings are fabricated from elbows and pipe and are intended primarily for the offshore industry. Applications are found on fixed platforms, jackup rigs, and construction barges. Possible locations of these types of joints are shown in figure 3.28.

The major objective of the preliminary analysis was to evaluate these special structural joints in light of current industry construction

TABLE 46
Nomenclature

Symbol	
A	Area, sq. in.
a, b	Major, minor diameters of ellipse, in.
B	Chord diameter, in.
C_D	Ratio of projected area
C_f	Flux factor
C_p	Ellipse function
D	Brace diameter, in.
e	Scallop of elbow, in.
F	Stress, ksi
F_S	Shear stress, ksi
$\bar{F}$	Stress corrected for flux effect, ksi
K_t, K_b	Uniaxial stress factors in tension and bending, respectively
$\bar{K}_t$, $\bar{K}_b$	Biaxial stress factors in tension and bending, respectively
R	Radius, in.
r	Radius of elbow (cross-section), in.
RKL	K brace with elbows, reinforced
RTYL	TY chord brace with elbows, reinforced
P_m	Maximum allowable load, kip
S	Perimeter, in.
t	Wall thickness, in.
t_W	Weld depth, in.
x, y	Orthogonal co-ordinates, in.
XL	XL vertical cross brace with elbows
θ	Angle of elbow intersection, degrees

practice. In simplest terms, the analysis indicated that the structural efficiency of the special reinforced K-brace joint is nearly two times that of a standard fabricated K-brace. The reinforced XL elbow joint is over three times as efficient.

The analysis also showed that the XL joint transfers loads with minimum welding requirements. These findings are based on certain assumptions and do not encompass all the variables inherent in any given construction project. A survey of these assumptions and the analytical framework form the basis for the above conclusions.

The stress concentration factor in fabricated tubular joints is decreased with smooth transitions and increases in the joint shear area. The new elbow type joints discussed here have improved strength to weight ratios because they possess both of these attributes.

Elbow Joint Geometry.

Figure 3.28 shows three new tubular joint designs which use elbows and pipe. There are other design combinations, but this analysis is limited to the RKL, RTYL, and XL joints.

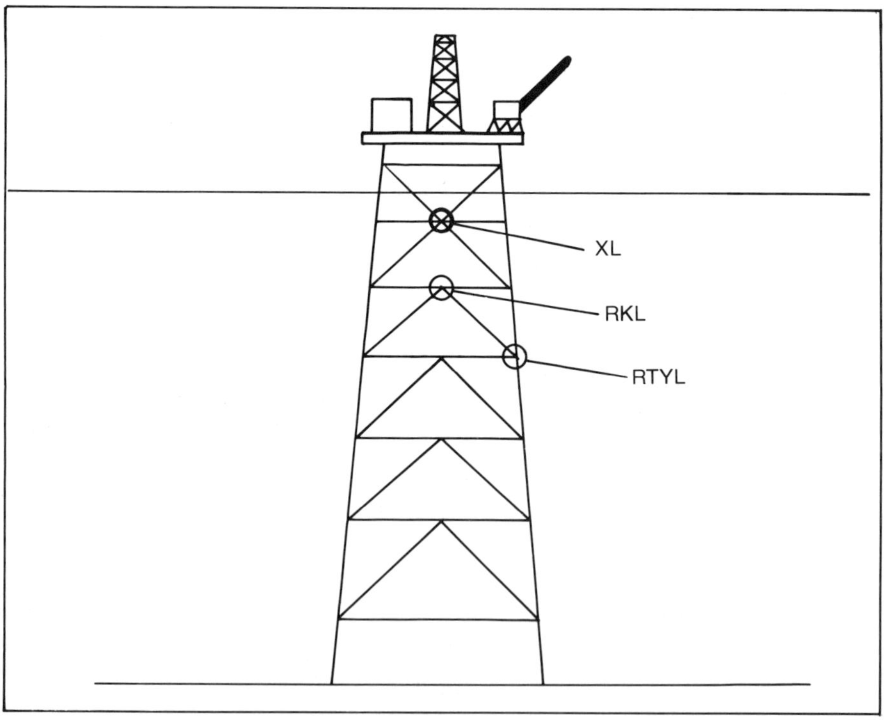

FIG. 3.28 Use of joint types

Since the jargon associated with these new joints may be unfamiliar, it is necessary to first define key terms. On each joint designation, the "R" represents plate reinforcement and the "L" indicates the use of an elbow. All secondary braces, including those used on the "XL" joint, were assumed to be inclined at an angle of 45°. The joint designations were:

RKL—K-brace with elbow, reinforced
RTYL—TY chord brace with elbow, reinforced
XL—XL vertical cross brace with elbows

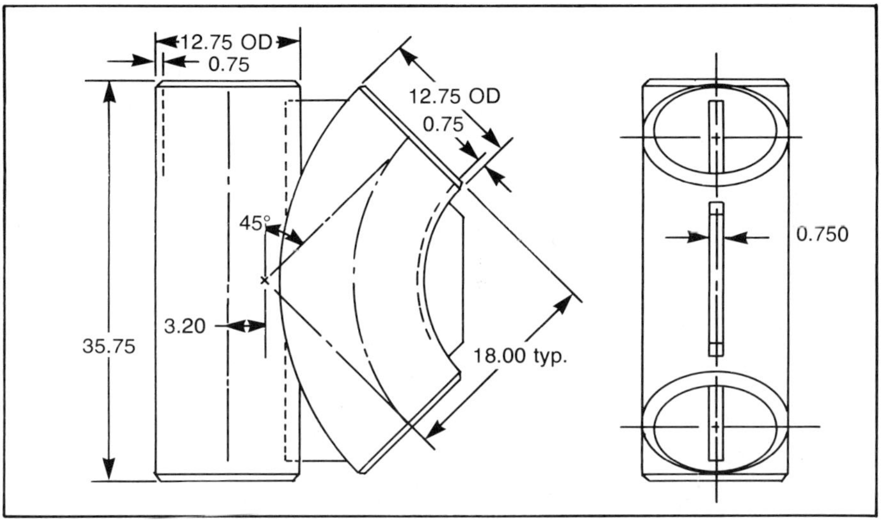

FIG. 3.29 RKL tubular joint geometry

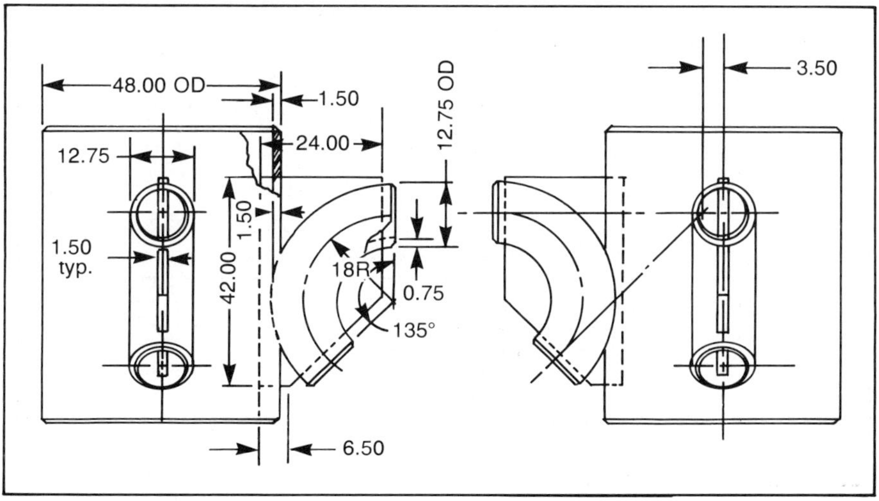

FIG. 3.30 RTYL tublular joint geometry

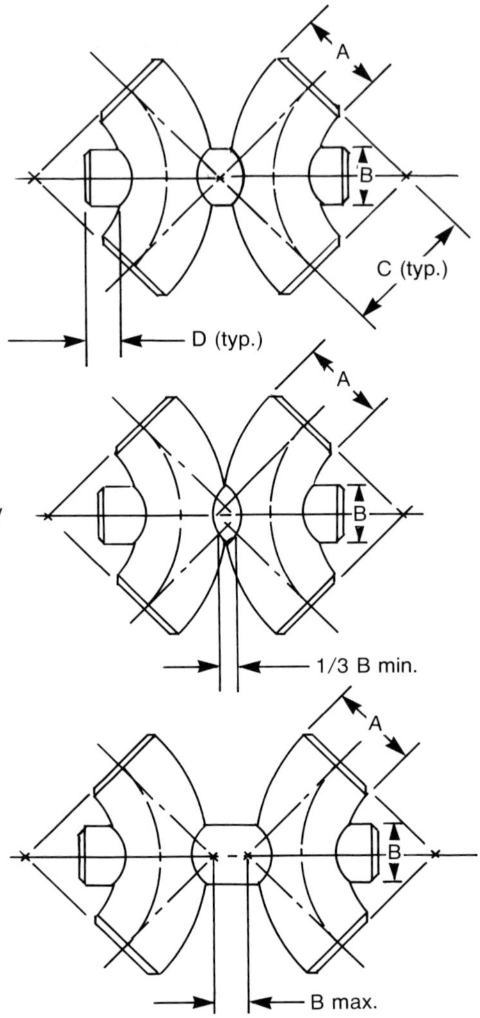

A	48-in.	90° L.R. elbow with 1.250 wall
B	36-in.	OD pipe with 1.250-in. wall
C	72-in	Radius of elbow
D	¼ B	Extension

FIG. 3.31 XL tubular joint

The specific geometry of the three joints is given in figures 3.29, 3.30, and 3.31.

Each of these joints is designed to accomplish a specific function within a given structural framework. The XL joint transfers forces through a lattice where a cross member is used to allow increased buckling loads. The RKL joint provides a connection at the midpoint of a lateral while the RTYL joint connects the laterals and diagonals to a platform column.

Joint eccentricity is the distance from the brace centerline intersection to the chord centerline divided by the chord diameter. The percent eccentricities of the new elbow joints discussed here are negative 25 for the RKL joint, negative 7.2 for the RTYL joint, and zero for the XL joint.

As noted earlier in the chapter, negative or zero eccentricity is a most desirable design feature and positive eccentricity should be avoided whenever possible. However, for certain situations in which the XL members are carrying minimal loads, it could be more cost effective to use a design of those joints having positive eccentricity.

Design and Analysis

The type of analysis used for tubular joints is dictated primarily by the complexity of the design conditions. A rather simplified approach is used in these preliminary analyses since the exact environmental conditions would be required to complete a detailed analysis. The final design should use either two-dimensional analysis,[(28,29)] or some type of three-dimensional finite element analysis[(30)]. In addition, an experimental approach such as strain gauges, birefringement coatings,[(31,32)] or brittle lacquers could also be used. The two-dimensional or three-dimensional methods of photoelasticity would also be an acceptable analytical approach.

A review of API specifications[(33)] was used as a reference point in these analyses. Generally, this specification provides procedures for checking punch shear but does not provide information for the overall analysis of tubular joints.

For comparison with existing designs, an analysis of a standard K joint, which is currently fabricated from pipe, was performed. The analysis provided the base data used to compare and evaluate the special elbow joint designs.

The results of this analysis were based on combined stresses obtained by AISC procedures.[(34)] The allowable stresses used were 0.4 of yield for shear and 0.6 of yield for axial and flexural loads. The analysis used for determining the appropriate biaxial stress concentration factors and derived force flux factors are also briefly discussed later. The primary load considered in each case was a tensile load of the 45° brace. Details of each step of the analysis are available in a Tube Turns Research Engineering report.[35]

Biaxial Stress Concentration Factors

Many of the design handbooks, such as Peterson[36] give results for uniaxial stress concentration factors which are not appropriate for three-dimensional joints.

Some of the earliest reported research findings for two-dimensional stress concentration factors were given by Bynum and DeHart.[37] These results suggest that uniaxial stress concentration factors are about 38% larger than the biaxial stress concentration factors. The uniaxial factors appropriate for the special joints analyzed in this study were selected and the biaxial correlation factor applied. The results of this exercise are plotted in figure 3.32.

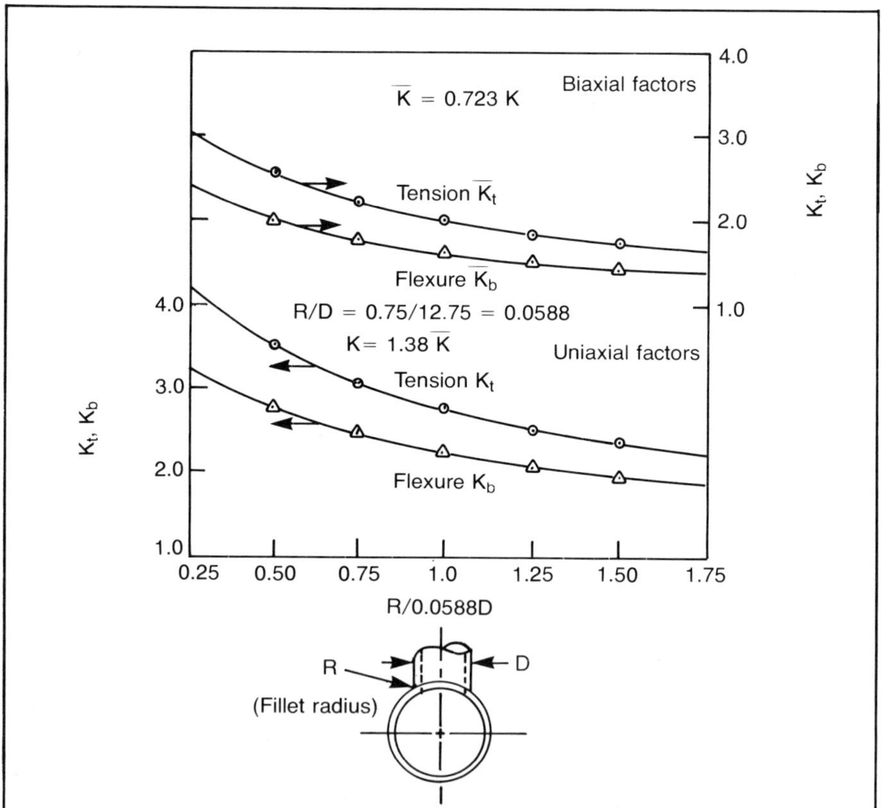

FIG. 3.32 Stress concentration factors

Joint Scallop Functions

The perimeter of the special RKL and RTYL joints attached to the straight circular member was calculated using the geometry given by the schematics in figure 3.33. It was assumed that the fitting was scalloped in a flat plane. The variation between the interface on a flat plane and the interface on a curved member is insignificant for the RTYL joint and gives a somewhat approximate result for the RKL joint.

The simplified relations given by Marks[(38)] were used in the analysis where the elliptical correction factor is given in figure 3.34. The results from the calculations for the shear area are geographically portrayed in figure 3.35, which displays the various functions versus the scallop ratio needed for the analysis.

The scallop ratio is defined as the radial distance removed from the curved member of the fitting divided by the radius of the

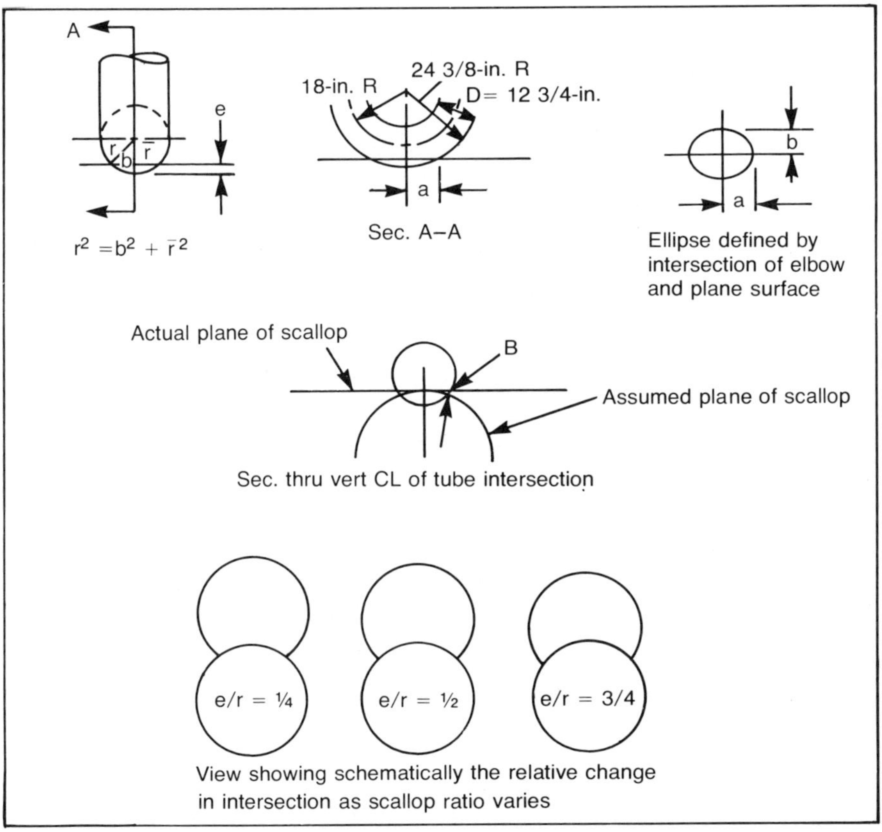

FIG. 3.33 Elbow joint geometry

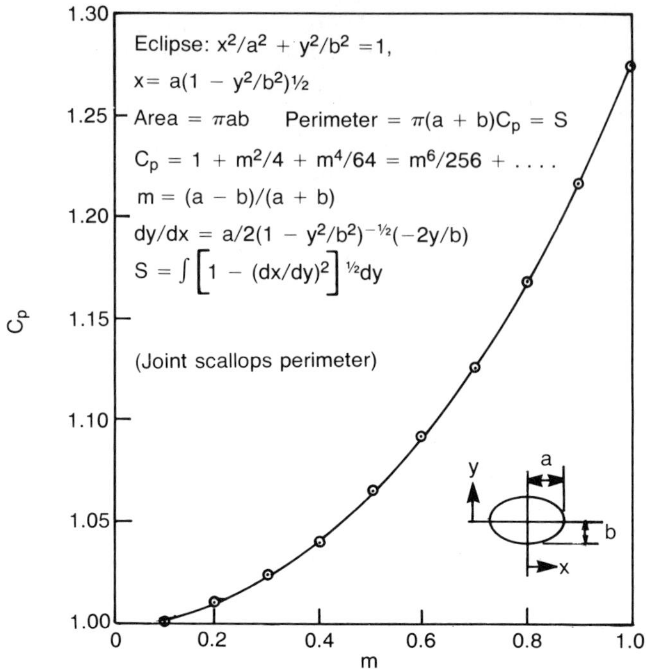

FIG. 3.34 Ellipse functions

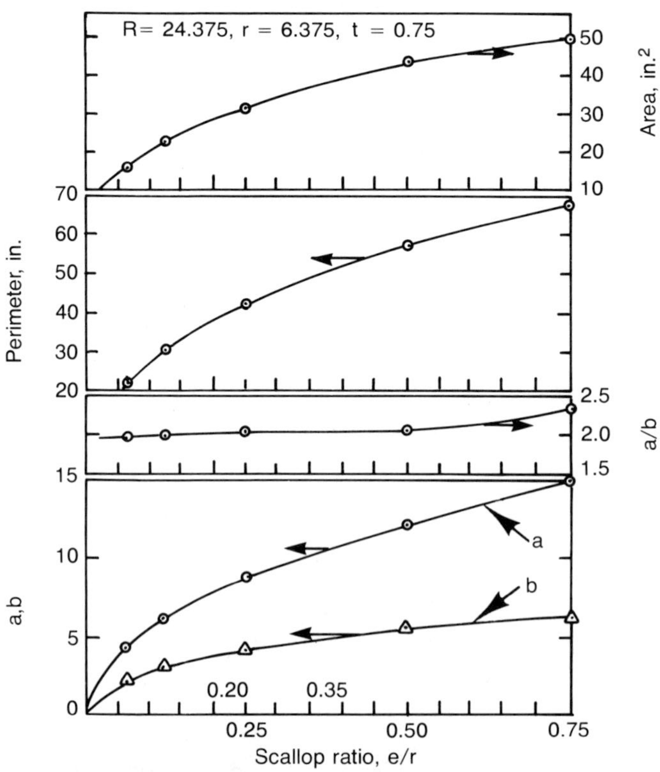

FIG. 3.35 Shear area of tubular joints

curved tubular member. It is interesting to note that for a wide range of scallop ratios, the ratio of the major to minor diameter of the ellipse varies from 2.0 to approximately 2.3 as shown in figure 3.35.

Force Flux Factors

The stress in a structure can be visualized as the force flux, normalized as the area perpendicular to the direction of the force. This

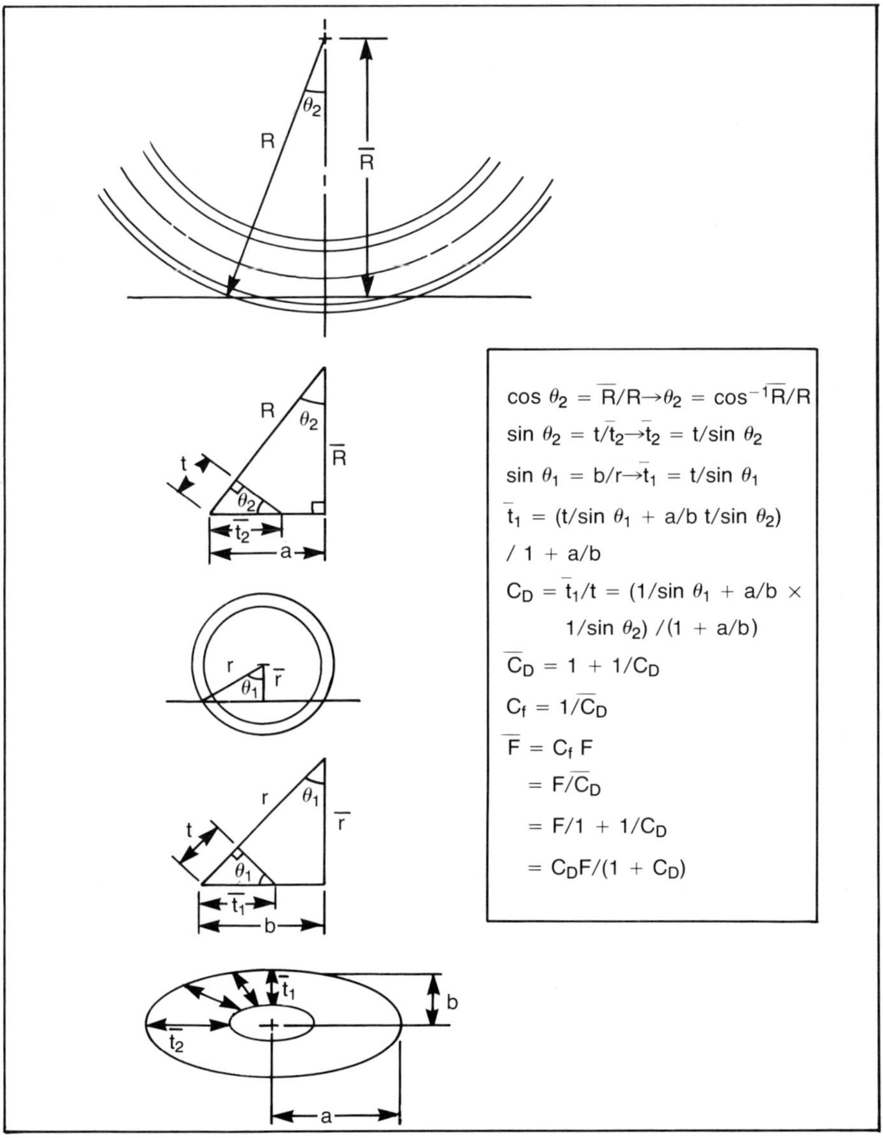

FIG. 3.36 Derivation of flux factors

is evident when viewing experimental hydraulic analogies, which are often used in theoretical structural formulations.

These analogies indicate that the force flux is better distributed during smooth transitions and is usually best achieved by a localized increase in the cross sectional area.

To account for the increased shear area of elbow joints, we have defined and derived the flux factors outlined in figure 3.36. The results of the calculations for the flux factors, as a function of the scallop ratio, are graphically illustrated in figure 3.37. This figure indicates, for example, that a curved member scalloped to one-half the radius reduces the force flux factor from 1.0 to 0.64, i.e., the efficiency of the joint is enhanced roughly 50%.

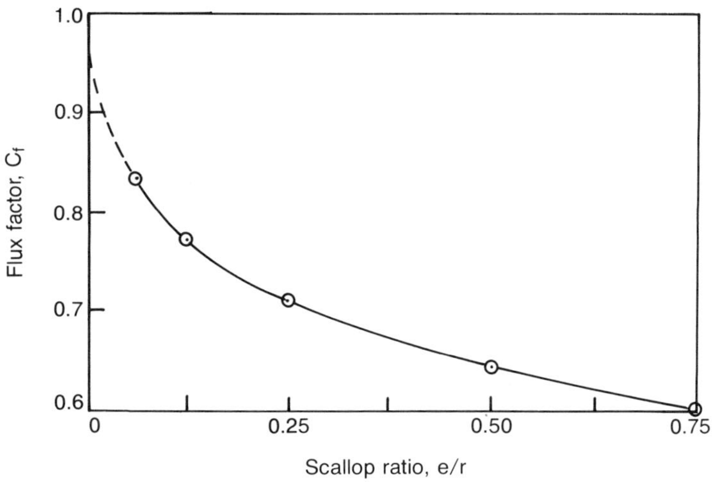

FIG. 3.37 Flux factor function of scallop ratio

RKL Tubular Joint Results

Analysis of the new RKL joint sketched in figure 3.29 was performed as described above. The parametric results for varying scallop ratio with constant reinforcement plate thickenss of 0.75 in., and for varying plate thickness with a constant scallop ratio of 0.35, are given in figures 3.38 and 3.39.

The load allowable factor is defined as the maximum brace stress based on the axial force applied to the nominal cross sectional area of the brace when the combined stress in the joint is at the yield point divided by the yield strength of the material.

The load efficiency factor is defined as the maximum allowable brace load of the RKL joint when the joint is at the yield point, compared to the maximum allowable brace load of a fabricated K joint when it is at the yield point.

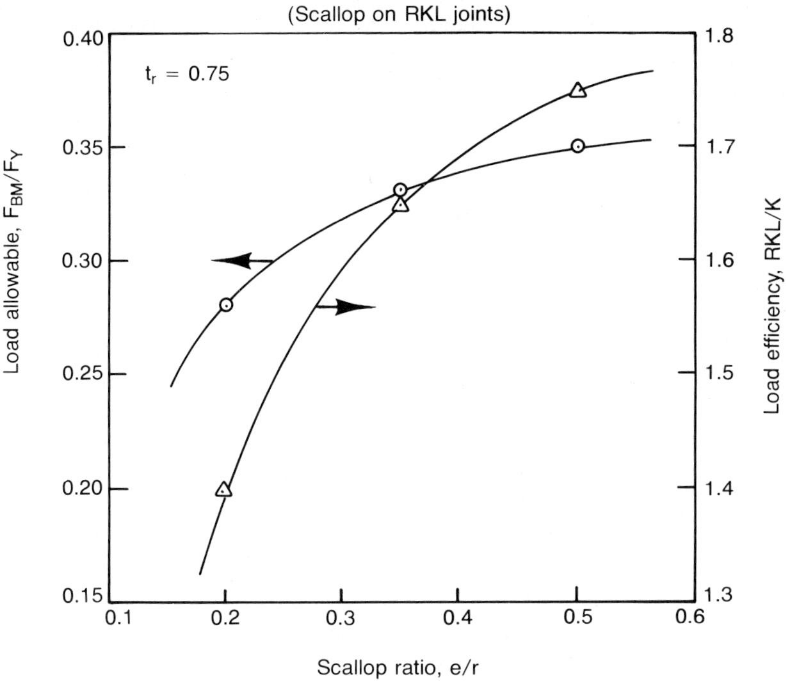

FIG. 3.38 Joint efficiency function

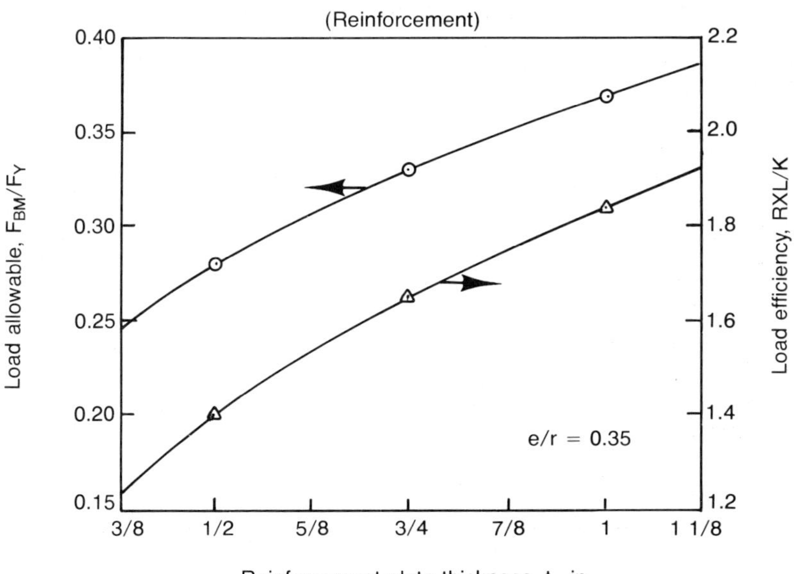

FIG. 3.39 Joint efficiency function

The results in figure 3.38 show that the RKL joint is better than the fabricated K joint by factors varying from 1.4 to 1.75 as the scallop ratio varies from 0.2 to 0.5 with a constant reinforcing plate thickness of 0.75.

The results in figure 3.39 show that the RKL joint out-performs the fabricated K joint by factors varying from 1.4 to 1.9 as the reinforcing plate thickness varies from 0.5 to 1.12 in. with a constant 0.35 scallop ratio.

RTYL Tubular Joint Results

Parametric results for the analysis of the RTYL joint sketched in figure 3.30 are given in figure 3.40 for varying reinforcement plate thicknesses with a constant 0.35 scallop ratio. As indicated above with the RKL joint, the load efficiency of the RTYL could be further increased by increasing the scallop ratio.

The results show that the load allowable factor increases from 0.33 to 0.69 as the reinforcement plate thickness increases from 0.25 to 1.75 in.

The load allowable factors were based on static AISC stress allowables. For dynamic load conditions, the load allowable factors given in figures 3.38 to 3.40 would increase by a factor of 33%. The load efficiency factors given in these figures are the same for both static and dynamic stress allowables.

XL Tubular Joint Results

The complete analysis of the XL joint [35] produced equations in dimensionless form so that various values of the axial force and the

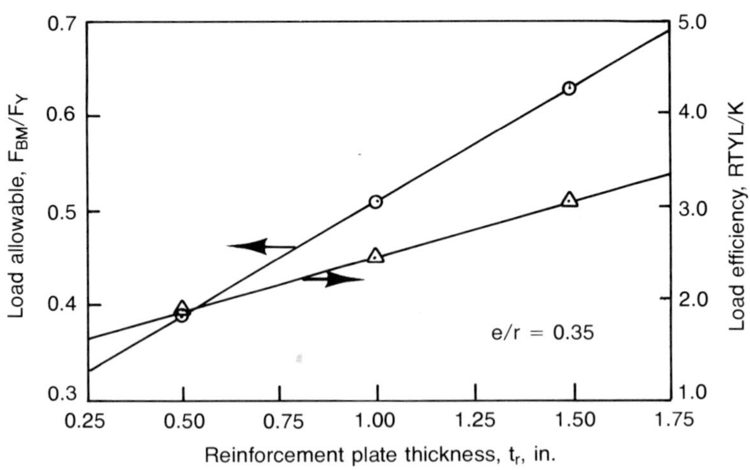

FIG. 3.40 Joint efficiency of RTYL joints

material yield strength could be easily substituted in the analytical format. The equations included the combined stress allowable with shear, axial and flexural loads.

The shear stress is the critical component of combined stress for the XL joint. For purposes of illustration, the load allowable on the XL joint sketched in figure 3.31A, as a function of the weld depth for a 60-ksi yield material, is shown in figure 3.41. The corresponding shear stress as a function of the weld size is also given in this figure.

This graph suggests that the weldment of the two curved members to the straight tubular section on four perimeters could be relatively small fillet weld and still be within allowable shear stress levels without affecting the overall efficiency of the joint.

For example, the maximum brace load varies from 550 to 760 kips as the weld depth varies from 0.12 to 0.25 in. The allowable load only varies from 760 to 1,100 kips as the weld depth varies from 0.25 to 1.25 in. In field applications, a full penetration weld is recommended.

Again, the maximum allowable load on the brace given in figure 3.41 is based on static AISC conditions. For dynamic conditions of wind and wave loads, the given allowable brace load can be increased 33%.

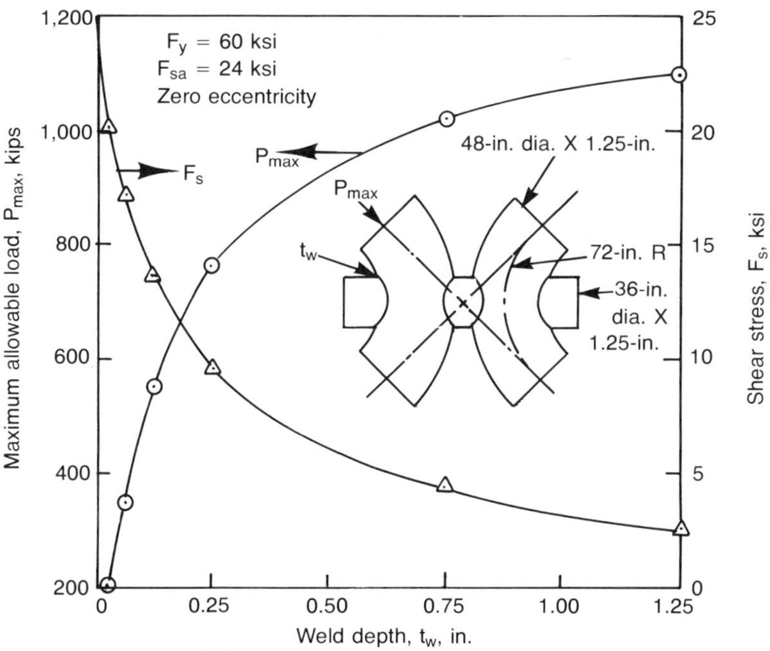

FIG. 3.41 Load ratings of XL joints

The analysis provided here is preliminary, but the results are encouraging. Static tests of these new types of tubular joints using structural elbows will be discussed in the next section. These results indicate improved designs can be achieved by using elbow type joints.

3.5 COMPARISON OF TEST STRENGTHS FOR NEW JOINT DESIGNS

The forged elbow tubular joints for offshore platforms offer better strength to weight ratios than the traditional fabricated joints and can be manufactured economically.

Six types of cross joints were tested for static tension and compression to give yield factors. The joint efficiency of each type of joint tested was then determined. The joint efficiency is considered to be the joint strength of each particular design divided by the joint strength of a fabricated cross joint.

TABLE 47
Nomenclature

Symbol	
D	Diameter, in.
F_y	Yield strength of material, ksi
F_u	Ultimate strength of material, ksi
F_j	Joint stress, ksi
T	Wall thickness, in.
F_x	Stress at failure, ksi

Geometry and Material Properties

The geometries of the tubular joints tested are shown in figure 3.42. The material properties of the six types of cross joints evaluated in the 13 tension and compression tests are given in table 48.

The joint diameter to thickness ratio was 36.5 for all the tests except for tests 5 and 6 where the wall thickness of the elbow cross was increased to give a 21.8 diameter thickness ratio. The yield strength from mill test reports varied from 37.9 to 44.4 ksi. The tensile strength varied from 57.9 to 74.2 ksi. The ultimate to yield strength factor varied from 1.38 to 1.79.

The standard fabricated cross was constructed according to API RP 2A guidelines and used for a reference.

All of the XL joints tested were made from long radius elbows and had zero eccentricity. In common piping terminology, a long

TABLE 48

Parameters in static tubular joint tests

Test no.	Load type	Test with legs	Joint description	Joint type	Diameter, D, in.	Wall, T, in.	D/T	Material strength: Yield, F_y, ksi	Material strength: Tensile, F_u, ksi	F_u/F_y	Max. joint yield stress, F_j, ksi
1	Tens	Yes	API fabricated	X	2.375	0.065	36.5	43.3	59.8	1.38	8.50
2	Comp	Yes	cross	X	2.375	0.065	36.5	37.9	57.9	1.53	6.63
3	Tens	No	Elbow cross	XLA	2.375	0.065	36.5	43.3	59.8	1.38	7.50
4	Comp	Yes		XLA	2.375	0.065	36.5	39.9	58.0	1.45	8.38
5	Tens	No	Elbow cross	XLB	2.375	0.109	21.8	39.9	58.5	1.45	19.50
6	Comp	No	thickened	XLB	2.375	0.109	21.8	39.9	58.5	1.45	19.00
7	Tens	No	Tube Turns	TSX	2.375	0.065	36.5	44.4	74.1	1.67	51.00
8	Comp	Yes	standard cross	TSX	2.375	0.065	36.5	41.5	74.2	1.79	15.47
9	Comp	No		TSX	2.375	0.065	36.5	41.5	74.2	1.79	42.50
10	Tens	No	Split cross	TRX	2.375	0.065	36.5	44.4	74.1	1.67	55.90
11	Comp	No	with plate	TRX	2.375	0.065	36.5	41.5	74.2	1.79	55.90
12	Tens	No	Cross w/double	TDRX	2.375	0.065	36.5	44.4	74.1	1.67	55.00
13	Comp	No	reinforcement	TDRX	2.375	0.065	36.5	41.5	74.2	1.79	61.80
							AVG.	41.8	67.0	1.60	

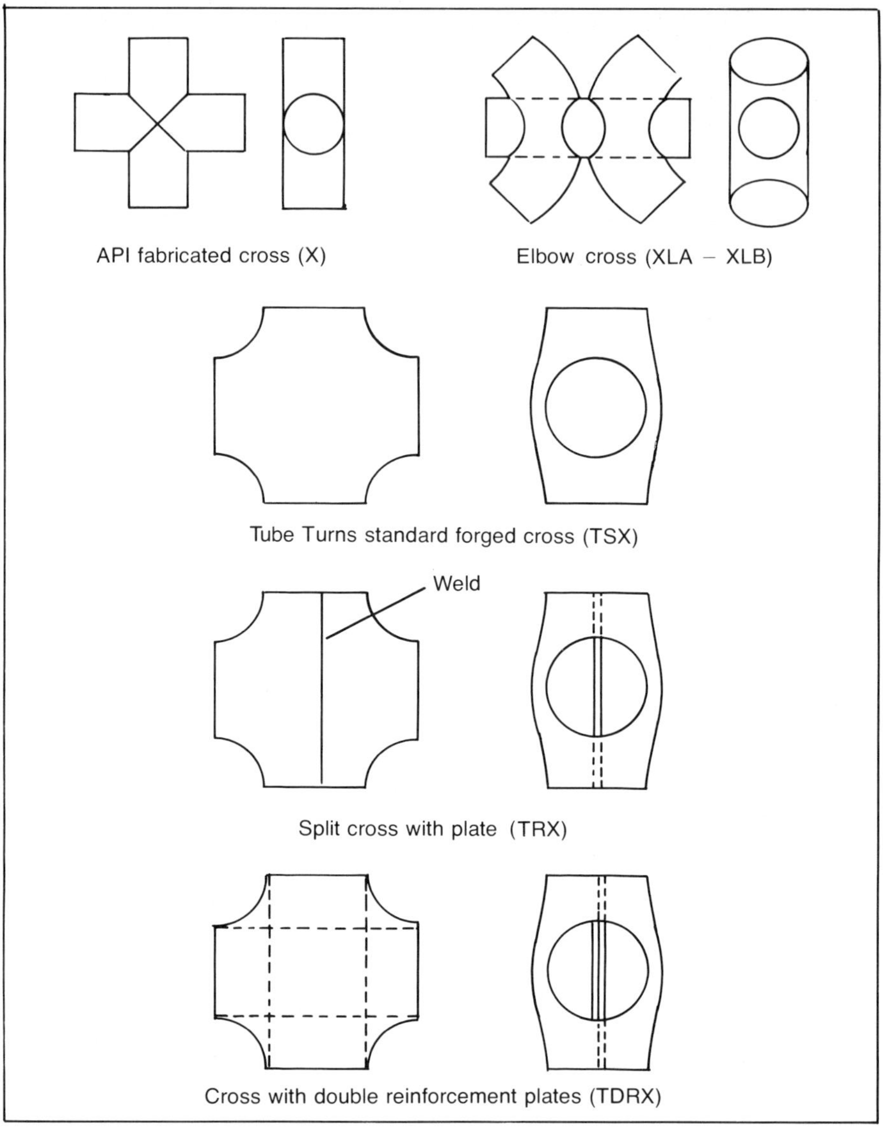

FIG. 3.42 Test model configurations

radius elbow is forged with the centerline radius equal to 1.5 times the nominal pipe diameter.

All of the models in tests 7 through 13 listed in table 48 used the Tube Turns standard forged cross, hereafter designated as joint TSX.

Two modifications of the TSX joint, the TRX and TDRX joints, were also developed and tested. These modifications used internal fillet welded reinforcing plates to stiffen the joint. In these test models, the stiffener plates were 0.104 in. thick and had properties of 36.6 ksi yield and 49.3 ksi tensile strength.

The TRX models consisted of splitting the basic TSX joint and fitting a single scalloped reinforcing plate in the joint body. This plate was centered in the joint and positioned 0.25-in. from ends where the pipe legs would be connected. The plate was welded to the inside of the joint and the split body welded together.

The TDRX models were not split but did have internal plate reinforcement. They consisted of fitting and internally welding two rectangular plates inside the joint with the plates again running to within 0.25-in. of the ends of each joint.

A quarter section plan view of the joint would then show a square cross with the reinforcing plates overlapping in the center of the joint.

Test Methods and Results.

All of the models were subjected to incremental static loads until a readily discernible permanent deformation of the joint was achieved. The maximum load was defined as the load level where deformations recorded on the deflection gauges began increasing for constant load increments.

The maximum load was then divided by the cross sectional area of the brace to get the joint stresses, (F_j) given in table 48.

The reported loads are, therefore, somewhat greater than the loads corresponding to exact initiation of joint yielding. This error appears to be of small consequence in comparing the relative efficiency of the various joints tested.

After the maximum elastic loads were determined, the joints were subjected to increased loads to observe the failure mechanism. The failure mechanism and the points, or areas, of failure with loads beyond yield varied considerably with the different types of joints. All tubular intersection welds in each of the test pieces were full penetration type. In the compression test on the API-type fabricated cross, the brace perpendicular to the axis of the applied load collapsed. In the tension test, failure of the welds was preceded by large, permanent ovalling of the joint.

During the tension tests on the elbow joints, the body ovalled slightly before failure. This initiated a tear in the heat affected zone near the welds which joined the elbow to the cross member.

Compression tests of the XL joints, resulted in buckling of the elbow on the side nearest the joint centerline and on both of the elbow sections having the applied load.

The TSX joint exhibited such strength that it was nearly rigid relative to the attached legs. Test 8 showed that the initiation of leg collapse in compression occurred at only 36% of the maximum joint load of the TSX model studied in test 9 without braces.

To introduce a failure load, further tests on the TX joints were conducted without braces. After making these adjustments in testing methodology, several TSX specimens were tested until observable ovalling was achieved.

Although dimensional deformations did occur in these joints, final failures were always limited to the heat affected zones of welded connections and never in the body of the cross. Attempts to force failures in the cross bodies were unsuccessful and suggest that this joint has far more static load carrying capability than the attaching pipe of equal wall thickness and strength.

Scale Effects

There has been considerable discussion in the literature on tubular joint design regarding the inherent size of scale effects of joint model tests. This has to do with the imperfect geometry of welds, such as ripple width and depth, pass width, and effective fillet radius.

These effects become more pronounced as the model scale decreases and the ripple depth itself becomes a stress riser in very small models.

It is felt that full-scale model tests and computer analysis should be seriously considered or implemented before finalizing a platform design. However, full-scale or even large scale model tests, as well as comprehensive computer analyses, are enormously expensive and would prohibit the investigation of all but a very limited number of parameters involved in new joint developments.

The scale effects in small model tests may lead to conservative results. More specifically, the stress concentration factor due to weld ripple depth decreases as the model size increases. This means that the measured maximum elastic stresses in a large model should be less than in a small model test.

It is suggested that the use of small model tests for preliminary design and evaluations be considered, especially when the results show a wide variation in joint stresses as a function of the type of joint, such as determined in these tests.

This view is tempered with the realization that parametric test results may show only small differences and the scale effects could very well mask any differences.

Joint Efficiency

Since the maximum joint stress is a function of the material strength, the joint stresses given in table 48 were normalized by dividing the joint stress by the yield strength of the particular model. The resulting yield factors are given in table 49.

TABLE 49
Yield factor and joint efficiency

Joint type	Yield factor, F_j/F_Y Tens	Comp	Avg.	Percent variation	Joint efficiency, $\frac{F_j/F_Y}{F_x/F_Y}$ Tens	Comp	Avg.	Percent variation
X	0.196	0.175	0.185	11	1.00	1.00	1.00	0
XLA	0.173	0.210	0.192	19	0.882	1.20	1.04	31
XLB	0.489	0.476	0.483	3	2.49	2.72	2.61	9
TSX	1.15	1.02	1.09	12	5.87	5.83	5.89	1
TRX	1.26	1.35	1.31	7	6.43	7.71	7.08	18
TDRX	1.24	1.49	1.37	18	6.33	8.51	7.41	29
AVG				12				15

The percent variation is one-half of the difference between the tension and compression test results divided by the tension-compression average. The average variation was only 12%.

The tension and compression tests would give identical results for perfect specimens provided the measurements were taken by sophisticated instrumentation. Admittedly, the specimens tested in this program are not perfect, and even with full scale models, one may expect a small experimental error.

Therefore, the scale effects could be relatively small. In view of this, the methods for determining elastic load limits provide satisfactory approximations.

The yield factors for the various joints studied were normalized against the yield factor for the fabricated cross joint, and the resulting joint efficiencies are given in table 49. The average variation in the joint efficiency was 15% and this can probably be viewed as an indication of the total inaccuracy from all sources.

The results show that the joint efficiency is about the same for the fabricated X and XL joints of the same diameter to thickness ratio, but when the elbows were thickened and the D/T for the joint was decreased from 36.5 to 21.8, the XL joint became 2.6 times stronger than the X joint.

The forged Tube Turns cross joint showed the greatest improvement over the fabricated X joint. The joint efficiencies were about 5.9, 7.1, and 7.4 for the TSX, TRX, and TDRX joints, respectively. This indicates that the difference between the joint efficiency of the split forged cross with a single stiffener plate and the forged cross with double internal reinforcement plates was not significant.

The result of tension and compression tests on tubular joints with six different geometries suggests that the vertical cross braces made from thickened forged elbows had a joint efficiency, or elastic load allowable, of roughly twice that of a field fabricated cross brace. The results achieved with the forged cross pipe fitting were even more encouraging.

The elastic load allowables for the standard forged cross were about six times greater than the regular fabricated cross brace, and the internally reinforced forged cross had an elastic load allowable nearly seven times that of a fabricated cross brace.

We have tried to provide platform designers with a simplified preliminary procedure for studying stresses in the joints of offshore structures and new approaches to the ever-present problem of efficiently completing tubular connections.

Earlier in the chapter, a capsulated overview of various design theories was shown. This information has been assimilated into concise formulas which should allow the platform designer to make quick and reasonably accurate assessments of a given structure's overall integrity.

It was shown that improvements in the basic building blocks of offshore structures, tubular joints, can be achieved with today's existing technology. The forged steel buttweld joints, developed and patented by Tube Turns, appear to offer several decided advantages over their fabricated counterparts. Their increased strength to weight ratio, improved compression and tensile properties, and reduced stress concentration characteristics suggest that these joints are technically feasible alternatives to fabricated joints.

The future areas of offshore exploration and development will require every effort to improve and refine design techniques. It is obvious that more information regarding the fatigue behavior of tubular joints would be helpful to anyone concerned with platform design.

3.6 FIXED-PLATFORM RISER DESIGN USING J TUBES

The advantages of the J-tube procedure is that it costs less in general, it can be installed very quickly, and is very reliable. The disadvantage of the J-tube method is that much more careful planning and organization is required compared to several other procedures for riser installation.

The principal methods for installing risers on fixed platforms are the L tubes, J tubes, reverse J, bending shoe, and special purpose procedures, such as flange and spool method, hydrocouple, and underwater welding. The L tube is formed by raising the end of the pipeline to or near the waterline using a derrick barge, and then lowering the pipeline as joints are stalked vertically.

The J tube is connected into the platforms during the time that the platform is being fabricated. To form the riser, a diver connects the messenger cable running through the J tube to the end of the pipeline lying on the mudline, or up onto the lay barge in the case of a pipeline originating at the platform. With a winch on the platform deck or by use of a derrick barge, the rope pulls the pipeline up through the J tube to deck level.

In the reverse J-tube procedure, the riser is formed by stalking pipe at the platform deck and letting the dead weight of the joint pipe initially push the pipe on through the curved portion of the J tube near the mudline. A crane barge or the lay barge pulls and supports the pipeline at or near the mudline until sufficient pipeline length is played out. The end of the pipeline is then raised to the waterline, joined to the pipe on the lay barge, and regular lay operations start.

The J-tube procedure can be used for initiating or for terminating lines. The reverse J is usually better for initiating lines, and it is sometimes better than the J for the large pipe sizes. For special situations, the J tube and the reverse J could be used for either initiating or terminating pipelines. However, with certain exceptions, special underwater couplings would have to be used to join the pipe ends or a derrick barge would lift the pipe ends, connect at the water line, and then lower back to the mudline.

The bending shoe method for forming a marine riser was developed by Shell and consists of fabrication of a curved guide installed on the platform jacket. The method is intended for terminating pipelines. Several ropes are attached to the pipeline at intermittent points, and the pipeline is raised from the mudline, made to conform to the shoe or guide, and then permanently coupled to the shoe.

There is no exact lower limit of water depth in using the J-tube procedure, but ordinarily it probably would not be considered for much less than 100 ft. Calculations show that the procedure could be used to advantage up to and over 1,200 ft water depth.

The J-tube riser procedure should be very well suited for use on gravity platforms where the riser must go through several bends from the mudline to the deck. Ordinarily, the pipeline goes through the first plastic deformation in the J tube at a short distance from the swedge on the lower end of the J tube. The pipeline then goes through a reverse plastic deformation at a point slightly above the tangent point of the straight and curved parts of the J tube. Since the work done in a plastic deformation is only one of many force components in the total winch force, extra bends can be accommodated with reasonable increases of required winch capacity.

Design Procedures

It would seem that pulling a pipe through a tube ought to be a simple matter. It is, but it usually takes a little more engineering to make it this way. A few of the considerations for the design of a J-tube type of riser are described in the next few paragraphs.

The riser portion of the pipeline is bare steel. With the relatively sharp radius in the J tube, a concrete coating would crack and spall, and a mastic coating would get rubbed off and gummed up in the J tube. Thin films would wear off in small areas giving rise to corrosion concentration points. A rubber stopper is sometimes installed between the bare pipe and the concrete coated pipe for the purpose of jointing in the swedge. This prevents oxygen rich water from circulating in the J tube which would cause excessive corrosion. The usual procedure then is to pull the pipeline until the concrete coated pipe enters the swedge. For corrosion control, the annulus between the riser pipe and the J tube is sometimes filled with a mud having suitable corrosion inhibitors after the pulling operation is completed.

If the diameter-to-thickness ratio, D/t, of the riser is very small, then excessive winch forces to plastically deform the pipe are required, and platform bracing of the J tube would also become a greater problem. If D/t is too large, the pipeline would float, and the probability of buckling the riser during plastic deformation becomes too great.

The best value of specific gravity, SG, of the pipeline to use is a matter of engineering judgment. The SG obviously should be greater than unity, but values very much larger than unity might be a waste of material and money.

Pipeline buckling strength is a function of: the combined effects of the bending strain or moment as the line is plastically deformed in the tube; the tensile axial forces in the line; the hydrostatic head of the water, especially in deep water (figure 3.43), the yield strength of the material; the D/t ratio; and the SG selected. While the tensile axial force in the line during pulling increases the buckling allowable, this decreases the energy required for plastic deformation of the pipeline in the J tube (figure 3.44). The energy reduction in plastic deformation during the bending of the riser due to the tensile axial force is a function of the material properties (ratio of ultimate strength to yield strength) and the geometrical properties (ratio of cross sectional area to moment of inertia) of the pipeline.

The pulling force to plastically deform the riser tube can be calculated fairly easily. This is done by relating the required bending moment to the plastic strain level, which is a function of the J tube curvature. Using the theory of plasticity for an idealized stress-strain curve, the moment to plastically deform the pipe over its whole cross

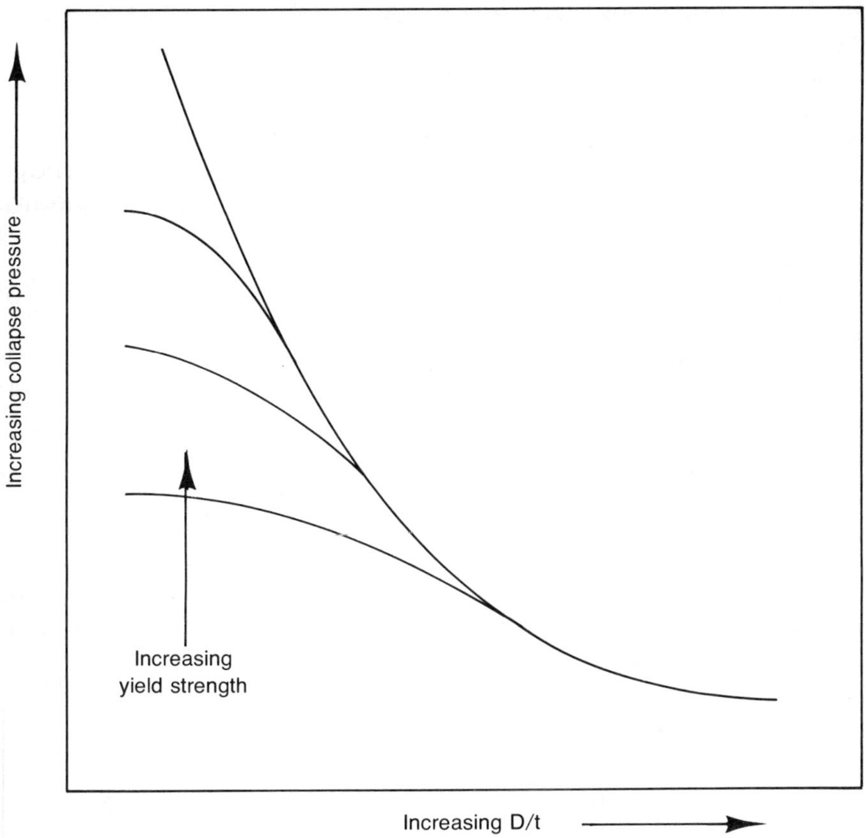

FIG. 3.43 Effect of yield strength and D/t on buckling due to hydrostatic pressure

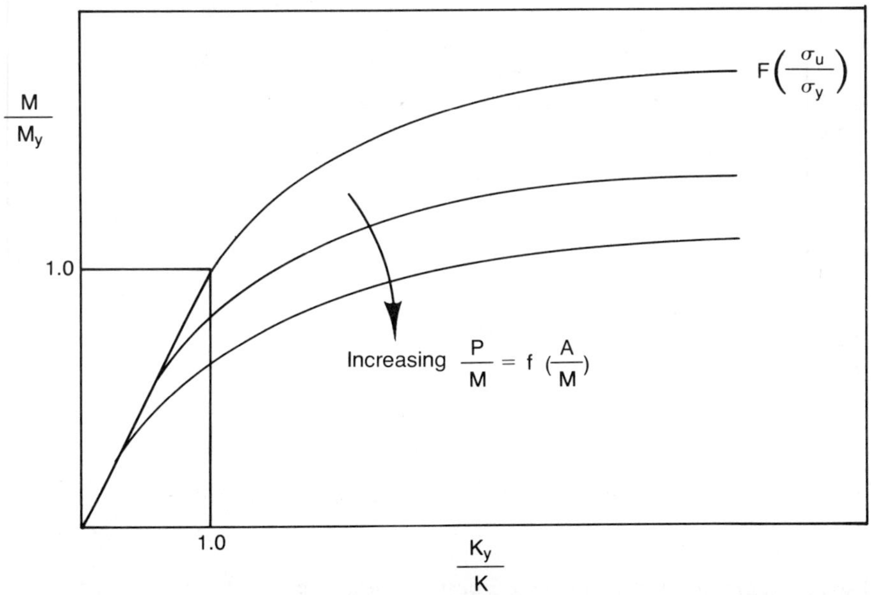

FIG. 3.44 Influence of axial force on plastic pipeline moment

section is only about 27% greater than the moment at the yield point, My, which is lower than for several other types of structural shapes (figure 3.45). This procedure is satisfactory for shallow to medium water depths. For deep water design, say over 400 ft water depth, the actual stress-strain curve of the material should be used to construct the design curve of plastic moment versus radius of curvature, K. Alternately, the design curves can be prepared using the curvature, ϕ, or strain, E. If these design curves are not used, especially with low carbon steels, the pulling force can be severely underestimated.

The radius of the J tube bend depends upon a number of factors such as buckling consideration of the pipelines, selected location of the upper end of the riser with respect to the deck layout, location of braces in the jacket, and other considerations. In general, the larger the radius of the J-tube bend, the less chance for buckling. Then, however, bracing a large bend J tube can become a problem.

All of the above information is consolidated, and the component forces during pulling are plotted to get the maximum required winch forces. The pipe might be pulled at anywhere between 1 and 50 ft/min. Since sliding friction is less than the static friction, an attempt

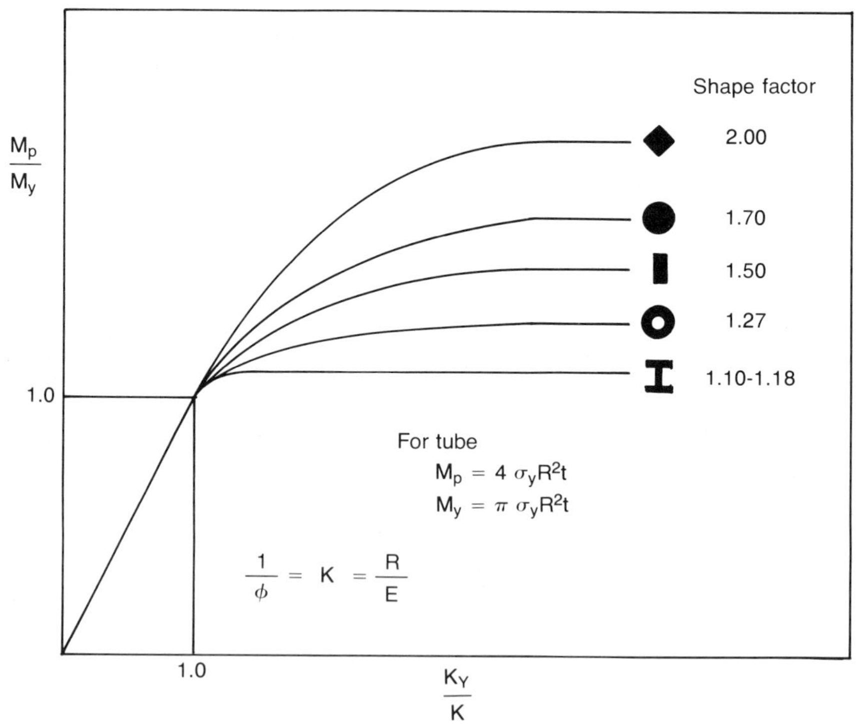

FIG. 3.45 Section shape factors

should be made to make the pipe pull as continuous as possible. Figure 3.46 shows the influence lines for the static and sliding conditions. This enables good control on pulling force when the pull is stopped for any reason and then resumed.

The method for analyzing the forces due to soil resistance depends on the particular situation. For hard bottom, a friction coefficient applied with the submerged weight per foot of the pipeline can be used. For very soft bottoms where the pipeline may have

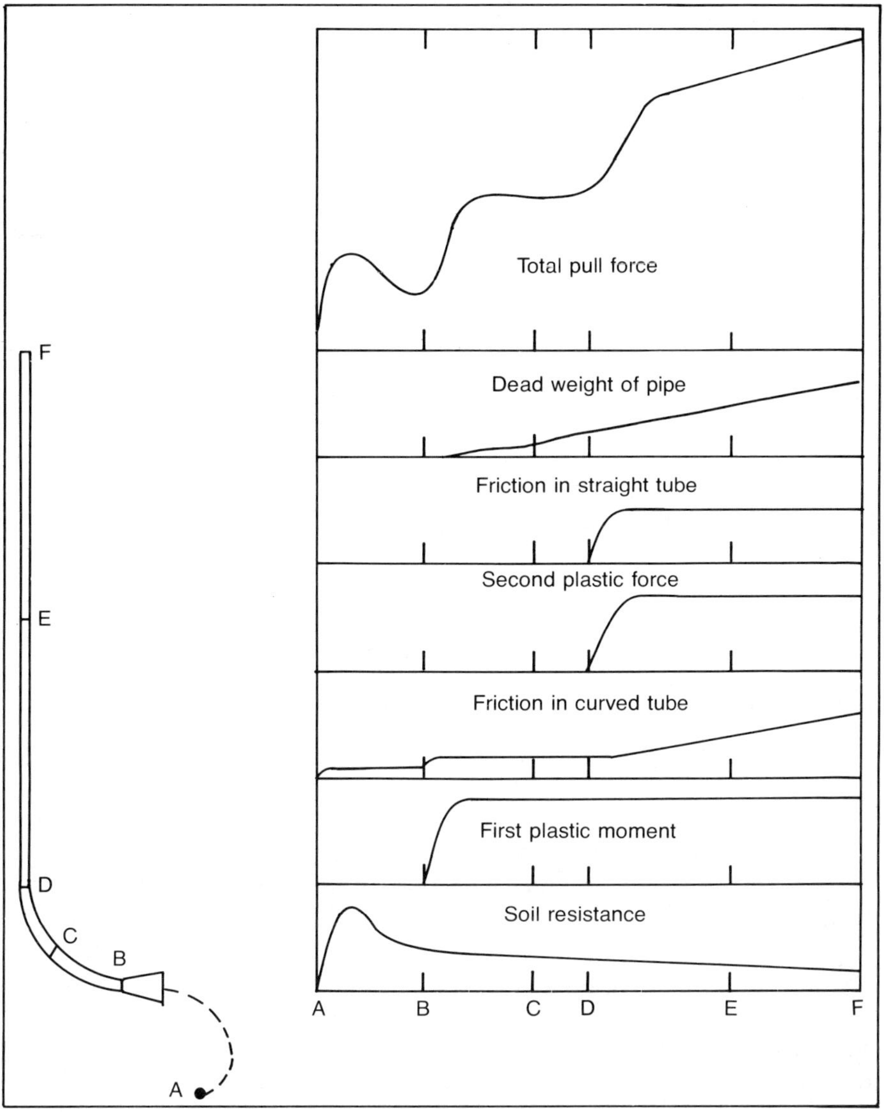

FIG. 3.46 Summary of pull forces as pipe goes through tube

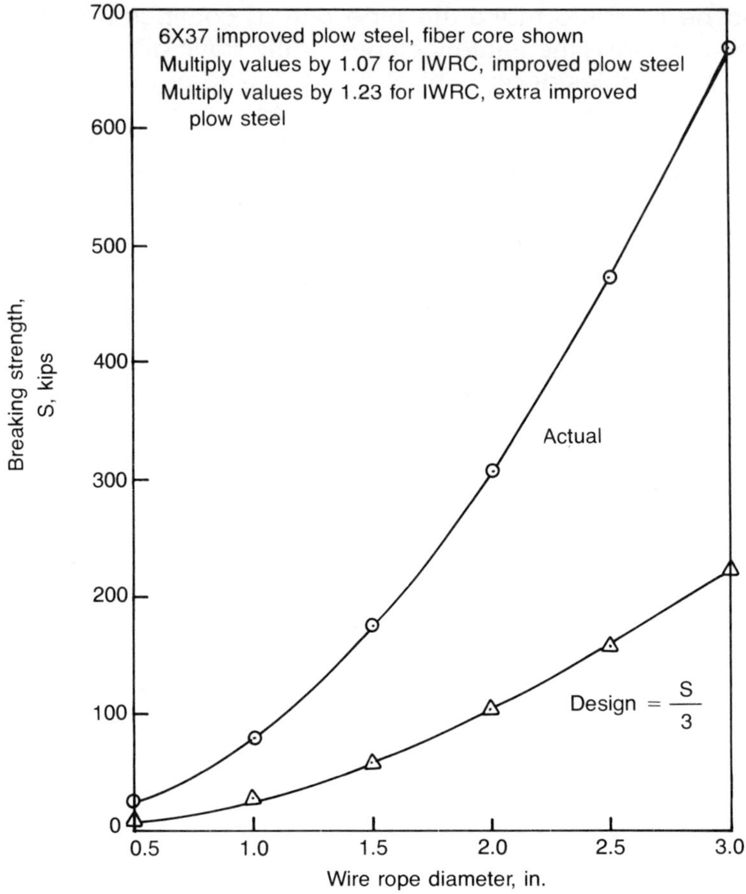

FIG. 3.47 Strength of pulling rope

settled considerably, the shear or plowing of the soil by the pipeline loop on the mudline should be calculated.

The final step is sizing the pulling winch and the wire rope (figure 3.47). To insure reliability, the rope should be sized so that the maximum force seen by the rope is roughly one-third of its breaking strength.

Operational Considerations

Operational factors, of course, dictate the direction to a large extent, and operational factors must be considered during the planning before the design work and during construction. As stated earlier, the disadvantages of the J-tube riser procedure is that so much planning is required. Then why bother?

In a nutshell, the J-tube riser can be installed in a few hours and in less than a day for water depths of 100 to 1000 ft with the proper planning. Installing an L tube or connecting underwater flanges or performing underwater welding in several hundred feet of water could require from several days to several weeks. Further, with proper design and planning, the J-tube riser is a lead pipe cinch, that is, as certain as anything can be.

But back to the planning. Some of the matrix of questions that must be answered during the planning stages follow:

Will there be short or long flowlines? For terminating lines, a loop of pipeline must be left on bottom. With many short lines, too much time and material could be wasted in the loops.

Which direction should the J tube be oriented? If this cannot be answered, can the cost of several J tubes with different orientations fabricated into the jacket be justified in a contingency plan?

What is the probability or availability of a derrick barge working nearby on the date of riser installation? If uncertain, is there sufficient deck area available for a pulling winch and other necessary supplies?

How will precautionary measures be taken to keep the J tube clear? If the header on the J tube is not sealed, junk might fall in and have to be fished out before pulling.

Who will be responsible for coordination between production and transportation, and how will it be handled? Will the activity associated with the installation of the riser effect drilling operations, and if so, how?

Should the J tube be sheltered by the jacket structure, and if so, how far below the waterline? (That is, how much of the J tube should be inside and outside of the jacket structure.) This means determining the remote probabilities of workboats, ships, or semisubmersibles plowing into the platform during storms.

What is the exact location and angle of the swedge with respect to the mudline? In determining this, possible shifting of the mudline, mudslides, and settling of the platform must be considered.

Since the J tube is built into the platform structure, the "breakaway" strength is usually very high and is in general better than with an L tube. If every law of the sea were violated, such as a ship dragging anchor on the pipeline at or near the J-tube swedge, then what should break out for minimum platform damage?

Conclusion

The J-tube riser procedure is extremely reliable when properly planned and executed.

As discussed and indicated above, it may not be the best procedure for some situations. The J-tube procedure among others should always be considered during the early planning stages of a new platform.

3.7 PLATFORM SAFETY BARRIER DESIGN

The barrier is used to protect marine offshore platforms from falling objects. It especially relates to protecting the drill floor by preventing falling objects such as drill collars, pipe, and the like from penetrating the drill floor and injuring people or damaging equipment in the substructure, and to protecting wellheads, especially Christmas trees over producing wells, from falling objects. The term "wellhead" as used herein includes the Christmas tree which also includes the master valves, tees, swab valves, and wing valves which sit on the casing spool and hanging equipment. The barrier particularly concerns placing of a safety barrier between the wellhead and/or interior area of an offshore structure so as to protect the offshore structure interior portions or wellhead from any falling debris.

As the world supply of hydrocarbons has dwindled, more and more attention has been given to searching for liquid hydrocarbons beneath marine bodies of water, necessitating the construction of larger and larger offshore platforms. In recent years especially, there has been considerable attention directed to the drilling and production of oil and gas wells located in water. These wells are drilled in the ocean floor from either fixed platforms or from floating structures or vessels. Currently, fixed platforms are the preferred means, which requires driving long piles into the ocean floor, erecting a working area above, drilling wells, and then producing through the wells from the oil deposits. When using such fixed platforms, the drilling rig is set on the platform above the surface of the water and the wells are drilled, normally using dry land techniques with some adaptations. After the well is drilled and completed, producing equipment separators, wellheads and the like are located on the platform itself above the surface of the water.

Usually, these structures contain many wells drilled from one offshore structure. As is well-known in the industry, it is a tremendous waste of time and capital to have a structure capable of having up to 30 wells and drilling all 30 wells before production is begun. Therefore, it is desirable that as soon as one well has been drilled to a reservoir, that it begin producing while other wells are concurrently drilled.

One problem with this desirable operation is the danger of damage to the wellheads by fire or by blowouts when drilling equipment is dropped on the Christmas trees at the head of producing wells. This is especially true where the Christmas trees are located on the platform itself or are only immediately below the surface in the area where much activity in the drilling phase of production is underway. In several instances, dropped objects have severely damaged producing wells, releasing an enormous amount of flammable hydrocarbons into the vicinity of the platform and adding to the danger of fire through sparks caused by falling metallic objects.

Fires which occur on platforms themselves will burn themselves out as soon as the fuel is exhausted unless the fire is fed by leaking wellheads. Such a state is obviously undesirable since damage to equipment and danger to personnel is enormous. In addition, the fire can result in wells producing in an uncontrolled state. Such fires can continue for weeks until the wells are killed by drilling directional wells for bottom hole control or are otherwise brought under control. Such well control is an enormous expense and waste of natural resources, in view of the time and expense it takes to stop the fire as well as the production lost during the fire.

At the present time there is a reluctance to include normal production equipment on platforms having producing wellheads; that is, there is a reluctance to drill and produce from the same platform. This reluctance is based on safety considerations, primarily to avoid exposing the producing wellheads to fire caused by falling objects or debris. This problem exists since drilling techniques require much overhead movement of heavy drill collars, pipe stems and other heavy objects. Since storage on an offshore platform is limited, these objects must constantly be moved over producing wellheads and personnel area. Lloyds of London and the United States Department of Energy have both consistently rated this activity as extremely high risk. The alternative to concurrent drilling and production is the use of two platforms side by side, with added expense and time or the cessation of production until all drilling is complete, again adding enormously to the time before production is begun, as well as lost income from the nonproducing wells.

One method of preventing falling debris from penetrating to the interior of a platform is simply to build the platform of materials so thick so as to be able to totally prevent penetration by falling objects. However, it is well known that on offshore platforms, excess weight is undesirable since added weight is added cost and expense of maintenance. Some form of protection is necessary, for example, in the North Sea, since codes and regulation require inherently safe simultaneous drilling and producing operations, which implies that inadvertant release of drill collars or other heavy objects of the like from the drilling rig not penetrate the drill floor or wellhead area of the platform. Since steel of sufficient thickness to prevent penetration is extremely heavy and very expensive, current methods of solving the problem primarily center on the use of wood timbers as a crash barrier. However, these wooden barriers are inadequate since they also add excessive weight to the structure. Wood timbers up to 18 inches in thickness are necessary. In addition, these timbers must be supported, which requires additional structure and thus more weight. As yet another negative aspect of wooden timbers, they tend to fuel a fire and hold glowing embers so that fighting a fire is more difficult.

It would be of great benefit to provide a lightweight, penetration resistant barrier which would protect the drilling area, wellhead area, and other interior portions of offshore platforms.

The object of the platform safety barrier is to provide a suitable safety barrier for offshore platforms. It provides a lightweight noncorrosive safety barrier having an extremely high capacity for shock absorption and high probability for precluding penetration of falling objects.

The barrier is an offshore structure safety barrier to prevent falling objects from penetrating horizontal surfaces comprising a penetration-resistant layer overlying an impact absorbing material of sufficient thickness to absorb the anticipated impact energy. Normally, this safety barrier will be inserted in sections beneath the horizontal surface to be protected, in the structural area normally between the I-beams and will be held in place either by bolting to the I-beams or by placing a second impact-containing layer beneath the I-beams so as to absorb the force of the blow when dissipated through the safety barrier. The safety barrier absorbs impact by deforming a rubber material or allowing a thermoplastic foam material to cave, while the penetration-resistant barrier at the upper edge of the horizontal safety barrier prevents penetration into interior spaces.

Detailed Description of the Drawings

Figure 3.48, an isometric view of an offshore structure shows the platform itself, which is supported on jacket legs with piles driven through the legs or through skirts into an ocean bed. The platform contains cranes which are used to move heavy objects from a storage area to a position where said objects can be utilized by a drilling tower. The drilling tower is movably supported on two sets of tracks which allows positioning of the tower and the substructure over any one of the multiple wellheads. As shown in the figures, several of the wellheads have Christmas trees indicating that they are producing wells, whereas some of the wellheads are undergoing drilling activity. Casing is shown descending into the sea floor and

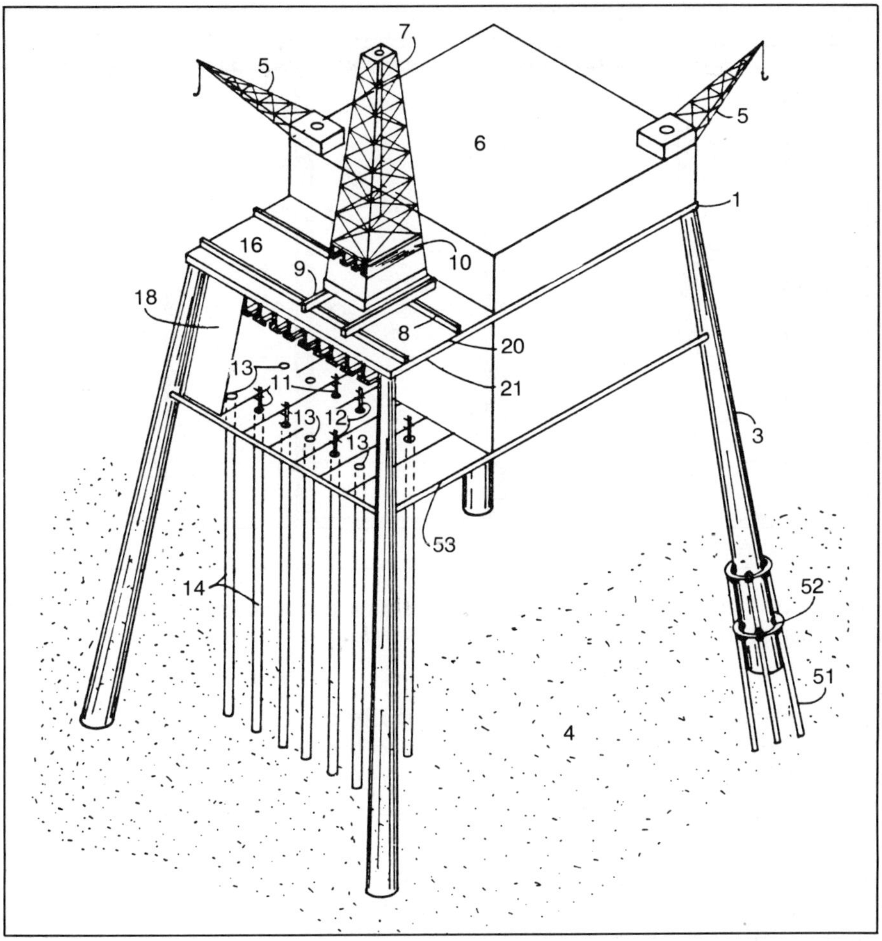

FIG. 3.48 Offshore structure with drilling activity and producing wells

thence into the oil reservoirs. Access from the drilling tower to the wellheads is provided through the horizontal deck. Usually such access is through removable panels, one for each wellhead so the other wellheads are protected during drilling activity. Vertical structural panels separate wells with panels on each of the four vertical sides around the wellhead. The height of the vertical walls around each wellhead from the cellar level up to the platform level typically varies from 18 to 38 ft.

Figure 3.49 is a cross-sectional view between two I-beams showing the upper deck and the lower level of the crash barrier. The steel plates both above and below the I-beam structure are held in

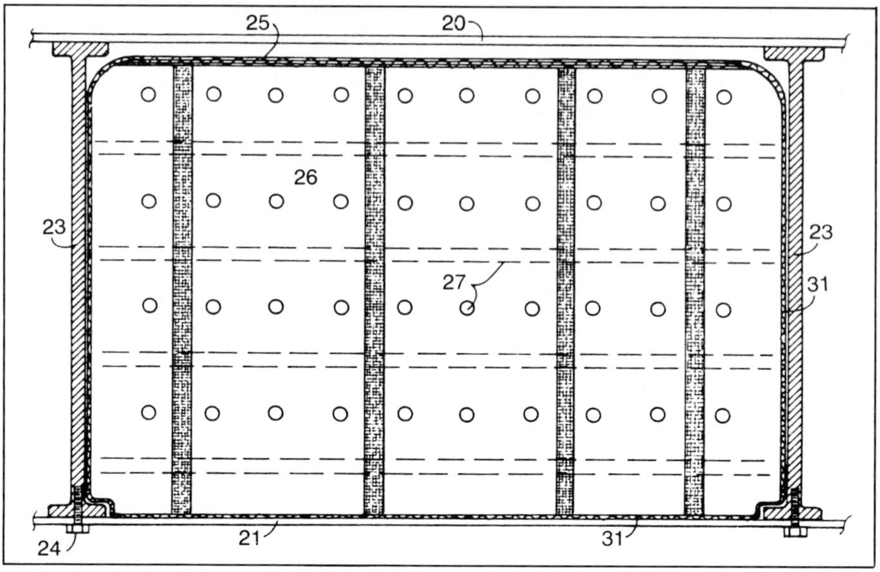

FIG. 3.49 Cross section of the interior of a horizontal surface between two structural members with safety barrier

place by any convenient means, usually by fastening bolts or welding or any other means well-known to those skilled in this art.

The safety barrier itself can be of any convenient size sufficient to withstand anticipated impact. For example, the I-beams, when in position as shown for offshore platform use, can be any depth up to 3 or 4 ft. Thus the thickness of the barrier contained therein need only be sufficient to withstand anticipated impact and need not necessarily fill the entire space.

In the particular mode shown in figure 3.49, a penetration resistant layer is placed over an impact absorbing layer. Apertures throughout the impact absorbing layer are provided to allow for some deformation of the absorbing layer. The penetration resistant layer if flexible, is optionally attached to the I-beams or if rigid, overlies the impact absorbing layer so as to spread the course of the impact over the entire layer. The entire safety barrier is supported by floor which easily contains the total force exerted on it since the force is spread by the safety barrier over the entire surface of the floor.

Figure 3.50 shows the safety barrier under maximum stress exerted by a falling heavy object. As the figure represents, the falling object has penetrated the upper level and has encountered the penetration-resistant barrier. In the figure, the penetration-resistance barrier is a flexible fabric which covers the entire upper surface of the impact absorbing layer and can be connected by straps to effec-

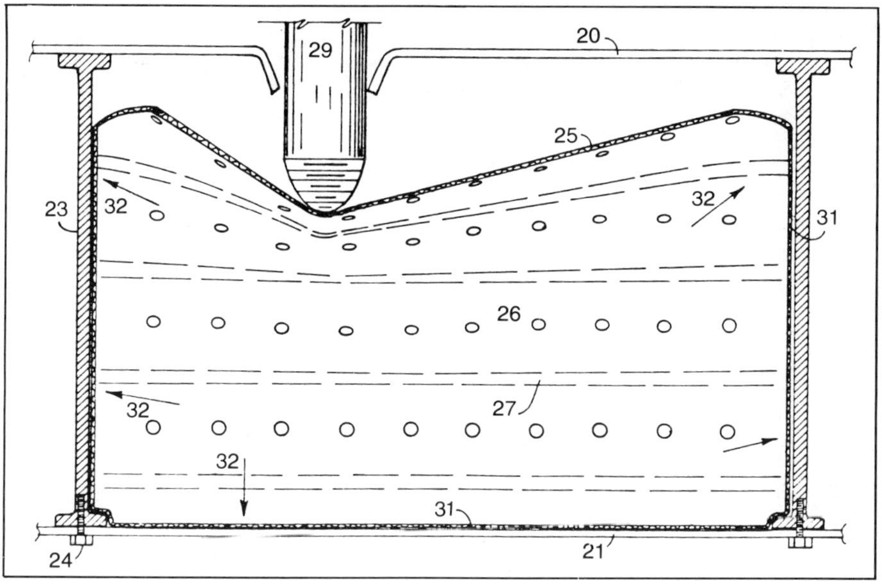

FIG. 3.50 Barrier in operation

tively surround the impact absorbent material. As the object descends further into the impact absorbing material, the penetration resistant material prevents the object from penetrating into the impact absorbing material. The deformation holes are compressed at the point of impact and the force of the falling object is transmitted equally in all directions away from the point of impact as indicated by arrows.

Once the falling object has been stopped and before it penetrates the lower level, it is removed and the upper deck is repaired. The safety barrier returns to its original configuration if it is an elastomer, or the penetration-resistant barrier is replaced if necessary, as when a thermoplastic foam is used.

If a foamed thermoplastic impact-absorbing layer is used or if a reinforced matting penetration-resistant barrier is used, replacement of the entire safety barrier would be necessary, since the thermoplastic foam successively caves as the object approaches the bottom level and no reuse is possible. This embodiment is shown in figure 3.51 wherein a rebar penetration-resistant barrier is shown overlying a fiber-reinforced rubber mat, which in turn overlies a foamed thermoplastic. The fiber reinforcing the rubber mat can be either wire or nonmetallic fibers. The foamed thermoplastic is lightweight and requires no deformation cavities. The rubber mat reinforced with fabric can be scrap sidewalls or scrap tires.

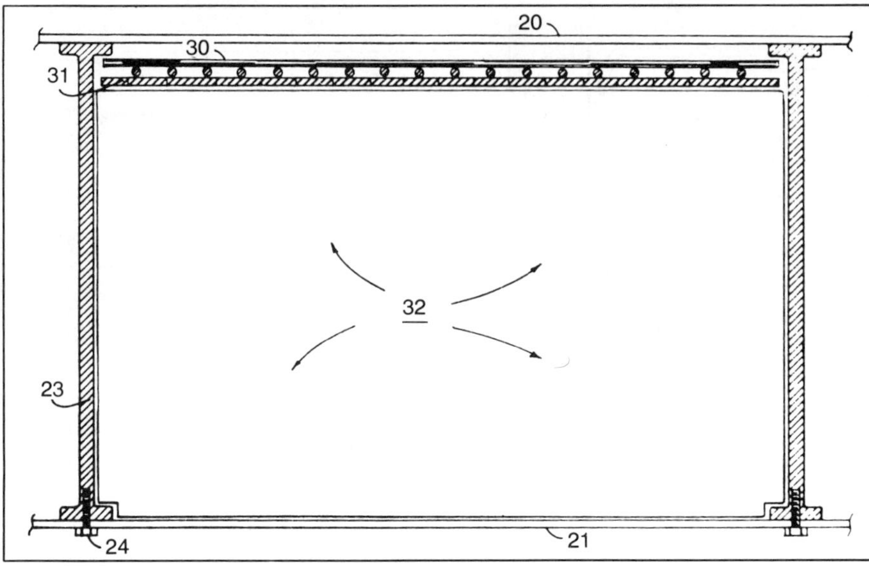

FIG. 3.51 **Use of readily available inexpensive materials**

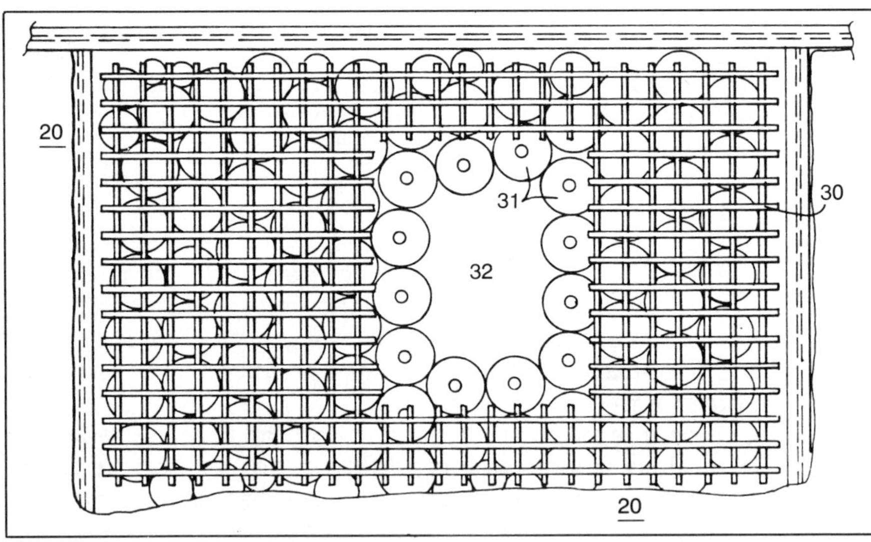

FIG. 3.52 **Top sectional view of figure 3.51**

Figure 3.52 is a top sectional view of figure 3.51, showing placement of the fiber reinforced rubber sidewalls situated so as to form additional penetration resistance and underlying the rebar grid. The sidewalls can be attached to the steel grid.

A fairly broad choice of materials is available for both the penetration-resistant barrier and for the impact absorbing layer. For example, the penetration-resistant barrier can be a network of rebar which overlies the impact-absorbing material. Since grids of steel reinforcement are so commonly used in construction with Portland concrete, the rebar is widely available and would be relatively inexpensive and its use could be immediate and easy to facilitate. However, more preferred but much more expensive than rebar would be materials specifically made for this purpose, such as a rubber mat reinforced with steel wire or high impact fibers such as fiberglass, nylon, polyester, steel, or aramid fibers or combinations of these. However, most preferred, but again more expensive, is a mat comprising several layers of penetration-resistant fabric already well known for use in safety vests or flack jackets. Such a penetration-resistant barrier would be comprised of several layers of aramid or nylon fiber which is extremely light in weight but resists penetration to a surprising extent.

Considering lowest cost, minimum weight, use of readily available materials, ease of fabrication, and installation with finished prefabricated panels, the preferred embodiment of this invention is a foamed plastic shock absorbing layer overlaid with one layer of sidewall from scrap tires. All tires consist of rubber reinforced with fiber such as nylon, rayon, polyester, or steel. This embodiment could be prefabricated sandwich panels with the penetration resistant layer of tire sidewalls and foamed plastic with steel plate over the tires and under the foamed plastic. Alternatively, the tire sidewalls can be tied in place, the steel plates welded in, and the plastic foamed in place in the field for use on existing installations.

The load can be made to spread over a wider area and crush more plastic by putting a grid of steel wire or rods over the scrap tire sidewalls. In either case, the scrap tire layer is required since without this layer, the drill collar would penetrate the upper steel cover plate, put the load on the steel grid, and the grid would simply shear through the foamed plastic or rubber without deforming the whole cross sectional area of the shock absorbing material. Figure 3.52 is a top view of figure 3.51 showing displacement of scrap tire sidewalls.

Designs for the safety barrier described here is based on the heaviest objects subject to drop and impact which must be absorbed. Usually the heaviest objects which are subject to dropping, with subsequent damage to the platform, are treble tiers of drill collars and casing slips. Although the size of such objects are subject to some variation, an average size for drill collars is the 9½ in. collar (3½-in. inside diameter, 7⅛ drill-collar pin, 217 lbs/ft, 31 ft long/collar), while casing slips average about 14 in. and weigh approximately 5000 lbs. These objects, when falling from maximum height, would approximate the data in table 50.

TABLE 50

	Collar	Slip
Weight, kip (1,000) lbs)	20	5
Drop height, feet	10	50
Impact velocity, feet/second	25.4	56.7
Momentum, mass × velocity	508	284
Barrier depth, feet	3	2
Barrier crush, feet	2	1
Impact deacceleration, ft/sec^2	161	1607
Dynamic force factor, numeric	5	50
Maximum dynamic force, kip	100	250
Deacceleration work, ft.-kip	100	125
Impact area, inches^2	70.8	1100
Absorber crush (true strain)	0.51	0.41
Minimum barrier modulus, psi	5400	1800

The impact absorbing layer can be any one of a number of readily available materials. Representative examples of such materials are natural rubber, synthetic rubber, mixtures of natural and synthetic rubber, polystyrene foam, polyurethane foam, and other foamed thermoplastic materials. These materials can easily contain flame retardant materials well-known to those skilled in thermoplastic art, which would prevent these lightweight safety barriers from adding any hazard to a fire encountered on an offshore structure.

The modulus of the barrier will vary with choice of materials used, but the material chosen should be in certain ranges for maximum effect. For collars, the modulus should be from about 4,000 to about 7,000, and for slips from about 1,000 to about 2,000, depending on the absorber thickness.

When the impact absorbing material is a foamed thermoplastic, polystyrene or polyurethane foams are preferred. These foams will normally be present in a layer of from about 0.5 to about 4 ft in thickness and have a density of from about 5 to about 15 lb/cu ft. When foamed thermoplastic materials are used as the impact absorbing layer, no void areas are necessary as the foamed areas themselves contain sufficient voids to allow compression of the foam during absorption of the impact. These materials successively "cave" as the falling object is slowed to a stop.

When natural or synthetic rubbers or mixtures of these are used as the impact absorbing material, it is necessary to provide void areas as shown in figures 3.49 and 3.50 to allow for deformation of the elastomer, since these materials are essentially incompressible. However, these materials have astoundingly high impact absorption properties. Comparing rubber to foamed plastic for the energy absorbing layer, foamed plastics would cost less and would weigh less than foam rubber. These materials must contain sufficient void area to allow for the deformation which would be caused by the anticipated impact. When these elastomers are used, the impact absorbing layer need not be as thick as when foamed thermoplastics are used and normally will range from about 2 in. to about 2 ft in thickness.

Thus, it can be seen that a lightweight, economical, easily replaceable safety barrier is provided for offshore structures. This safety barrier is easily placed so as not to require valuable floor space and prevents falling objects from penetrating horizontal surfaces. The invention is applicable not only to drilling floors, wellhead areas, but also personnel and maintenance areas such as turbines, water and food supplies, and so forth.

While certain embodiments and details have been shown, it will be apparent that various changes and modifications may be made without departing from the spirit or the scope of the invention.

3.8 PLATFORM CRANE SYSTEMS DESIGN

Procedures for crane design for offshore fixed platforms are difficult to cope with because of the various factors which must be considered. The objectives are to serve the function, to do this at economical costs, and to make the operation safe from the standpoints of both life and preservation of the capital investment.

The integrity of all of the crane components taken as components and as a unit or of the system design must be carefully evaluated. The system design gets complicated because several groups or companies are ordinarily involved in design and manufacturing of the systems components. The major components include the rope, the boom, the crane structure, crane controls, the pedestal, and the pedestal foundation on the platform.

Controls for Safe Design

A number of platform cranes have torn loose at the pedestal or at the crane foundation with overloads and have fallen into the sea with subsequent loss of life and equipment. Besides these losses, the logistics at the platform and continuing operations can be jeopardized.

The overloads are generally caused by attempts to lift weights exceeding the rated crane design capacity. Less frequently (but still often enough to warrant consideration), the ropes or the hook can get snagged on the platform itself as the line is reeved in during hook retraction. There have also been cases where the boat was inadvertently snagged resulting in a crane overload. And there have been reports of the boat running out to sea while unaware of a crane wire snag.

These situations have led to better and safer controls than before, but the controls are still inadequate. Besides the overload alarm systems, some cranes now have an automatic clutching mechanism that engages when the dynamic capacity of the crane is reached. This causes a release of the load at a speed in proportion to the percent of overload.

These safety systems, however, are of questionable value for a snagged boat running out to sea. When all of the wire is finally unwound from the drum, there is a high probability that the rope end will whiplash causing the rope to fix itself to the structural components of the boom.

To handle all of these contingencies, a safety link should be installed above the hook. The reliability of such a passive system would also be greater than the more expensive available active systems.

Sources of Critical Failure

The common crane failure modes are illustrated in figure 3.53 where the order of failure is in the preferred sequence. That is, the component to fail first should be the rope breaking, then the boom yielding, then the pedestal yielding, then the boom buckling, and so on. The boom still has strength after buckling. Boom failure here indicates the loads when the boom actually comes apart after buckling. This assumes a normal design of buckling factors greater than the yield strength but less than the ultimate strength of the materials.

The bull ring usually has adequate safety factors in the design since the amount of steel in the tie back is small relative to the other crane components. The cost is therefore of little consequence. Except for the bull ring, the pedestal foundation should have the largest design factor so that the foundation is the least likely and is the last component to experience a possible failure in the system.

The basic problem in the safe design of a crane system is the applied structural design factors and their relation to the usable strength of the wire rope. For a given maximum design load, static hook capacity is normally taken as one-third of the breaking

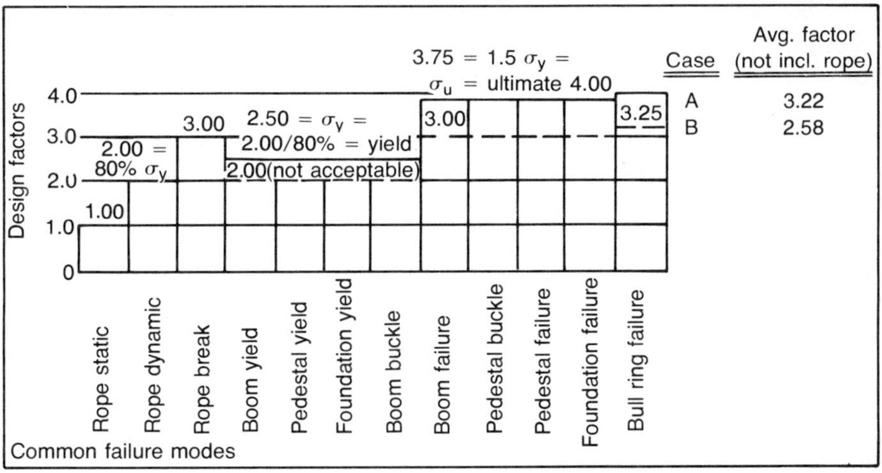

FIG. 3.53 Factors in crane design practice

strength of the rope. The dynamic factor of two is usually considered in the design. So the rope strength is still one and a half times greater than the maximum dynamic hook load ever expected.

Crane Design Practice

The 12 primary load factors for consideration in the crane design system are illustrated in figure 3.53. This shows that for minimum crane design, and the least expensive, one would take the maximum structural stress equal to the material yield strength with the maximum dynamic load causing stresses at the verge of yield. Low strength steels are usually used for these structures so that welding repairs and installation can be easily accommodated in the field. The ultimate strength for many low carbon steels is very close to one and a half times the yield strength. Theoretically, the safety factor during plastic response is then 50%. The problem with this is that ultimate strength of the structure equals the breaking strength of the rope giving an equal probability for rope break or ultimate failure of the boom, the pedestal, or the foundation.

Kicking the design factor up to 2.5 on the rated static crane capacity gives an ultimate factor on the structural members of 3.75, or 25% greater than the rope strength. The problem is that such an apparently safe structure may not truly have the expected theoretical margin of safety. The plastic response of a boom framework or the pedestal between the yield and the ultimate, especially after buckling, is extremely difficult to approximate and may be impossible to accurately predict. Therefore, graduated safety should be reviewed.

Since the safety factor of the structure is roughly proportional to weight, which is, in turn, proportional to cost, the average of the 9 primary design factors (excluding the rope) can be used as a good indication of crane cost. Thus, the average factors for the above 2 cases were 3.22 and 2.50, that is, raising the design factor 25% increased the cost by about the same percentage. But this linearity does not always hold as we will show.

Graduated Safety Design

The inherent difficulties in reliable enough theoretical procedures for plastic design with buckling leads to a consideration of graduated safety. That is, a fixed safety factor could be applied to each component in the crane system. This is shown in figure 3.54 and vividly illustrates the compounding effect of even small fixed design factors.

More explicitly, our definition of fixed applied safety factors means that with a factor of 1.25, for example, the boom yields at 25%

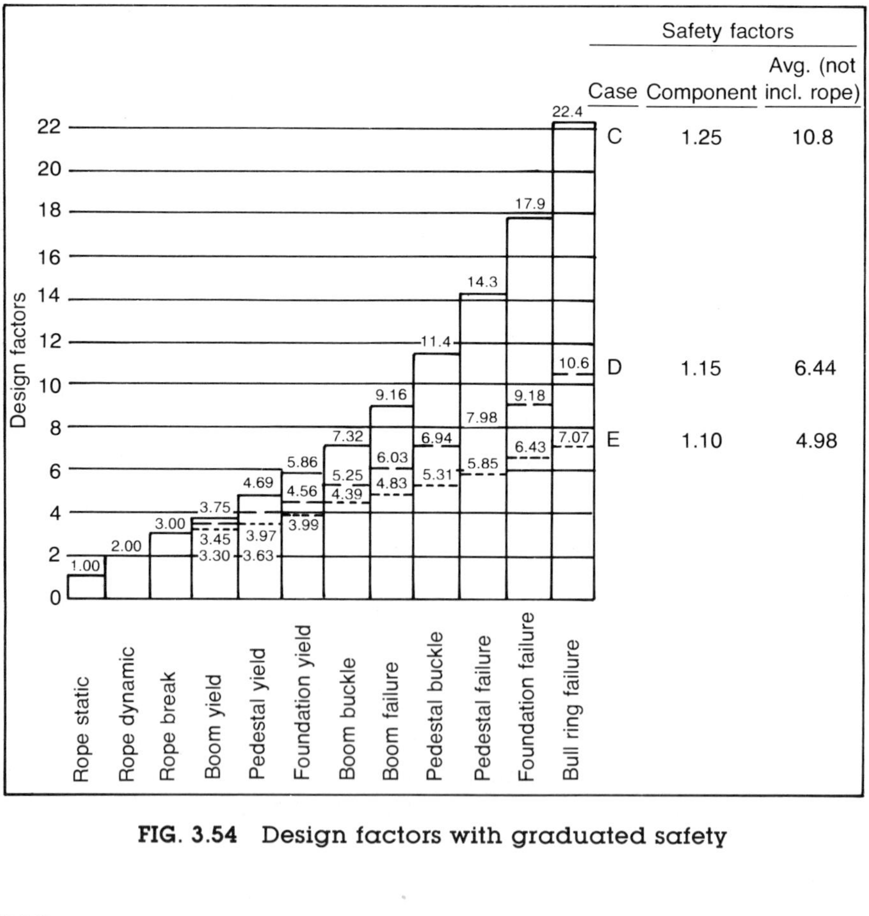

FIG. 3.54 Design factors with graduated safety

greater load than rope break, the pedestal yield at 25% greater load than the load at which the boom starts to yield, and so on. Three cases were studied using fixed component factors of 1.25, 1.15, and 1.10. This shows how the resulting weight can become unfeasible even with small fixed factors. For example, with the successively applied 15% factor, the safety factor on foundation failure would be 9.18 times the rated static hook load.

For the fixed factors of 1.25, 1.15, and 1.10, the resulting average system factors were 10.8, 6.44, and 4.98, respectively. Comparing the lowest average of 4.98 with a common design practice using 25% as in figure 3.53 with an average of 3.22 shows that the minimal practical graduated safety of 10% costs about 4.98/3.22 = 1.55, or 55% more. Use of a 15% graduated safety would cost about 6.44/3.22 = 2.00, or roughly twice the cost of using the 25% factor in figure 3.53. The results of these unfavorable cost comparisons provide incentives for reviewing other design approaches.

Systems Design with Safety Link

Rather than compound the safety factors on each component of the crane system, the optimum approach would be to use a safety link above the crane hook designed in such a way that the safety link would be weaker than the other components. The effects of this are illustrated in figure 3.55 for the safety link set variously at values of 2.0,

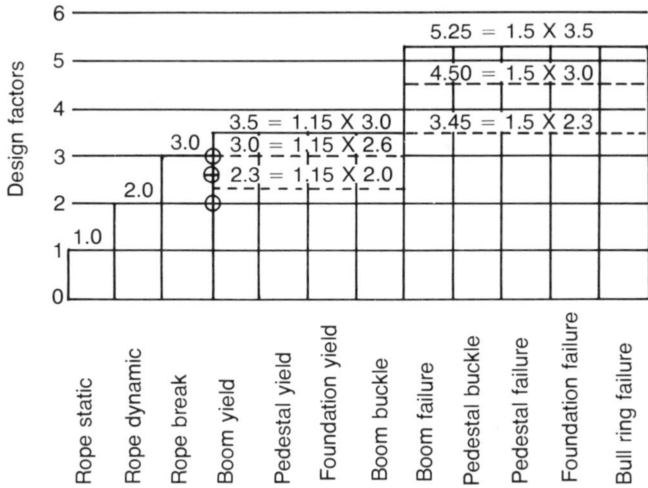

Safety factors		
Case	Set Safety Link	Average (not incl. rope)
F	3.0	4.47
G	2.6	3.83
H	2.0	2.94

FIG. 3.55 Design with safety link and fixed safety factors

2.6, and 3.0 times the rated static hook capacity. With a 15% margin of safety on yield strength applied to the structural members, the corresponding structural safety factors would be 2.3, 3.0, and 3.5. The structural safety factors on ultimate would be 3.45, 4.50, and 5.25, respectively.

The chart shows that the three corresponding average factors would be 2.94, 3.83, and 4.47. Comparing these averages with Case A in figure 3.53 shows that the system with a safety link could cost 13% less when set equal to the dynamic rating, about 39% more when set equal to the breaking strength of the rope, and 19% more when set at 2.6 times the static hook rating. It is quite significant that the overall structural integrity of the system and the operating safety could be improved at equal or reduced costs, depending upon the extent above dynamic crane rating chosen.

Aspects of Safety Link Design

We are not aware of commercially available, passive safety links for the crane hook. However, it would seem that a mechanical link should use a pin in shear rather than a so called weak link in tension because most of the allowable elongation is between the yield strength and the ultimate strength with no significant permanent deformation up to yield strength. Further, the fatigue strength is drastically reduced for loadings between the yield and the ultimate strength. On the other hand, shear failures are characterized by very

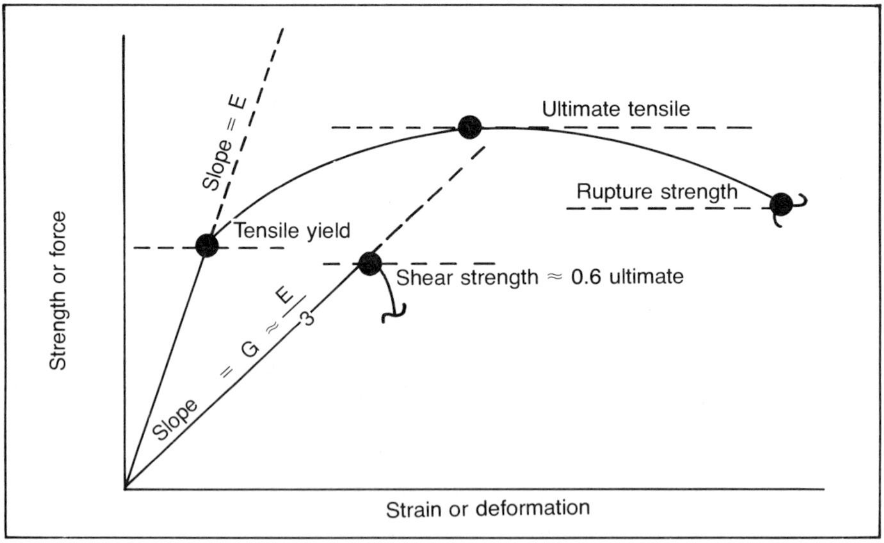

FIG. 3.56 Typical failure functions of loading modes

small deformation up to ultimate shear with the deformation between shear ultimate and rupture strength being very small. This is illustrated in figure 3.56, which has not been drawn to scale for purposes of clarity.

General Crane Requirements

The preceding discussions indicate the complexity and difficulty in choosing the appropriate design factors for the primary components in an offshore platform crane system. However, the illustrations show how the general requirements of crane systems can be specified for improved safety with small additional costs and even reduced costs with selected designs.

REFERENCES

1. Bryant, J. E., "Circular Tubular Joint Design," Thesis, Tulane Univ., 1962.
2. "Planning, Designing, and Constructing Fixed Offshore Platforms," Bul. RP 2A, 9th ed., November 1977, API.
3. Noel, J. S., Beale, L. A., and Toprac, A. A., "An Investigation of Stresses in Welded T Joints," SFRL Rep. 550-3, March 1965, Univ. of Texas.
4. Barnoium, B., and Thebault, J., "In Situ Fatigue Analysis of a Tubular Joint: Hydrodynamic and Mechanical Aspects," Paper 3301, Proc. OTC 1978, pp. 2185–96.
5. Andrian, L. E., Sewell, K. A., and Womack, W. R., "Partial Investigations of Directly Loaded T Joints," Thesis, Southern Methodist Univ, 1958.
6. Horton, S. B., and Toprac, A. A., "Investigation of Elastic Stresses in Welded Tubular Steel K Joints," SFRL Rep. 550-6, January 1966, Univ. of Texas.
7. Toprac, A. A., and Chen, J., "Stresses in Steel Tubular Y Joints," SFRL Rep. 550-7, January 1966, Univ. of Texas.
8. Toprac, A. A., and Brown, R. C., "An Experimental Investigation of Tubular T Joints," SFRL Rep. 550-8, January 1966, Univ. of Texas.
9. Bouwkamp, J. G., "Behavior of Tubular Truss Joints under Static Loads," Rep. SESM 67-33, January 1968, Univ. of California (Berkeley).
10. Marshall, P. W., and Toprac, A. A., "Basis for Tubular Joint Design," Welding Research Supplement, May 1974.
11. Marshall, P. W., "Basic Considerations for Tubular Joint Design in Offshore Construction," Bul. 193, WRC.

12. Bouwkamp, J. G., "Behavior of Tubular Gusset Plate Joints," Paper 1821, Proc. OTC, 1973.
13. "Planning, Designing and Constructing Fixed Offshore Platforms," Bul. RP 2A, 9th ed., November 1977, API.
14. Bynum, Jr., D., "Deepwater Pipelines," *Offshore*, February 1977, pp. 81–95.
15. Reber, Jr., J. B., "Ultimate Strength Design of Tubular Joints," Paper No. 1664, OTC Proc., 1972.
16. Morrow, J. D., "Cyclic Plastic Strain Energy and Fatigue of Metals," Spec. Tech. Pub. No. 378, ASTM, 1965.
17. Halford, G. R., "The Fatigue Toughness of Metals: A Data Compilation," TAM Rep. No. 265, Dept. Theoretical and Appl. Mech., Univ. of Ill., May 1964.
18. Miner, M. A., "Cumulative Damage in Fatigue," *Trans. ASME*, Vol. 67, 1945, p. A159.
19. Kwan, C. T., "Fatigue Analysis for Lattice Legs of Jack-Up Drilling Platforms," Paper No. 77-Pet-11, ASME.
20. Bynum, Jr., D., "How to Calculate and Evaluate Rig Fatigue," *Offshore*, Dec. 1976, pp. 212–215.
21. Marshall, P. W., and Toprac, A. A., "Basis for Tubular Joint Design," *Welding Res. Sup.*, May 1974.
22. Grover, H. J., Gordon, S. A., and Jackson, L. R., "Fatigue of Metals and Structures," NAVWEPS 0025534, Rev. June 1960, Supt. Doc.
23. Bouwkamp, J.G., "Tubular Joints Under Static and Alternating Loads," Rep. SESM, June 1966, Univ. of Calif. (Berkeley).
24. Kurobane, Y., and Konomi, M., "Some Simple S-N Relationships in Fatigue of Tubular K Joints," Trans. AIJ, No. 212, October 1973, pp. 9–22.
25. Kawahara, M. and Iwasaki, T., "Analyses of Fatigue Crack Growth Behavior in Welded Tubular T Joints," Paper No. 3251, OTC Proc., 1978, pp. 1719–1728.
26. Ohtake, F., Sakamoto, S., Tanaka, T., Kai, T., Nakazato, T., and Takigawa, T., "Static and Fatigue Strength of High Tensile Strength Steel Tubular Joints for Offshore Structures," Paper No. 3254, OTC Proc., 1978, pp. 1747–1755.
27. Bynum, Jr., D., "Establishing Multi-Operational Capabilities for Rigs," *Petroleum Times*, July 25, 1975.
28. Bynum, D., Jr., and Chen, C. C., "Structural feasibility study of forged joints," Report No. 18240, July 1974, prepared for Tube Turns Co.
29. Bynum, D., Jr., "Fabricated tubular joints for jack-up legs," Report No. 9450, prepared for Engr. Technology Analysts, Inc., Sept. 1974.
30. Bynum, D., Jr., "Computer analyses of 650 kip mast," Final

Report 21, prepared for Universal Drilling Co., June 1978.
31. Bynum, D., Jr., "Birefringent coating analysis of a tensile plate containing a heterogeneous inclusion using fringe multiplication," MS thesis, Univ. Wash. library, May 1961.
32. Bynum, D., Jr., and Harris, T. A., "Stress analysis of rocket motor cases using Zandman method," Report No. DDM6678, prepared for Rocketdyne Co., October 1966.
33. "Planning, designing, and constructing fixed offshore platforms," Bul. RP 2A, 9th ed., API, November 1977.
34. Manual of steel construction, 7th Ed., AISC, 1973.
35. Bynum, D., Jr., "Preliminary stress analysis of reinforced K-brace, leg chord, and XL-brace fittings fabricated from elbows," Report No. 70.146, (not published), Tube Turns Div., Chemetron Corp., June 14, 1976.
36. Peterson, R. E., "Stress concentration factors," John Wiley Pub., 1953.
37. Bynum, D., Jr., and DeHart, R. C., "Fillet and groove stress concentration," *Exp. Mech.*, June 1964.
38. Baumeister, T., and Marks, L. S., editors, Standard Handbook for Mechanical Engineers, 7th ed., McGraw Hill pub., 1967.

CHAPTER

4 MARINE PIPE-LAY DESIGN

4.1 MARINE PIPELINE DESIGN CHARTS

Design and construction of marine pipelines is a complicated business because of all the parameters that must be considered. The best approach for pipeline analyses is to solve a large number of simple problems, and then relate these solutions into a systems analyses. Two of these simple problems are the selection of suitable and economical pipeline coating weights and the determination of appropriate pipe wall thickness to preclude buckling during construction. Design charts are given for simplifying analyses on these two problems, and example results are given.

Some of the other considerations in pipeline design include determination of method for platform connection, evaluation of weather data, use of weather data to assess operational risks regarding postponement of pipeline abandonment in severe weather conditions. Additional aspects of pipeline design include selection of available stingers, appropriate setting of tensioner, pipe buckle propagation during construction, etc.

Economics

Because of the current economic situation, it is worth a lot of time to optimize the pipeline wall thickness as well as the weighting material. The dollar costs for increasing the negative buoyancy one-tenth of a point on large pipelines are increased about 9.5%. For example, in raising the SG from 1.05 to 1.15 on a bare high strength steel pipe, the steel costs alone increase about \$50 K/mile with a 3/32 in. increase of wall thickness on a 30-in. line, and about \$90 K/mile with a ⅛ in. increase of wall thickness on a 40-in. line.

Material Properties

To do the juggling act on selection of pipeline wall thickness and weighting material, it is necessary to know the mechanical properties of the material candidates which will usually be to API

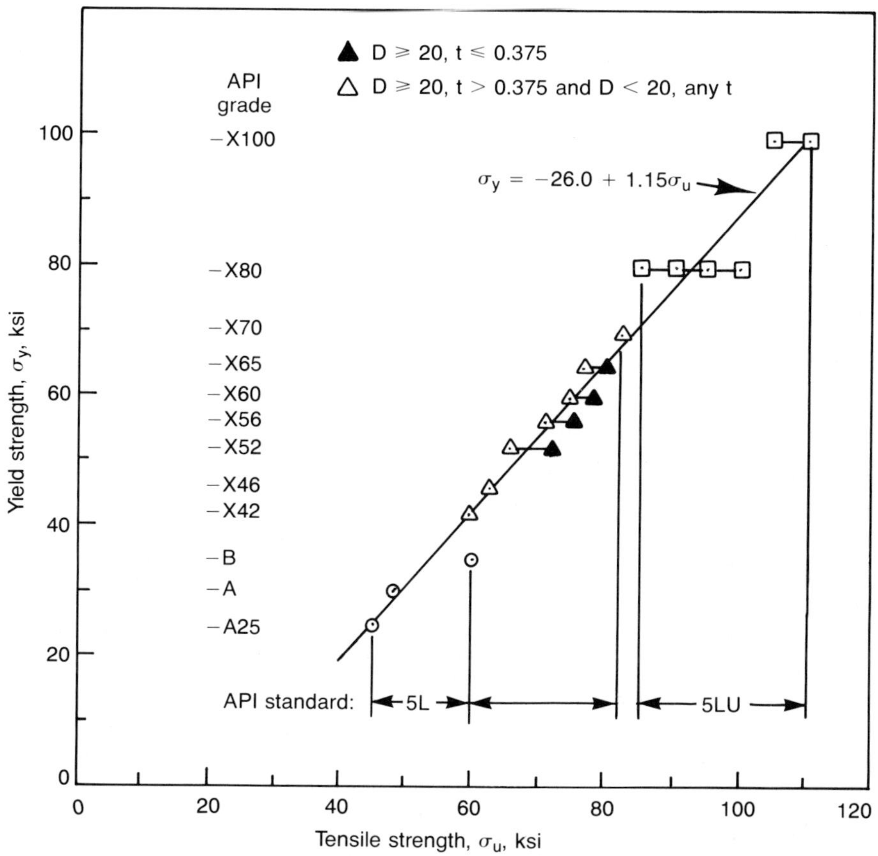

FIG. 4.1 Tensile strength of API-grade pipe

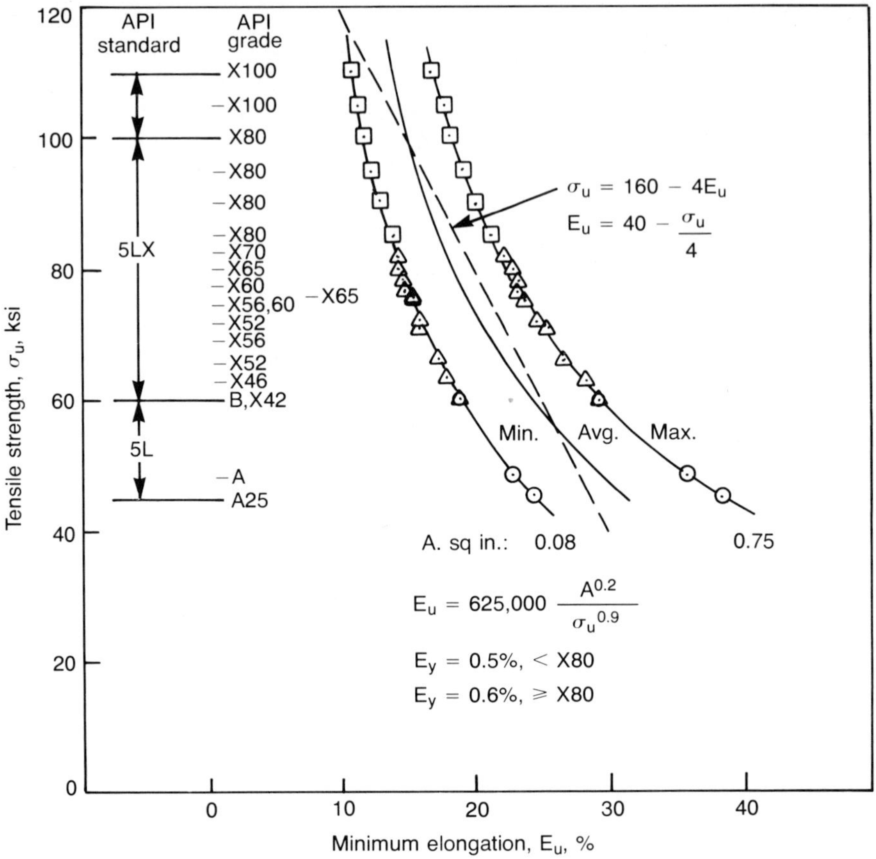

FIG. 4.2 Variation of minimum elongation with tensile strength

specifications. Material properties given in API Specs 5L, 5LX, and 5LU are summarized in figures 4.1–4.3.

A rough fit of the data for σ_y vs. σ_u was derived and is shown on figure 4.1. Use of the equation simplifies the plastic analyses involved in buckle propagation studies.

For thin walled pipes, (i.e., high D/t ratio), load is limited by buckling. For thick walled pipes, load is limited by plastic strain allowable which is related to the strength factor, SF. The equation for the strength factor, defined as the ratio of tensile to yield strength, is given in figure 4.3.

Safety Factors

For thick walled pipe it is suggested that the factor of safety be tied to ductility which is indicated by the strength factor. Using the

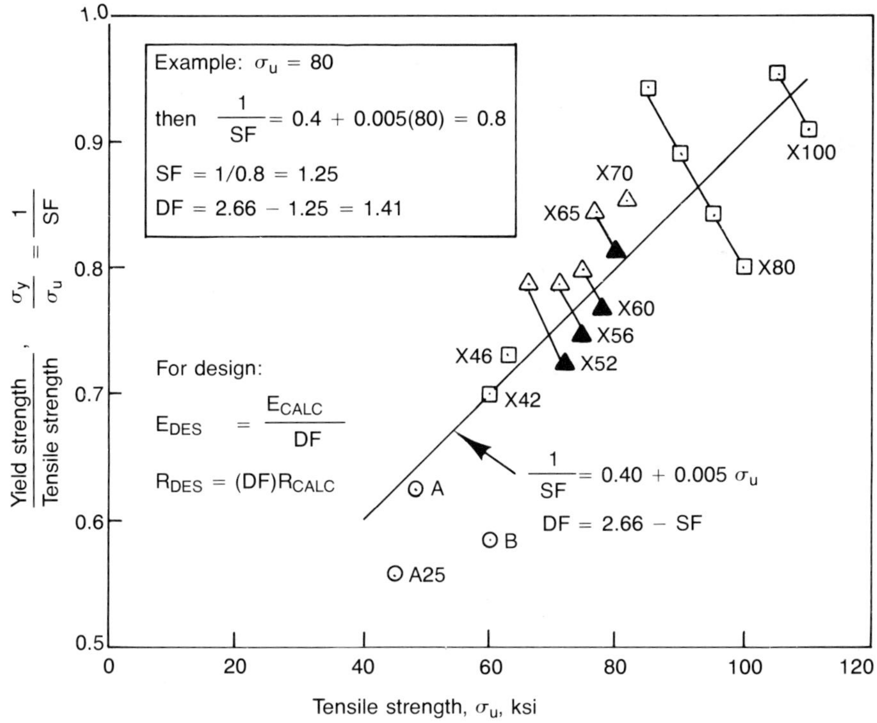

FIG. 4.3 Strength factor for API-grade pipelines

equation in figure 4.3, the ultimate strength is about 1.66 times greater than the yield strength for the low strength steels and approaches the value of one as a limit for high strength steels. This suggests a design factor, DF, that increases with a decreasing strength factor. The resultant DF equation and design use are given in figure 4.3.

The factor of safety, FS, for thin wall pipes can be defined as the reciprocal of the sum of the load factors for pressure, Fp, and for bending, Fb. Determination of these two load factors is described later in this paper.

SG Design Charts for Bare Pipe

Optimization of pipeline specific gravity (SG) for best economics is complicated by possible variation in the density and the thickness of both the protective coating to inhibit corrosion and the weighting coating to decrease pipe wall thickness and thereby lower steel costs. Useful equations are summarized in table 51. But parametric analyses in preliminary designs are quicker by using design charts rather than the equations, so few charts are given here.

TABLE 51
Formulas for specific gravity calculations

Bare Pipe

$$w_a = w_p = \frac{\pi}{4}(D^2 - d^2)\frac{490}{144} \approx 10.68\ D_p t_p \qquad (4.1)$$

$$w_d = \frac{\pi}{4} D_p \frac{64}{144} = 0.349\ D_p^{\,2} \qquad (4.2)$$

$$w_w = w_a - w_d \qquad (4.3)$$

$$SG = \frac{w_a}{w_d} = \frac{w_a}{w_a - w_w} = \frac{1}{1 - \dfrac{w_w}{w_a}} \qquad (4.4)$$

$$SG \approx \frac{10.68\ D_p t_p}{0.349\ D_p^{\,2}} = 30.6\ \frac{t_p}{D_p} \qquad (4.5)$$

$$NB \approx 30.6\ \frac{t}{D} - 1 \qquad (4.6)$$

Somastic Coated Pipe

$$w_a = w_p + w_s \qquad (4.7)$$

$$W_s = \frac{\pi}{4}\left[(D_p + 2t_s)^2 - D_p^{\,2}\right]\frac{135}{144} \approx 2.95\ D_p t_s \qquad (4.8)$$

$$w_d = 0.349\ D_s^{\,2} \qquad (4.9)$$

Weighted Pipe

$$w_a = w_p + w_s + w_c \qquad (4.10)$$

$$w_c = \frac{\pi}{4}(D_c^{\,2} - D_5^{\,2})\frac{p_c}{144} = \left.\begin{matrix}0.763\\0.899\\1.036\end{matrix}\right\}(D_c^{\,2} - D_5^{\,2})\ @\ p_c = \left\{\begin{matrix}140\\165\\190\end{matrix}\right. \qquad (4.11)$$

$$w_d = 0.349\ D_c^{\,2} \qquad (4.12)$$

Specific gravity for various combinations of the diameter and wall thickness of bare pipe is given in figure 4.4. This graph could be used with designs of nonweighted pipe having thin film (epoxy) protective coatings.

SG Design Charts for Coated Pipe

General SG design charts for 6–50-in. OD; 0.25–2.00-in. wall thickness, with and without a corrosion protection coating; and wall thickness of weighting material 0–4 in., in 1-in. increments, are given in figures 4.5–4.10. Note that curves extend into range of positive buoyancy. This data is given so that the curves could be used in

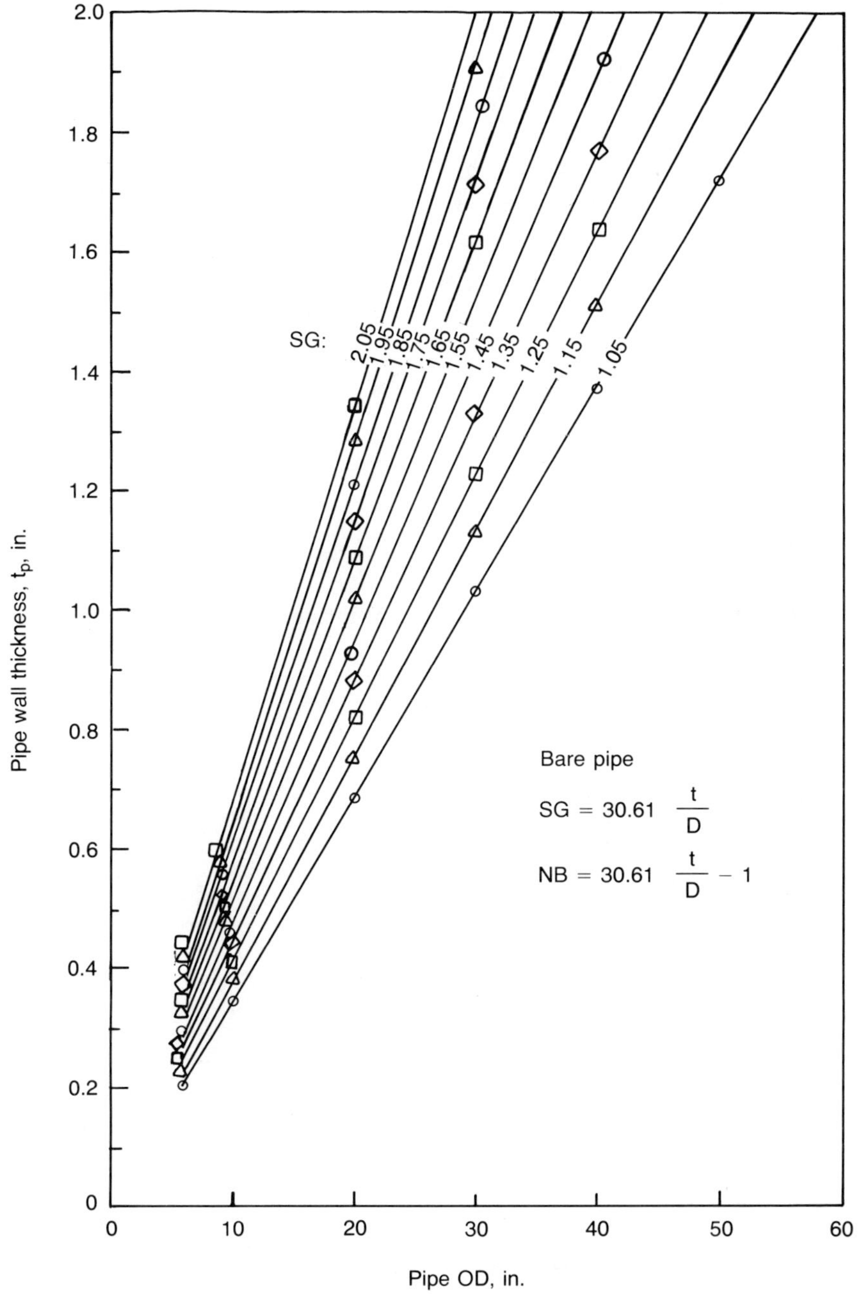

FIG. 4.4 Specific gravity design chart for bare pipe

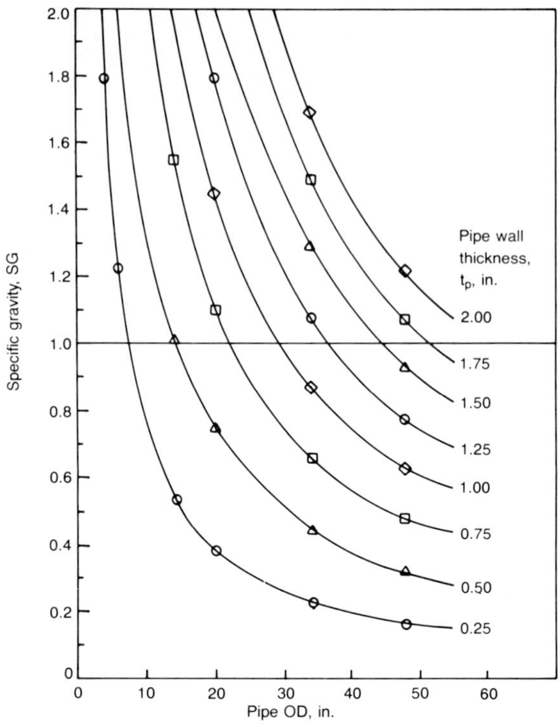

FIG. 4.5 Specific gravity without somastic corrosion-protective coating

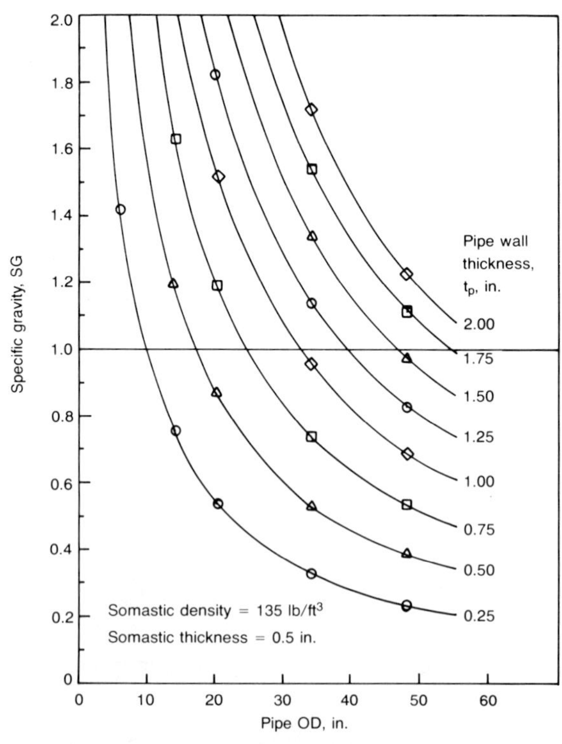

FIG. 4.6 Specific gravity with somastic corrosion-protective coating

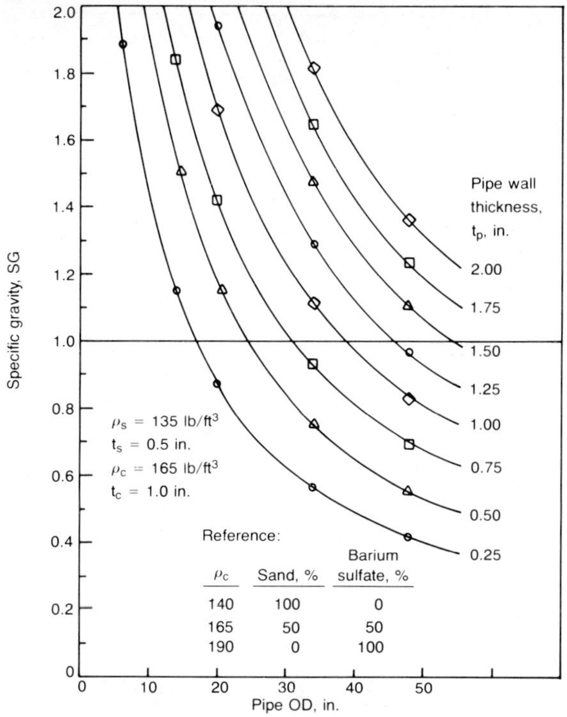

FIG. 4.7 Specific gravity with 1-in. concrete

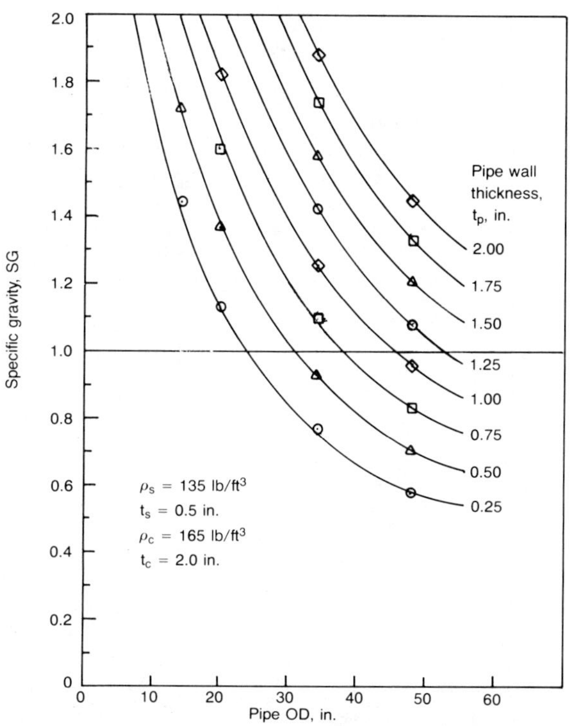

FIG. 4.8 Specific gravity with 2-in. concrete

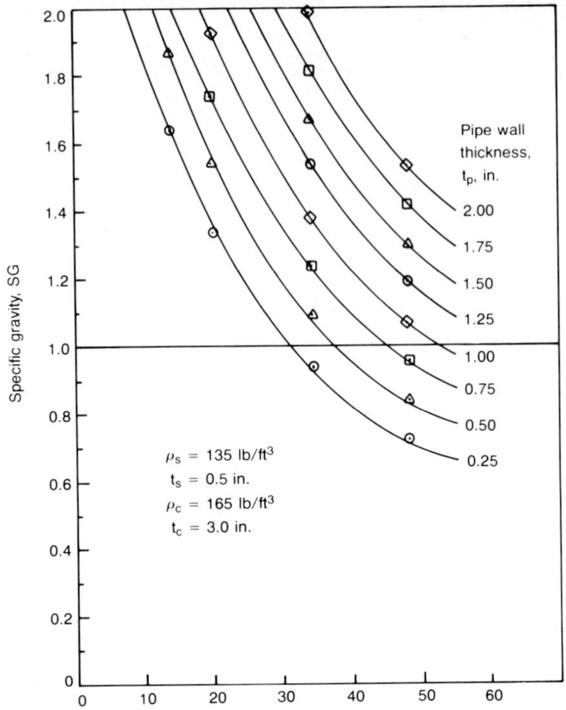

FIG. 4.9 Specific gravity with 3-in. concrete

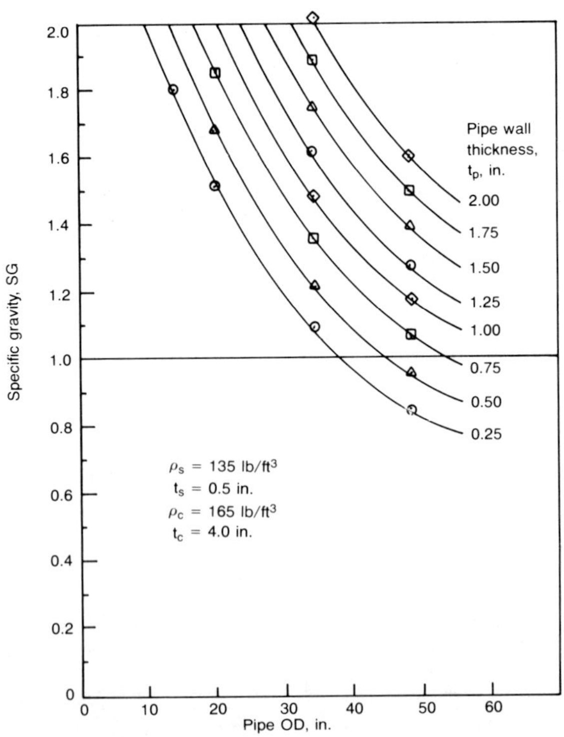

FIG. 4.10 Specific gravity with 4-in. concrete

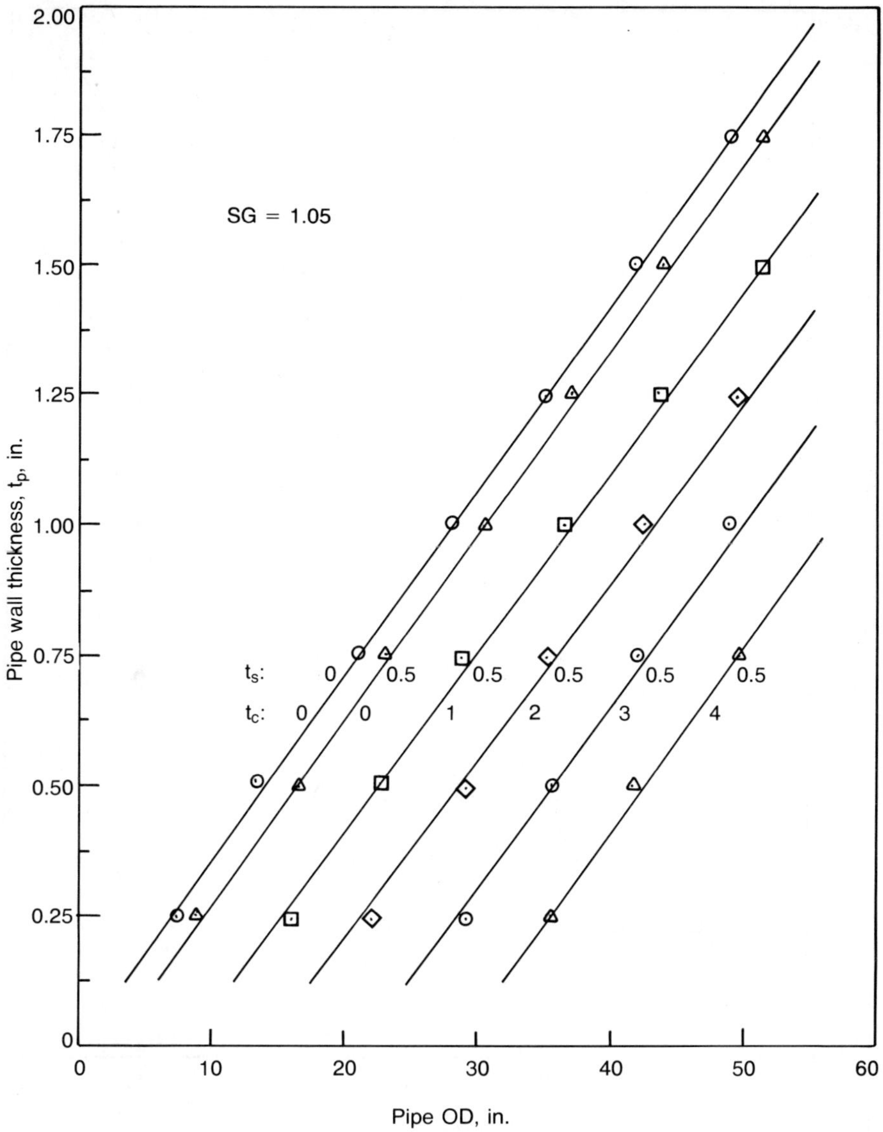

FIG. 4.11 Pipe parameters at 1.05 sp gr

special anchor designs. Intermittent special weighting anchors on the line instead of uniform coatings would seldom even be considered for offshore operations, but they are not infrequently used for certain onshore operations such as in swamps and marshy areas.

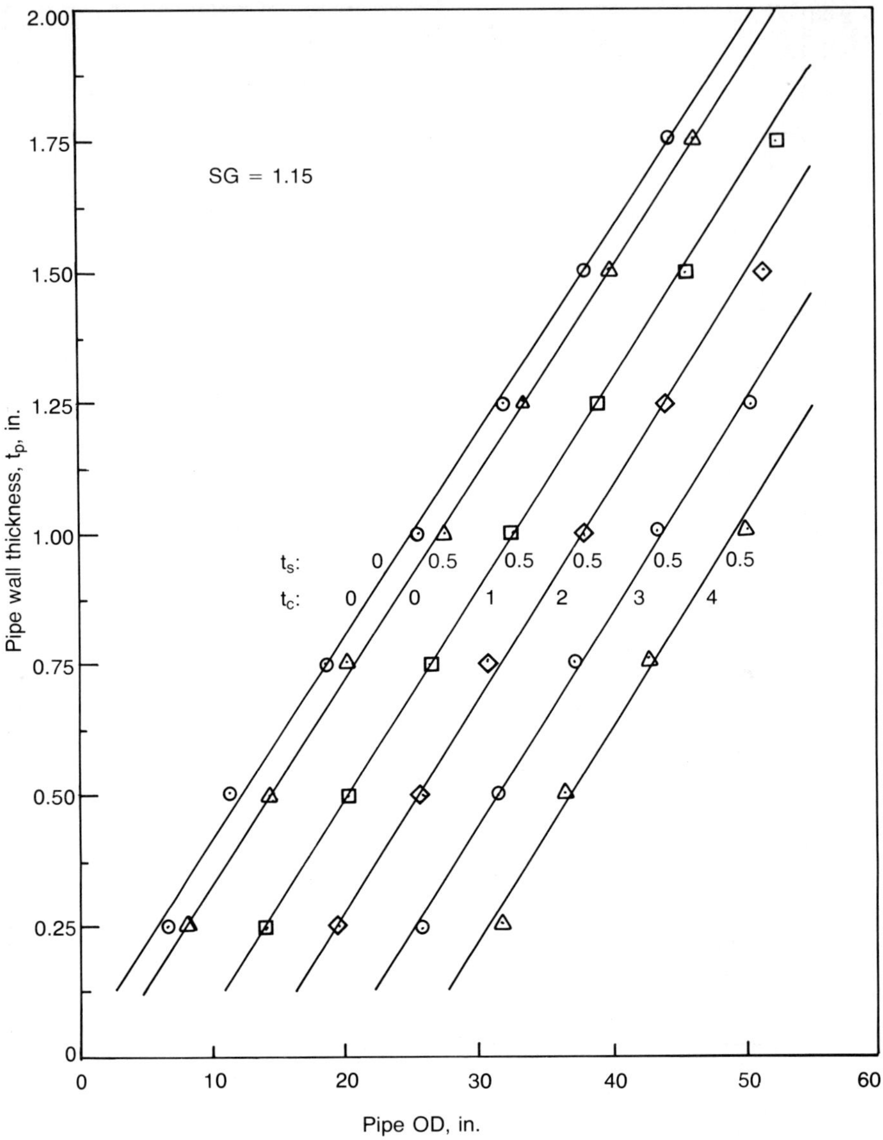

FIG. 4.12 Pipe parameters at 1.15 sp gr

Using the SG vs D plots in figures 4.5–4.10, the design charts of t vs D with various coatings were cross plotted for selected SG and are given in figures 4.11–4.14. The required SG is determined from a study of water currents at the mud line.

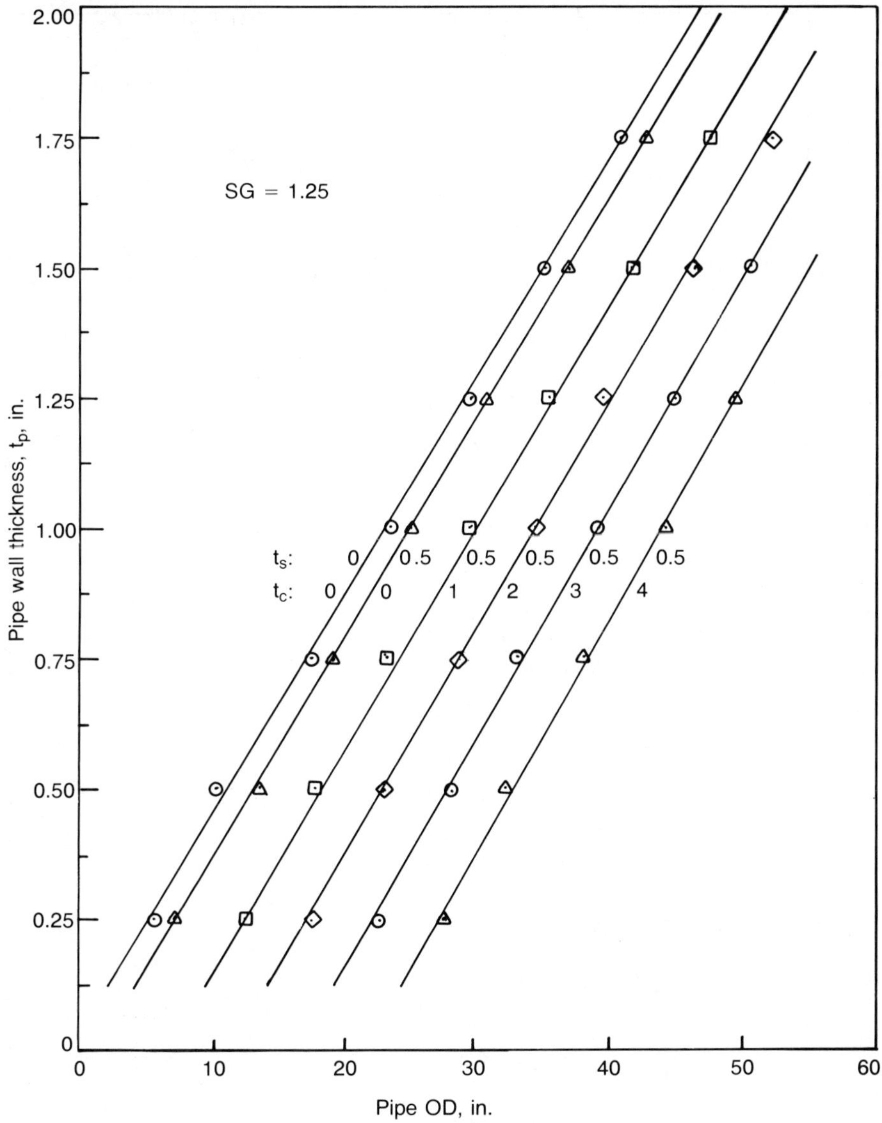

FIG. 4.13 Pipe parameters at 1.25 sp gr

Parametric SG Design Charts

Note that the possible combinations of parameters were restricted to limit the number of SG charts presented in this paper. In all of the design charts given, the corrosion protective coating was

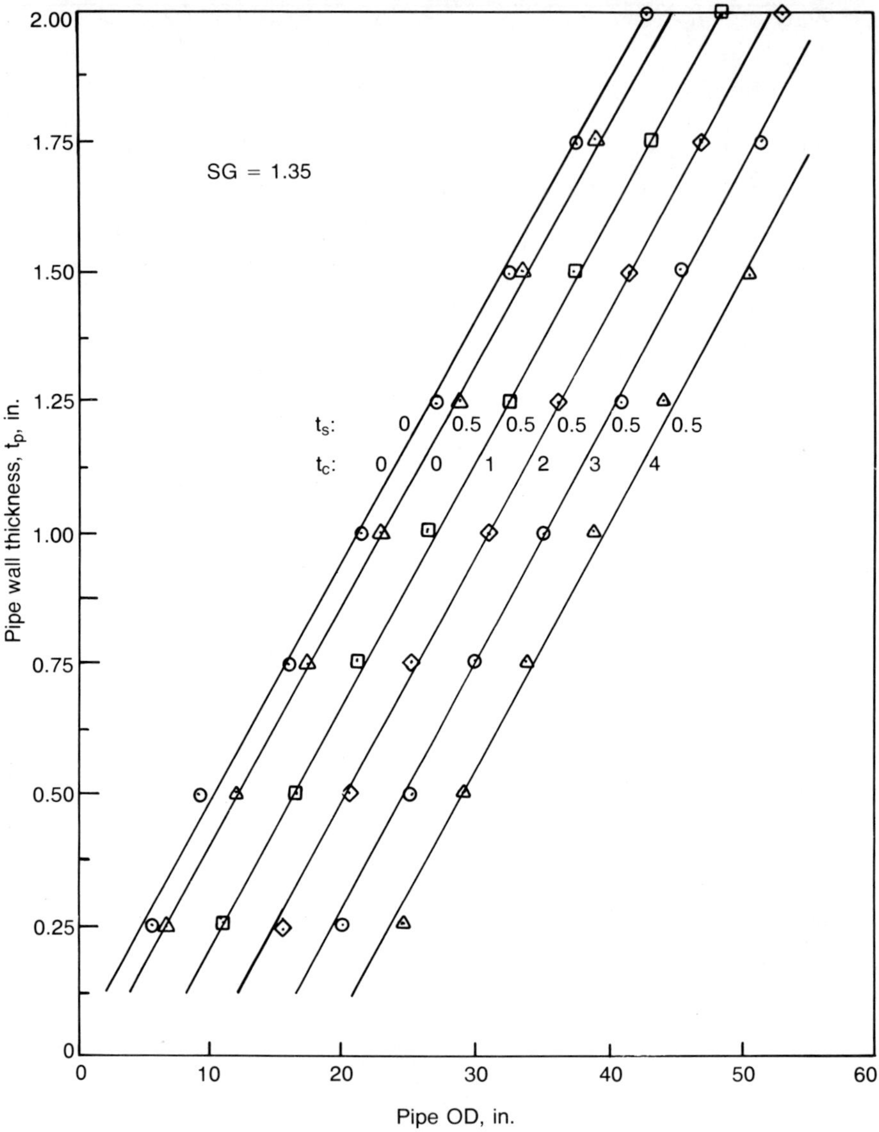

FIG. 4.14 Pipe parameters at 1.35 sp gr

restricted to a nominal 0.5-in. wall thickness and a density of 135 lb/cu ft, even though other values are used in practice. Also a 50–50 mix of sand and barium sulfate was considered as the materials composition in the weighted coating for a density of 165 lb/cu ft. For reference, densities of weighted coating material are 140 for 100% sand and 190 for 100% barium sulphate.

Ten SG design charts were presented here with the above restrictions on the variables. A design handbook would give data for three different thicknesses of protective coating, all at the same density, and three different densities of weighting coating with four thicknesses of weighting coating. Such a parametric exercise would result in 66 charts in addition to those presented here.

General Pipeline Buckling Analyses

The pipeline can buckle two ways. In pressure buckling due to excess water depths for a given line, the pipe collapses. Under certain conditions, the buckle can propagate along several miles of the line to a lesser water depth where the pressure is less than the calculated propagating pressure. Buckling arrestors are sometimes used to mitigate pipeline damage.

The second mode of buckling failure is the bending buckle. This can happen in the sag-bend region with insufficient tension or in the overbend region with improper stinger design.

Each type of buckle failure has a distinctive characteristic shape. Pressure buckle shape is due to a plastic collapse, and pipe cross section looks like an 8. Bending buckles look like a pattern of diamond shapes in the thin wall pipes. Bending buckle in thick wall pipe is a single plastic crimp.

There is not complete agreement in the literature regarding the most correct or best equations to use for calculating buckling of cylinders. The problem is complicated by several variables such as D/t ratio, material properties, out-of-roundness, nonuniform wall thickness, surface imperfections, and elastic vs. plastic buckling modes. For pipeline design, the Southwell theoretical relation for pressure buckling and the Timoshenko empirical relation for bending buckling are usually applicable.

Design Charts for Pressure Buckling

Southwell equation for pressure buckling with material constants for steel plugged in is given in table 52. The water depth parameter, $\overline{H}$, was tabulated for various values of D/t and yield strengths corresponding to API grades of steel with the results given in table 52 and figure 4.15. Note that the water depth parameter, $\overline{H}$, equals water depth, H, times coating area factor, C_p, divided by pressure factor, F_p.

The concrete weighting coating has negligible strength but causes an increase in effective pressure on the pipe. The coating area factor, C_p, used to correct for this is given in figure 4.16 as a function of pipe diameter and concrete coating thickness.

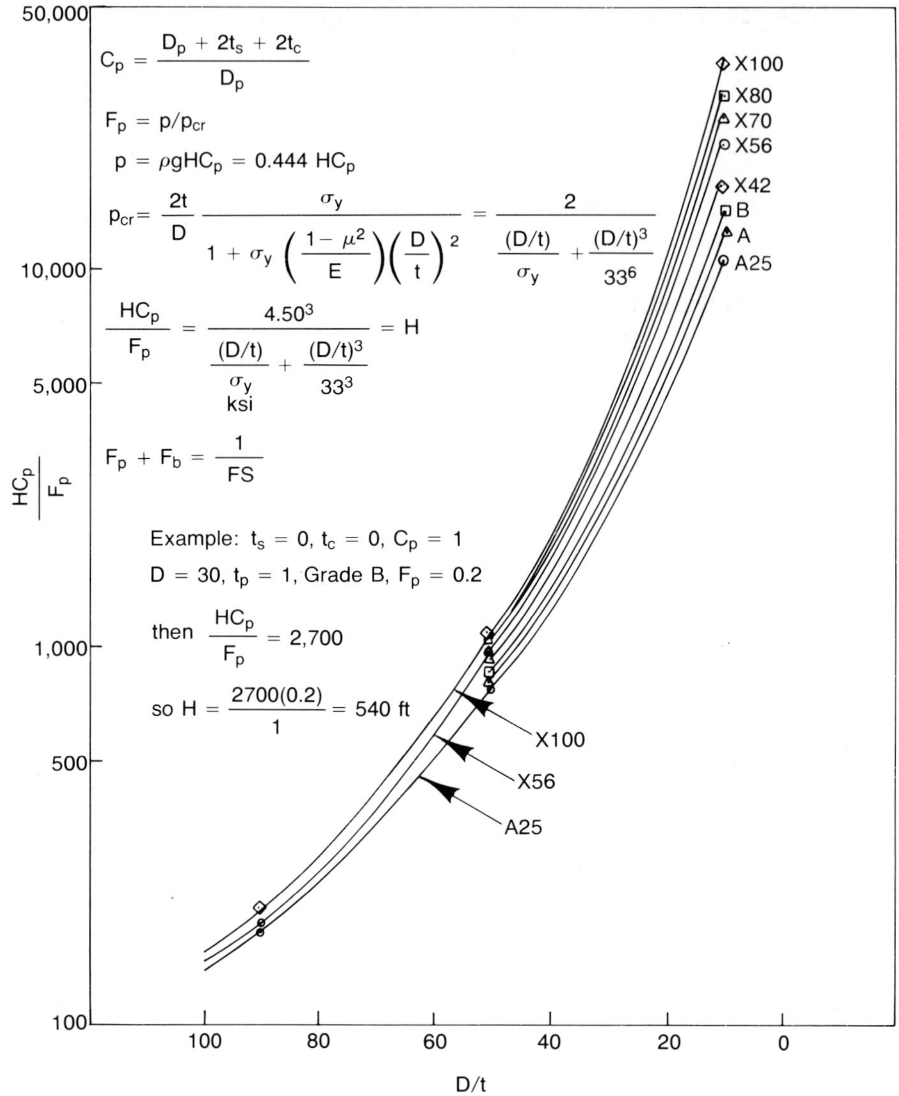

FIG. 4.15 Design chart for pressure buckle analysis

The pressure factor, F_p, is the ratio of the working water depth divided by the water depth where the pipe collapses. The sum of the pressure factor and the bending factor, F_b, must not exceed unity.

Results in figure 4.15 show how allowable water depths increase with decreasing diameter-to-thickness ratios of pipe. The design chart is somewhat deceptive at first because the results are

TABLE 52
Pressure buckle results

Grade	API Spec	σ_y	D/t: 10	30	50	70	90
			Water Depth Parameter, $\overline{H} = \frac{H\,C_p}{F_p}$				
A25	5L	25	10500	2230	777	341	175
A		30	12400	2480	825	354	179
B		35	14200	2690	863	363	182
	5LX	42	16800	2940	904	373	186
		56	21500	3320	961	386	190
		70	26000	3610	1000	395	193
	5LU	80	29000	3770	1020	399	194
		100	34500	4020	1050	406	196

$$p_{cr} = \frac{2t}{D}\left[\frac{\sigma_y}{1 + \sigma_y\left(\frac{1-\mu^2}{E}\right)\left(\frac{D}{t}\right)^2}\right] = \frac{2}{\frac{(D/t)}{\sigma_y} + \frac{(D/t)^3}{33^6}}, \text{ Southwell} \tag{4.13}$$

$$p = p_g\,H\,C_p = 0.444\,H\,C_p \tag{4.14}$$

$$F_p = p/p_{cr} \tag{4.15}$$

$$\frac{H\,C_p}{F_p} = \frac{4.50^{\underline{3}}}{\frac{(D/t)}{\underset{\text{ksi}}{\sigma_y}} + \frac{(D/t)^3}{33^{\underline{3}}}} \tag{4.16}$$

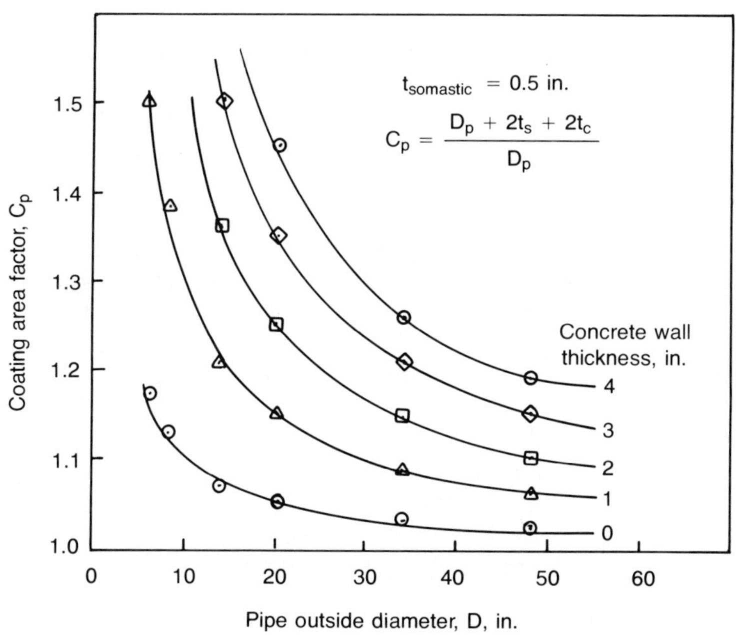

FIG. 4.16 Pressure increase due to coatings

plotted on log scale. For comparison at D/t of 30, for example, water depth allowable for X100 is only about 80% greater than for material grade A25. Water depth allowable varies much more greatly with D/t than with yield strength. With Grade B, for example, allowable water depth increases by a *factor* of 78 as D/t decreases from 90 to 10.

Design Charts for Bending Buckling

Relation for bending parameter, $\bar{B}$, uses Timoshenko's empirical bending buckle equation, as shown in table 53. Strictly speaking, the Timoshenko equation is for symmetric axial compression, and test values show that the buckling load allowable for pure bending is roughly 1.4 times greater than for pure compression. It is more reasonable to assume that buckling would occur at the same value of compressive strain regardless of whether the compressive strain is caused by pure bending or by symmetrically applied axial loads.

By similiar reasoning, the final design should take the tensioner force into account. The allowable bending strain would increase and the allowable radius of curvature would decrease with increasing tensioner force.

Neglecting the influence of the tensioner force, which gives conservative results, the bending parameter, $\bar{B}$, is plotted in figure 4.17 as a function of material yield strength and D/t radio. $\bar{B}$ includes a factor of 12 so that the radius of curvature, R, will come out in units of feet instead of inches.

To compare results of bending buckling values in the figure 4.17 design chart, at D/t of 30 for example, the minimum radius of curvature for Grade A25 is 2.6 times greater than for Grade X100 steel. With Grade B, for example, minimum radius of curvature increases by a factor of 9 as D/t decreases from 90 to 10.

Example Pressure Buckling Design

The preceding design charts are especially useful in preliminary design. However, once the pipe diameter and SG are determined from required flow rates, bottom currents, and other factors, the procedures are greatly simplified. With a 30-in. pipe, for example, some of the various combinations of pipe wall and coating thickness as a function of SG, without regard to buckling considerations, are given in table 54.

To further simplify these discussions, it is assumed that an SG of 1.15 is adequate. Resulting allowable water depths as a function of pressure function, coating thickness, and maximum and minimum yield strengths of API grade 5LX steels are given in table 55.

TABLE 53
Bending buckle results

API Spec	σ_y	D/t: 10	20	30	40	50	60	70	80	90	100
		Bending Parameter, $\overline{B} = \frac{R\,F_b}{D/12}$									
5L	25	24.2	48.4	72.6	96.8	121	145	169	194	218	242
	30	20.9	41.8	62.7	83.6	105	125	146	167	188	209
	35	18.5	37.0	55.5	74.0	92.5	111	130	148	167	185
5LX	42	16.1	32.2	48.3	64.4	80.5	96.5	113	129	145	161
	46	15.0	30.0	45.0	60.0	75.0	90.0	105	120	135	150
	52	13.8	27.6	41.4	55.2	69.0	82.8	96.6	110	124	138
	56	13.1	26.2	39.3	52.4	65.5	78.6	91.7	105	118	131
	60	12.5	25.0	37.5	50.0	62.5	75.0	87.5	100	113	125
	65	11.9	23.8	35.7	47.6	59.5	71.4	83.3	95.2	107	119
	70	11.3	22.6	33.9	45.2	56.5	67.8	79.1	90.4	102	113
5LU	80	10.4	20.8	31.2	41.6	52.0	62.4	72.8	83.2	93.6	104
	100	9.17	18.3	27.5	36.7	45.9	55.0	64.2	73.4	82.5	91.7

$\underset{\text{Bending}}{\sigma_{cr}} \approx C \underset{\text{Compression}}{\sigma_{cr}}$, $C \approx 1.4$ cut use 1.0 for safety

$$\sigma_{cb} = \frac{E\left(1.2\ t/D - 10^{-7}\ \frac{D}{2t}\right)}{1 + 0.004\ \frac{E}{\sigma_y}} \approx \frac{36^{6}t/D}{1 + \frac{120}{\sigma_{y\ ksi}}},\ \text{Timoshenko Empirical} \quad (4.17)$$

$$\sigma_b = \varepsilon_b E = \frac{D}{2R} E \quad (4.18)$$

$$F_b = \sigma_b/\sigma_{cb} \quad (4.19)$$

$$\frac{R\,F_b}{D/12} = 0.417\left(\frac{D}{t}\right)\left(1 + \frac{120}{\sigma_y}\right) = \overline{B} \quad (4.20)$$

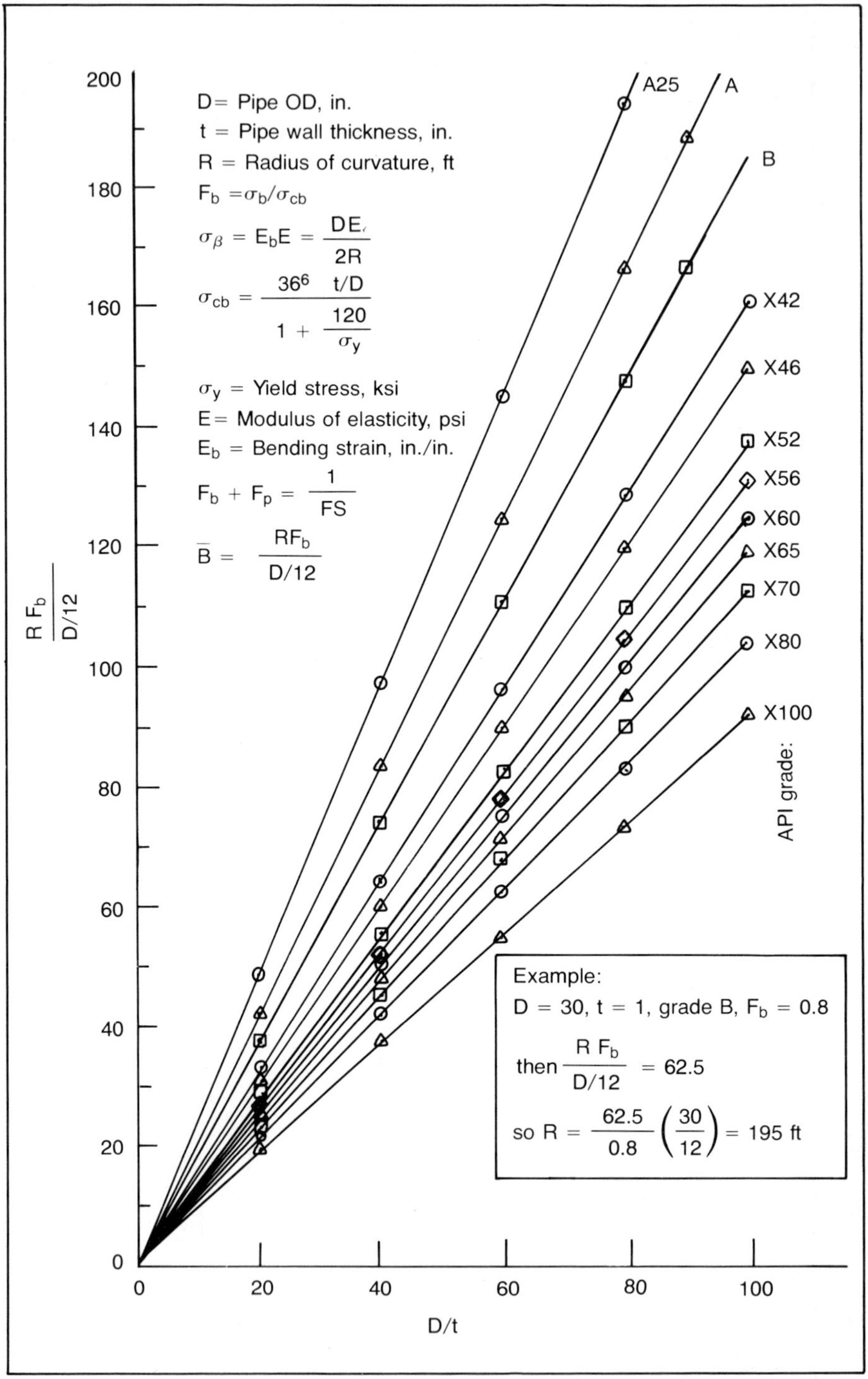

FIG. 4.17 Design chart for bending buckle analysis

Results are plotted on log graph in figure 4.18. This vividly shows that allowable water depth is a weak function of yield strength and a strong function of diameter to thickness ratio of the pipe.

Example Bending Buckle Design

Continuing with the example of a 30-in. pipe and a required SG of 1.15 and using the equations given in table 52 results were calculated for various values of the bending factor, coating thickness, and minimum and maximum grades of steel in API Standard 5LX. Minimum radius of curvatures is given in table 56 and maximum allowable bending strain is given in table 55.

Assume that maximum water depth is 400 ft. From figure 4.18 we see that a 1-in. concrete on a 15⁄16 pipe with a 0.2 pressure factor,

TABLE 54
Example coating combinations on 30 in. pipe

		Pipe Wall Thickness, t_p, in. (to nearest higher 1⁄16 in. increment)					
	t_c:	0	0	1	2	3	4
SG	t_s:	0	0.5	0.5	0.5	0.5	0.5
1.05		1 1⁄8	1	3⁄4	9⁄16	15⁄16	1⁄8
1.15		1 3⁄16	1 1⁄8	15⁄16	11⁄16	7⁄16	1⁄4
1.25		1 15⁄16	1 1⁄4	1	15⁄16	5⁄8	3⁄8
1.35		1 7⁄16	1 3⁄8	1 3⁄16	1	3⁄4	9⁄16
	C_p:	1.00	1.04	1.10	1.17	1.24	1.30

TABLE 55
Maximum water depths, for example D = 30, SG = 1.15

				Water Depth, H, ft.					
			$F_p = p/p_{cr} =$	0.2		0.3		0.4	
t_p	t_s	t_c	D/t σ_y:	42	70	42	70	42	70
1 3⁄16	0	0	25.3	826	1070	1240	1610	1650	2040
1 1⁄8	0.5	0	26.7	714	906	1070	1360	1430	1810
15⁄16	0.5	1	32.0	468	564	702	846	936	1130
11⁄16	0.5	2	43.6	216	246	324	369	432	492
7⁄16	0.5	3	68.6	64	68	96	102	128	136
1⁄4	0.5	4	120	13	13	20	20	26	26

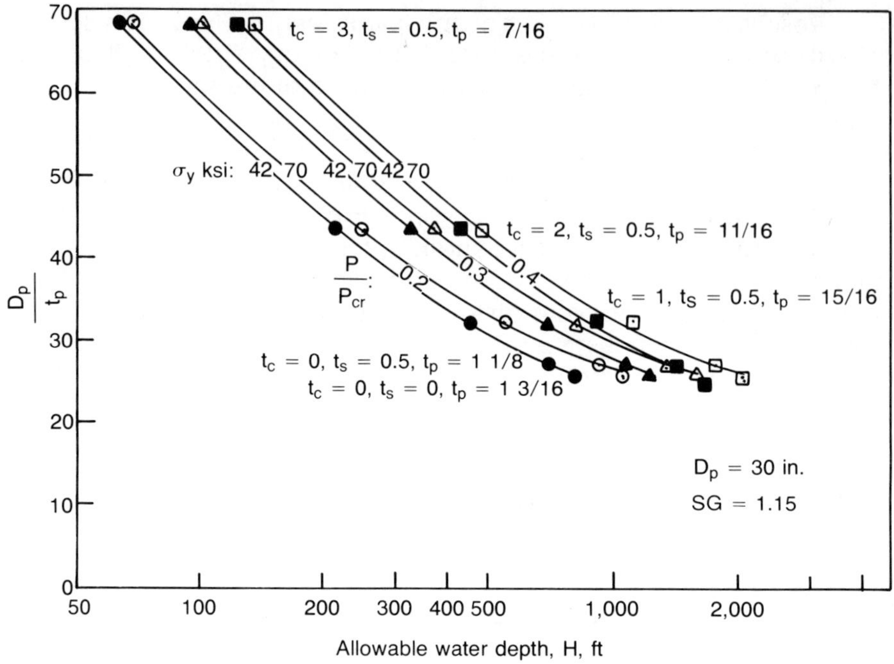

FIG. 4.18 Results for allowable water depth

TABLE 56
Minimum radius of curvature, for example D = 30, SG = 1.15

					Radius of Curvature, R, ft.					
			$F_b = \sigma_b/\sigma_{cb}$ =		0.6		0.7		0.8	
t_p	t_s	t_c	D/t	σ_y:	42	70	42	70	42	70
1 3/16	0	0	25.3		170	119	146	102	127	89.5
1 1/8	0.5	0	26.7		179	126	154	108	135	94.5
15/16	0.5	1	32.0		215	151	184	129	161	113
11/16	0.5	2	43.6		293	206	251	176	220	154
7/16	0.5	3	68.6		461	324	395	277	346	243
1/4	0.5	4	120		806	566	591	485	605	425

or a 2-in. concrete on 11/16 pipe with a 0.4 pressure factor is adequate using X42 steel in either case. Use the 11/16 pipe since costs would be lower.

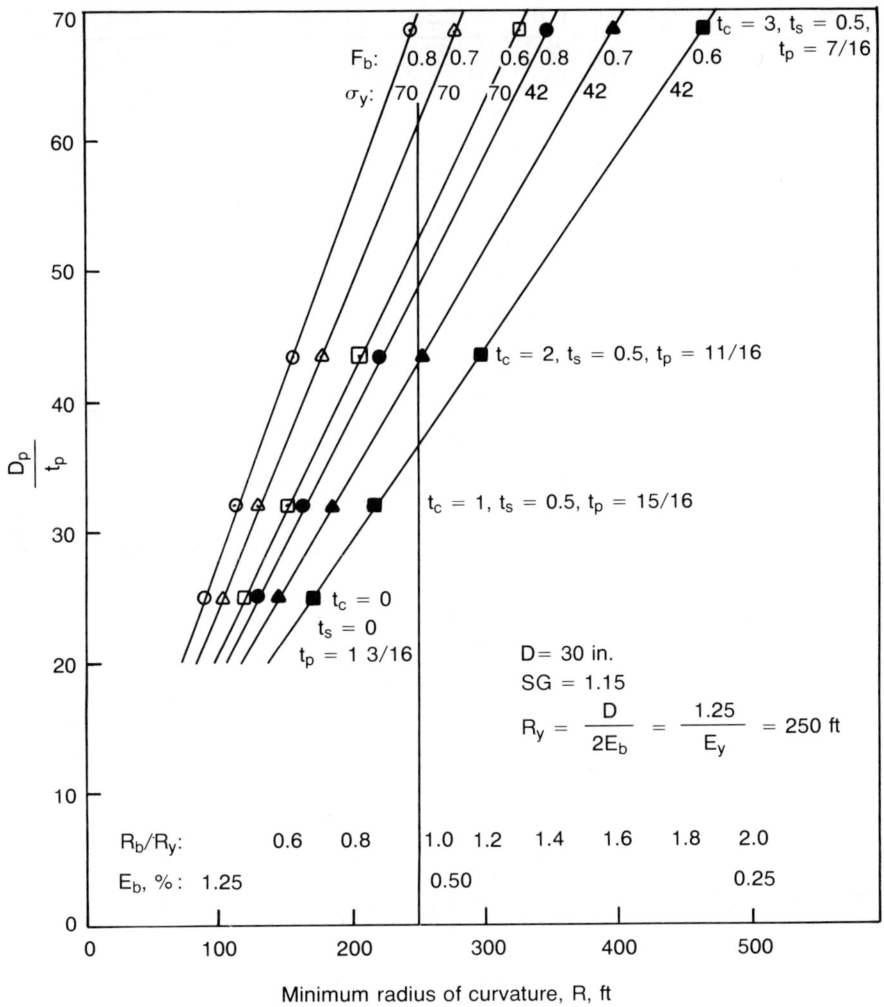

FIG. 4.19 Minimum radius of curvature based on buckling

With a 0.4 pressure factor, the bending factor is 0.6. The minimum radius of curvature for the 11⁄16 × 42 pipe with a 2-in. coating is about 290 ft (figure 4.19). This minimum radius of curvature corresponds to a maximum strain of about 0.43 (figure 4.20). The computed strain level is a little high for most pipeline construction jobs. The strain level would be decreased by using a greater radius of curvature on the stinger configuration and by increasing the tensioner setting.

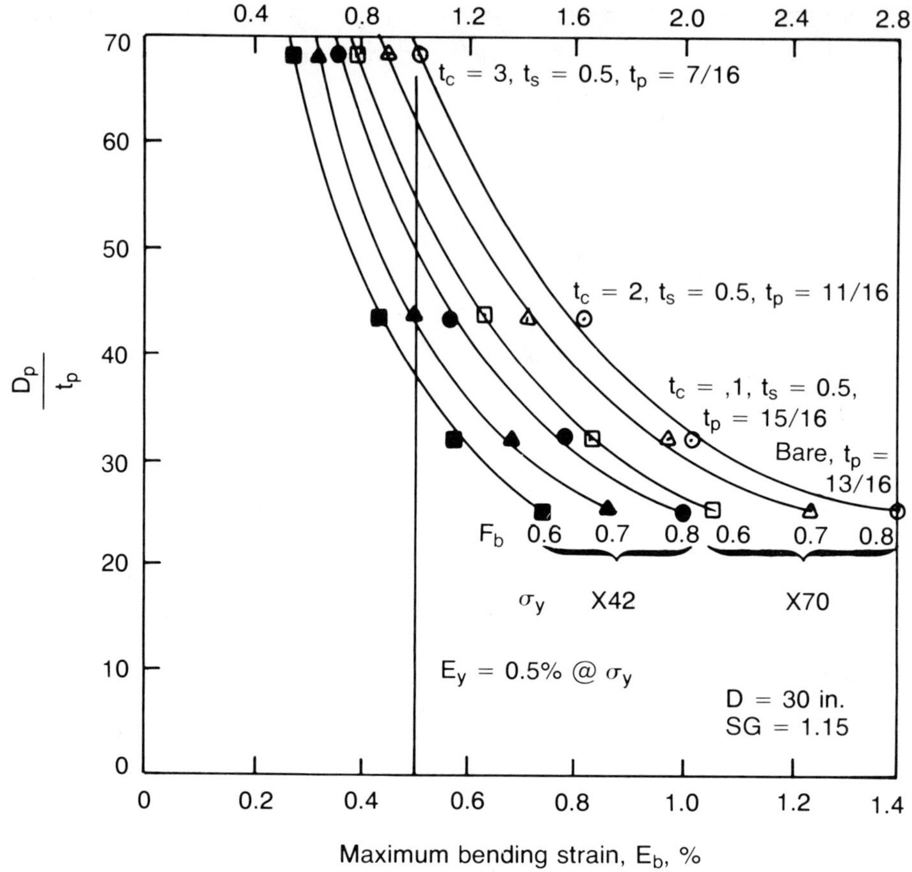

FIG. 4.20 Maximum bending strain based on buckling

TABLE 57
Maximum bending strain, for example D = 30, SG = 1.15

			$F_b = \sigma_b/\sigma_{cb}$ =	Bending Strain, ε_b, in./in.					
				0.6		0.7		0.8	
t_p	t_s	t_c	D/t σ_y:	42	70	42	70	42	70
1 3/16	0	0	25.3	0.74	1.05	0.86	1.23	0.98	1.40
1 1/8	0.5	0	26.7	0.70	0.99	0.81	1.16	0.93	1.32
15/16	0.5	1	32.0	0.58	0.83	0.68	0.97	0.78	1.12
11/16	0.5	2	43.6	0.43	0.62	0.50	0.71	0.57	0.81
7/16	0.5	3	68.6	0.27	0.39	0.32	0.45	0.36	0.51
1/4	0.5	4	120	0.16	0.22	0.21	0.26	0.21	0.29

Conclusion

Obviously many combinations of coating thickness and density, pipe wall thickness, and yield strength could be selected to perform adequately from the structural standpoint. Final design would be based on analyses as described here plus economic considerations. Final design would also include influence of manufacturing tolerances on thickness, out-of-round and other factors.

4.2 BUCKLING CONSIDERATIONS FOR THE DESIGN OF MACHINE PIPELINES

It is important to be able to predict the response of a pipeline to external loads during marine pipe-laying operations. Failure, due to excessive stress or buckling, is very costly, particularly in deep water. The open literature contains reports of many investigations on buckling of circular rings and tubes subject to different kinds of loadings, which often consider geometric imperfections such as out-of-roundness.

The problem primarily involves a combination of circumferential buckling due to hydrostatic pressure and longitudinal bending. In order to achieve an understanding of each individual mode of collapse, a review of the open literature was made. The gist of these studies is then reduced to one governing equation which is used in examples of practical design problems, and recommendations are made for use of this method by the pipeline industry.

As early in 1858, Fairbairn recorded results and developed empirical equations for the collapse of riveted circular tubes subject to uniform radial pressure. In 1885, Levy developed the theoretical buckling equation used today for perfectly circular cylinders. In 1886, Bresse developed a method to determine circumferential bending stress due to elliptic eccentricities. Bryan presented experimental results and a corresponding theoretical relationship in 1888. In 1900, Hovgaard worked with the collapse of elliptic cylinders. In 1906, Stewart investigated steel lap welded tubes from 2 to 10 in. in diameter and showed that Fairbairn's empirical equations were sometimes off by a factor of 3 or 4. Studies of long circular cylinders subject to uniform radial pressure were continued by Carmen and Carr in 1906, by Stewart in 1907, Lilly in 1910, and Cook in 1913. The following year, Cook reported an investigation on short tubes. Southwell developed an empirical equation in 1915 which agreed with experimental results up to the yield point of the material. Studies on short tubes were continued by Carmen in 1917. In 1931, the teams of Sanders and Windenburg and of Jasper and Sullivan further studied thin circular cylinders subject to radial pressure.

In 1932, Flugge added another dimension by studying combined axial and bending forces, followed by Lundquist in 1933 and Donnell in 1934. Out-of-roundness was investigated in 1933 by Timoshenko, by Windenburg and Trilling in 1934, and by Sturm in 1936. Also in 1936, Jasper studied the effect of yield strength on seamless circular cylinders used for casings. Similar studies were performed by Frame in 1938, and in 1939, by Holmquist and Nadai. In 1939, a survey of previous works including combined circumferential bending and uniform radial loading was presented by Main. Also in 1939, Edwards and Miller studied the problem of combined axial and radial loads. Improvements were made in subsequent years for circular cylinders under uniform radial pressure by Clinedinst in 1939 and by Sturm in 1941.

The significance of out-of-roundness on load carrying capability was realized and investigated by Holt in 1952 and by Galletly and Bart in 1956. It was studied for more general imperfections by Donnell in 1956. By this time, it was recognized that buckling could occur when the cylinder was in the plastic range. In 1957, Ades studied this problem considering ovalization during bending. Axial and radial loads were again studied by Stuiver and Tomalin in 1959. In 1962, Kennedy and Venard presented a summary of developments to include radially loaded circular cylinders. In 1967, the effect of longitudinal bending on out-of-roundness was reported by Broussard, Barry, and McCarron. In 1970, Heise and Esztergar studied a host of codes related to elasto-plastic buckling response to uniform radial pressure and suggested improvements in allowable stresses. In 1971, two API reports showed results for casings subject to axial and circumferential loads. In 1973, Mesloh, Sorenson, and Atterbury showed nondimensional, quantitative results for circular cylinders subject to bending, and uniform radial pressure as applied to marine pipe-lay problems.

The problems caused by buckling in the circumferential direction due to hydrostatic pressure and longitudinal bending as independent factors are of a simple nature compared to buckling under a combination of these loadings. The following sections discuss each mode of buckling separately. Emphasis is placed on the various parameters affecting buckling, such as out-of-roundness, length-to-diameter ratio, diameter-to-thickness ratio, safety factors, yield strength and modulus, and the nonlinearity of the stress-strain curves. Once these factors are established for circumferential and longitudinal buckling separately, the problem becomes one of relating the two modes in a meaningful fashion to obtain the combined effect. A method is given here for combining these loadings. Graphical results for use in design and in marine pipe-lay operations are also presented.

TABLE 58
Nomenclature

D	= Outside diameter, in.
E	= Young's modulus, psi
E′	= Local slope of stress-strain curve, psi
f( )	= Mathematical representation for a function, __ __ __
I	= Moment of inertia at buckling, in.^{4}
K	= Ratio of inside to outside diameter, numeric
M_{cr}	= Longitudinal bending moment at buckling, in-lb
M	= Longitudinal bending moment, in-lb
N	= Ratio of D/t, numeric
P_{cr}	= Uniform radial pressure at buckling, psi
P	= Uniform radial pressure, psi
P^*	= Out-of-round uniform radial failure pressure, psi
Q	= Ratio of maximum to minimum diameters at place of greatest distortion, numeric
R	= Outside radius, in.
R^*	= Radius to maximum eccentric point, in.
S	= Maximum circumferential stress at failure with no bending, psi
S_B	= Circumferential bending stress at failure, psi
S^*	= Maximum axial stress at failure, psi
S.F.	= Safety factor, numeric
t	= Wall thickness, in.
W	= Water depth to pipeline section with maximum moment, ft
μ	= Poisson's ratio, numeric
σ_Y	= Yield stress, psi

Buckling Due to Hydrostatic Pressure

Published literature for the most part has been directed at circumferential elastic buckling of long circular cylindrical shells subject to uniform radial pressure. Some of these theoretical elastic buckling formulas based on perfectly circular cross sections are given in table 59. These derivations based on thin shell theory do not apply to large thickness-to-diameter ratios.

To consider buckling in the plastic range, Holmquist and Nadai developed a modified modulus for use in Timoshenko's elastic buckling equation. They also developed an approximate relation for plastic buckling due to hydrostatic pressure, and compared their work with the equations developed by Clinedinst. A summary of these relations are given in table 61.

The obvious question asked in many of the investigations is how close is the comparison between experimental results and theoretical predictions for buckling. Many investigators found that discrepancies were considerable for the lower diameter-to-thickness ratios. Figure 4.21 presents results of Jasper and the elastic theory of Timoshenko. Stewart ran experiments on 10-in. diameter steel tubes hav-

TABLE 59
Elastic buckling formulas with hydrostatic pressure, no out-of-roundness

Eq. No.	Equation	Source	Date
(4.21)	$P_{cr} = \dfrac{2E}{(1 - \mu^2)\left(\dfrac{D}{t} - 1\right)^3}$	Levy	1885
(4.22)	$P_{cr} = \dfrac{2E}{(1 = \mu^2)\left(\dfrac{D}{t}\right)\left(\dfrac{D}{t} - 1\right)^2}$	Clinedinst	1939
(4.23)	$P_{cr} = \dfrac{2E}{(1 - \mu^2)\left(\dfrac{D}{t}\right)^3}$	Timoshenko-Gere	1961
(4.24)	$P_{cr} = \dfrac{2E}{(1 - \mu^2)}\left[\left(\dfrac{t}{D}\right)^3 - \left(\dfrac{t}{D}\right)^4\right]$	Lame	—
(4.25)	$P_{cr} = \dfrac{2E(3 + f(D/L))}{3\left(\dfrac{D}{t}\right)^3(1 - \mu^2)}$	Sturm	1936

TABLE 60
Plastic buckling formulas with hydrostatic pressure, no out-of-roundness

Eq. No.	Equation	Source	Date
(4.26)	$P_{cr} = \dfrac{8EE'}{(1 - \mu^2)\left(\dfrac{D}{t}\right)^3(\sqrt{E} + \sqrt{E'})^2}$	Holmquist-Nadai (Timoshenko)	1939
(4.27)	$P_{cr} \approx \dfrac{8EE'\left(1 + \dfrac{3t}{D}\right)}{(1 - \mu^2)\left(\dfrac{D}{t}\right)^3}$	Holmquist-Nadai	1939
(4.28)	$P_{cr} = \dfrac{8EE'\left(1 + \dfrac{2t}{D}\right)}{(1 - \mu^2)\left(\dfrac{D}{t}\right)^3}$	Clinedinst	1939

ing 0.15–0.20-in. wall thickness and reduced his work to empirical equations for design. Other relations were developed by Bresse and Timoshenko, and a summary of these relations are given in table 61.

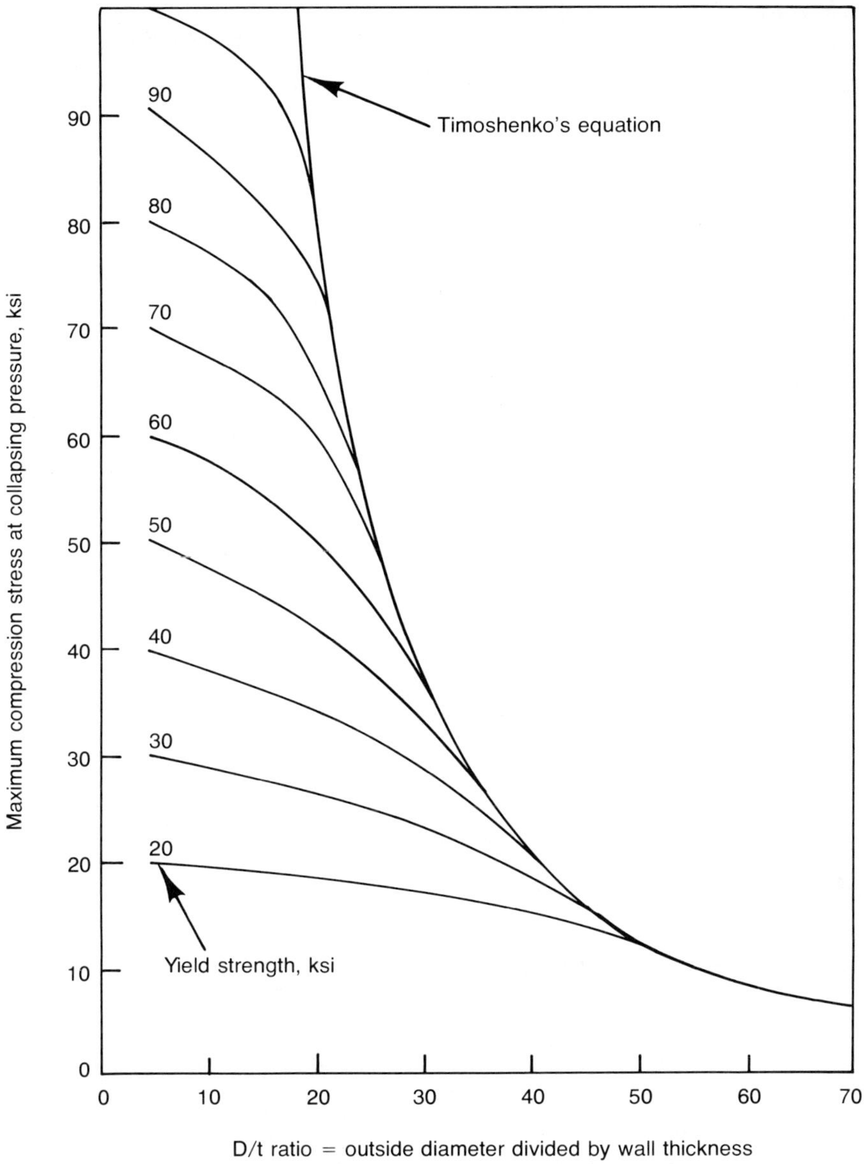

FIG. 4.21 Influence of compression yield point on the collapse resistance of pipelines

When the actual eccentricity cannot be evaluated, these formulas lose their significance. Many investigators have developed curves which fit experimental data up to the yield stress of the material. These empirical formulas seem to be the most reliable for

TABLE 61
Elastic buckling with hydrostatic pressure with out-of-roundness

Eq. No.	Equation	Source	Date
(4.29)	$P^{*} = \dfrac{(0.0926)(P_{cr} - 47.6)}{(Q - 0.874)^{1.25}} + 47.6$	Stewart	1907
(4.30)	$S_B = \dfrac{3P_{cr}ED^2}{8t^2\left(1 - \dfrac{P_{cr}t^3}{2ED^3}\right)} \left(\dfrac{R^{*2} - R^2}{R^{*2}}\right)^2$	Bresse	1886
(4.31)	$p^{*2} - \left[\dfrac{2\sigma_Y t}{D} + \left(1 + \dfrac{6e}{t}\right) P_{cr}\right] P^{*} + \dfrac{2\sigma_Y t}{D} P_{cr} = 0$	Timoshenko	1933

design purposes for the whole range of diameter-to-thickness ratios. A host of empirical equations are available and are summarized in table 62.

The discussion thus far has made no mention of the effect of length-to-diameter ratios on buckling. Jasper and Sullivan have shown that the radial collapse pressure is essentially the same for all length-to-diameter ratios over eight. Cylinders in this range are referred to as long cylinders.

Buckling with Longitudinal and Axial Forces

The problem of buckling in a suspended pipeline involves longitudinal bending and axial forces as well as uniform radial pressure. It will expedite matters to treat axial forces and longitudinal bending together rather than to cite the numerous references available in the open literature on axial buckling of noncurved shells. Flugge has shown that a satisfactory approximation of the safe side can be obtained by assuming that buckling occurs when the maximum compressive stress becomes equal to the critical stress calculated for symmetrical shell buckling (table 63).

Experimental tests by Lundquist and by Donnell showed a large discrepancy between theoretical values predicted by Flugge's equation and test results. This discrepancy is primarily due to imperfections (out-of-roundness and longitudinal dimples). The tests also showed that cylinders in either pure bending or pure axial

TABLE 62
Empirical buckling formulas for hydrostatic pressure with out-of-roundness

Eq. No.	Equation	Source	Date
(4.32)	$S = \sigma_y - \dfrac{\sigma_y{}^2 \left(\dfrac{D}{t}\right)^2 (1 - \mu^2)}{4E}$	Johnson	—
(4.33)	$S = \dfrac{37000}{1 - 0.00084 \left(\dfrac{D}{t}\right)^2}$	Stewart	1907
(4.34)	$S = \dfrac{\sigma_y E}{E + \sigma_y(1 - \mu^2) \left(\dfrac{D}{t}\right)^2}$	Southwell	1915
(4.35)	$S = \dfrac{\sigma_y E(1 + K)(K - 1)^2}{2K^2E(K - 1)^2 + 4\sigma_y(1 - \mu^2)(K^3)(K + 1)}$	Rhys	—
(4.36)	$S = \dfrac{52000}{1 + 0.0011 \left(\dfrac{D}{t}\right)^2}$	Jasper-Sullivan	1931
(4.37)	$S = \dfrac{\sigma_y}{2t}(2.71t - 0.043D)$	Werke	—
(4.38)	$S = \dfrac{\sigma_y}{2t}(2.45t - 0.045D)$	Dunn	—

TABLE 63
Buckling with longitudinal bending and axial force

Eq. No.	Equation	Source	Date
(4.39)	$S^* = \dfrac{Et}{R\sqrt{3(1 - \mu^2)}}$	Flugge	1932
(4.40)	$S^* = \dfrac{E\left(0.6\,\dfrac{t}{R} - 10^{-7}\,\dfrac{R}{t}\right)}{1 + 0.004\,\dfrac{E}{\sigma_y}}$	Timoshenko	1961

compression buckle at a much lower load than predicted by Flugge's formula. Furthermore, the tests revealed that the lower values of buckling pressures for both pure bending and pure compression were nearly the same at all diameter-to-thickness ratios. Based on these results, Timoshenko developed an empirical equation for pure axial compression (table 63). Since the lower bounds for pure axial compression can be equated to pure longitudinal bending, the equation is applicable for pure longitudinal bending as well. Values using Timoshenko's equation are plotted in figure 4.22.

Effect of Diameter-to-Thickness Ratios on Marine Pipe-lay

More recently Mesloh, Sorenson, and Atterbury found that the longitudinal bending moment causing buckling varies as an inverse power of the diameter-to-thickness ratios. Ades used a more rigorous approach to evaluate the bending strength of circular tubes considering ovalization during bending in the plastic range. Broussard, Barry, and McCarron reported the amount of ovalization realized for 30-inch diameter pipes with 0.5- and 0.562-inch wall thicknesses using marine pipe-laying operations. They showed that only about one-tenth of one percent out-of-roundness occurred at the yield stress.

In order to predict whether a pipeline will fail due to stress or to elastic buckling during marine pipe-laying operations, it is advantageous to categorize pipelines according to diameter-to-thickness ratios. These ratios are important because they obviously influence

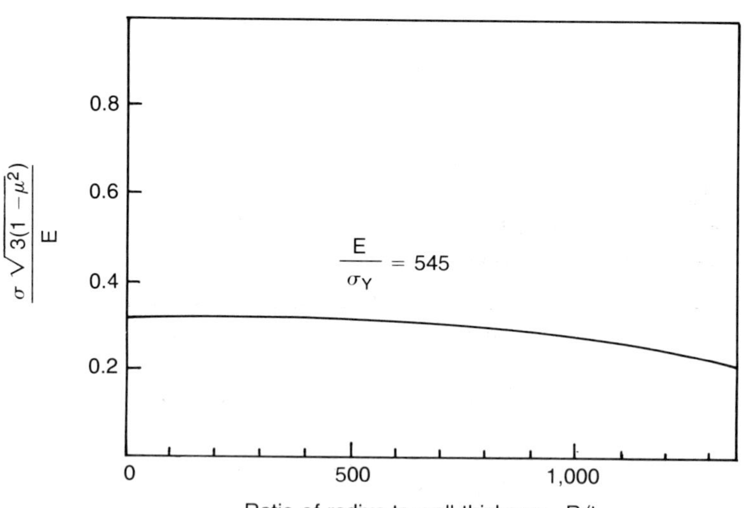

FIG. 4.22 Buckling stress due to bending

the stress level in the pipeline wall. The higher the stress, the more the out-of-roundness. Also, as the stress increases, the stress strain curve may become more nonlinear even below the yield stress.

The diameter-to-thickness ratio of which these factors influence pipeline response has been shown to be around 25 for both circumferential buckling and for pure longitudinal bending. This ratio also depends on the material properties. The effect of diameter-to-thickness ratios on buckling will contribute to the best choice of reliable governing equations for prediction of pipeline buckling.

Formulation of Elastic Buckling Analysis in Marine Pipe-lay

Buckling in a pipeline during the marine pipe-lay operation can present many problems. The pipeline is not straight and is subject not only to uniform radial forces but also to axial forces and longitudinal bending. When considering uniform radial forces, out-of-roundness must be taken into account as well.

It has been shown by Mesloh, Sorenson, and Atterbury that the two major factors influencing buckling of offshore pipelines are bending moment and uniform radial pressure. To determine the critical hydrostatic pressure at which buckling occurs with out-of-roundness and without any longitudinal bending, we recommend that the empirical equation of Southwell be used. The maximum amount of ovalization realized at the yield stress is about one-tenth of one percent, as shown by Southwell and others.

With larger eccentricities due to fabrication or other factors, one should consider another of the reliable out-of-roundness equations.

It is recommended that the critical bending moment at which buckling occurs, without any hydrostatic pressure, be calculated by Timoshenko's empirical equation. Using these two governing equations for critical hydrostatic pressure without longitudinal bending and critical bending moment without hydrostatic pressure, buckling with hydrostatic pressure and longitudinal bending acting simultaneously can be predicted. Figure 4.23 shows a straight design line, and every point above the line will possibly lead to buckling. The area below the line is the safe region. The curved line in figure 4.23 is the actual buckling envelope.

Solving the bending moment, M, in the first equation listed in table 64 by substitution of the variables from the next three equations results in the design equation for allowable bending moment (also given in table 64). From this equation, the bending moment, M, becomes the critical buckling moment, Mcr, at any water depth, W, for a given pipeline. Figure 4.24 shows a plot of maximum allowable bending moment vs. water depth for pipeline diameters of 12, 24,

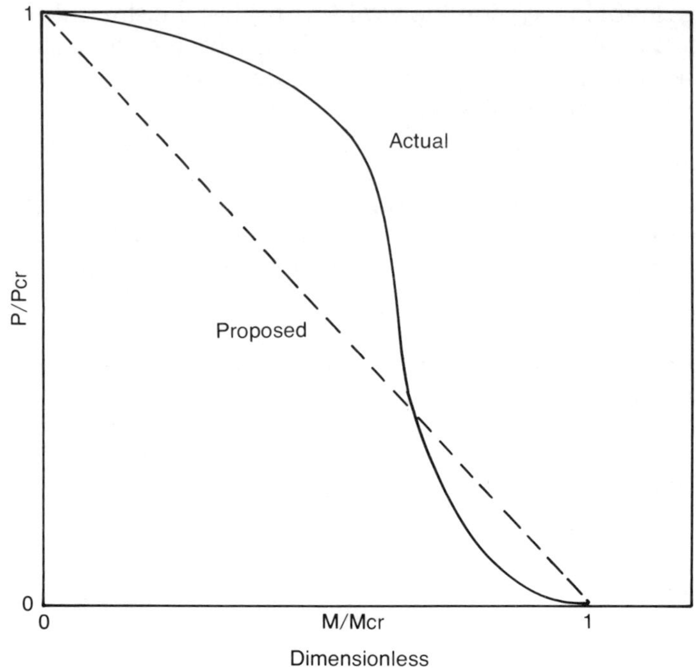

FIG. 4.23 Failure envelope for pipelines with uniform pressure and longitudinal bending forces

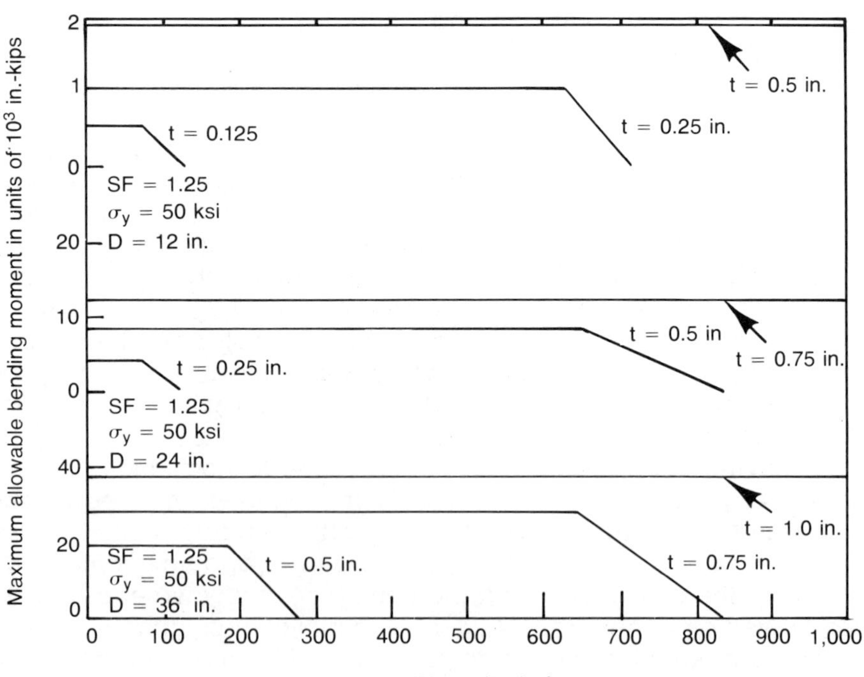

FIG. 4.24 Maximum allowable bending moment vs. water depth for various pipeline diameters

TABLE 64
General elastic buckling criteria for marine pipelay

Eq. No.	Equation	Source	Date
(4.41)	General Failure Criteria $\frac{1}{SF} = \frac{P}{P_{cr}} + \frac{M}{M_{cr}}$	Mesloh, Sorenson, and Atterbury	1973
(4.42)	Hydrostatic Buckling Pressure without Longitudinal Bending $P_{cr} = \frac{2\sigma_y E}{NE + \sigma_y N^3(1 - \mu^2)}$	Southwell	1915
(4.43)	Longitudinal Bending without Hydrostatic Pressure $M_{cr} = \frac{\sigma_y EI \left(\frac{2.4}{N} - 10^{-7}N\right)}{D(\sigma_y + 0.004\ E)}$	Timoshenko	1961
(4.44)	Hydrostatic Pressure $P = \frac{64W}{144}$	—	—
(4.45)	Allowable Bending Moment $M = \frac{\pi t D^2 \left(\frac{2.4}{N} - 10^{-7}N\right)}{8(\sigma_y + 0.004E)} \times \left[\frac{\sigma_y E}{SF} - 0.222\ W\ (NE + 0.91\sigma_y N^3)\right]$	—	—

and 36 inches with a safety factor of 1.25. The oblique lines predict buckling failure below the yield stress, and the horizontal lines predict the moment at which the yield stress is reached. Figure 4.25 shows the effect of different yield stresses on the maximum allowable bending moment, again with a safety factor of 1.25.

From the open literature, three equations were selected which govern circumferential buckling with out-of-roundness, buckling due to longitudinal bending, and hydrostatic pressure. These three equations were combined into a single equation to determine the maximum bending allowable, and the results were plotted. The results showed that buckling in a marine pipeline can definitely occur below the yield stress in a practical pipe-laying operation. It can also be the governing mode of failure in deep water for realistic combinations of diameter and wall thicknesses.

When buckling is considered to be a possible mode of failure, buckling arrestors should be employed. When such discrete stiffeners are used, they must be considered in the dynamic analysis of the

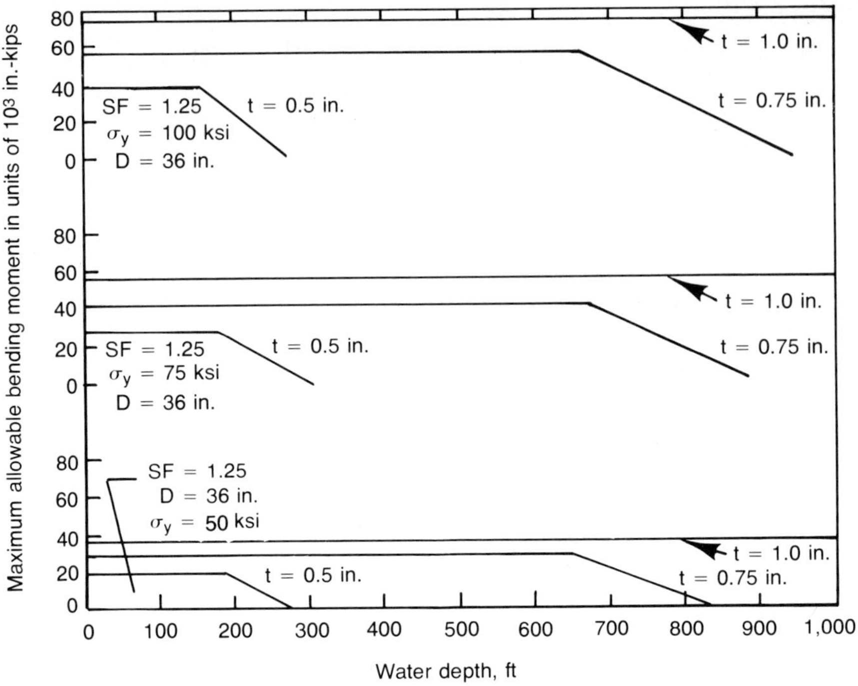

FIG. 4.25 Maximum allowable moment vs. water depth for various yield stress values

pipeline. Computerized analytical techniques can be used expeditiously to accurately determine bending moments and the deflected shape and to account for point loads due to discrete stiffeners.

4.3 COMPUTER-SIMULATED SUBSEA PIPE-LAY PROBLEMS

Pipeline abandonment and recovery in rough-water areas such as the North Sea can occur as often as once a month. Moreover, costs to the operator for marine pipe-lay operations can cost as much as $100,000/day.

Therefore, downtime during pipe-lay and recovery operations must be reduced. To this end, a variety of new design developments and pipe-lay conditions can now be modeled with analytical computer tools. One such tool is ETA, a digital-computer routine developed by Engineering Technology Analysts, Inc.

In this program, input includes the pipeline geometry and mechanical factors such as barge tension, water depth, wave height, and current speed. Computer output gives resultant pipeline coordinates, forces, moments, stresses, strains, and curvature at any prescribed point.

Pipe-lay Considerations

Table 65 gives a capsule view of pipe-lay methods in use today. The curved-stinger method using tensioners is more versatile than other methods. But the larger the stinger, the more apt it is to be damaged or lost in deepwater operations.

To reduce either the length or size of the stinger, curved instead of straight stingers are now commonly used. But short, curved stingers usually require the use of tensioners.

Marine pipelines have been laid primarily by four general methods: pipe-lay barges, reel barges, flotation, and bottom pull.

A list of the design procedures leading up to marine pipe-lay operations is shown in figure 4.26. This chart can only indicate the varied factors involved in the early planning stages of the job. Types of subbottom profiles normally used in the surveys are given in table 66.

For pipe-lay, the operator and contractor must work in a small area of critical allowables. For example, if rough water comes up and the pipeline is abandoned too soon, valuable time is lost and costs are escalated.

On the other hand, waiting too long to abandon the pipeline during large vessel motions could create excessive pipeline tensions or damage the stinger. These tensions could cause a propagating buckling of the pipeline.

Computer Analyses

Use of experimental-model tests is often required to verify design of new developments. This is not easily done parametrically. Improved theoretical methods of pipe analysis during laying and recovery operations should be used.

TABLE 65
Pipe-lay methods

Method	Advantage	Limitations
Nontensioned pipe lay	Simple, few mechanical problems	Long stinger required, not practical for deep water
Tensioned, straight stinger	Deepwater capability	Economics of stinger design
Tensioned, curved stinger	Deepest-water capability	Tensioner hardware
Reel barge	Fast, economical	Smaller-size pipe
Pull from shore	Onshore pipeline assembly	Relatively short distances from shore

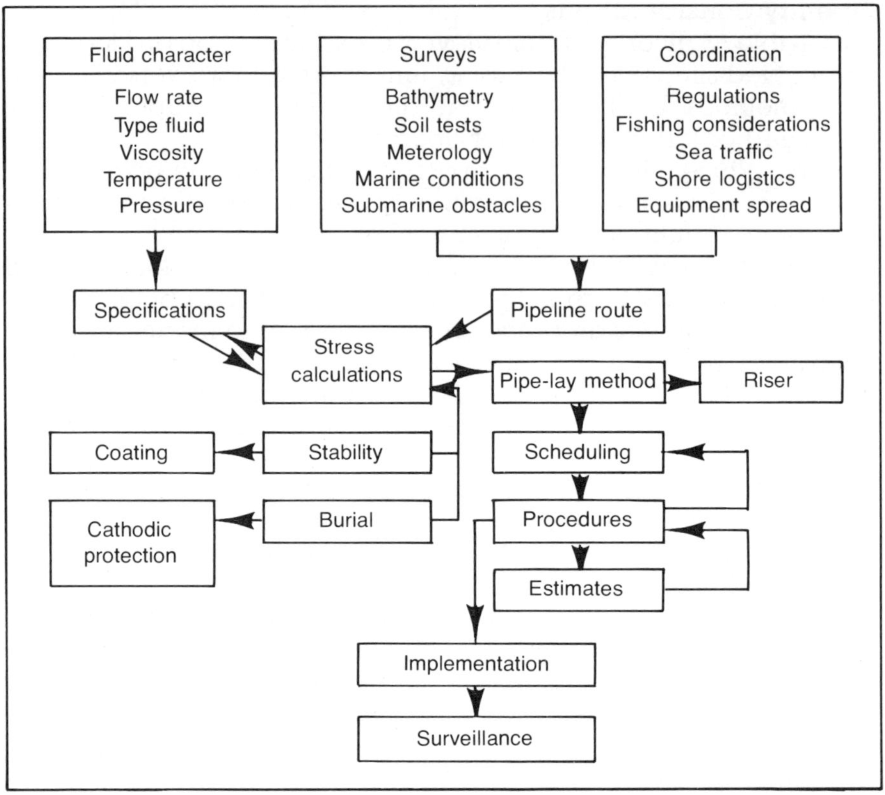

FIG. 4.26 Design procedures for marine pipe-lay operations

TABLE 66
Summary of subbottom profiles

System	Resolution, m	Penetration, m	Sound source
Sparker	4	400	Capacitance discharge
Boomer	1.5	100	Electromechanical
Ecosounder	0.1	5	Electromechanical
Subbottom profiler	0.5	70	Fixed-frequency transducer

A wide variety of new designs and pipe-lay conditions can now be modeled with analytical computer tools. The ETA/Pipelay computer routine is used to analyze such marine pipe-laying problems as the Forties field to Cruden Bay pipeline in the North Sea.

There, 32-in. pipe in up to 600-ft water depths under severe criteria was analyzed. The program determined effects of wave, current, and barge motion on stresses in the stinger and pipe as it was being laid.

The program considers large deflections during pipe-lay. It can handle any pipeline size, shape, and stiffness, variable tensioner forces, and configurations. It also handles different water depths, wave and current forces in any direction, and various bargemooring systems.

ETA/Pipelay considers the effects of the dynamic response of the pipe-lay barge on the pipeline through a quasistatic technique.

Program capability exists to model-buckling arrestors, valves, and various types of connectors. Results have closely compared experience with strain-gauge measurements taken during a series of actual pipe-laying situations.

Different Approaches

Studies have compared various procedures for designing and analyzing marine pipelines. The approaches that have been used, and their relative merit are as follows.

Perhaps the simplest and most direct method would be the use of catenary equations. This technique, however, gives only "ballpark" types of answers and is inadequate for any changes of section properties. It cannot be used in analyses to account for pipeline valves or buckling arrestors.

Also, closed-form solutions of differential equations have been used. Here, however, the solution becomes unwieldy if more than one or two changes in section are considered.

As a third technique, the program uses a nonlinear, finite-element program which provides solutions with no restrictions. Comparison of ETA/Pipelay results with linear, finite-element programs shows that the linear approaches overestimate stresses by about 50%. Catenary equations underestimate the stresses by about 50%.

Thus, use of accurate methods to estimate stresses becomes critical in operations which are near design limits.

Analyses using catenary equations would be in error for large-diameter or thick-walled pipes since the catenary equations neglect pipe rigidity.

Linear, finite-element analyses could also cause considerable error. This is because large displacements of small diameter pipe are not properly accounted for.

ETA computer input, on the other hand, includes pipeline geometry and mechanical considerations. It also includes barge tension, water depth, wave height, and current speed.

Computer output provides resultant pipeline coordinates, forces, moments, stresses, strains, curvature, and design factors at any point along the line.

Input results are three-dimensional so the pipeline can have a varying coating or wall thickness. With this method, currents and waves can be applied from any direction.

Pipeline behavior can be simulated when it is relatively rigid—48-in.-OD pipe in 200 ft of water—or when it is relatively flexible—12.75-in.-OD pipe in 1,500 ft of water.

Example Analysis

Sample-problem variables are given in table 67 for both pipe-lay and recovery. The problem portrays a pipeline pulled by wire rope up to 35 ft below the end of the stinger.

Coordinates of the pipe recovery configuration are 20 ft or less below the pipe-lay curve for the overbend region (figure 4.27). They are about 10 ft maximum above the pipe-lay curve for the sag-bend region.

It was found that 125 kips at the tensioner is about right for this pipe-lay problem. This value is, therefore, assumed for recovery operations.

TABLE 67
Variables for sample problem

410 ft	Water depth
15 ft	Distance from tensioner to main waterline
125 kip	Barge tension
1,000 ft	Curved stinger radius
30×10^6 psi	Young's modulus of pipe
0.3	Poisson's ratio of pipe
36 in.	Pipe outside diameter
0.75 in.	Pipe wall thickness
3.5 in.	Pipe coating thickness
50 ksi	Pipe stress allowable
734 lb/ft	Weight of coated pipeline in air
89.3 lb/ft	Weight of coated pipeline in water
62.4 lb/ft^3	Weight of water
10×10^6	Effective Young's modulus of recovery wire rope
3 in.	Diameter of recovery wire rope
7.66	Specific gravity of steel
2.33	Specific gravity of coating
3.45	Specific gravity of coated pipe in air
1.17	Specific gravity of coated pipe in water

Distribution of forces, moments, and flexural stress along the pipeline for pipe-lay and recovery are then compared in figures 4.28, 4,29, and 4.30.

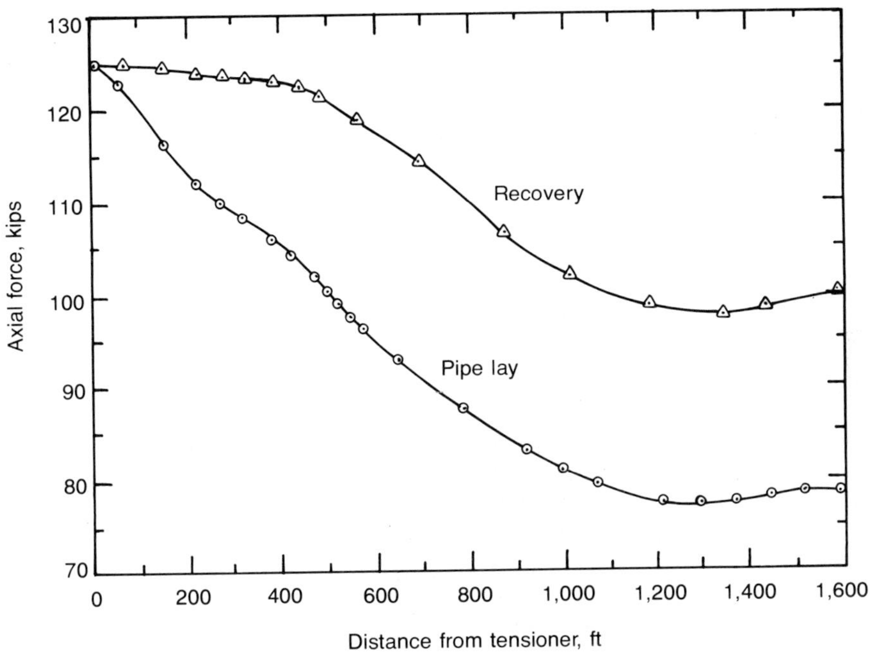

FIG. 4.27 Pipeline coordinates

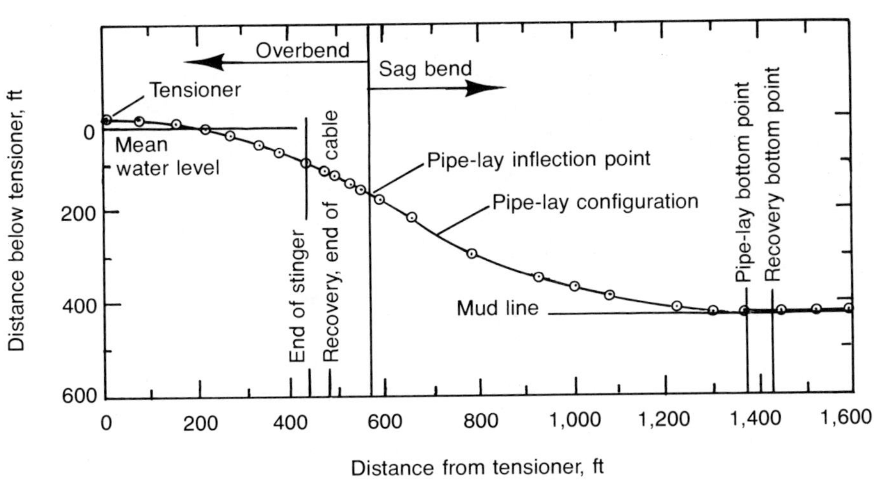

FIG. 4.28 Axial forces

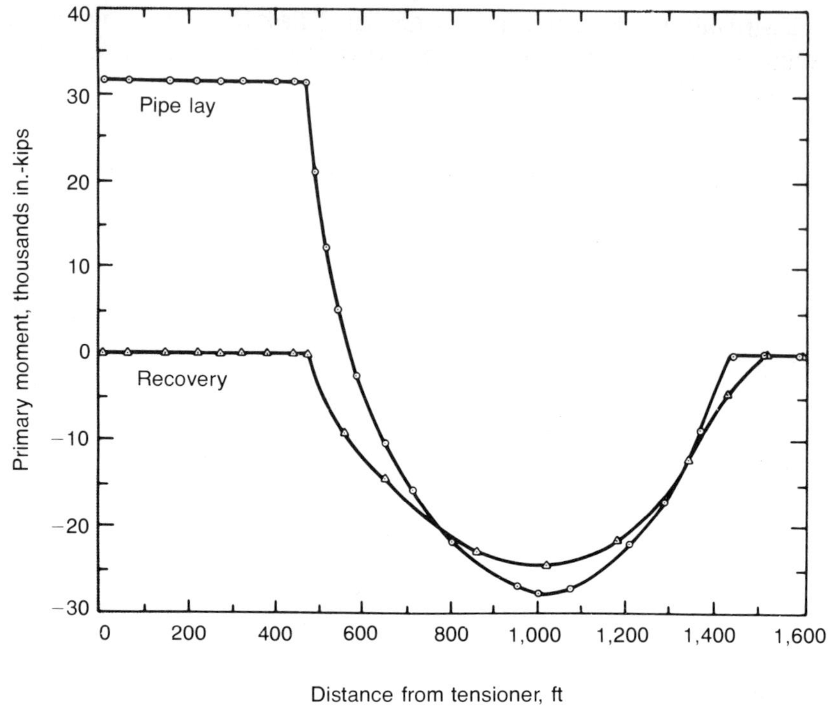

FIG. 4.29 Primary moments

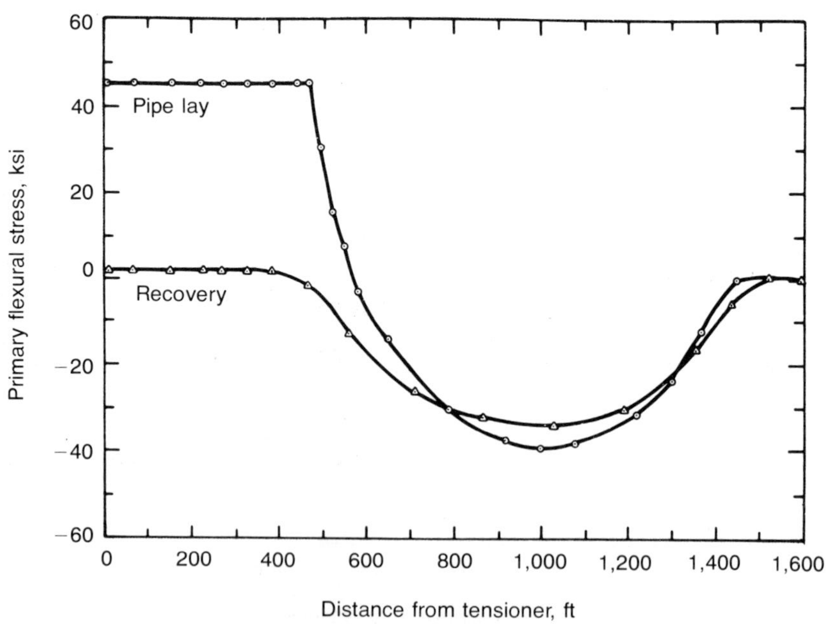

FIG. 4.30 Primary flexural stress

Combined stress or design factors are shown in figure 4.31. This indicates that pulling with the wire rope in recovery while the force is set at the winch-tensioner valve gives a satisfactory operation.

This is true even though pipe-lay has an overbend and sag bend while the recovery operation puts only a sag bend in the pipe. The combined strains are compared in figure 4.32.

In the sample problem for 410-ft water depths, the horizontal distance from the tensioner to the point of mud line contact is about 1,400 ft (figure 4.27). This is for both the pipe-lay and the recovery operation.

The aspect ratio of this horizontal distance divided by the water depth here is about 3½. This is an average value for most pipe-lay operations.

In both pipe-lay and recovery, the axial force is a maximum at the tensioner and decreases with distance below the mean waterline (figure 4.28).

In the sag-bend area, the axial force is about 60% of the set tensioner force during pipe-lay. The axial force is about 80% of the winch-tensioner force during recovery.

During the pipe-lay operations in the sample problem, the maximum moment of 32,000-in. k is constant along the stinger. It then decreases to a maximum negative value of 28,000-in. k in the sag-bend area of the pipeline (figure 4.29).

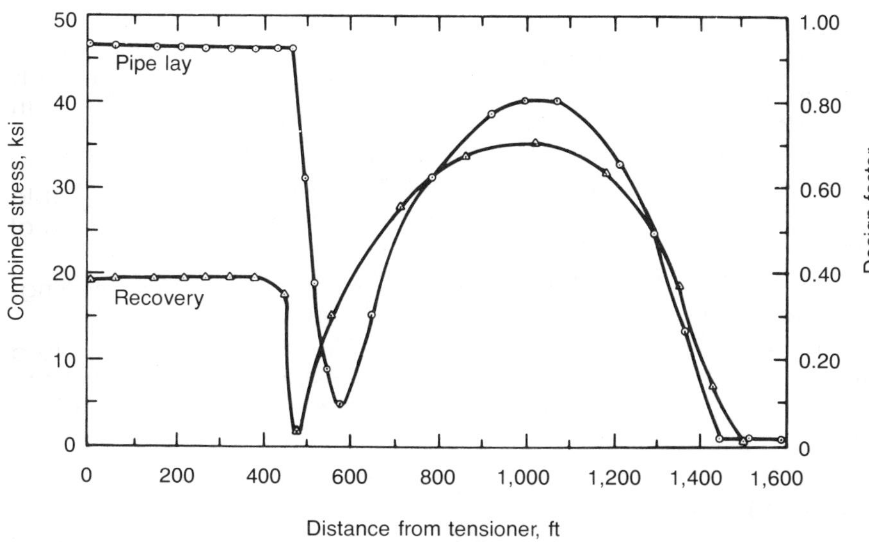

FIG. 4.31 Combined stress and design factor

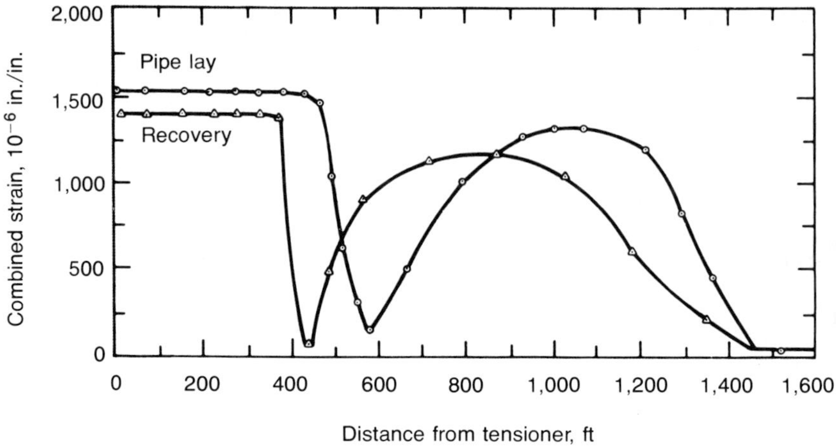

FIG. 4.32 Combined strain

A wire rope runs from the capped end of the pipeline up to a winch during recovery, and there is no moment taken by the rope. In the sag-bend area, the negative moment during recovery is only slightly less than the negative moment during the pipe-lay operation.

Since the primary stress is directly proportional to the primary moment, the primary stress response is similar (figure 4.30).

Combined stress (figure 4.31) is taken here as the sum of the absolute values, not the algebraic sum, of axial stress and flexural stress. The maximum combined strain is handled in a similar manner (figure 4.32).

Actually, the flexural stress on the top side of the pipeline is tension in the overbend area up to the inflection point. It is then in compression from the inflection point down to the mud line.

Flexural stress on the bottom side of the pipeline is negative in the overbend and positive in the sag-bend area. A stress allowable of 50 ksi (table 67) is used which is normally taken as about 85% of the tensile yield strength.

With this, the maximum combined stress is highest on the stinger; it decreases and then peaks again in the sag-bend area.

The peak combined stress in the sag-bend is equivalent to a design factor of about 0.8 for the pipe-lay operations (figure 4.29). This means that the peak stress is about 80% of the 50 ksi allowable.

Water-depth Considerations

Present water-depth capability for marine pipe-lay is only slightly greater than about 1,200 ft. As recent as 15 years ago, the limiting water depth for 30-in. marine pipelines was 200 ft. It is now possible to lay 30-in. lines in 800-ft water depths.

A number of schemes have been proposed for extending pipe-lay water-depth capability such as the use of inclined ramps on semisubmersibles.

Although there are operational problems to overcome, the semisubmersible is considerably more stable in rough seas than a conventional barge. And an inclined ramp restricts high stresses to only the sag-bend region.

With the conventional stingers, however, pipeline stresses peak both in the overbend and in the sag-bend regions.

Capability limits of inclined ramps are expected to vary from about 800-ft water depths with 36-in. lines up to perhaps 1,400 ft with 6-in. lines.

A major problem is that considerably more sophisticated equipment will be required with inclined ramps. For example, almost completely automated welding, X-ray, and coating-doping stations will be required.

An example of current water-depth requirements can be found in the Frigg field. A part of the field is in Norwegian waters only about 100 miles from the coast. But the 1,000-ft-deep Norwegian trench has detered pipe-lay operations with existing equipment.

It is reported, however, that the third-generation pipe-lay vessels will be able to lay pipelines at this great water depth. These vessels include:

- Saipem's Castoro VI, semisubmersible pipe-lay vessel.
- The Viking semisubmersible.
- Santa Fe's new reel barge.

Sample Problem

Using the ETA program, the forces, moments, stresses, and strains were predicted for laying pipe in 235 and 410-ft water depths (table 68).

Wave and current forces were not applied to this particular study. Only barge tension, gravity, and friction forces at the mud line were considered for each water depth. Results show the small effect that variations in moderate water depths have on pipelines laid in calm seas.

Tensioner force set for 235-ft depth was 120 k whereas the tensioner force for 410 ft was 124 k. This difference in tensioner force is negligibly small for a practical situation. Thus, the tensioner force can be considered constant in this example.

TABLE 68
Variables for water depth

234 and 410 ft	Water depth from mean water level to mud line
120 and 124 kips	Barge tension
15 ft	Distance from tensioner to mean waterline
1,000 ft	Curved stinger radius
36 in.	Pipe outside diameter
0.75 in.	Pipe wall thickness
3.5 in.	Pipe coating thickness
734 lb/ft	Weight of coated pipeline in air
89.3 lb/ft	Weight of coated pipeline in water
7.66	Specific gravity of steel
2.33	Specific gravity of coating
3.45	Specific gravity of coated pipeline in air
1.17	Specific gravity of coated pipeline in water
30×10^3 ksi	Young's modulus of the pipe
0.3	Poisson's ratio of the pipe
50 ksi	Allowable stress in the pipe

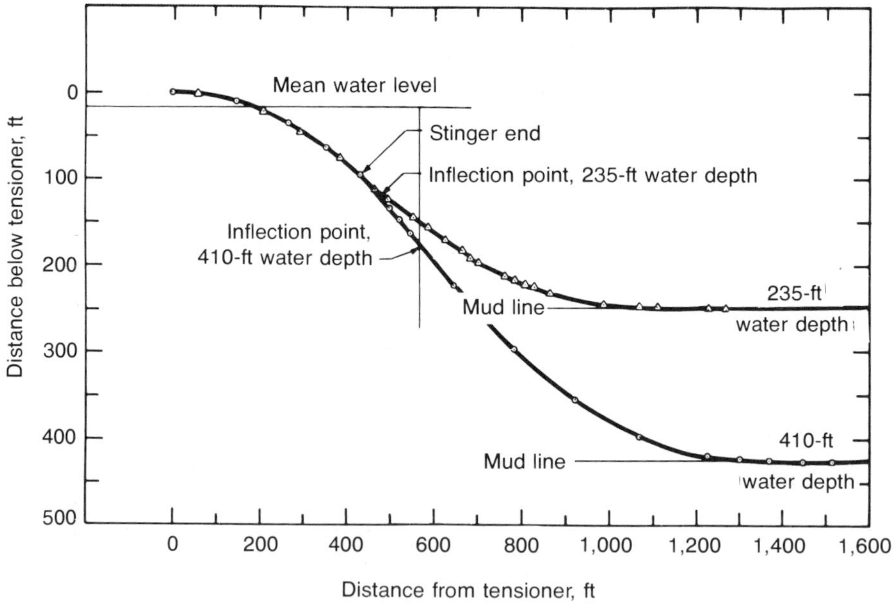

FIG. 4.33 Pipeline coordinates

At 235 ft, the distance from the tensioner to the inflection point in the pipeline was 452 ft, vs. 570 ft for 410-ft water depth (figure 4.33).

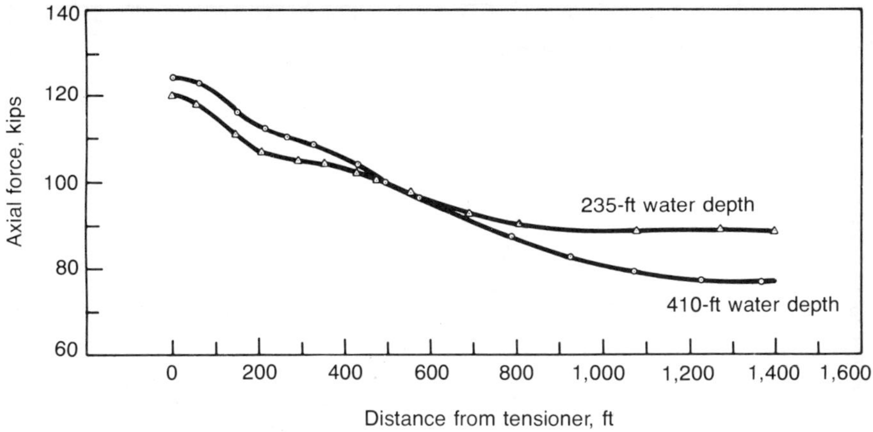

FIG. 4.34 Axial forces

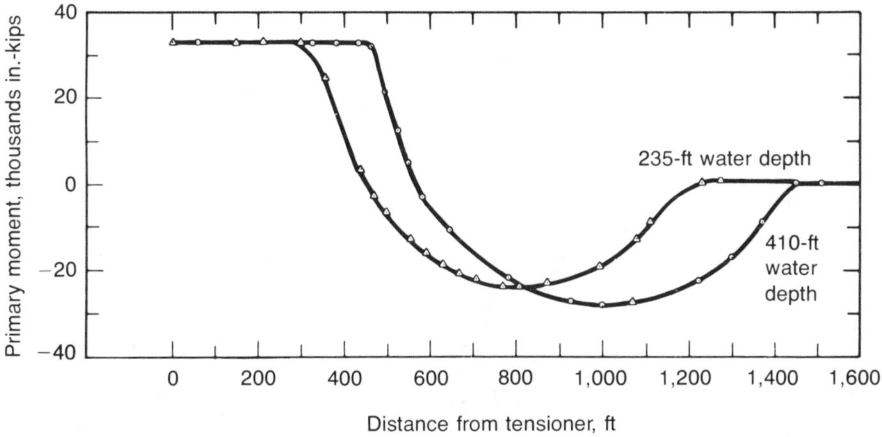

FIG. 4.35 Primary moment

Horizontal distance from the tensioner to the mud line tangent point was 1,151 ft for 235-ft water depth and 1,370 ft for 410-ft water depth.

Distance from the tensioner along the pipeline to the mud line tangent point was 1,188 ft for 235-ft water depth and 1,457 ft for 410-ft water depth.

Axial force in the pipeline decreases with distance below the mean waterline (figure 4.34). The axial force in the sag-bend region decreases to 101 kips for 235-ft water depth and to 96 kips for 410-ft water depth.

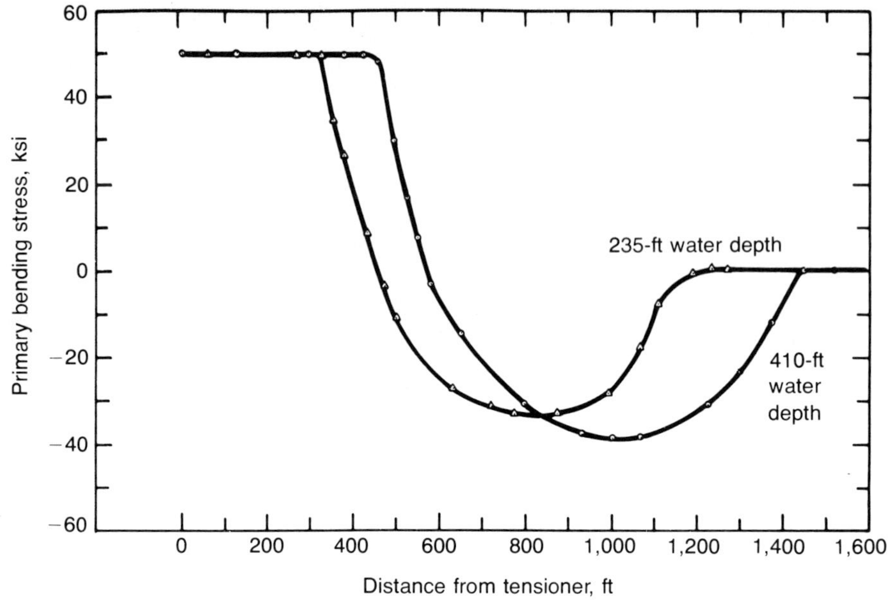

FIG. 4.36 Primary bending stress

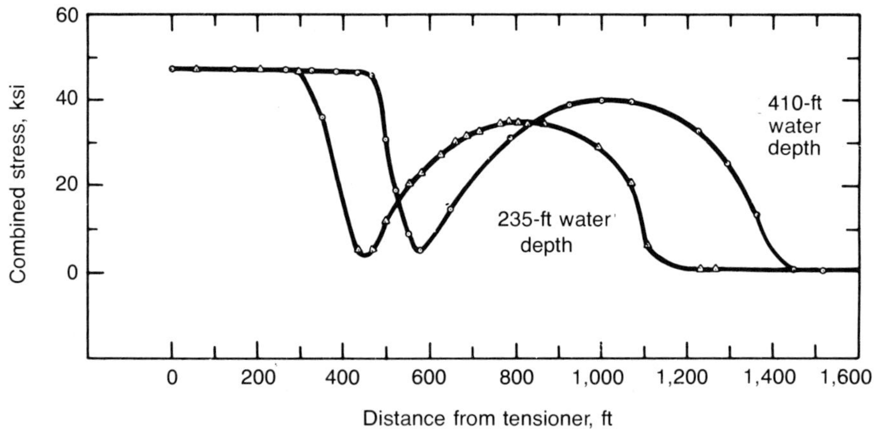

FIG. 4.37 Combined stress

The primary moment in the section of the pipeline on the stinger was 32,000-in. kips for both water depths (figure 4.35). The same stinger curvature was used in both cases.

Maximum primary moment in the sag-bend area was 24,000-in. k for 235-ft and 28,000-in. k for 410-ft water depth.

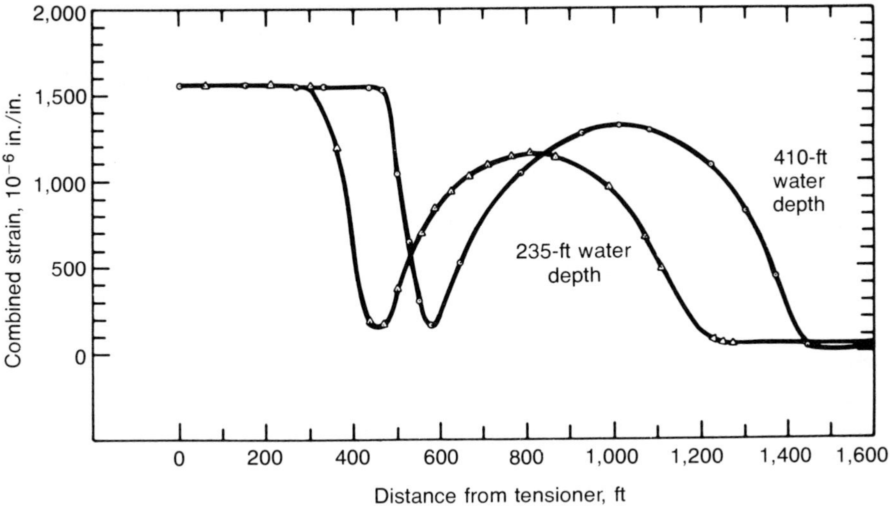

FIG. 4.38 Combined strain

Primary bending stress in the sag-bend area of the pipeline peaked at a negative 33.5 ksi for 235-ft and a negative 38.9 ksi for 410-ft water depth (figure 4.36).

Maximum combined stress in the pipelines on the stinger was 46.4 k for both water depths (figure 4.37).

Maximum combined stress in the pipeline in the sag-bend area was 34.6 ksi for 410 ft. Based on a 50-ksi design stress allowable, this gave sag-bend stress factors of 0.69 for 235-ft and 0.80 for 410-ft water depth.

As discussed earlier, the combined stress is the sum of the absolute values, not the algebraic sum, of axial stress and flexural stress. Maximum tensile stresses occur on the top side of the pipeline in the overbend and on the bottom side of the pipeline in the sag bend.

Maximum combined strain in the pipeline on the stinger was 0.15% for both water depths (figure 4.38). The maximum combined strain in the sag-bend region of the pipeline was 0.13% for 235 ft and 0.12% for 410-ft water depth.

We would expect considerably different results when wind, waves, current, and barge motion are taken into account.

4.4 TENSIONER VARIATIONS AFFECT STINGER DESIGN

Constant-force tensioners are seen as a means of increasing pipe-laying capability in deep water depths under severe conditions. But it is essential to be able to predict pipeline response at different tension values.

Sample cases show that when barge tension is increased from 80 to 120 k deflection is reduced by about 10 ft.

Reduction in deflection also causes reduction in curvature and hence in the bending stresses. Combined stresses are shown to be reduced by about 25% in the sag-bend region with a 50% increase in tensioner force.

Stingers

The types of equipment available greatly affect overall pipe-lay operation. The type of stinger used, for instance, affects the required tensioner force. Four types of stingers have been used thus far:

- Rigid straight.
- Flexible straight.
- Curved.
- Articulated.

Most pipelines to date have been laid with the rigid straight stingers.

Severe limitations can be placed on the contractor by lack of suitable equipment. Consider a 1965 pipe-lay operation in the Gulf of Mexico.

With a 20-in.-OD, 0.406-in.-W.T. pipeline having a 0.625-in. somastic coating plus a 3-in.-thick, 140-lb/cu ft concrete coating, the sp gr was 1.35. With a 2,000-ft sag-bend radius and a 292 ft-k moment, the sag-bend stress was 24 ksi and the overbend stress was 33 ksi.

Calculations showed that for the maximum 100-ft depth, the required length would be 2,400 ft for a straight stinger and 1,400 ft for a single curvature stinger.

With the use of a flexible stinger which allowed for double curvature (figure 4.39), required stinger length was reduced to 807 ft. This short stinger was used in the pipe-lay operations just described.

The first pipe-lay barges did not have tensioners. Thus, very long stingers were necessary to keep the pipeline slope near the bottom low enough to prevent overstress in the pipeline.

However, without tensioners, the required length was four to five times water depth. So, pipe-lay was limited to relatively shallow water.

Water-depth capability, however, has been increased by use of tensioners with curved stingers which can be much shorter than straight stingers. However, the rigid, curved stinger is limited to a relatively small water-depth range since the slope of the sag bend changes with water depth. But rigidity of curved or straight stingers makes them susceptible to damage due to vessel motion.

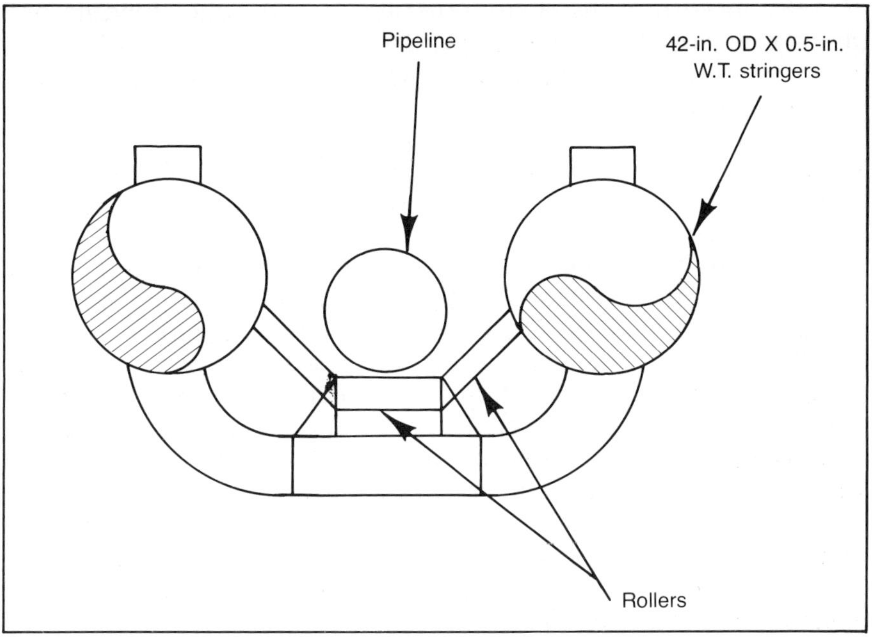

FIG. 4.39 Flexible stinger

Articulated stingers provide many more operational capabilities for pipe-lay operations. An articulated stinger consists of a series of connected segments, each having buoyancy chambers. Special hinge joints between segments allow for vertical flexibility.

With a different number of segments connected and with the segment chambers having adjustable buoyancy, pipe can be laid in a wide range of water depths.

Tensioner and Stinger Design

As pipe diameter increases, stinger length and required tensioner forces increase.

There are, however, technical and economical limits on stinger lengths. For example, as a rule of thumb, stingers over 300 ft long cost about $1,000/ft. This means that a 400-ft stinger would cost about $400,000.

The greater the stinger length, the higher the probability that the stinger will be damaged or lost. In addition, there are the costs of pipeline abandonment and recovery.

In the North Sea, severe storms have risen in as short a time as 5 hours. Such a short time precludes abandonment and shelter in time to prevent damage.

For these reasons, it appears that conventional stingers are almost pressed to their technological limits. This indicates that thicker pipe and larger-capacity tensioners will probably be used on new pipe-lay barges.

Required tension force on a pipe-lay barge is directly proportional to the thickness of concrete coating on the pipe. With the 32-in., Frigg field pipeline, for example, most of the line has a 1.16 sp gr with 1⅞-in. coating.

At the short end, the line starts out at 1.6 sp gr with 4⅞-in. coating of 190 lb/ft. It then decreases to 1.16 sp gr with four incremental changes—each change occurring about 8 miles apart.

Two tandem track-type Rasco tensioners designed to develop a total of 80 k tension and 88 fpm maximum payout are on Santa Fe's *Choctaw 1*. During the vessel's first operations in the Bass Strait, tensions as high as 75 kips were experienced. Normally, a 35- to 45-kip tension was set with surge fluctuations of 8 k.

A 50-ton constant tension winch is available for use in pipeline abandonment and recovery operations. The stinger hitch on the *Choctaw II* is designed for 1.66-million-lb horizontal reaction and 1-million-lb vertical reaction.

Tensioning capability on pipe-lay vessels varies much. With two Western Gear 100-k units, the ETPM 1601 has a tension capability of 200 k.

The Viking semisubmersible has two tandem track-type tensioners and a capacity of 300 kip at 175 fpm. Brown & Root's Bar 323 has three 75-kip track-type tensioners.

During actual marine pipe-laying operations, it is often desirable to adjust barge tension to reduce either sag or lateral deflection caused by waves and current. Too much sag will cause excessive bending stresses in the pipeline.

Axial Stress

Axial force increases with increased barge tension. But the overwhelming stress-producing factor is bending. Thus, overall combined stresses are reduced by increasing barge tension.

There is a point of diminishing return where an increase in tensioner force will not yield a corresponding reduction in deflection and stress. The capability now exists to determine the minimum required barge tension and the value at which a diminishing return is realized.

We have found that, for a given pipeline size and water depth, an optimum tension exists to give minimum pipeline stress (figure 4.40). This is because at low tensions, flexural stress is high. As tension increases, the pipeline straightens, flexural stress decreases, and axial stress increases.

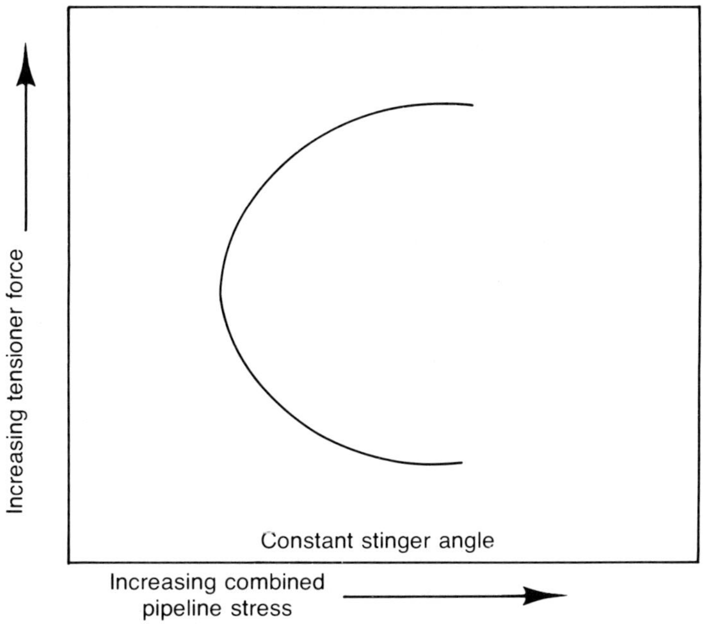

FIG. 4.40 Pipe-stress change with tensioner force

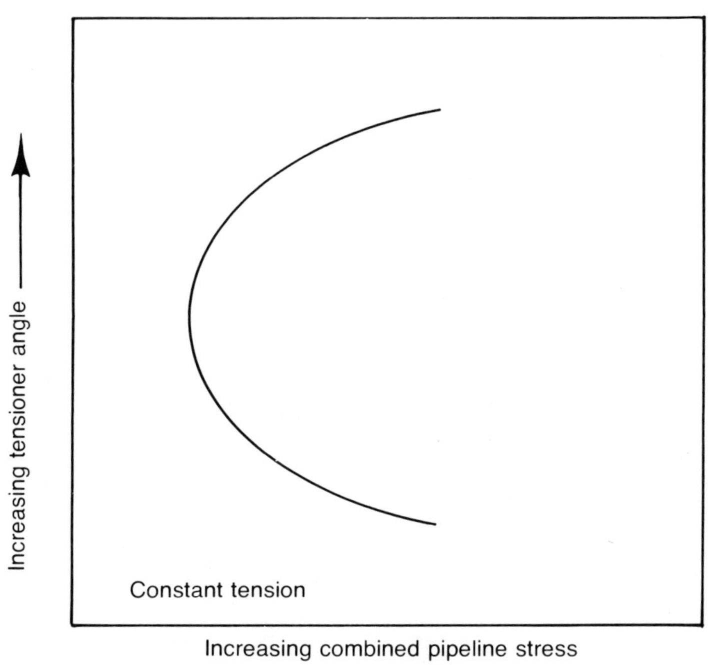

FIG. 4.41 Pipe-stress change with tensioner angle

Net effect is that the combined stresses decrease. Flexural stress keeps on decreasing with increased tension. After a while, however, the tensions become high enough that the axial stress is significant.

For the case of constant tension, pipeline size, and water depth, it can also be shown that there is an optimum angle at the end of the stinger.

At the optimum angle, there are minimum flexural stresses in the pipeline. As the stinger angle deviates from optimum, flexural stresses increase, giving an increase in combined stress (figure 4.41).

Stingerless Pipe-laying

A promising method for extending water-depth capabilities of marine pipe-lay operations appears to be use of a stingerless barge.

With this method, the pipeline slides off the ramp at an angle adjusted so the pipeline has only axial force. There are no flexural stresses in the normal overbend region.

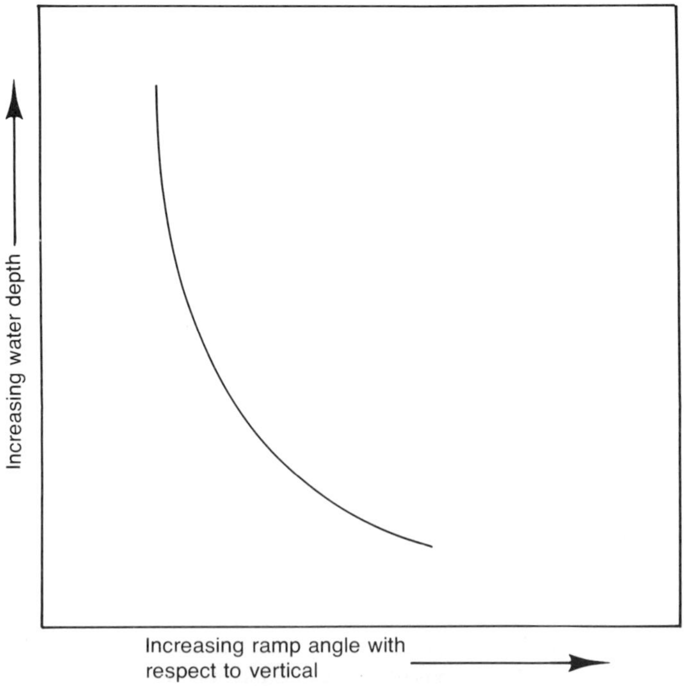

FIG. 4.42 How water depth affects ramp angle

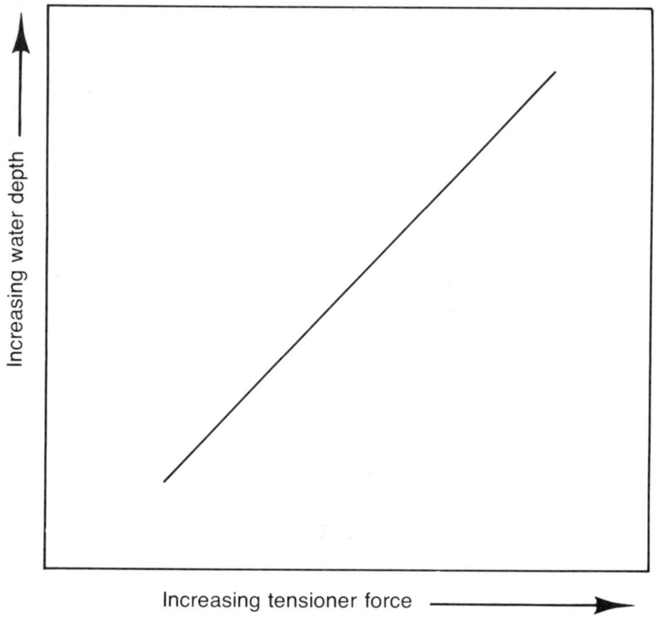

FIG. 4.43 How water depth affects tensioner force

For stingerless pipe-lay barges, the optimum angle decreases with increasing water depth (figure 4.42). The optimum tensioner force increases with increasing water depth (figure 4.43).

Regarding stingerless pipe-lay barges, Dixon and Rutledge have developed nondimensional curves of maximum strain in the pipeline. It is a function of pipeline stiffness for varying tensioner forces (figure 4.44) and for varying angles of the line at the barge with respect to the vertical (figure 4.45).

Their procedures, using stiffened catenary equations, provide good first approximation for preliminary design of inclined ramps for deepwater operations.

Example Problems

Two tensioner forces, 80 and 120 k, were studied to determine response of pipelines when there are no wave and current forces present. Only the tensioner force, gravity, and soil friction at the mud line were considered.

To show the influence of tensioner force levels on pipe-lay operations, detailed analyses were made for the two sample problems. For a 36-in. pipeline in 235 ft of water, the tensioner force was set at 80 and at 120 k. All other possible variables were held constant (table 69).

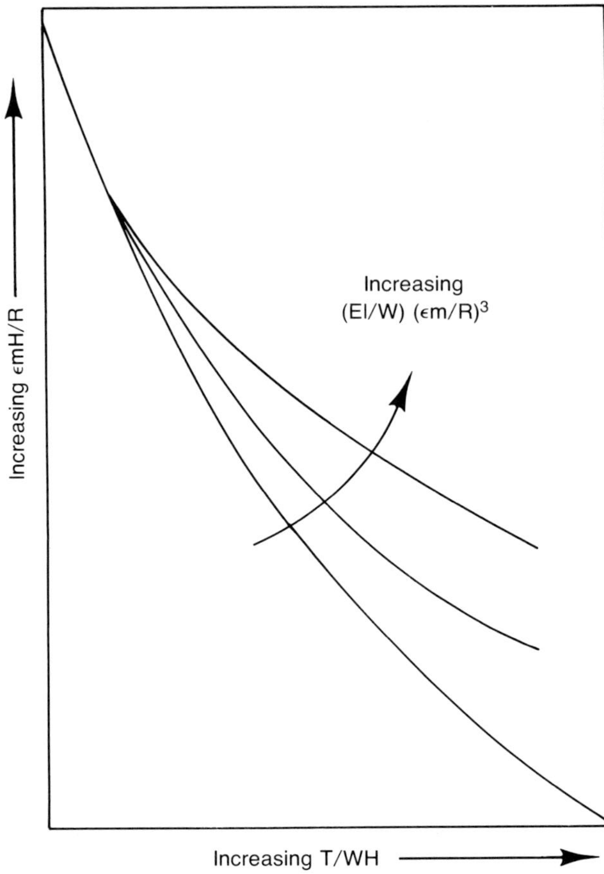

FIG. 4.44 How pipe strain varies with barge tension

Vertical distances from the tensioner to the inflection points were determined to be 121 ft for 80 k and 104 ft for 120 k (figure 4.46). The horizontal distances from the tensioner to the inflection points were 467 ft for 80 k and 452 ft for 120 k.

Distances along the pipeline from the tensioner to the inflection points were 485 ft for 80 k and 468 ft for 120 k.

Horizontal distance from the tensioner to the mud line tangent point was 1,068 ft for 80 k and 1,151 ft for 120 k. In the sag-bend area, the higher tensioner force pulled the pipeline up about 10 ft.

Distances along the pipeline from the tensioner to the mud line tangent point was 1,108 for 80 k and 1,188 ft for 120 k.

At the mud line, axial forces in the pipeline were 48 k for 80-k tensioner force and 88 kip for 120-k tensioner force (figure 4.47). The

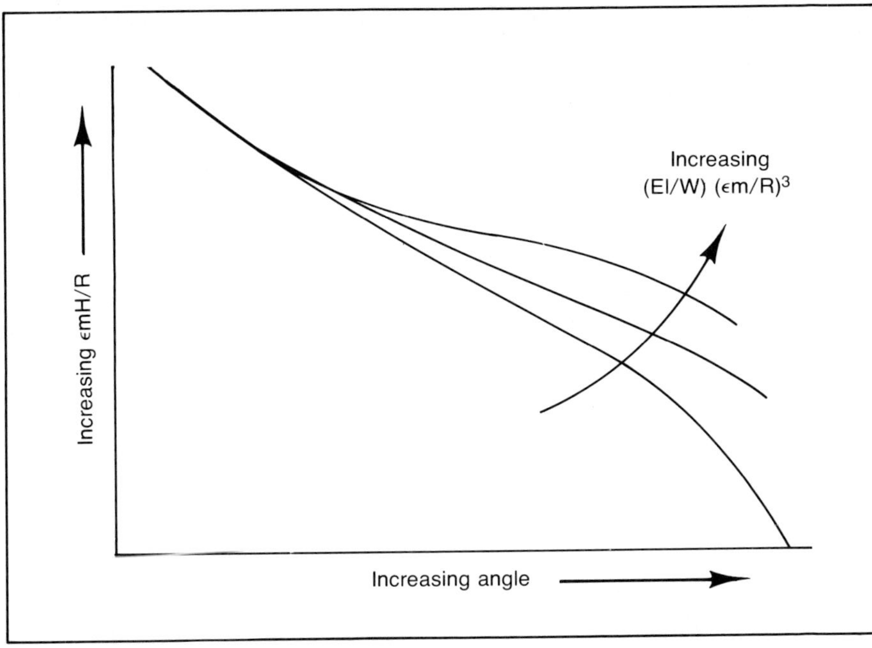

FIG. 4.45 How pipe strain varies with angle at lay barge

TABLE 69
Summary of operating conditions

Parameter	Quantity
Water depth, ft	235
Distance from tensioner to mean waterline, ft	15
Curved stinger radius, ft	1,000
Pipe outside diameter, in.	36
Pipe wall thickness, in.	0.75
Pipe coating thickness, in.	3.5
Weight of coated pipeline in air, lb/ft	734
Weight of coated pipeline in water, lb/ft	89.3
Specific gravity of steel	7.66
Specific gravity of coating	2.33
Specific gravity of coated pipeline in air	3.45
Specific gravity of coated pipeline in water	1.17
Young's modulus of pipe, ksi	30×10^3
Poisson's ratio of pipe	0.3
Allowable stress in pipe, ksi	50
Barge tension, kip	80, 120
Surface water current, knot	0
Wave height, ft	0
Wave length, ft	0

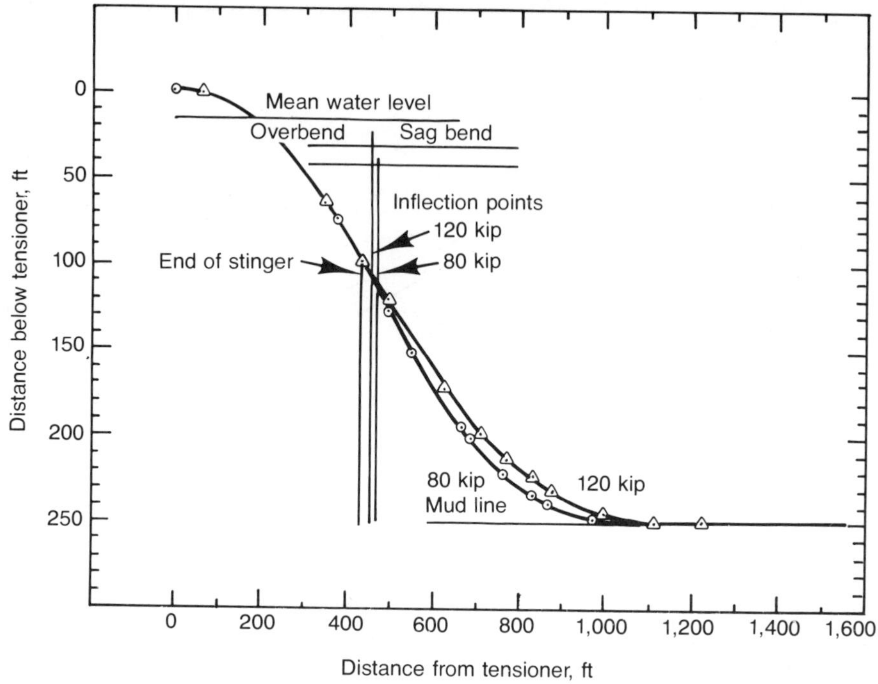

FIG. 4.46 Inflection points

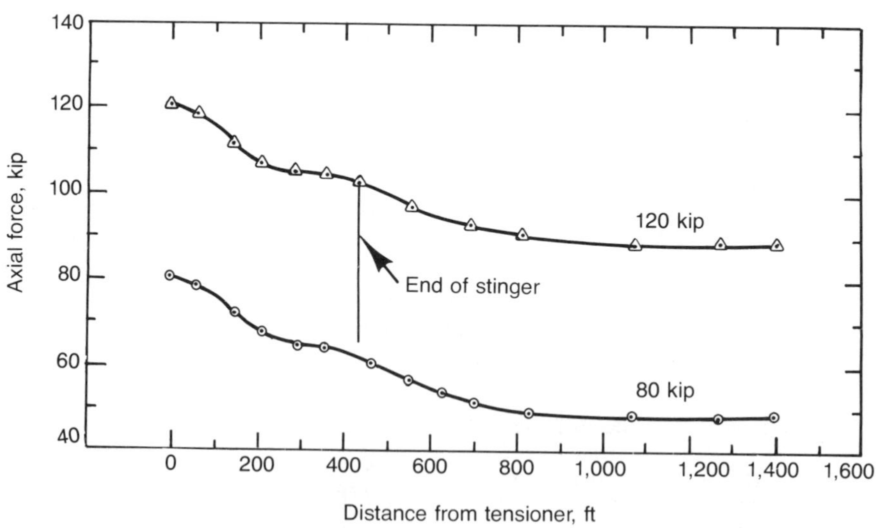

FIG. 4.47 Axial forces

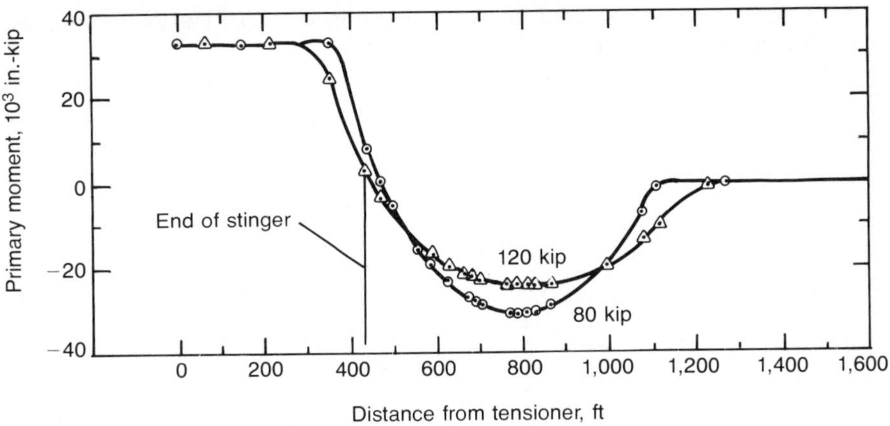

FIG. 4.48 Primary moment

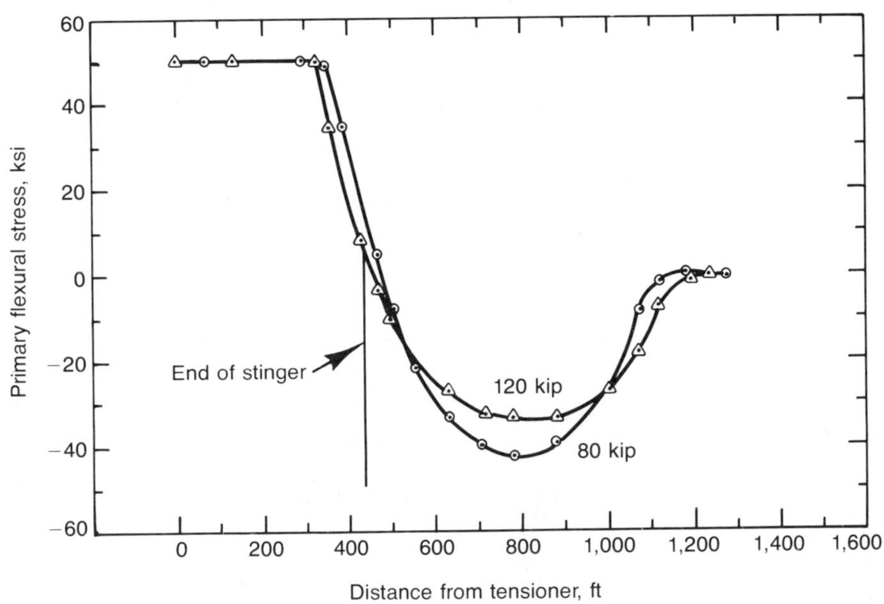

FIG. 4.49 Primary flexural stress

primary flexural moment was the same at 32,300-in. k in the over-bend region for both.

In the sag-bend region, primary flexure was 30,200-in. k for 80 kip and 24,000-in. k for 120 k (figure 4.48). In the sag-bend region, flexural stress was 42 ksi for 80 k and 34 ksi for 120 k (figure 4.49).

Combined axial and flexural stress in the sag bend was 43 ksi for 80 k and 35 ksi for 120 k (figure 4.50).

Combined strain in the overbend was 0.15% for both cases (figure 4.51). Combined strain in the sag bend peaked at 0.14% for 80 k and 0.11% for 120 k.

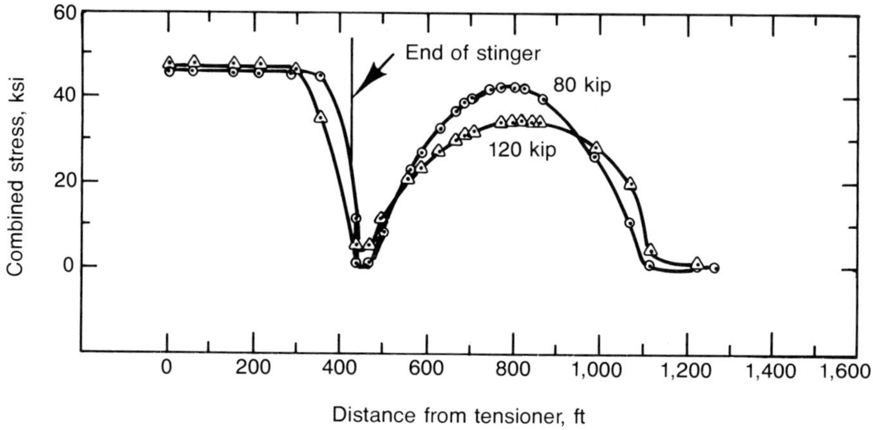

FIG. 4.50 Combined stress

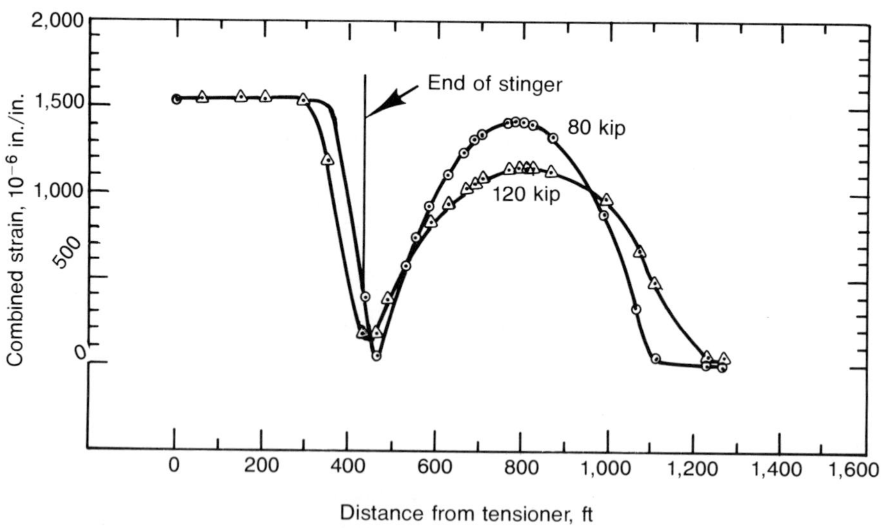

FIG. 4.51 Combined strain

4.5 PREDICTING PIPE-LAY BARGE MOORING ORIENTATION

Optimum orientation for combined effects of waves, winds, and currents from any direction can be easily simulated with computer procedures. Results should help reduce downtime due to weather conditions during actual pipe-laying

It may be necessary to lay pipe along a narrowly prescribed path that has been preswept to eliminate ridges and subsea sand dunes. But procedures described here would allow the contractor to accurately simulate procedures for specific conditions and to stay within the prescribed route.

Three sample cases are investigated. Results show that barge mooring during continuous pipe-laying and during holding conditions is important.

For a continuous pipe-laying operation, the lateral-slack mooring case is the best choice. This is because stresses are greatly reduced even though lateral deflections are increased.

Orientation

When storms suddenly develop, mooring orientation becomes very important during pipe-lay operations.

For example, storms with 80-mph winds and 20-ft waves have been generated within a 4.5-hour period in the Frigg field in the North Sea. Under these conditions, capability of the mooring system and vessel orientation directly affects the upper limit of wave and current which a pipeline can withstand.

Static offset and orientation of the pipe-lay vessel are a direct function of forces due to winds and currents. The transient dynamic offsets are due to the wave forces.

Tabulation on the 10% wind exceedance by season for 19 areas worldwide is shown in table 70 for comparisons. Of these areas studied, offshore eastern Canada seems to have the highest winds.

Mooring Systems

Mooring equipment on board a pipe-lay vessel varies much. *Choctaw I*, for example, has eight mooring lines and four diesel-driven, 42-in. double-drum winches. Winches carry 3,000 ft of 2-in. wire rope for the 20-k anchors. The wire ropes have a breaking strength of 190 tons.

The ETPM 1601 pipe-lay and crane ship has 6 forward and 4 aft-positioning electrohydraulic winches with 150-ton pull capacity. Each winch has 6,560 ft of 2.5-in wire rope and an 18.5-ton anchor.

The Viking semisubmersible pipe-lay vessel has fourteen mooring winches instead of the usual 4 to 8 winches. Two winches

TABLE 70

Average annual 10% wind-exceedance data by season

Area	Coordinates		Wind speed in knots			
	Latitude	Longitude	Spring	Summer	Fall	Winter
Southern North Sea	52-07N	05-11E	26	23	32	33
Adriatic Sea	43-31N	16-26E	23	17	23	27
Gulf of Guinea	00-42S	08-45E	10	15	10	11
Mozambique	25-55S	32-34E	26	28	31	28
Persian Gulf	26-16N	50-37E	18	10	15	19
Arabian Sea	24-48N	66-59E	9	35	14	20
Yellow Sea	37-29N	126-38E	24	19	27	28
Sea of Japan	39-43N	140-06E	20	17	18	22
Malaysia	03-07N	101-33E	12	17	16	22
Kalimantan	0	109-00E	11	24	17	22
Timor Sea	12-26S	130-52E	18	23	17	19
Bass Strait	37-52S	144-45E	31	30	33	24
Canadian Arctic	74-43N	94-59W	24	26	31	32
Eastern Canada	43-56N	60-01W	33	26	32	35
Gulf of Alaska	61-10N	150-01W	25	20	29	28
California	32-44N	117-10W	22	22	20	21
Gulf of Mexico	29-59N	90-15W	27	18	28	20
Caribbean Sea	18-30N	77-55W	22	29	20	22
Argentine	38-44S	62-11W	31	34	29	25

Note:

1. The exceedance data mean that in the southern North Sea, the average annual wind speed is greater than about 26 knots for 10% of the time in the spring—March, April, and May.
2. Coordinate 52-7N—southern North Sea—for example, represents 52 degrees and 7 minutes North.
3. Source: derived from several issues of Ocean Industry (data from Storm Research Institute, Houston).

are located at each of the four corner columns and three winches at each of the midship columns.

With 8,000-ft lines at the breast columns and 11,000 ft lines at the bow and stern, the vessel can advance 3,000 ft before resetting anchors.

A computer constantly monitors and calculates required winch torque and speed. Calculations are based on measured line force and line angle at the fair-leads, with manual, semiautomatic, or completely automatic adjustment.

Saipem *Castoro VI* semisubmersible pipe-lay vessel has 12 mooring winches. This vessel also has provisions for fully automatic mooring operations.

By using automated mooring systems, proper barge position and orientation can be maintained to achieve the best operating results.

Mooring can be controlled by a computerized system. The system compares planned and actual barge anchor tensions and angles of departure at the fair-leads and makes the necessary corrections.

Example Problems

In this series, we use a 3.5-fps (about 2 knots) surface current with a linearly decreasing distribution to 1.5 fps at the ocean bottom.

There are no restrictions, however, on the type of current distribution which can be considered in computer simulation of pipe-lay operations. It would have been just as easy to use the parabolic, the one-seventh rule, a current-direction reversal at some depth or any other unusual distribution of current.

Three sample problems were analyzed for comparison. A 36-in. pipeline in 235-ft water depth was considered.

All of the possible variables were held constant except vessel offset and vessel orientation (table 71).

An extremely severe environment of 20-ft waves and 2-knot winds were acting in a direction perpendicular to the barge heading for the sample cases.

The three cases of mooring orientation investigated were station keeping, offset mooring, and slack mooring.

Station keeping represents a holding condition. There, the vessel is allowed to rotate but has no offset with respect to a line drawn from the barge heading to the mud line tangent point (figure 4.52).

In all three cases, the pipeline is displaced from this line due to the current and the waves.

TABLE 71
Summary of operating conditions

Title	Description
Station keeping	The barge is fixed laterally but is free to rotate. This represents a holding condition.
Offset mooring	The barge is free to move laterally but does not rotate. Position is held by mooring system.
Slack mooring	The barge is free to rotate and is free laterally but mooring lines are oriented to allow sufficient tensioner forces.

Note: Waves and current act simultaneously and perpendicular to the barge heading.

Parameter	Quantity
Water depth, ft	235
Distance from tensioner to mean waterline, ft	15
Curved stinger radius, ft	1,000
Pipe outside diameter, in.	36
Pipe-wall thickness, in.	0.75
Pipe-coating thickness, in.	3.5
Weight of coated pipeline in air, lb/ft	734
Weight of coated pipeline in water, lb/ft	89.3
Specific gravity of steel	7.66
Specific gravity of coating	2.33
Specific gravity of coated pipeline in air	3.45
Specific gravity of coated pipeline in water	1.17
Young's modulus of the pipe, ksi	30×10^3
Poisson's ratio of the pipe	0.3
Allowable stress in the pipe, ksi	50
Barge tension, kip	120
Surface water current, knots	2.07
Wave height, ft	20
Wave length, ft	200
Wave period, sec	6.3

In what is called the offset-mooring problem, the vessel is restrained by the mooring system from rotation and from translation in the direction of the barge heading.

The vessel is allowed to offset laterally until equilibrium with the tensioner force is achieved; then the mooring system fixes this orientation. A constant tensioner force of 120 k was used in all three sample problems.

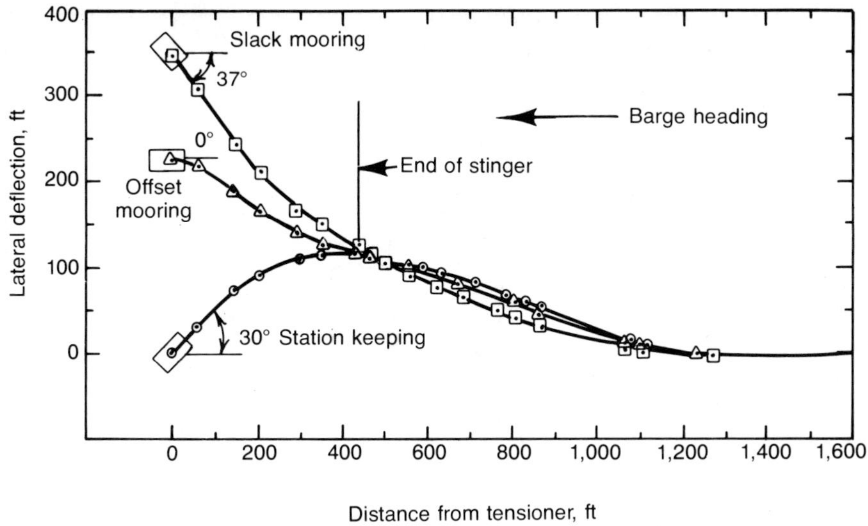

FIG. 4.52 Distance from tensioner vs. lateral deflection

In the slack-mooring condition, the vessel is restrained from surge. The vessel is allowed to offset laterally and rotate until equilibrium with the pipeline is achieved. Then this orientation is fixed by the mooring system.

Results

With the given conditions, the computer-simulated pipe-lay operation gave a lateral translation of 229 ft for offset mooring and 350 ft for slack mooring (figure 4.52).

Distance along the pipeline from the tensioner to the mud-line tangent point was 1,218 ft for station keeping, 1,212 ft for offset mooring, and 1,256 ft for slack mooring. The lateral moment went up to 211,600-in kip for the offset mooring condition (figure 4.53).

Maximum calculated lateral bending stresses in the overbend were 76 ksi for station keeping. They were 47 ksi for slack mooring and 294 ksi—or overstressed by a factor of nearly 6—for the offset mooring condition (figure 4.54).

Maximum lateral-bending stresses in the sag bend were 45 ksi for station keeping, 43 ksi for offset, and 30 ksi for slack mooring.

In the overbend region of the pipeline, the maximum combined stresses were 89 ksi for station keeping and 47 ksi for the slack-mooring condition (figure 4.55). In the example offset-mooring condition, the pipeline failed.

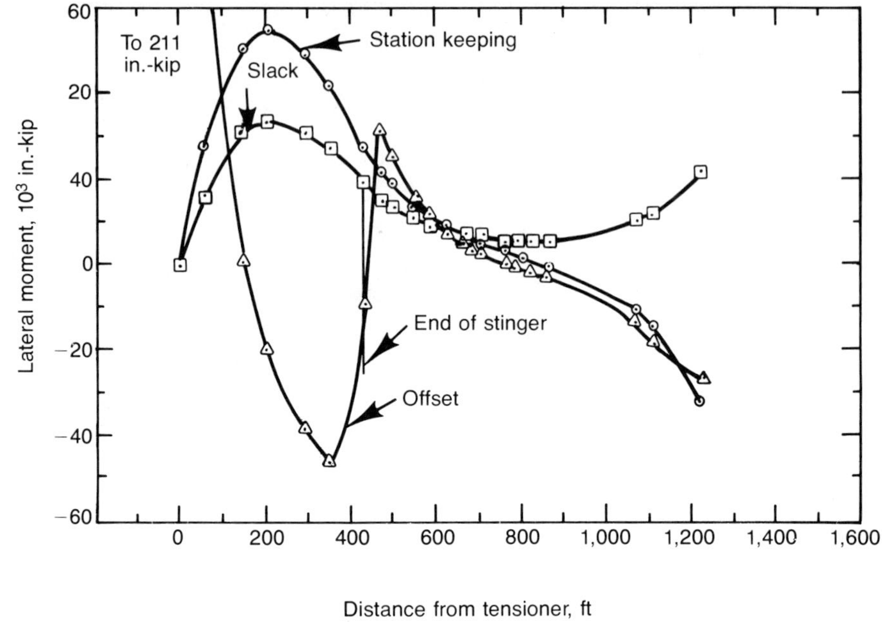

FIG. 4.53 Distance from tensioner vs. lateral moment

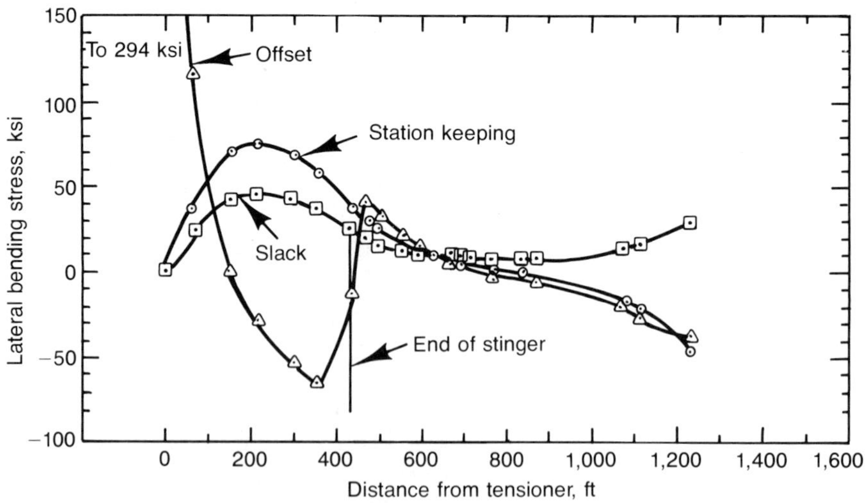

FIG. 4.54 Distance from tensioner vs. lateral bending stress

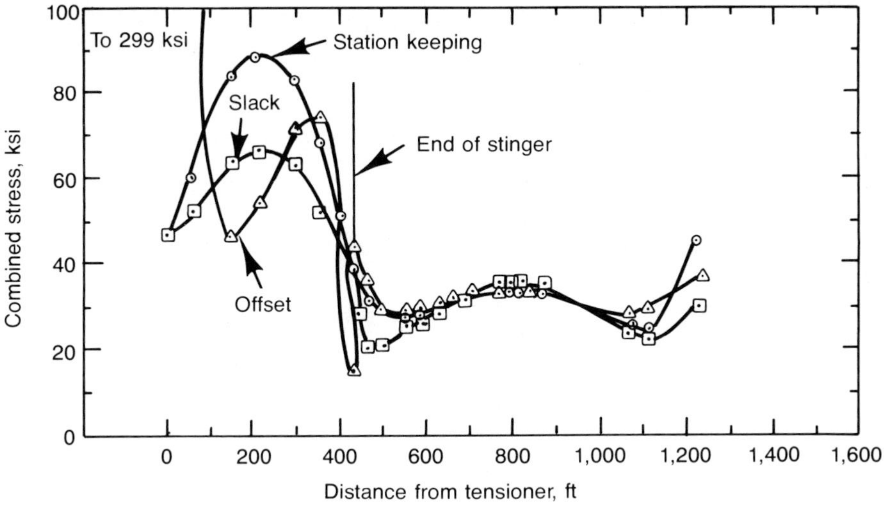

FIG. 4.55 Distance from tensioner vs. combined stress

Note that the combined stress and combined strain in the middle area of the sag-bend region peak at about the same value for all three cases.

There is, however, another inflection point near the mud line, and the combined stress kicks up to different values in each case at the mud-line tangent point. This is a result of the mud-line boundary conditions used in the computer simulations.

Other Conditions

Boundary conditions were assumed to be complete fixity against translation and rotation at the tangent point of the pipeline with the ocean bottom.

For hard bottoms, the coefficient of friction would be low, corresponding to free rotation. If the ocean bottom had boulders or rock formations on each side of the pipeline route, this would correspond to zero lateral allowable rotation.

Intermediate conditions of very low-strength soils can be simulated. This is done by using calculated effective spring constants to partially resist rotation and translation at the ocean bottom.

In very shallow water depths—100 ft—bottom conditions strongly influence stress and strain levels in the pipeline at and about the mud-line tangent.

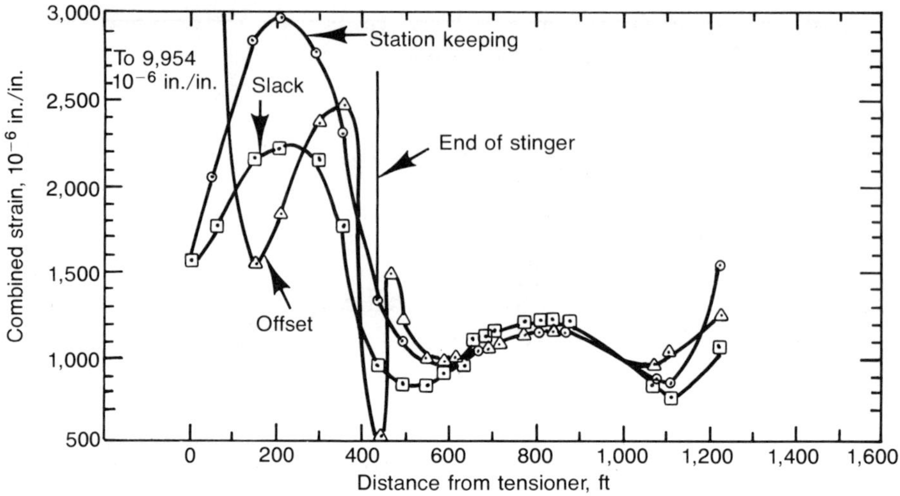

FIG. 4.56 Distance from tensioner vs. combined strain

Then, consider very deepwater depths—over 600 ft. There, bottom conditions have a very small influence on pipeline stress at the mud line, if currents near the ocean bottom are zero or very small.

In the overbend region, the maximum combined strains were 0.29% with station keeping, 0.99% with offset, and 0.22% with slack mooring (figure 4.56).

In the sag bend, maximum combined strains were 0.12% with station keeping, 0.15% with offset, and 0.12% with slack mooring.

Maximum values of the different strain components given here occur at different locations along the pipeline. Thus, the combined strain or stress will not necessarily occur at the same location as the maximum value of any strain component.

The slack-mooring case was the only one out of the three studied which had stresses less than the assumed design allowable of 50 ksi. The design factor was 0.94 with this case.

Station-keeping mooring case had maximum stresses 90% greater and maximum strains 32% greater than the slack-mooring problem. The offset-mooring case failed on the basis of stress and experienced plastic deformation of 1%.

4.6 COMBINED LOADS AFFECT MARINE PIPE-LAYING

Weather conditions in a specified offshore area can critically affect marine pipe-lay operations. Thus, pipe-lay procedures must be adjusted to meet varying weather conditions in the area.

Wave and current forces act simultaneously on the pipeline. These must be analyzed to plan marine pipe-lay operations to get accurate results.

Computer-simulation procedures can be used to determine operation limits of various combinations of environmental loads. Severe conditions of 20-ft waves and a 2-knot surface current are used to show how such environmental loads influence pipeline stresses.

This makes it possible to predict when an operation can begin or when it must be abandoned. Downtime or unnecessary damage to the pipeline or pipe-lay vessel can thus be avoided.

North Sea Problems

Summer months are often referred to as the weather window for North Sea pipe-lay operations. Even so, midsummer storms in 1973 severely hampered operations on BP's 115-mile, 32-in. crude line to the Forties field.

Phillips' 220-mile, 34-in. crude line to the Ekofisk field was also affected. The weather was so bad in 1973 that pipe-laying at Ekofisk field was shut in for 27 days in December.

Weather conditions can cause long downtimes, even for semi-submersible type pipe-lay vessels.

Bad weather conditions were combined with both labor and technical problems. With pipeline buckling and damaged stingers, an abnormally low rate of about 3 miles of line were laid in about 7 months in 1970.

Working time for the vessel in Bass Strait weather conditions would normally be about 60%. This is 50% greater than for the conventional flat-bottomed type of pipe-lay barges. But the vessel still set a number of new records.

Future pipe-lay barges will need to be designed to operate in specific sea conditions. Pipe-lay operations must also be adjusted to meet environmental conditions. Viking Jersey's pipe-lay barge, for example, was designed to operate in 15-ft wave heights and to withstand a "100-year" North Sea storm.

Four example cases are presented here and comparisons are drawn. As a control, an analysis is made for a pipeline being layed in calm seas. In each case, effects of current, of waves, and of current and waves acting together and perpendicular to the pipeline are studied.

It was found that overbend stresses with given waves and current conditions are about 95% greater than with calm seas. This is about 36% greater than with waves alone.

Ideal Vessel

In any marine pipe-laying operation, the environment decides allowable water depths, stinger shape and length, and tensioner force. Environmental conditions dictate when the operation can begin or be abandoned due to excessive pipeline stresses or deflections.

Cost of constructing and laying pipelines under more-severe criteria continues to rise. And the risks of working in unknown environmental conditions is intensified. Costly downtime must be avoided as more miles of greater diameter pipelines are being laid.

Many studies have been made to find the ideal pipe-lay vessel for severe weather conditions like those in the North Sea, but this ideal vessel does not exist.

Flat-bottomed or shipshape pipe-lay barges are too often forced to drop the pipeline due to severe weather conditions. Large semisubmersibles have adequate stability in heavy seas. But, without the shelter of the leeward side, it is not possible for support vessels to resupply in heavy seas.

A summary of some useful worldwide weather data on sea states with 10% exceedance values is useful for early design (table 72). This is not a complete picture, however. Two different areas could have the same 10% exceedance values with different exceedance conditions.

Example Problems

Examples of computer-simulated marine pipe-lay operations with various environmental conditions show effects of weather on the pipeline. All possible variables except weather states were held constant to simplify the comparisons (table 73).

A 36-in., 0.75-in. W.T. pipeline was laid in 235-ft water depths with a constant tension of 120 k. A curved stinger having a 1,000-ft radius was extended 430 ft horizontally from the pipe-lay vessel.

In the four cases of weather states considered, current and waves acted perpendicular to the pipeline:

- It was assumed there were no currents or waves during the pipe-lay operations.
- A 2-knot surface current linearly varying to 43% of this value at the mud line was applied.
- Twenty-ft waves, 200 ft long and with a 6.3-sec period were applied to the pipeline.
- Pipe-lay operations were simulated for combined 2-knot current and 20-ft waves.

TABLE 72

Average annual 10% wave exceedance by season

Area	Coordinates		Sea height, ft				Swell height, ft			
	Latitude	Longitude	Spring	Summer	Fall	Winter	Spring	Summer	Fall	Winter
Southern North Sea	52-07N	5-11E	7	6	9	10	12	11	12	13
Adriatic Sea	43-31N	16-26E	6	5	7	9	14	8	12	14
Gulf of Guinea	0-42S	8-45E	6	5	6	6	10	8	9	11
Mozambique	25-55S	32-34E	9	10	12	11	8	8	10	6
Persian Gulf	26-16N	50-37E	2	5	4	7	1	7	4	8
Arabian Sea	24-48N	66-59E	6	8	6	5	6	14	8	7
Yellow Sea	37-29N	126-38E	6	4	5	10	8	8	3	2
Sea of Japan	39-43N	140-06E	6	5	7	9	12	4	5	10
Malaysia	3-07N	101-33E	3	5	3	7	12	11	12	13
Kalimantan	0-0	109-00E	4	5	4	7	3	6	6	10
Timor Sea	12-26S	130-52E	2	5	6	7	11	7	5	5
Bass Strait	37-52S	144-45E	6	7	8	8	11	12	11	12
Canadian Arctic	74-43N	94-59W	—	6	—	—	—	10	—	—
Eastern Canada	43-56N	60-01W	10	7	9	12	13	12	13	14
Gulf of Alaska	61-10N	150-01W	5	4	9	6	na	na	na	na
California	32-44N	117-10W	7	5	6	6	10	8	11	8
Gulf of Mexico	29-59N	90-15W	7	5	7	6	11	1	10	11
Caribbean Sea	18-30N	77-55W	6	8	7	6	9	12	8	9
Argentina	38-44S	62-11W	6	6	7	5	7	12	9	10

Notes:

1. Exceedance data mean, for example, that in the Southern North Sea the average annual seas are greater than about 7 ft during 10% of the time in the spring (March, April, and May).
2. Coordinates 52-7N, for example, means 52 degrees and 7 min north.
3. Source: Derived from several issues of Ocean Industry (data from Storm Research Institute, Houston).

TABLE 73
Operating conditions

Condition	Description
Calm	Calm seas with no waves or currents.
Current only	Only the current acts on the pipeline and in a perpendicular direction.
Waves only	Only the waves act on the pipeline and in a perpendicular direction.
Combined	Waves and current act simultaneously and perpendicular to pipeline.

Parameters	Quantity
Water depth, ft	235
Distance from tensioner to mean waterline, ft	15
Curved stinger radius, ft	1,000
Pipe outside diameter, in.	36
Pipe wall thickness, in.	0.75
Pipe coating thickness, in.	3.5
Weight of coated pipeline in air, lb/ft	734
Weight of coated pipeline in water, lb/ft	89.3
Specific gravity of steel	7.66
Specific gravity of coating	2.33
Specific gravity of coated pipeline in air	3.45
Specific gravity of coated pipeline in water	1.17
Young's modulus of pipe, ksi	30×10^3
Poisson's ratio of pipe	0.3
Allowable stress in pipe, ksi	50
Barge tension, kips	120
Surface water current, knots	2.07
Wave height, ft	20
Wave length, ft	200
Wave period, sec	6.3

Note: The barge is fixed laterally but free to rotate.

For all four cases, structural responses in the vertical plane were the same (table 74). Axial stress decreased from 120 k in the overbend to 101 k in the sag bend.

The sag-bend region runs from the inflection point to the mudline tangent point. The primary flexural stress in the vertical plane varied from 45 ksi maximum in the overbend to 34 ksi maximum in the sag bend.

TABLE 74
Maximum values in vertical plane for analyses

Variable	Overbend	Sag bend
Vertical distance from tensioner to:		
Inflection point, ft	104	—
Mud-line tangent point, ft	250	—
Horizontal distance from tensioner to:		
Inflection point, ft	104	—
Mud-line tangent point, ft	—	1,151
Axial force, kips	120	101
Strain, micro-in./in.		
Axial	48	41
Primary flexure	1,501	1,116
Stress, ksi		
Axial	1.4	1.2
Primary flexure	45	34

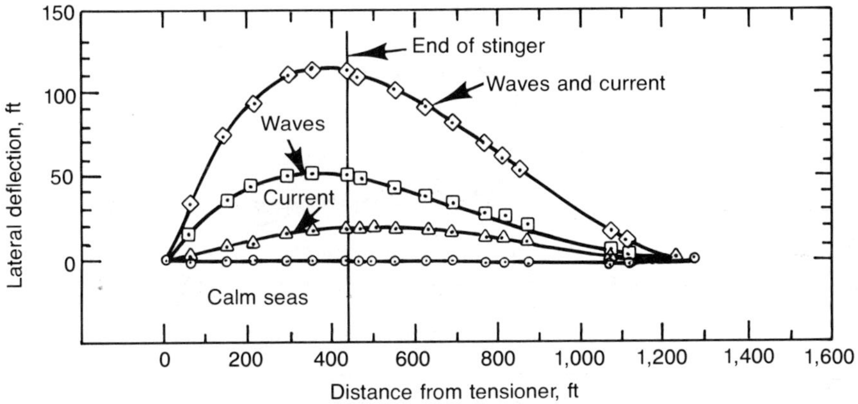

FIG. 4.57 Lateral deflection in pipeline

As expected, maximum lateral deflection of the pipelines in the four cases was greatest and was 115 ft from the vertical plane with combined waves and current (figure 4.57). Maximum lateral deflection was 19 ft with current alone and 52 ft with waves alone.

Maximum lateral moments and stresses occurred in the case of combined environment in the overbend region. Maximum values in the overbend for combined environment were 54,300-in. k of

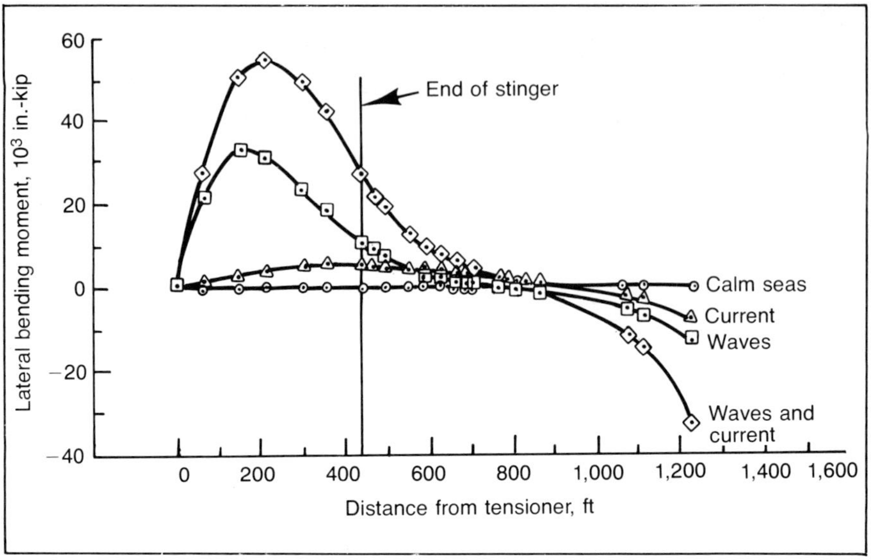

FIG. 4.58 Lateral bending moment in pipeline

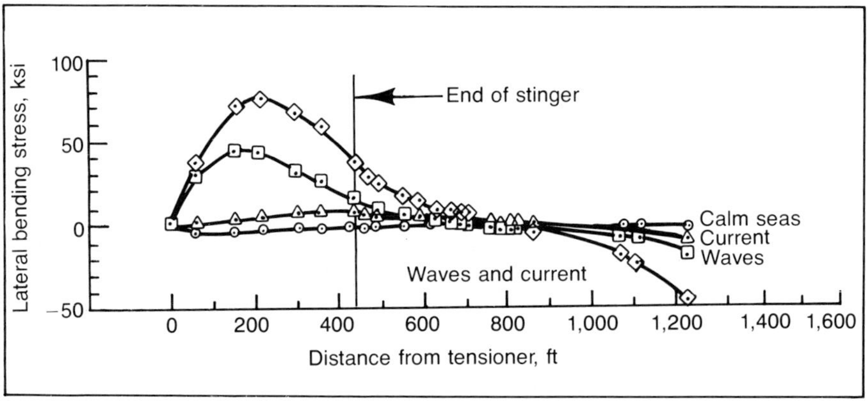

FIG. 4.59 Lateral bending stress in pipeline

moment (figure 4.58) and 75.7-ksi flexural stress (figure 4.59).

Effects of current acting alone and waves acting alone cannot be added together directly to determine safe operations in areas with a severe environment because of the nonlinear behavior of pipeline response.

Results of adding structural responses in the lateral plane only for current and wave loads and comparing this sum with the combined case proves this point (table 75).

Simulation using combined environment gave a maximum pipeline deflection. This was 61% greater than the algebraic sum of the deflections due to current and waves acting alone.

TABLE 75

Nonlinear results in lateral plane for computer-simulated pipe-lay operations

Variable	2-knot current	20-ft waves	Combined current & waves	Sum of current & waves	Combined divided by sum, %
Maximum deflection, ft	19.2	52.1	115.0	71.3	1.61
Maximum overbend response					
Flexural moment, 10^3 in.-kips	5.9	33.1	54.3	39.1	1.39
Strain, micro-in./in.	278.0	1,541.0	2,523.0	1,819.0	1.39
Stress, ksi	8.3	46.2	75.7	54.5	1.39
Maximum sag-bend response					
Flexural moment, 10^3 in.-kips	5.0	12.9	32.4	17.9	1.81
Strain, micro-in./in.	386.0	472.0	1,505.0	858.0	1.75
Stress, ksi	7.0	18.0	45.2	25.0	1.81

Structural responses for the combined environment cases were about 39% greater in the overbend. And they were about 79% greater in the sag bend than estimates reached by taking the sum of the separate responses.

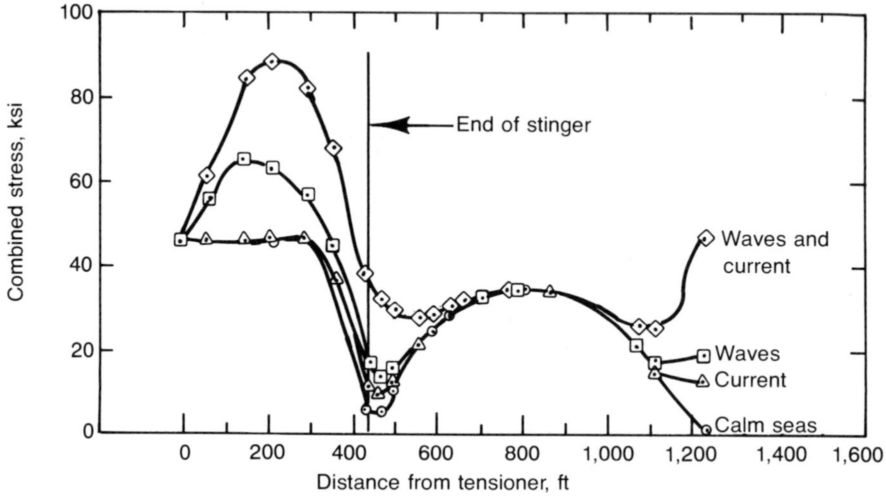

FIG. 4.60 Combined stress in pipeline

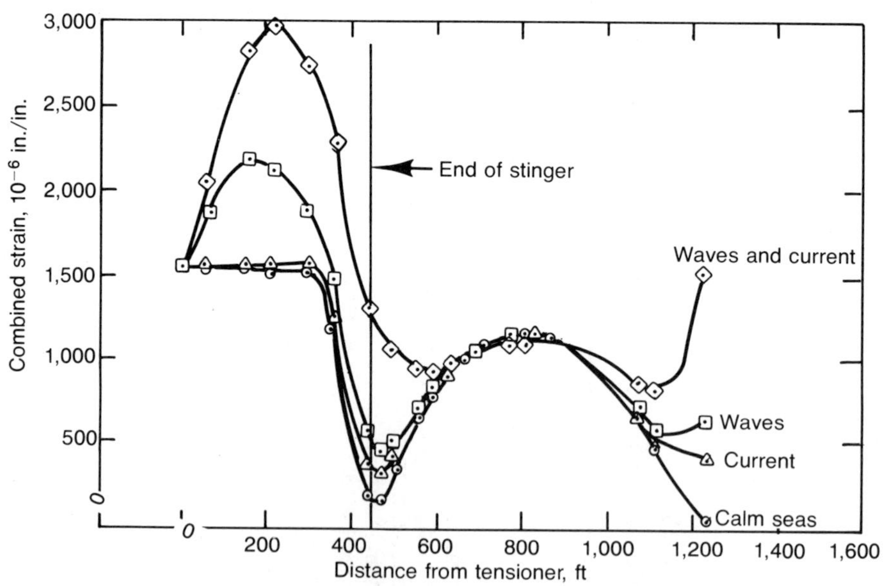

FIG. 4.61 Combined strain in pipeline

Vector Summation

For design purposes, limits of wave and current forces beyond which pipe-lay operations should be abandoned must be determined. To do this, the vector sum of the stress and strain due to axial force, primary flexure, and lateral flexural moment should be evaluated.

Maximum combined stress for waves and current acting simultaneously was 89 ksi in the overbend and 46 ksi in the sag bend (figure 4.60). Maximum strain for combined environment was 0.30% in the overbend and 0.16% in the sag bend (figure 4.61).

With each condition simulated, design factors based on a 50-ksi stress allowable were greater in the overbend than in the sag bend (table 76). The design factor in the overbend was 1.79 for combined conditions vs. 0.92 for calm seas.

The 2-knot current did not have much effect, but the design factor in the overbend for 20-ft waves acting alone was 43% greater than for calm seas. The design factor for combined environmental loads was 96% greater than for calm seas.

A rough estimate on limiting condition for the given pipeline design would be 1.1-knot surface current and 11-ft waves. This is based on the 1.79 design factor with combined environmental loads.

However, to set the actual limits, an additional computer simulation would be required since the pipeline response is nonlinear with wave height.

4.7 HERE'S HOW WIND AND WAVE DIRECTIONS AFFECT SUBSEA PIPE-LAYING

Marine pipe-laying operations must be abandoned with severe weather conditions, so it is essential to simulate safe operational conditions and determine the exact combined conditions which signal pipeline abandonment.

By understanding the limits of environmental conditions as a function of wave and current directions, efficiency of pipe-lay vessels can be improved. Downtime because of too early abandonment can also be decreased or avoided. Sample problems were computer-simulated to determine conditions in the pipeline. One case involved an operation in which waves and current forces acted perpendicular to pipeline. The other case involved waves and current forces acting parallel to pipeline, in the same direction as the barge heading.

TABLE 76
Key values for design and evaluation

Variable	Weather state			
	Calm	2-knot current	20-ft waves	Combined
Distance along pipeline from tensioner to inflection point, ft	467.00	468.00	473.00	490.00
Distance along pipeline from tensioner to mud-line tangent point, ft	1,187.00	1,188.00	1,194.00	1,218.00
Overbend region				
Combined strain, %	0.15	0.16	0.22	0.30
Combined stress, ksi	46.00	47.00	66.00	89.00
Design factor	0.92	0.93	1.32	1.79
Sag-bend region				
Combined strain, %	0.12	0.12	0.12	0.16
Combined stress, ksi	35.00	35.00	35.00	46.00
Design factor	0.69	0.69	0.69	0.93

All possible variables were held constant except wave and current direction. When waves and current were perpendicular to pipeline, stresses were 124% above those where waves and currents were parallel with pipe-lay vessel heading.

Sea Changes

Changes in sea conditions naturally occur during marine pipelaying operations. Wave height and current velocity may drastically and suddenly increase or decrease and lead to necessary alterations in pipe-lay procedure.

When changes in sea states and directions are not accounted for, costly downtime or damage can result. A change in angle between the waves and current relative to barge heading can be very detrimental to the operation.

Magnitude of responses due to changes in the angle of waves and current in relation to the pipeline has been examined. Capability exists to run parametric studies for the complete range of angular headings. This includes headings parallel to the pipeline and headings opposite in direction to the barge heading.

Pipeline response highly depends on this angle. Typical weather states in the area are also instrumental in setting operation guidelines. Stresses and lateral deflections can thus be precisely predicted for the expected angles of attack.

The prevailing direction of waves and current forces in relation to the submerged pipeline during pipe-laying operations is also important. The Santa Fe *Choctaw*, for example, can continue to lay pipe in 10 to 14-ft seas, depending on wave and current direction.

Riser Tie-in

One of the most critical aspects of pipe-lay operations is the platform connection of the pipeline with the riser.

Direction of wind, wave, and current forces must be considered when planning location of the tie-in. For safer operations, the riser should be installed on the sheltered leeward side of platform.

For preliminary design, data on directions of wind, wave, and swell (table 77) as well as planned month of installation should be carefully considered.

There are three main types of riser connections. Conventional riser-connection procedure involves using the pipe-lay barge davits to pull up the line and then to weld on an ell section. The pipeline is gradually lowered as successive vertical sections are welded together to form the riser. When the line reaches bottom, the riser is clamped to a platform leg.

TABLE 77
Most frequent directions of wind, waves, and swell

Area	Condition	Jan.	Feb.	Mar.	Apr.	May	Jun.	Jul.	Aug.	Sep.	Oct.	Nov.	Dec.
Gulf of Mexico	Wind	N	E	SE	E	E	SE	E	E	E	E	N	N
	Wave	N	SE	SE	E	E	SE	E	E	E	NE	NE	N
	Swell	SE	SE	SE	SE	E	SE	SE	SE	E	NE	NE	NE
Persian Gulf	Wind	NE	NW	NW	N	NW	NW	W	W	W	NW	NE	NE
	Wave	NW	NW	NW	NW	NW	NW	NW	NW	NW	NW	NW	NW
	Swell	NW	NW	NW	NW	NW	NW	NW	NW	NW	W	NW	NW
Southern N. Sea	Wind	SW	SE	SW	SE	SW	SW	SW	SW	NW	SW	SW	SW
	Wave	SW	E	SW	SW	SW	SW	SW	SW	SW	SW	SW	SW
	Swell	NW	SE	N	W	SW	SW	N	N	N	N	N	SW
Malaysia	Wind	NE	NE	NE	NE	S	S	S	SW	SW	SW	NE	NE
	Wave	NE	NE	NE	NE	S	S	SE	SW	SW	SW	NE	NE
	Swell	NE	NE	NE	NE	S	SW	S	SW	S	NE	NW	NE
Bass Straits	Wind	W	N	W	W	NW	N	NW	NW	W	NW	W	W
	Wave	SW	W	W	NW	NW	W	SW	SW	W	W	W	W
	Swell	SW	W	W	SW	SW	SW	SW	SW	W	SW	SW	SW

Source: Several issues of Ocean Industry (data from Storm Research Institute, Houston).

The second method is the pull tube system where a J section of pipe is welded to the platform during fabrication. The pipeline is pulled up through the J tube with a winch on the platform deck.

A third procedure for forming a riser is the bending-shoe method. The shoe is fabricated along with the platform. An extra length of pipeline is laid so a deck winch can pull the line up around the shoe to form the riser.

Normally, 3 to 20 days are required to make a pipeline riser tie-in. Cost can run from $250,000 at 150-ft water depths to $1.2 million at 300-ft water depths for 30-in. pipelines. It has been estimated, for instance, that a tie-in will cost about $1.3 million on the Frigg field line.

Type of riser and location should be considered in early stages of platform design, especially if the J tube or bending shoe is to be used.

It is also important that field personnel thoroughly understand the planned riser-connection procedures. And sufficient data and instrumentation with which to work should be available.

Pipeline Bury

Pipe-lay bury requirements vary considerably with the area. These requirements are influenced by bottom conditions, current levels, shipping and fishing traffic, water depth, etc.

Wind, wave, and current directions also affect orientation of the bury barge. It is essential to determine correct tensions in the mooring line to keep the barge in the same vertical plane as the pipeline.

Stingray natural-gas-pipeline system, for example, had 228 miles of 12- to 36-in. line laid in 1974 in the Gulf of Mexico. Bury requirement was a 3-ft depth in less than 200-ft water depths. Lines were not buried in over 200-ft water depths.

The 32-in., 230-mile line laid from Total Oil's North Sea Frigg field to St. Fergus, Scotland had 80% laid in greater than 360-ft water depths. It was designed for 1⅞-in., 190-lb concrete over ⅜-in. rebar cages. Specific gravity runs from about 1.16 to 1.6 near shore. Bury requirements were 3 m near shore, and the rest were 1 m, done by jetting.

An anticipated problem with the Frigg field line was trawler fishing-net weights striking the line. The design took into account 6-ton weights striking the line 18 times in 20 years, on the average, with the boat traveling 4 to 6 knots.

Actual jetting power on bury barges varies but is about 20,000 gpm at 2,500 psi. Such equipment is on the Brown & Root bury barge *Bar 331*, built at Levingston Shipbuilding in Orange, Tex. *Bar 331* has gas-turbine-powered jet pumps totaling 40,000 hp.

TABLE 78
Summary of API 1104 welding code

Defect classification	Individual weld length	Continuous 12-in. weld length
Undercutting	Depth up to lesser of 1/32 in. or 12.5% W.T. and length up to lesser of 2 in. or 1/6 of weld length.	
Class 1		
Gas pockets (porosity)	Spherical: lesser of up to 1/8 in. or 25% W.T.	Variable per code
	Cluster: up to 1/2 in. (1/16 max. pocket)	Up to 1/2 in. total length
	Piping (wormhole): lesser of up to 1/8 in. or 25% W.T.	Variable per code
	Hollow beads: up to 1/2 in.	Up to 2 in. total length with adjacent discontinuities separated with minimum 2-in. sound metal

Class 2		
Inadequate weld-root penetration	Up to 8%	Up to 1 in.
Inadequate penetration due to high-low	Up to 2 in.	Up to 3 in.
Internal concavity	Any length, if radiographic image density less than adjacent base metal. If greater, burn-through spec applies.	
Incomplete fusion (surface defect)	Up to 8%	Up to 1 in.
Incomplete fusion due to cold lap (internal defect)	Up to 2 in.	Up to 2 in.
Burn through	Lesser of up to 1/4 in. or W.T.	Up to 1/2 in.
Wagon tracks (elongated slag inclusions)	Up to 2-in. length or 1/8-in. width	Up to 2 in. at intervals greater than 1/32 in.
Isolated slag inclusions	Up to 1/8-in. width	Up to 1/2 in. and 4 or less inclusions with 1/8 in. maximum width
Class 3		
Cracks	Up to 5/32-in. length star cracks or shallow crater cracks.	

On Santa Fe's *Cherokee,* conversions at the NDSM yard in Amsterdam changed the vessel from a pipe-lay to a bury barge. The diesel pumps now deliver 16,000 gpm at 2,500 psi.

Codes and Regulations

Additionally, a number of material requirements, codes, and regulations have to be met in any particular area of operation. Codes of these regulatory agencies specify maximum allowable stresses for the pipelines. To satisfy these codes, wave and current directions must be considered to determine forces on the line caused by the directions and their effects on the pipeline.

Pipeline-welding requirements can vary with the type of steel used and the area of operation. API requirements are the most universally used. Qualifications for marine-pipeline welders are outlined in API 1104, Standards for Welding Pipelines and Related Facilities (table 78).

Most marine-pipeline welding operations are done by welders using stick electrodes. However, Brown & Root has had good results with their use of the CRC Crose automatic welding equipment.

Grade of steel for the pipeline depends on water depth and operating pressure. Trade-offs of concrete coatings with wall thickness, method of pipe-lay operations, and ocean bottoms conditions, etc., are also included. A few examples follow:

- Stingray line used grade API 5LX-60 in the 30-in. diameter, 0.625-in. W.T. main lines. Operating pressure is 1,200 psi. Gathering lines were API 5XL-52.
- The 32-in., 0.75-W.T. Frigg field line used grade API 5XL-65 for an operating pressure of 2,160 psi in 400-ft water depths.
- The 1974 Cinta field, 18-in. OD, 0.5-in.-W.T. lines in the Java Sea in 75 to 130-ft waters, used grade API 5XL-42 with an 1,850-psi test pressure.

Example Problems

For the case studies, influence of the direction of the waves and current on the pipeline operations was computer-simulated.

From other work, it has been determined that the worst case was both waves and current acting perpendicular to the pipeline. The least-damaging case involved a direction coincident with the barge heading.

In the example problems discussed here, a 36-in., 0.75 W.T. was considered with 1.17 sp gr of the coated line in water. The barge tension was 80 k.

An extreme environmental condition of a 2-knot surface current acting simultaneously with a 20-ft wave was considered. In both cases, all conditions were the same except for current and wave heading (table 79).

Results of the study are summarized in figures 4.62 to 4.65. In the discussion of these results, primary moments and stresses act in the vertical plane.

Lateral moments are in a plane perpendicular to the plane of the pipeline. A parallel wave means the wave direction is coincident with the barge heading.

For parallel waves and current, the distances from the tensioner to the inflection point were 113 ft vertically, 469 ft horizontally, and

TABLE 79
Summary of operating conditions

Title	Description
Perpendicular	Simultaneous waves and current act perpendicular to the pipeline.
Parallel	Simultaneous waves and current act parallel to the pipeline and in the same direction as the barge heading.

Note: In both cases the barge is fixed laterally and is free to rotate.

Parameter	Quantity
Water depth, ft	235
Distance from tensioner to mean waterline, ft	15
Curved stinger radius, ft	1,000
Pipe outside diameter, in.	36
Pipe wall thickness, in.	0.75
Pipe coating thickness, in.	3.5
Weight of coated pipeline in air, lb/ft	734
Weight of coated pipeline in water, lb/ft	89.3
Specific gravity of steel	7.66
Specific gravity of coating	2.33
Specific gravity of coated pipeline in air	3.45
Specific gravity of coated pipeline in water	1.17
Young's modulus of the pipe, ksi	30×30^3
Poisson's ratio of the pipe	0.3
Allowable stress in the pipe, ksi	50
Barge tension, kips	80
Surface water current, knots	2.07
Wave height, ft	20
Wave length, ft	200
Wave period, sec	6.3

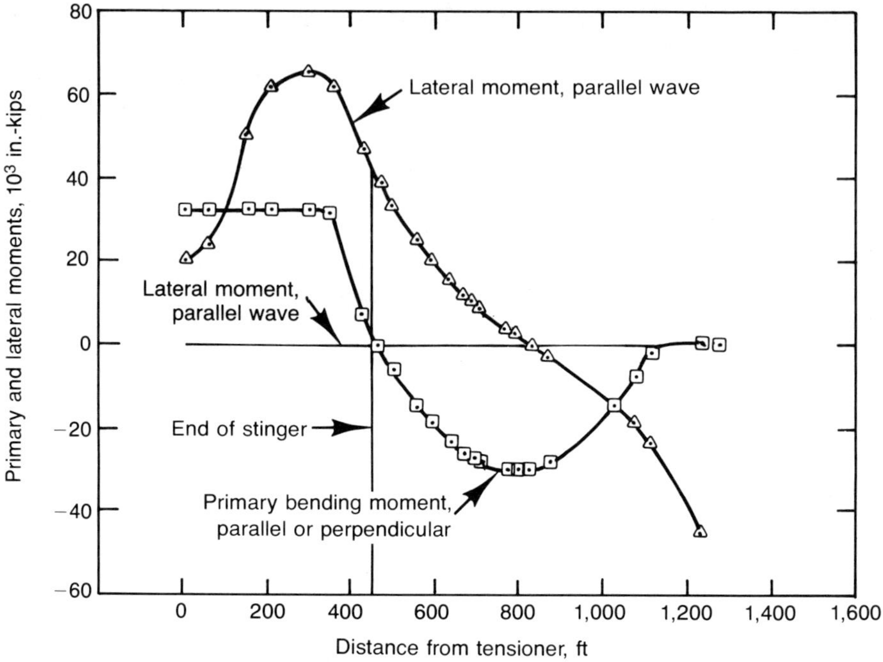

FIG. 4.62 Primary and lateral moments

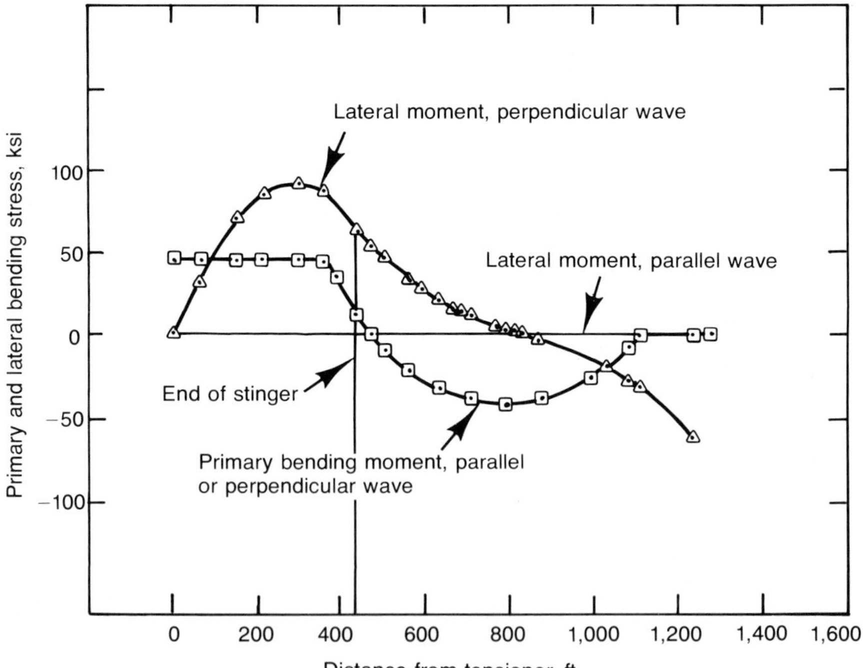

FIG. 4.63 Bending stresses

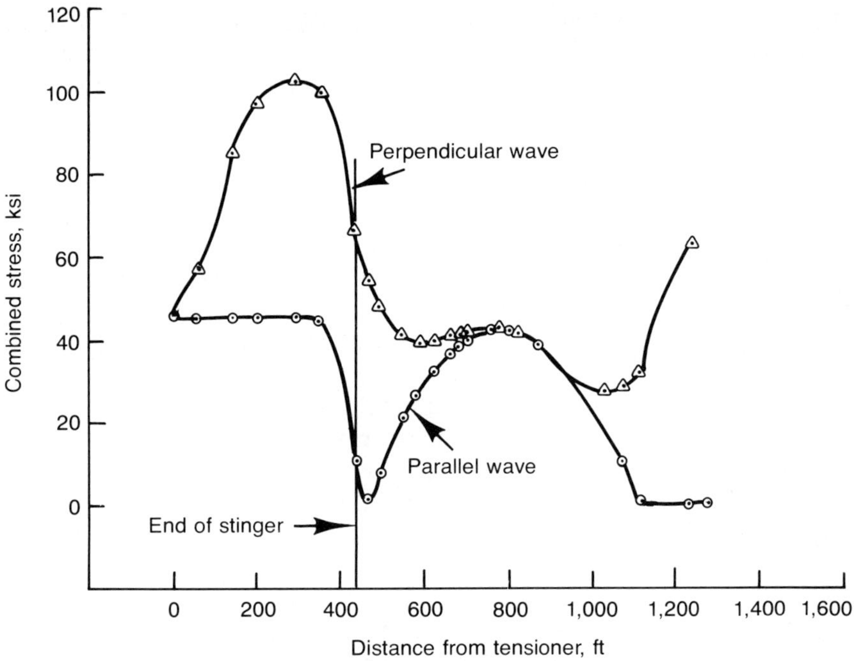

FIG. 4.64 Combined stress

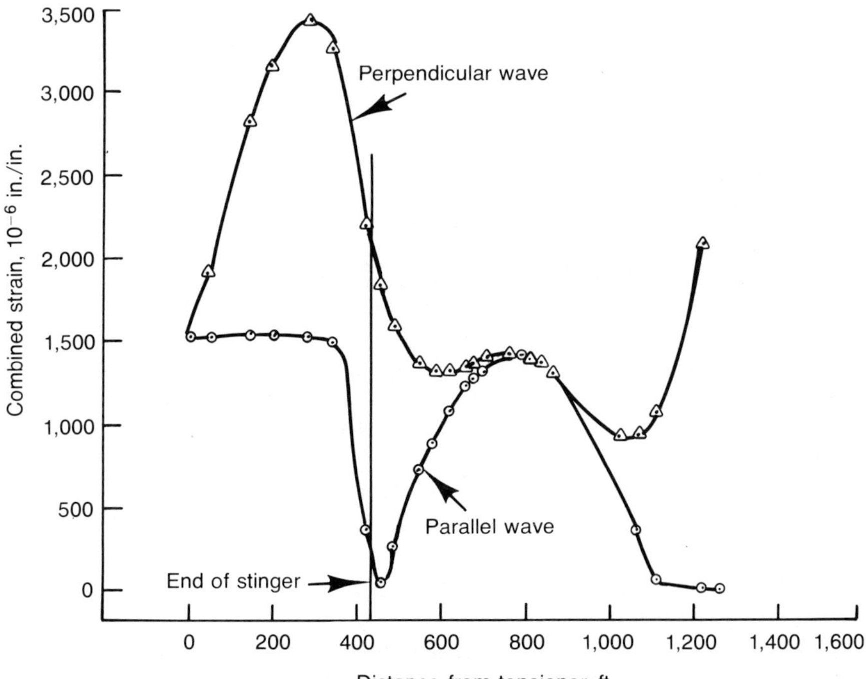

FIG. 4.65 Combined strain

486 ft along the pipeline. Corresponding distances for perpendicular waves and current were 121 ft, 455 ft, and 515 ft, respectively.

With parallel waves, the distances from the tensioner to the mud-line tangent point were 1,068 ft horizontally and 1,108 ft along the pipeline. Corresponding distances for perpendicular waves were 1,068 ft and 1,168 ft, respectively.

Maximum axial force in the sag bend was 60 k for parallel waves and 61 k for perpendicular waves. For both directions, the maximum primary flexure was 32,300-in. k in the overbend and 30,000-in. k in the sag bend (figure 4.62).

For perpendicular waves, the maximum lateral moments were 65,600-in. k in the overbend and 45,100-in. k in the sag bend.

Maximum primary bending moment for both cases was 45 ksi in the overbend and 42 ksi in the sag bend (figure 4.63). There was, of course, no lateral stress for parallel waves. Maximum lateral stress for perpendicular waves was 92 ksi in the overbend and 63 ksi in the sag bend.

Maximum total combined stress is the number used for design and assessment. For parallel waves, the maximum combined stress was 46 ksi in the overbend and 43 ksi in the sag bend (figure 4.64). For perpendicular waves, maximum combined stress was 103 ksi in the overbend and 64 ksi in the sag bend.

For the given severe environmental conditions, parallel waves gave a maximum design factor of 0.92. With perpendicular waves, the pipeline was overstressed by a factor of 2.06.

Combined strains for parallel waves were 0.15% in the overbend and 0.14% in the sag bend (figure 4.65). The combined strain for perpendicular waves was 0.34% in the overbend and 0.21% in the sag bend.

4.8 VESSEL MOTIONS ADD PIPE-LAY STRESS

During subsea pipe-laying, severe environmental conditions may arise to cause high waves and strong currents. Depending on barge size and type of pipeline being laid, barge-response amplitude can become high compared with the incident wave amplitudes.

Dynamic stresses due to vessel motions added to normal pipe-lay stresses could cause severe damage to the pipeline and to the stinger.

Sample cases discussed here show that the increase in sag-bend design factor for heave is about five times greater than the increase for surge. Roughly, about an 8% increase in the design factor per foot of vessel motion could be used for preliminary design in relatively shallow water.

Computer-simulated pipe-lay operations could be used to set realistic operational limits regarding vessel motions and environmental conditions. The procedures can be used to reduce downtime due to premature pipeline abandonment. Pipeline damage is reduced by calling for abandonment before the pipeline is overstressed.

Vessel Motion

Motions of a pipe-lay vessel are, of course, greatly influenced by vessel shape.

For instance, the least motion and, hence, the most operating time is achieved with the large semisubmersible. This is followed by the shipshape vessel and by the conventional pipe-lay barge.

Response operators of motion divided by wave height for any particular type of pipe-lay vessel are a direct function of the wave period.

With semisubmersibles, the heave response typically varies between 0.1 at 5-sec wave periods up to unity at about 22-sec wave periods (figure 4.66). Surge or sway operators for semisubmersibles

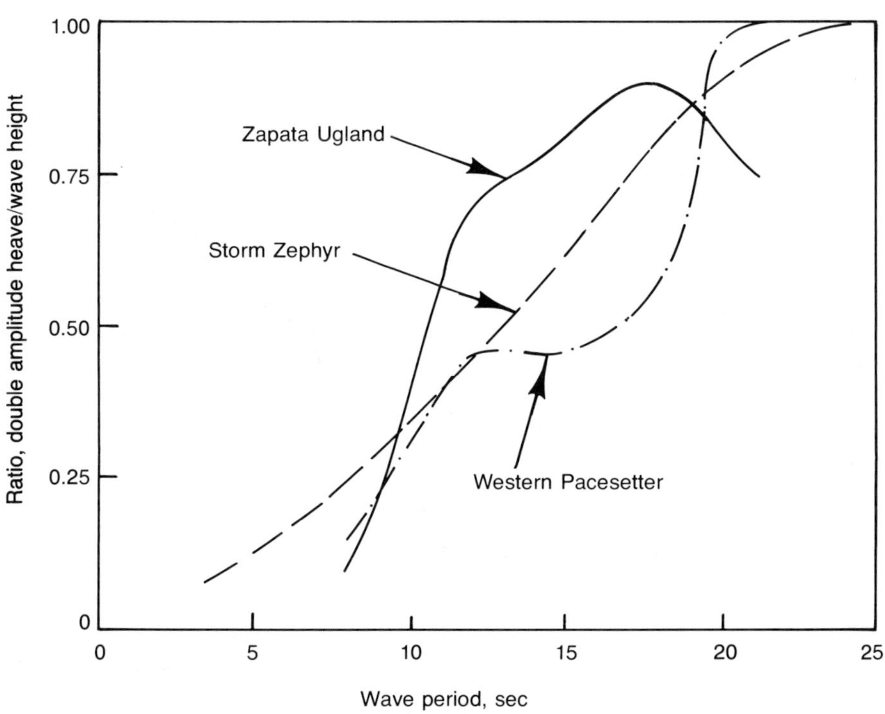

FIG. 4.66 Semisubmersible heave motion vs. wave period

typically vary between 0.1 for 9-sec periods to about 0.85 at 22-sec periods (figure 4.67).

It is generally believed that the northern North Sea presents the world's most severe environment. This, however, is not always true.

Consider a 50% probability that a significant wave height, H_s, will not be exceeded. Weather data for this probability show that H_s is 6.3 ft for the Grand Banks of eastern Canada in the summer. H_s is only 4.8 ft for the northern North Sea in the summer and 2.5 ft for the Gulf of Mexico annually.

Operating efficiencies for these areas have been compared. And for a conventional pipe-lay barge, the efficiency is about 80% in the Gulf of Mexico vs. 45% in the northern North Sea.

With large semisubmersible types of pipe-lay barges, the operational efficiency is about 90% for the Gulf of Mexico compared with 80% for the northern North Sea.

These figures assume maximum significant wave heights of 8 ft for conventional pipe-lay barges and 15 ft for large semisubmersibles.

High winds also affect pipe-lay operations. Winds create pressure on the side of the pipe-lay barge that causes the vessel to surge or sway. This surge can cause excessive tension in the pipeline and

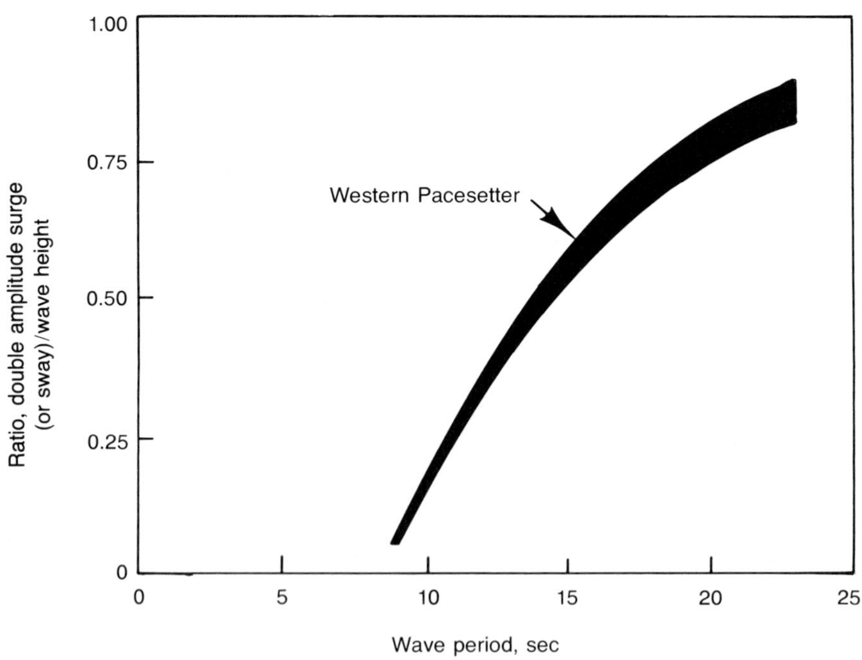

FIG. 4.67 Semisubmersible surge motion vs. wave period

possible buckling. High winds also cause wind-driven currents at a rate of about 2% on the wind speed.

To keep discussions simple, winds and currents were not applied in the computer-simulated pipe-lay operations reported here. Influence of waves as well as winds and currents would, however, be taken into account when planning actual pipe-laying.

Design allowable for combined stresses during pipe-lay is taken as 75–90% of the yield stress. This means that, with dynamic stresses being up to one-third of the design allowable, dynamic stresses must be considered in the design and analyses.

Pipe-lay capability is a function of the combined seaworthiness of the barge and the stinger. The articulated stinger can increase the sea state in which a pipe-lay barge can operate.

Normally, when the weather becomes so severe that pipe-lay vessel motions are intolerable, the pipeline is lowered to the bottom. The stinger is abandoned and the vessel seeks shelter. Damage to the stinger often occurs during abandonment operations. A lot of time is then lost in recovery and repair of the stinger and in getting back on station.

To help alleviate this problem, for example, the Viking semisubmersible pipe-lay vessel has a retractable stern ramp instead of the conventional stinger.

Dynamic Conditions

To determine the dynamic forces, it is necessary to know the velocity and acceleration at each section along the pipeline.

Velocity and acceleration of the barge due to a steady-state harmonic sea wave were calculated. Neglecting friction along the ramp and stinger, the end of the pipeline must move with the barge. Therefore, resulting velocities and accelerations at each pipe section are a function of barge motion.

Morrison's equation was used to obtain the maximum forces realized at each node along the pipeline by using the calculated velocities and accelerations. Drag and inertia coefficients and particular response operators of the barge based on experience and conservation were assumed.

These forces were then used as input data in the computer simulation to obtain the total response of the pipeline to dynamic and static loadings.

The pipeline was assumed to act as a rigid pendulum and to rotate about the point of contact on the mud line. This was done to determine velocity and acceleration at each pipe section.

Velocity or acceleration is the product of the instantaneous barge velocity. Acceleration is multiplied by the ratio of the straight-line distance from the point of the pipeline contact with the mud line to a particular pipe section. This is then divided by the distance from the mud line to the tensioner.

Examples

To determine the influence of dynamic stresses on marine pipe-lay operations, computer-simulated investigations involving vessel motions were performed. Three case studies are presented here.

Other studies show that the dynamic effects on the marine pipeline due to vessel sway within reasonable limits is negligible.

In the three case studies presented, all variables were held constant except for vessel motion (table 80). For comparative purposes, the first case involved pipe-lay operations with no winds, waves, or currents.

The other two cases considered 20-ft wave heights of 6.3-sec period and no wind or currents. A response operator of 0.5 was assumed.

TABLE 80
Operating conditions

Condition	Description
Heave	Upward motion of barge 5 ft from mean water level.
Surge	Aft motion of barge 5 ft from motionless position.
Motionless	No motion relative to mud-line plane.

Parameter	Quantity
Water depth, ft	235
Distance from tensioner to mean waterline, ft	15
Curved stinger radius, ft	1,000
Pipe outside diameter, in.	36
Pipe-wall thickness, in.	0.75
Pipe-coating thickness, in.	3.5
Weight of coated pipeline in air, lb/ft	734
Weight of coated pipeline in water, lb/ft	89.3
Specific gravity of steel	7.66
Specific gravity of coating	2.33
Specific gravity of coated pipeline in air	3.45
Specific gravity of coated pipeline in water	1.17
Young's modulus of the pipe, ksi	30×10^3
Poisson's ratio of the pipe	0.3
Allowable stress in the pipe, ksi	50
Barge tension, kips	80
Surface water current, knots	0
Wave height, ft	20
Wave length, ft	200
Wave period, sec	6.3

In the second case, the single-amplitude upward heave of the vessel was 5 ft. The third example case considered a 5-ft single-amplitude aft surge of the pipe-lay vessel.

These three case studies assume a distance of 15-ft from the tensioner to the mean waterline.

The horizontal distance from the tensioner to the end of the constant curvature stinger, having a 1,000-ft radius, was 430 ft.

The 36-in. diameter, 0.75-W.T. pipe had a sp gr of 1.17 in water. In the three cases discussed here, the tensioner was set at 80 k. The allowable maximum stress in the pipeline was taken to be 50 ksi.

Deflections of the marine pipeline were computed from the mean position of the vessel excursions. Distances along the pipeline from the tensioner to the inflection point were 485 ft for the case of normal operations with no motion. It was 511 ft for the case of vessel heave, and 492 ft for the case of vessel surge (table 81).

Dynamic effects of vessel surge compared with vessel heave were relatively small. And distances along the pipeline for the motionless case and for aft surge were about the same.

Results

Computer-simulated pipe-lay operations showed that the 80-k axial force set at the tensioner decreased along the pipeline. The decrease in the sag bend was to about 48 k for motionless operation, 30 k for heave, and 32 k for surge (figure 4.68).

The primary flexural moment was 32 million in.-lb maximum in the overbend for all three cases (figure 4.69). In the sag bend, the primary moment peaked at 30 million in.-lb for the motionless case, 41 million in.-lb for heave, and 32 million in.-lb for surge.

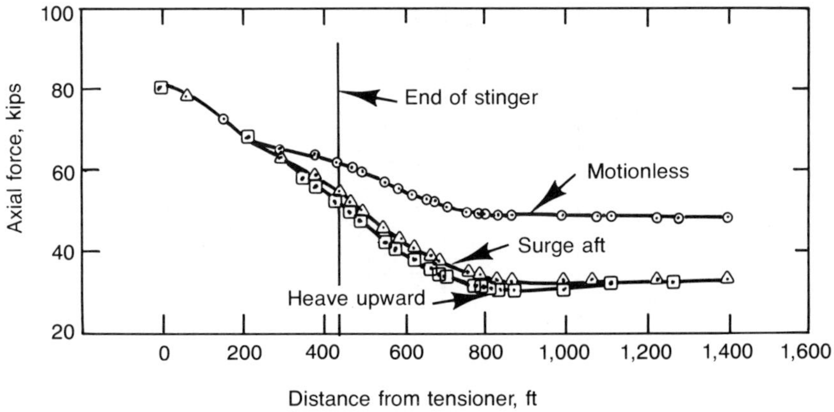

FIG. 4.68 Axial forces

TABLE 81
Summary of deflections

Variable	Distance, ft		
	Motionless	5-ft vessel heave, up	5-ft vessel surge, aft
From tensioner to inflection point			
Horizontal distance	468	490	474
Vertical distance	121	127	116
Distance along pipeline	485	511	492
From tensioner to mud-line tangent point			
Horizontal distance	1,068	985	1,068
Vertical distance referenced to mean waterline	250	250	250
Distance along pipeline	1,108	1,028	1,108

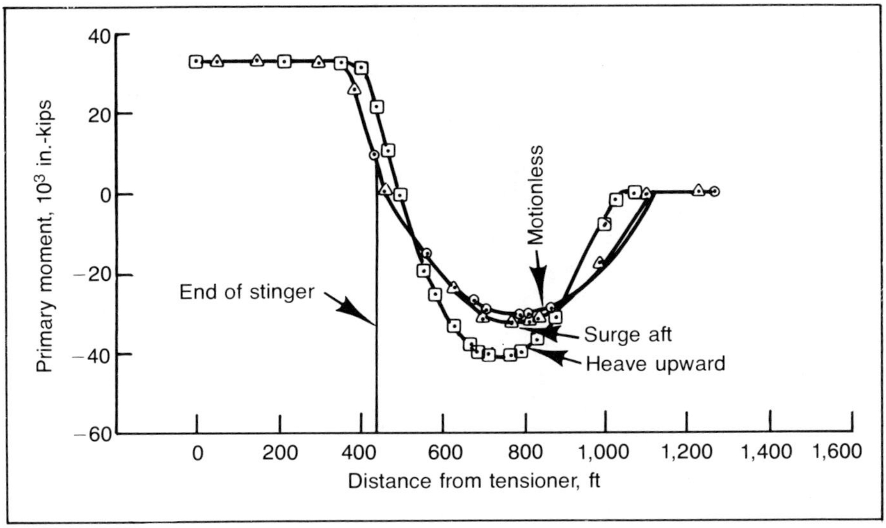

FIG. 4.69 Primary moment

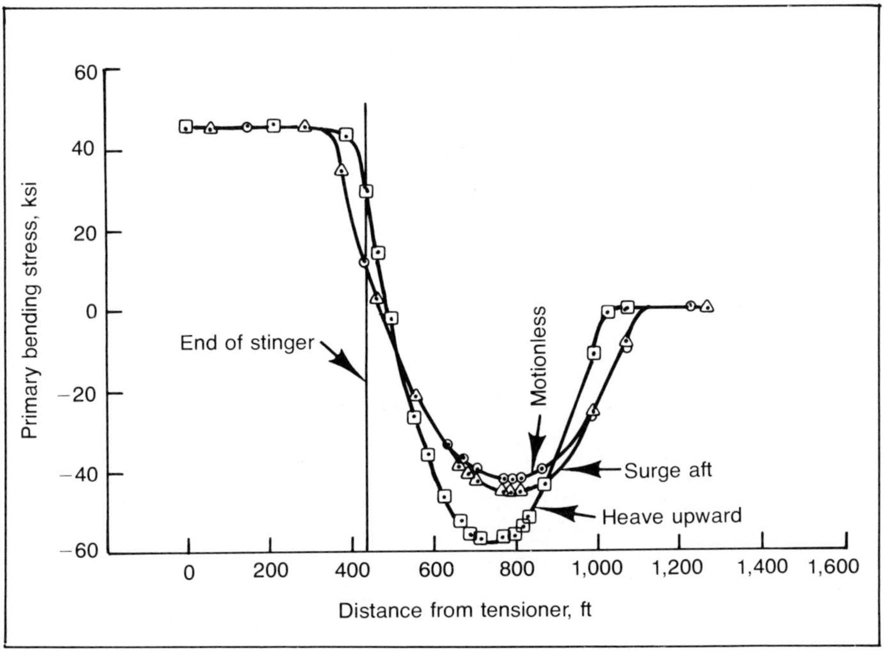

FIG. 4.70 Primary bending stress

These moments gave a maximum stress in the overbend of 45 ksi. Peak stresses in the sag bend were 42 ksi for the case of motionless, 57 ksi for heave, and 45 ksi for surge (figure 4.70).

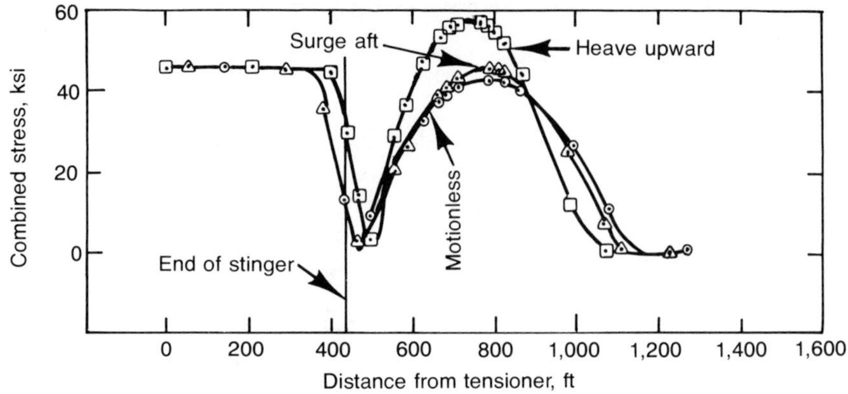

FIG. 4.71 Combined stress

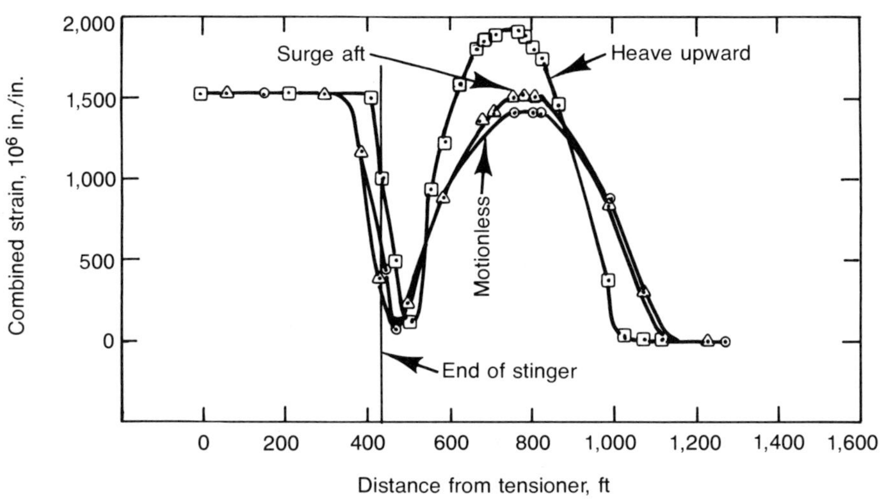

FIG. 4.72 Combined strain

Combined axial and flexural stress in the overbend was 46 ksi for all three cases. The maximum combined stresses in the sag-bend area were 43 ksi for motionless, 57 ksi for heave, and 45 ksi for surge (figure 4.71). The combined strains varied in a similar manner (figure 4.72).

The design factor is the ratio of maximum combined stress to the design allowable stress and is usually used for evaluation. Results showed that the sag-bend factor was less than the overbend factor with no motions, as would be expected.

Vessel motions increased the design factor in the sag bend from 0.85 to 1.15 for 5-ft upward heave (table 82). This was a 35% increase

TABLE 82

Maximum values of structural responses

Variable	Overbend region			Sag-bend region		
	Motionless	Heave	Surge	Motionless	Heave	Surge
Loads						
Axial force, kips	80.4	80.5	80.4	59.6	47.2	49.6
Primary flexure, 10^6 in.-lb	32.3	32.3	32.3	30.2	40.8	32.2
Strain, micro-in./in.						
Axial	32	32	32	23	32.	20
Primary flexure	1,500	1,501	1,501	1,403	1,915	1,495
Combined	1,532	1,532	1,532	1,423	1,909	1,509
Stress, ksi						
Axial	1.0	1.0	1.0	0.7	0.6	0.6
Primary flexure	45.0	45.0	45.0	42.1	56.9	44.9
Combined	46.0	46.0	46.0	42.7	57.3	45.3
Design factor	0.92	0.92	0.92	0.85	1.15	0.91

or about 7% dynamic stress per foot of vessel heave. Vessel motions increased the sag-bend design factor from 0.85 to 0.91 or about 7% for 5-ft aft surge.

Net result of the vessel motions for this particular case was essentially no change in the overbend region. Vessel motions gave dynamic stresses in the sag bend that amounted to a total of about 8.4%/ft of simultaneous surge and heave. It is unlikely that either surge or heave would be acting alone.

4.9 RIG MIX OF OFFSHORE CONSTRUCTION FLEET

Generally, when performing stress and buckling analyses of pipelines, a trade-off results. For example, as pipeline wall thickness decreases, the buckling-safety factor decreases, and tensioner and stinger requirements are also reduced.

As wall thickness increases, stress and buckling-safety factors increase as well as equipment needs and material costs. Increasing wall thickness by 1/16 in. in a 100-mile-long 30-in. pipeline increases material costs by about $1.8 million.

Major factors must be considered in the design stage to prevent buckling. They are the ratio of diameter to wall thickness, material properties of the pipe, and manufacturing tolerance on roundness and wall thickness.

For large diameter-to-thickness ratios (D/t), the buckling mode is instability and collapse. With small D/t ratios, the buckling mode is instability due to plastic strain.

Buckling Analysis

The literature shows that problems involving buckling of cylindrical shapes have been investigated for some time.

In 1858, Fairbairn developed empirical equations for the collapse of riveted circular tubes subjected to uniform radial pressure. His results were used extensively until 1906 when Stewart published his work and empirical equation for pressure on steel lapwelded tubes.

Theoretical and empirical equations developed by Levy in 1884, Carman and Carr in 1906, Southwell in 1915, Flugge in 1932, and Timoshenko in 1934 are still used today.

For marine-pipeline buckling design and analysis, we developed relationships for combined flexure, axial loads, external pressure, and accounted for pipeline flattening. Design curves for use in parametric analyses were also developed (figure 4.73).

Our procedure relates allowable external pressure and bending moment with the equation of Mesloh, Sorenson, and Atterbury, which is:

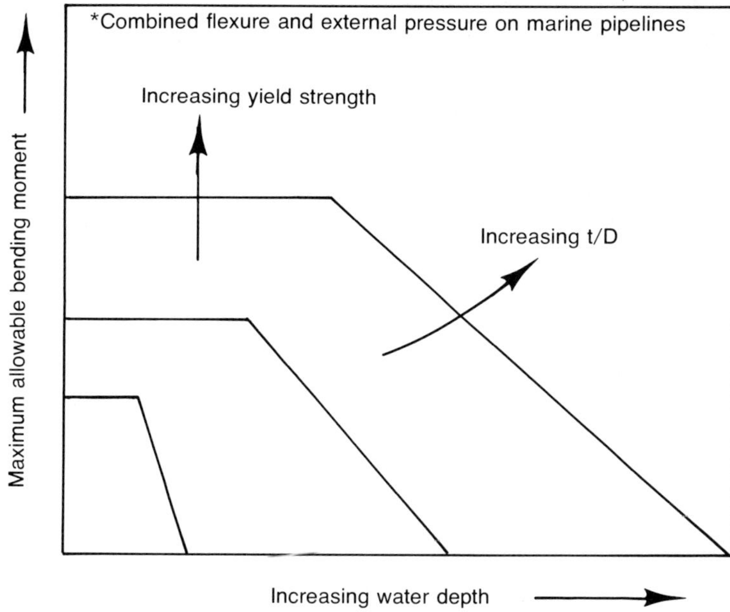

FIG. 4.73 Design curves

$$[P(SF_p)/P_{cr}] + [M(SF_m)/M_{cr}] = 1 \qquad (4.48)$$

The maximum allowable moment, M, depends upon the tensioner force. The moment values can be determined by the ETA/ Pipelay computer program.

The critical-collapse pressure was determined from the Southwell equation where the maximum flattening is 0.1% at the yield stress, $\delta\ \gamma$.

$$P_{cr} = [2\ (t/D)\ \delta\ \gamma]/[1 + (0.9\ \delta\ \gamma/E)\ (D/t)^2]$$

The critical buckling moment was determined from Timoshenko's equation. (4.49)

$$M_{cr} = (EI/D)\ [(2.4\ t/D) - (10^{-7}\ D/t)]/[1 + (0.004\ E/\delta\gamma)]$$

These equations do not account for stiffness of the concrete coating on marine pipelines. This is not necessary, unless the pipe has a large OD with thick coatings.

When necessary to correct for concrete coating, a substitution is made with a calculated effective modulus of elasticity. This is the same approach used for sizing portland-concrete beams with steel reinforcement.

Once the pipeline is set on bottom, primary loading is external pressure, and compression stress at collapse pressure is easily calculated (figure 4.74).

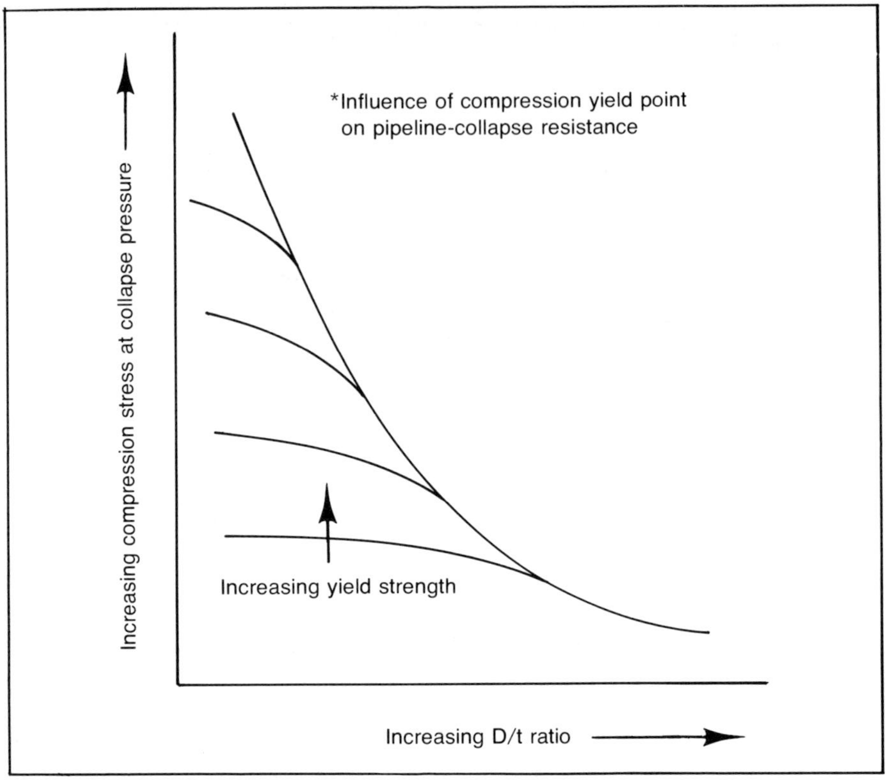

FIG. 4.74 Compression stress

However, the most critical situation is when the pipeline approaches bottom. There, sag-bend flexural stresses are high in addition to the hydrostatic head of the particular water depth.

For this reason, the combined influence of flexural stress and the collapse mode should be considered in design and analysis procedures.

Safety Factors

Code ASA B31.8 gives a permissible bending strain of 1.3%, which is satisfactory for onshore pipelines. Because of the great costs for repairing offshore, however, greater factors of safety must be used for marine pipe-lay.

With maximum stresses in marine pipelines limited to about 80% of yield stress of the material, this amounts to a maximum bending strain of about 0.16%. Bending criteria for marine pipelines are, therefore, about eight times more conservative than onshore criteria.

Normally, strain is limited to 0.2% in the overbend and 0.15% in the sag bend during pipe-lay operations.

However, subsea lines can fail during laying due to buckling at much lower values of stress or strain than those predicted from theoretical stress analyses.

Three modes of buckling failure must be considered.

The Euler type of pipe buckling usually starts in the sag bend during extreme conditions such as excessive vessel motions. The factor of safety should be at least 1.75 for the Euler buckling mode.

A second failure mode is residual flattening. This should be limited to about 0.5% to prevent interference when running pigs. The safety factor should be at least 1.75 on 0.5% flattening. Figure 4.75 shows the flattening function of bending strain.

For a 30-in. OD line with 0.5-in. W.T., the percent of flattening is about 0.5% at 0.12% bending strain and about 2% at 1% bending strain.

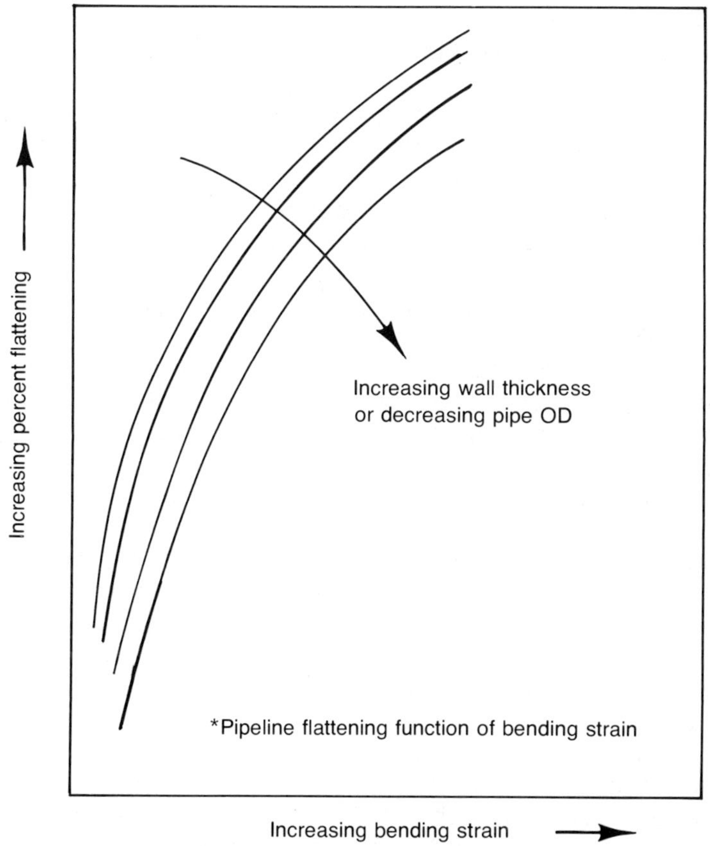

FIG. 4.75 Pipeline flattening

The third failure mode to consider is collapse due to external pressure. The safety factor on collapse should be at least 3.5 since failures sometimes propagate along the pipeline a distance of one to three times water depth.

Scour Considerations

The ocean bottom is sometimes first swept before laying the pipeline when the bottom is extremely irregular.

After laying the line, procedures then vary according to water depth, type of soil, bottom currents, and susceptibility to mud slides. After the line is laid and if it is laid in less than about 200-ft water depths, a sled with high-powered jets is pulled along the line with a bury barge.

Alternately, in less than about 100-ft water depths, a trench can be dredged before laying the line. As the line settles into the trench, sediment eventually covers the line and precludes snaring by fishing trawlers.

In shallow water of up to 75 ft, sediment seems to develop uniform grading with some resistance to subsequent currents and resultant scouring.

Currents are usually low or nonexistent in greater than 300-ft water depths, but damaging mud slides have been known to occur at this depth.

Critical depths regarding pipeline susceptibility to storm damage and scouring are then between 75 and 300 ft.

To prevent scouring, the line is held either gravitationally or mechanically. Gravitational protection is usually provided by increasing the specific gravity, that is, thicker concrete coating on the line.

Mechanical protection is provided by grouted auger-type anchors or imbedded fluted anchors at various intervals along the line. The pipeline is then secured to the anchors with steel bands.

In 1956, Hurricane Flossie first showed the effects of storms on unburied pipelines. It was reported that two 10-in. flow lines were laterally displaced up to 1,500 ft.

If a current runs over an unburied line, a hydrodynamic lift occurs. This lift phenomena can be visualized as being similar to the aerodynamic lift of the air envelopes about an airplane wing.

Experiments have shown that the lift coefficients can be as much as 50% greater than the drag coefficient. This periodic pipe lift and resultant scour can lead to severe pipeline damage or failure if left unchecked.

Postpipe-lay Buckling

Analyses to preclude buckling must be considered during both pipe-lay operations and after pipe-lay. For pipe-lay operations, preclusion of buckling is taken into account in the design by proper pipe diameter and thickness. Also to be considered is pipe coating, stringer configurations, and tension level.

If the route has not been preswept, weight of a trenching machine on the pipe can buckle a pipe running over a bottom precipice.

Depth at which a pipeline is buried is, of course, a matter of economics.

Recent estimates for a 10-ft-deep trench in 150-ft water depths are about $1,000/running ft for blasting with shaped charges. It costs about $35/ft for jetting sand. The desirable or required burial depth varies widely with conditions.

Normal jetting procedure is to pump water from the jet barge to the burial sled under high pressure. However, an innovation has been recently introduced. This method uses four submersible pumps, each delivering 9,000 gpm.

Each pump is directly coupled to submersible electric motors rated at 450 hp at 812 rpm. Power to the motors is supplied to two 1,300-hp diesel engines driving two 960-kva generators at 2,300 v on the support barges.

Economics

In a wet buckle where the pipeline breaks open and is flooded, 1 to 3 months can be required to dewater, recover, and repair the lines.

This costs $3 million to $11 million in deep, rough waters. A dry buckle requires 1 to 2 weeks and costs $750,000 to $2 million to repair.

Diver costs for repair and inspection in the North Sea, for example, are now expensive. Costs there for top diving personnel are currently running about $4,700/day.

Water-depth records for marine pipelines to platforms show water depths have varied from 20 ft in 1947 to 410 ft in 1974. Maximum water-depth capability is also a function of pipe diameter.

The 32-in. pipeline in 360-ft water depths installed in the North Sea Frigg field cost about $1 million/km. Normal tie-ins were estimated to cost $1.3 million. So, a buckling job would cost more.

Stingray pipeline, offshore Louisiana, installed in 1974, had average costs per mile varying from $267,000/mile for a 12¾-in. line to $557,000/mile for a 36-in line. Average cost for replacing one joint of 36-in. line in 70-ft water depths was estimated in one report to be $900,000.

In a January 1974 report submitted to the U.K. House of Commons, it was estimated that 10 advanced pipe-lay barges would be required from 1975 to 1980. They would be used to lay the needed 1,500 miles of new pipelines in the U.K. sector of the North Sea.

Based on 30–36-in. lines in 400 to 550-ft water depths, estimated capital cost for barges was $283 million. This included $7 million for repair and maintenance. Pipeline construction was estimated to cost $1,128 million which included $48 million for repair and maintenance.

As a rule of thumb, the direct pipeline costs from this report will therefore be about $750,000/mile.

There were four major pipelines started and two major lines finished in the North Sea in 1974. The combined 1,000-mile length of those six lines cost about $1 billion or roughly $1 million/mile.

Total costs in the North Sea are now running about $7,500 for capital investment per new barrel of oil per day (b/d). About $1,700 per b/d capital investment is required for a pipeline and delivery system.

For example, the Brent system, which services five fields, will have an initial cost of $500 million to carry 300,000 b/d. This line now carries oil from Cormorant to the Shetland Islands.

Present-day costs for first class vessels run about $50–60 million for pipe-lay vessels, $30–40 million for a derrick barge, and $20–30 million for a trenching barge. The cost just 10 years ago was about $8–10 million for a pipe-lay barge and $6–8 million for a derrick barge.

Offshore Rig Mix

Counting the rigs in operation and under construction, at the end of 1975, there were about 80 U.S. rigs and 39 foreign rigs for a total of 119 rigs. However, 37 out of the 119 rigs were multipurpose units.

In terms of rig mix, this is 22% pipe-lay, 34% derrick-crane barges, 13% bury barges, and 31% combined derrick-lay barges.

This rig count is for the first-class fleet of rigs for offshore construction and excludes inland and swamp barges. Pipe-lay vessels of less than 250-ft length and lift vessels with less than 150-ton cranes are generally excluded unless already proven for moderate seas.

A few vessels doing harbor work but capable of offshore operations are included. About 10 of the various types of vessels are marginal.

If all of the rigs for a particular purpose were operating at one time, the maximum worldwide vessel capability would be 63 for pipe-lay. For lifting operations, 77 vessels would be used, while 16 would be used for bury operations (table 83).

In percentages, the capability will be 53% for pipe-lay, 65% for lift, and 13% for bury operations. This adds up to 131, not 100%, because of the double counting of multipurpose rigs.

These percentages mean, for example, that if all vessels capable of pipe-laying were engaged in pipe-lay work, 63 vessels or 53% of the fleet of 119 rigs could lay pipe.

A more likely equipment-usage spread can be determined. One could assume the multipurpose rigs will be deployed for particular operations in the same distribution as the single-purpose rigs.

TABLE 83
Maximum capability with specified use of multipurpose rigs

	U.S.	Foreign	Total
Total vessels	80	39	119
Maximum capability for specified use			
Lay	38	25	63
Lift	47	30	77
Bury	16	0	16
Total	101	55	156
Percent maximum capability			
Lay	48	64	53
Lift	59	77	65
Bury	20	0	13
Total	127	141	131

Probable composition of the fleet deployment then is 41 pipe-lay vessels, 62 vessels for lifting, and 16 vessels for bury operations, totaling 119 rigs.

On a ratio basis, we would expect a long-term average of about 1.5 times as many vessels engaged in lifting operations compared with pipe-lay. About 0.4 times as many vessels would be engaged in bury operations as in pipe-lay.

Worldwide Buckling Costs

At the beginning of 1973, the extent of marine pipelines was estimated by one source as 15,000 miles worldwide. We estimate that present offshore pipeline mileage is about 17,500 miles.

Because of lack of readily available data, it is necessary to make some rather broad assumptions. Assume a starting date for purposes of this exercise of about 1955 for marine pipe-lay operations. This means that about 410 rig-years of work with pipe-lay vessels has been finished to date.

This estimate mileage of marine pipelines and rig-years gives an overall average work record of about 43 miles of pipeline per pipe-lay vessel per year.

As explained earlier, by the end of 1975, roughly 41 vessels were engaged in pipe-lay operations. This gives an annual capacity of about 1,760 miles of new marine pipelines.

It is understood that the amount of line laid depends on weather conditions, political factors, construction schedules, etc. This makes the business extremely cyclical. However, we are more interested in trends and long-term averages here.

Cost of repairing buckled pipelines can run from $100,000 for small pipe in calm, shallow waters to over $2 million for large pipe in rough, deep waters. Thus, we will use an average cost of $500,000/buckle.

On the average, a marine-pipeline buckle occurs about every 50 miles or so. However, the number of buckles varies widely with the environment.

A buckle might occur every 15 to 25 miles during bad weather in the North Sea. Experienced contractors have laid up to 200 miles of line in the Gulf of Mexico in single season with no buckles.

Putting these rough numbers together means that, in the 3-year period, 1975–1977, average annual worldwide cost of marine-pipeline buckles were about $17.5 million.

Steel sleeves to serve as buckling arrestors are sometimes put on pipelines at intervals of 300 to 1,200 ft apart, depending upon job conditions.

This expedites repair procedures and reduces pipeline damage due to buckle propagation. It is a matter of economics to weigh repair costs with spacing and type of buckling arrestors.

Computer-simulated pipe-lay operations have been described previously. These, with the buckling procedures described here, can be used to determine critical buckling moments as a function of particular job conditions.

These procedures and economic analyses could be used to optimize the type and spacing of buckling arrestors. This would result in the cost of marine-pipeline buckles being greatly reduced.

4.10 GENERAL CONSIDERATIONS BEFORE LAYING OFFSHORE PIPELINES

Water depth greatly affects sag-bend stresses in shallow water, but not as severely as one might expect.

Increasing barge tension reduces stresses to a point of diminishing return where a further increase in tension will not decrease the curvature or the combined stresses.

Waves and current heavily affect pipeline response. They must be analyzed as acting simultaneously and cannot be added separately to obtain the total effect. Combined waves and current have a greater influence than when the two loads are added separately.

Further, the direction of waves and current with respect to the barge heading influences the pipeline curvature and combined stresses.

Stresses can also be reduced by proper positioning of the barge during hang-on mooring conditions or while laying pipe during hostile weather.

If abandonment is inevitable, recovery then becomes the immediate problem. Pipe can be successfully recovered by using the same barge tension as in the pipe-lay operations and by using the recovery cable.

In severe weather, barge response may increase and cause considerable dynamic forces. Quasistatic analysis shows that deflections and stresses due to this excessive motion can be accurately determined.

Special Considerations

Considering cost and complexity of marine pipe-laying and risks that contractors face, overall cost effectiveness should be determined.

In general, use of marine pipeline transportation is the most efficient with regard to operating costs. In fact, the energy cost for pipeline transportation is about two-thirds that of floating storage and tanker transport (table 84).

However, both initial and long-term operating costs must be studied on a case-by-case basis. This is because of the varying influence of a knowledge of reservoir, water depth, distance considered, and other factors. Reliable economic analyses to determine the optimum marine transportation system are therefore complex.

Marine pipelines must maintain structural integrity in corrosive seawater for long periods. So, special considerations must be given to inhibiting corrosion.

This corrosion is inhibited by a thin, epoxy coating and, more commonly, by a thick mastic coating. Mastic consists of about 87%

sand, limestone, and fiber aggregate with about 13% asphalt by weight. This is a more flexible coating than asphaltic concrete pavement (about 5% asphalt.)

TABLE 84
Energy efficiency for transport

Mode	BTU/ton-mile
Pipeline	450
Railroad	670
Waterway	680
Truck	3,800
Airplane	42,000

Portland concrete coating over mastic is commonly weighted with iron ore. The pipeline is further protected by cathodic protection. This is done by either impressing a direct current in the pipeline or by connecting sacrificial metal anodes to the pipeline. Zinc bracelets are usually placed about 0.25 miles apart along the line.

We have discussed marine pipeline requirements in terms of allowable stress or strain and design factors during pipe-lay operations. The radius of curvature of the pipeline in the vertical plane corresponding to the discussed strains is usually about 1,000 to 1,200 ft.

There is usually a minimum curvature specified for the as-laid condition also. Consider the Cinta field marine pipelines laid in the Java Sea in 1974 by McDermott for Pertamina.

There, the minimum on-bottom pipeline radius was specified as 5,000 ft. The minimum radius in the overbend and in the sag bend during pipe-lay operations at Cinta field was specified as 1,100 ft.

Computer Simulation

Petroleum deposits are often found in deep water where seas are severe and inconsistent. To prevent a costly mission failure, many techniques have been developed to predict pipeline response during actual pipe-laying.

Analytical and experimental methods are possible. But experiments are always costly. Some analytical methods, however, are only valid for stiff pipelines in shallow water. Others apply only to quite flexible pipelines in deep water.

There are other methods which are more general. But they do not allow for the large deflections typical of many marine pipe-laying operations. Since bending is the primary source of stress and buckling failures, an improper S shape in the pipeline during laying can be disastrous.

Overall, prediction of pipeline response for a host of situations has been shown to be possible with the use of analytical computer programs.

Investigations

In all investigations, many of the possible variables were held constant (table 85). Also, each discussion was overly simplified for purposes of illustration.

Actually, all of the variables have to be simulated and handled together for guidance in actual pipe-laying.

Fifty computer runs were made as background for these reports. Some were preliminary and were used to set variables appropriate to the specific effects. Results of 23 of these runs are given here.

Cross correlations were made between 13 of these runs for this series (table 86). Results of the other 10 runs are given to substantiate some of the general conclusions provided earlier in the chapter. These 10 runs show:

- Effects of nonlinearity as a function of wave height.
- The small effects on pipeline stresses when wave and current directions are parallel with the barge heading.
- Negligible effects of reasonable values of vessel sway, etc.

TABLE 85
Summary of pipeline properties

Distance from tensioner to mean waterline, ft	15
Curved stinger radius, ft	1,000
Pipe outside diameter, in.	36
Pipe-wall thickness, in.	0.75
Pipe-coating thickness, in.	3.5
Weight of coated pipeline in air, lb/ft	734
Weight of coated pipeline in water, lb/ft	89.3
Specific gravity of steel	7.66
Specific gravity of coating	2.33
Specific gravity of coated pipeline in air	3.45
Specific gravity of coated pipeline in water	1.17
Young's modulus of the pipe, ksi	30×10^3
Poisson's ratio of the pipe	0.3
Allowable stress in the pipe, ksi	50

TABLE 86
Summary of example pipe-lay conditions

Case	Comparison title	Water depth, ft	Tension kips	Environment: Surface current, knots	Environment: Wave height, ft	Environment: Direction with respect to pipeline, degrees	Vessel motion: Distance, ft	Vessel motion: Direction	Vessel motion: Free laterally	Maximum calculated pipeline offset, ft
1	Water-depth recovery	410	125	0	0	—	0	—	no	0
2		410	125	1.25	10	0	0	—	no	0
3	Recovery	410	125	0	0	—	0	—	no	0
4		410	125	1.25	10	90	0	—	no	38.1
5	Tensions & motions	235	80	0	0	—	0	—	no	0
6		235	80	2.07	0	0	0	—	no	0
7		235	80	0	20	0	0	—	no	0
8	Direction	235	80	2.07	20	0	0	—	no	0
9		235	80	2.07	0	90	0	—	no	27.9
10		235	80	0	20	90	0	—	no	77.2
11	Direction	235	80	2.07	20	90	0	—	no	169.6
12	Water-depth									

	tensions & environment	235	120	0	0	—	0	—	no	0
13	Environment	235	120	2.07	0	90	0	—	no	19.2
14	Environment	235	120	0	20	90	0	—	no	52.1
15	Environment moorings	235	120	2.07	20	90	0	—	no	115.0
16		235	120	2.07	0	90	0	—	yes	42.8
17		235	120	0	20	90	0	—	yes	111.4
18	Moorings	235	120	2.07	20	90	0	—	yes	228.7
19	Motions	235	80	0	20	0	5	up	no	0
20		235	80	0	10	0	2.5	down	no	0
21		235	80	0	20	0	5	forward	no	0
22	Motions	235	80	0	20	0	5	aft	no	0
23	Moorings	235	120	2.07	20	90	0	—	yes	349.6

Key deflections (table 87) and maximum structural responses in the overbend (table 88) and in the sag bend (table 89) were tabulated for comparisons.

Note that the computer gives the distribution of component values for stress and strain. In addition, it gives the maximum values of the different components as they occur at different locations along the pipeline.

For example, the maximum combined strains given in the tabulated summaries are not necessarily equal to the sum of the components.

Recovery

When a pipeline is abandoned, the problem of recovery must be faced. This can prove to be time-consuming and costly. An example problem was investigated.

Parameters included a barge tension of 125 k, 410-ft water depth, and a 3-in. diameter wire recovery cable with a Young's modulus of 14×10^3 ksi. Same barge tension used for the pipe-lay operation was used for recovery operations.

It was found that when the pipeline was pulled up to 35 ft below the end of the stinger, the recovery configuration was 20 ft or less below the pipe-lay curve in the overbend. It was also about 10 ft above the pipe-lay curve for the sag-bend region.

Combined stresses realized during recovery were generally less than those in the pipe-laying operation and were thus well below the allowable limits.

Water Depths

A comparison was made between two different water depths, 235 ft and 410 ft, with no external wave or current forces considered. This comparison revealed the basic difference in pipeline response due to increased water depth which is a vital concern during pipe-lay.

With a tensioner force of 120 k the maximum sag-bend stress was only 15% higher for 410 ft of water than for 235 ft of water. This difference is significant but is not as large as expected.

Tensioner Force

A comparison was made to see how much increasing the tensioner force would decrease the deflection and hence the combined stresses.

Two tensions, 80 k and 120 k were compared with no wave or current forces. The pipeline with the higher tensioner force realized 25% less combined stress in the sag-bend region.

TABLE 87
Summary of key dimensions

Case	depth	Distance from tensioner to inflection point, ft			Distance from tensioner to mud-line tangent point, ft		
number	ft	X	Y	Along pipe	X	Y	Along pipe
1	410	570.1	176.6	605.6	1,370.3	425	1,457.1
2	410	570.1	176.6	605.6	1,370.3	425	1,457.1
3	410	458.7	117.3	479.6	1,429.6	425	1,517.6
4	410	457.4	116.3	478.4	1,429.6	425	1,518.2
5	235	467.5	120.8	485.4	1,068.3	250	1,107.5
6	235	467.5	120.8	485.4	1,068.3	250	1,107.5
7	235	468.9	113.3	485.8	1,068.3	250	1,107.5
8	235	468.6	113.4	486.4	1,068.3	250	1,107.5
9	235	468.0	113.3	486.7	1,068.2	250	1,108.9
10	235	467.5	113.1	495.8	1,068.3	250	1,121.3
11	235	454.5	120.8	515.4	1,068.3	250	1,168.4
12	235	452.2	104.4	467.7	1,150.7	250	1,187.5
13	235	452.2	104.4	468.2	1,150.7	250	1,188.2
14	235	452.2	104.4	473.1	1,150.7	250	1,194.3
15	235	452.2	104.4	490.3	1,150.7	250	1,217.9
16	235	452.2	104.4	468.5	1,150.7	250	1,188.5
17	235	452.2	104.4	471.0	1,150.7	250	1,193.3
18	235	452.2	104.4	483.6	1,150.7	250	1,212.0
19	235	489.6	126.7	511.1	985.2	250	1,028.1
20	235	461.9	106.9	477.6	1,151.0	250	1,187.5
21	235	462.5	110.7	479.7	1,108.8	250	1,147.4
22	235	473.6	116.0	491.9	1,067.8	250	1,107.5
23	235	452.2	104.4	526.9	1,150.7	250	1,256.1

TABLE 88
Maximum overbend responses

	Loads			Strain, 10^{-6} in./in.				Stress, ksi				
Case	Axial force, F_x, kips	Primary flexure, M_x, 10^6 in.-lb	Offset flexure, M_y, 10^6 in.-lb	Axial ε_x	Offset ε_y	Primary ε_z	Combined ε_c	Axial σ_x	Primary σ_y	Offset σ_z	Combined σ_c	Stress factor
1	124.0	32.2	0	50	0	1,503	1,550	1.50	45.0	0	46.4	0.930
2	125.2	32.3	0	50	0	1,503	1,550	1.51	45.0	0	46.5	0.930
3	125.0	15.4	0	1,276	0	414	1,672	1.78	5.79	0	23.4	0.468
4	125.0	15.4	0.46	1,276	6	414	1,673	1.79	5.79	0.172	23.4	0.468
5	80.4	32.3	0	32	0	1,500	1,532	1.00	45.0	0	46.0	0.919
6	80.4	32.3	0	32	0	1,500	1,532	1.00	45.0	0	46.0	0.919
7	80.4	32.3	0	32	0	1,500	1,532	1.00	45.0	0	46.0	0.919
8	80.4	32.3	0	32	0	1,500	1,532	1.00	45.0	0	46.0	0.919
9	80.4	32.3	8.9	32	415	1,500	1,579	1.00	45.0	12.4	47.4	0.947
10	80.4	32.3	38.8	32	1,827	1,501	2,372	1.00	45.0	54.0	71.2	1.423
11	80.4	32.3	65.6	32	3,052	1,501	3,428	1.00	45.0	91.6	102.8	2.057
12	120.2	32.3	0	48	0	1,501	1,548	1.40	45.0	0	46.4	0.929
13	120.2	32.3	5.9	48	278	1,501	1,565	1.40	45.0	8.3	46.9	0.939
14	120.2	32.3	33.1	48	1,541	1,501	2,195	1.40	45.0	46.2	65.9	1.317
15	120.2	32.3	54.3	48	2,523	1,501	2,978	1.40	45.0	75.7	89.3	1.787
16	120.2	32.3	21.0	48	977	1,501	1,838	1.40	45.0	29.3	55.1	1.103
17	120.2	32.3	107.6	48	5,002	1,501	5,270	1.40	45.0	150.1	158.1	3.162
18	120.2	32.3	210.6	48	9,792	1,501	9,954	1.40	45.0	293.8	298.6	5.973
19	80.5	32.3	0	32	0	1,501	1,532	1.00	45.0	0	46.0	0.919
20	80.5	32.3	0	32	0	1,501	1,532	1.00	45.0	0	46.0	0.919
21	80.4	32.3	0	32	0	1,501	1,532	1.00	45.0	0	46.0	0.919
22	80.4	32.3	0	32	0	1,501	1,532	1.00	45.0	0	46.0	0.919
23	120.2	32.3	33.8	48	1,573	1,501	2,217	1.45	45.0	47.2	66.5	1.330

TABLE 89

Maximum sag-bend responses

Case	Loads: Axial force, F_x, kips	Loads: Primary flexure, M_z, 10^6 in.-lb	Loads: Offset flexure, M_y, 10^6 in.-lb	Strain, 10^{-6} in./in.: Axial ε_x	Strain: Offset ε_y	Strain: Primary ε_z	Strain: Combined ε_c	Stress, ksi: Axial σ_x	Stress: Primary σ_y	Stress: Offset σ_z	Stress: Combined σ_c	Stress factor
1	96.3	27.9	0	39	0	1,298	1,330	1.16	38.9	0	39.9	0.798
2	96.8	27.9	0	39	0	1,295	1,328	1.17	38.8	0	39.8	0.797
3	122.9	24.3	0	1,255	0	1,131	1,434	17.6	33.9	0	35.1	0.703
4	123.0	24.3	1.6	1,255	73	1,131	1,682	17.6	33.9	2.2	35.2	0.704
5	59.6	30.2	0	24	0	1,403	1,423	0.7	42.1	0	42.7	0.854
6	59.7	30.1	0	24	0	1,398	1,418	0.7	41.9	0	42.5	0.851
7	59.7	30.2	0	24	0	1,403	1,423	0.7	42.1	0	42.7	0.854
8	59.7	30.1	0	24	0	1,398	1,418	0.7	41.9	0	42.5	0.851
9	59.6	30.2	7.6	24	352	1,403	1,429	0.7	42.1	10.6	42.9	0.857
10	59.6	30. 2	13.6	24	863	1,403	1,423	0.7	42.1	25.9	42.7	0.854
11	61.2	30.2	45.1	24	2,095	1,403	2,115	0.7	42.1	62.9	63.5	0.857
12	101.4	24.0	0	41	0	1,116	1,152	1.2	33.5	0	34.6	0.691
13	101.4	24.0	5.0	41	386	1,116	1,155	1.2	34.5	7.0	34.7	0.693
14	101.4	24.0	12.9	41	472	1,116	1,152	1.2	33.5	18.0	34.6	0.691
15	101.4	32.4	32.4	41	1,505	1,116	1,554	1.2	33.5	45.2	46.2	0.925
16	101.4	24.0	3.2	41	149	1,116	1,158	1.2	33.5	4.5	34.7	0.695
17	101.4	24.0	20.1	41	935	1,116	1,153	1.2	33.5	28.0	34.6	0.692
18	101.4	24.0	31.0	41	1,443	1,116	1,489	1.2	33.5	43.3	44.7	0.893
19	47.2	40.8	0	32	0	1.915	1,909	0.6	56.9	0	57.3	1.145
20	66.1	24.0	0	27	0	1,163	1,182	0.8	34.8	0	35.5	0.709
21	70.2	28.2	0	28	0	1,313	1,339	0.8	39.4	0	40.2	0.804
22	49.6	32.2	0	20	0	1,495	1,509	0.6	44.9	0	45.3	0.905
23	101.6	24.0	21.5	41	1,000	1,116	1,181	1.2	33.5	30.0	35.4	0.709

The implication is that stresses are primarily due to bending since the large tensioner force increases axial forces but decreases bending moments. Obviously, there is an upper limit on the tensioner force that will not yield any further significant reduction in sag-bend stress.

Mooring

Pipeline response, which governs the success of a given operation, is also greatly influenced by barge positioning. Three mooring configurations—station-keeping mooring, slack mooring, and offset mooring—were investigated:

- Station-keeping mooring occurs when severe weather interrupts laying of pipe in calm seas forcing operations to halt and perhaps be abandoned.
- Offset mooring is a sway condition when barge operation continues in the face of adverse sea conditions where the mooring system precludes barge rotation.
- Slack mooring is the same as offset mooring but with the barge free to rotate to the least stress orientation.

Overbend stresses are the most highly influenced by offset mooring, station-keeping mooring, and slack mooring respectively. Station-keeping mooring had maximum stresses 90% greater and maximum strains 32% greater than the slack-mooring problem.

The highest to lowest stresses occurring for these three cases show that the proper mooring configuration will allow optimization of combined stresses.

Waves and Current

In addition to tensioner forces and water depths, the environment has a large effect on pipe-laying operations. Extreme waves and current require that pipe-laying be abandoned.

Forces on a pipeline in 235 ft of water with a barge tension of 120 k and severe sea conditions acting perpendicular to the pipeline were evaluated. The evaluation compares effects of waves, effects of current, and effects of waves and current acting together vs. a calm sea condition.

Both lateral deflections and combined stresses prove to behave nonlinearly. That is, the response to waves and current acting together do not equal the sum of waves and current acting separately. The combined lateral deflection is as much as 60% greater than the sum of waves and current acting separately.

Results also showed that combined stresses in the sag bend were about 80% greater than results obtained from the sum of individual wave and current effects. Overbend stresses for waves and current acting simultaneously were about 95% greater than for calm seas and about 36% greater for waves acting alone.

To show the differences that wave and current directions have on pipe-laying, a comparison was made between pipeline responses for a problem with 80 k of tensioner force. Two directions were investigated:

- Waves and current acting simultaneously and perpendicular to the pipeline.
- Waves and current acting parallel to and in the direction of the barge heading.

Combined stresses for perpendicular waves and current were around 124% greater in the overbend region and at the mud line than for waves and current acting parallel to the pipe.

In some cases of severe weather, the response of the pipe-lay barge to waves becomes considerable. When this occurs, the dynamic forces due to motion response of the barge are transferred to the attached pipeline.

Forces due to drag and inertia are then calculated and superimposed on the forces realized if the barge and pipeline are quasistatic. The upward heave increased the combined sag-bend stresses by 35%. The aft surge increased the combined sag-bend stresses by 7% over those found for a calm sea.

Barge Motion

Case studies on the effects of vessel motions on marine pipe-lay operations showed that reasonable values of sway caused negligible changes in pipe stress. Vessel motions caused by wave forces increased the sag-bend stresses by about 7% per foot of heave and about 1.4% per foot of surge for shallow water.

Vessel motions caused by wind and current forces would cause the dynamic stresses to increase further. Valves and buckling arrestors would also cause increased dynamic stresses, and all of these effects can be easily simulated with computer procedures.

Buckling

Early in this series, methods were presented to determine the critical moment as a function of barge position, critical collapse pressure, and hydrostatic pressure. The procedures were used to develop design curves.

Insufficient data are available precisely to determine the annual number of buckles and cost with marine pipe-laying. Using some approximate methods, it was shown that about 35 buckles per year will probably be experienced in the near future.

This, of course, is strongly influenced by the weather conditions of the particular areas having the most demand for pipelines. The average cost to repair pipe-lay buckles was estimated at about $17.5 million per year.

Future Fleet

We have repeatedly heard discussions concerning the need for long-term projections of the required offshore construction fleet. After much study, we have formulated some preliminary construction rig forecasts.

One problem is that the offshore construction business is quite risky for operators and contractors. The business is constantly affected by various governmental actions and unmeasurable political factors.

Demand for offshore construction work is therefore cyclical. Some of the imponderable elements can be bypassed by using 5-year averages.

When an operator considers expanding his fleet, he must consider the current leased acreage in a given area. He must also consider rig-building programs of competitors and the oftentimes stringent construction schedules. With these uncertainties, it is still useful to have an idea of the direction the industry is going.

ETA published a 50-year forecast on offshore drilling-rig requirements. Study was based on published estimates of worldwide undiscovered recoverable reserves. This means the ultimate recoverable resources less the sum of the accumulative offshore production and proven reserves.

Offshore production follows offshore exploration in a quantifiable manner. It thus seems that pipe-lay operations and other offshore construction requirements should be correlated with offshore exploration.

The first-class fleet of offshore exploratory rigs reached about 376 in late 1975. First-class rigs exclude workover vessels, coring vessels, and submersibles. The number of construction vessels would then be 31.6% of the number of first-class exploratory rigs.

Taking into account the normal time lags and a large number of other factors along with expected exploration demands, a 25-year forecast was derived. It was estimated that the construction fleet will increase from the 119 vessels in 1975 to about 265 vessels in the year 2000.

Construction vessels are built as single purpose and sometimes for multipurpose operations. Ratios of single-purpose vessels were used in this study to classify multipurpose rigs so that a rig mix could be determined.

This procedure gives the rig mix as 35% pipe-lay vessels, 52% lift vessels (derrick or crane barges), and 13% bury barges (table 90). This shows that the effective pipe-lay fleet will grow from about 41 vessels to 101 vessels by the year 1995.

Single-purpose-capability distribution is 53% pipe-lay, 65% lift, and 13% bury, which adds up to 131% of the fleet in 1975 (table 91). This means that 37% of the classified pipe-lay vessels are multipurpose. About 35% of the classified derrick and crane barges are multipurpose.

Taking an average 25-year life for the construction rigs, 90 construction rigs would have to be built in the period 1975–1990 just for replacement.

By 1995, the fleet expansion would be 170 rigs and 90 replacements for retirement to give a required new construction of 260 rigs in the period 1975–1995.

TABLE 90
Construction-fleet mix

Year	Pipe-lay vessels	Lift vessels	Bury vessels	Total vessels
1975	41	62	16	119
1980	56	84	21	161
1985	71	106	27	204
1990	86	128	32	246
1995	101	150	38	289
2000	93	138	34	265
percent mix	35	52	13	100

TABLE 91
Construction-fleet capability

Year	Pipe-lay operations	Lift operations	Bury operations
1975	63	77	16
1980	85	105	21
1985	108	133	27
1990	130	160	32
1995	153	188	38
2000	140	172	34
Percent capability	53	65	13

TABLE 92
Building-rate demand of new construction vessels

Period	Pipe-lay vessels	Lift vessels	Bury vessels
1975–1980	15	22	5
1975–1985	30	44	11
1975–1990	45	66	16
1975–1995	60	88	22
1975–2000	60	88	22

Period	Total expansion	Retirement replacements	Total new rigs
1975–1980	42	0	42
1975–1985	85	30	115
1975–1990	127	60	187
1975–1995	170	90	260
1975–2000	170	95	265

TABLE 93
Accumulative expected capability of new construction vessels

Period	Pipe-lay operations	Lift operations	Bury operations
1975–1980	22	27	5
1975–1985	61	75	15
1975–1990	99	122	34
1975–1995	138	169	34
1975–2000	140	172	34

Classified capability of new rigs to be built can be determined in the same manner as for the total fleet shown here. Results show that between 1975 and 1995, 138 new pipe-lay equipment spreads will be required. Fleet expansion would include 60 new pipe-lay vessels and 90 new replacement rigs (table 93).

4.11 MARINE PIPELINE ROLL PARAMETERS

It is important to be able to quantify the magnitude of pipeline roll during marine pipe-lay operations, especially when planning for the wet buckle contingency. Valves are often installed in the line.

It would, of course, defeat the purpose if the pipe rolled 180 degrees with the valve operators ending upside down and buried in the mud. Pipeline roll is caused by several sources, but the most significant is the dynamically induced plastic strains in the suspended pipeline when laying pipe in bad weather conditions. Procedures to determine pipeline roll are discussed and the results are given in a design chart.

Sources of Pipeline Roll

There are two primary sources that will cause a suspended marine pipeline to roll. The first and more obvious source is lateral forces on the pipeline. These lateral forces could be due to sideways currents, waves, barge yaw or sway, and the influence on the pipeline as the barge goes through a turning radius. Each of these individual effects were studied. Analyses showed that combining all of these sources and taking the worst case conditions, the total induced roll would be on the order of 20–30 degrees during lay of one suspended pipeline span.

The second and more important source of induced pipeline roll is due to plastic overbend strains. This occurs in bad weather conditions and is due to barge heave and pitch. This can occur with none of the above lateral forces. With all elastic and no plastic overbend strains and no lateral forces on the pipeline, there would be no pipeline roll.

Fabrication procedures used to make pipe might effect the roll during lay operations. This would be the result of residual strains in the pipe set-up by the plate rolling process and by the pipe welding process in the steel mill to form the pipe. For a spiral weld pipe, all of these residual strains would be distributed around the pipe along the suspended pipeline span in such a way that the roll in pipe-lay operations would not be effected by these residual strains due to the averaging effect over the pipeline span.

Phenomenon of Marine Pipeline Roll

The pipe roll in laying operations with plastic overbend strains is due to the double curvature in the suspended pipe span. That is, if the pipe was suspended in a single curvature, then it would not roll regardless of how much the plastic overbend strains were.

Pipe roll with plastic overbend strain and double curvature can be reasoned as follows. Assume that a high value of plastic strain is put into the overbend region and that this plastic strain, induced midway in the stinger, is constant down to the inflection point of the suspended span. If the pipe did not roll, then from the inflection point down to the mud line, the pipe has to work to reverse the positive plastic overbend strain into negative elastic sag-bend strains. If a swivel is imagined at each 40-ft. joint of pipe, then the work done by the sag-bend portion of the pipe in reversing curvature would be a minimum if each joint of pipe rotated 90 degrees as it passed through the inflection point. Without the swivels, each joint rotates a little bit. The amount of rotation is determined from minimizing the elastic strain energy of torsion and the plastic strain energy of plastic flexure reversal in the sag bend.

The above arguments are illustrated in an elementary way in figure 4.76 for no roll, 90-degree roll, and 180-degree roll when the curvature in the suspended line is the same in all 3 cases. The influence lines of induced dynamic plastic strains are illustrated for the vertical plane. The shaded portion of the plastic reversal figures corresponds to the plastic work in reversing the plastic positive strain in the overbend to correspond with the required curvature in the sag bend. This shows how the required work is reduced drastically with pipeline roll.

Elastoplastic Material Properties

The exact shape of the stress-strain curve is normally of little consequence since most structural analyses is calculated for elastic condition. But the exact stress-strain relations greatly influence elas-

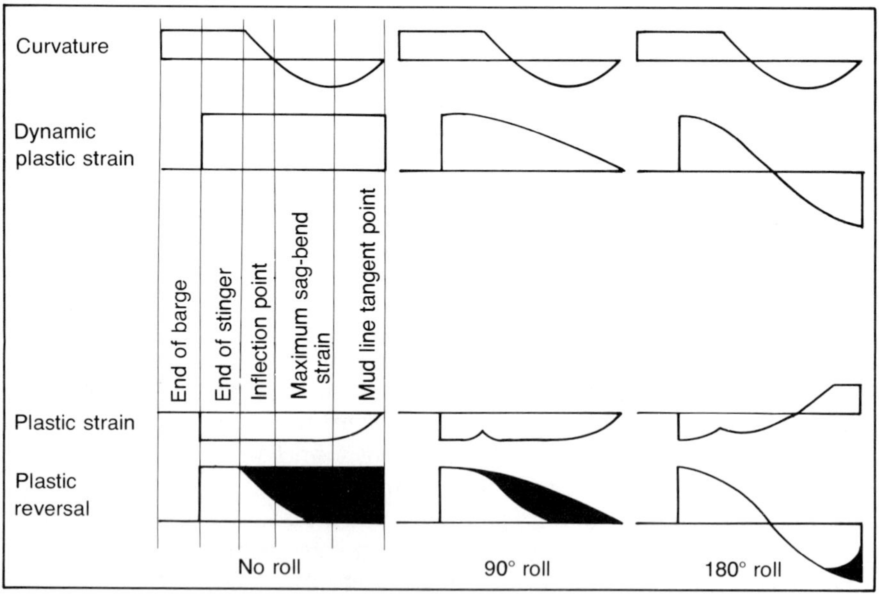

FIG. 4.76 Plastic work in suspended pipeline after dynamically induced plastic strain at stinger

toplastic analyses, and these exact load diagrams are difficult to find. The exact shape of the true stress-strain curve for pipe with 0.2 to 0.5 percent strain would require special successive tests on pristine material samples to ascertain the proportional limit (extent of linear stress-strain), the elastic limit (no plastic strain but nonlinear stress-strain), and the true character of the stress-strain curve between the elastic limit and the yield stress. This data varies with steel mill, grade of steel, temperature during steel rolling, and other variables. However, sufficiently reasonable properties can be obtained by approximation, and this was done by two methods with the average values from the two methods used in the analyses.

First, the API definition of 0.5% strain at the yield stress defines one point. It was assumed that the proportional limit was the same as the elastic limit and was 40 ksi or two-thirds of Grade × 60 yield stress. These two points were then connected with a parabolic curve. The second method used the theory of materials science. This theory assumes a proportional limit of one-third of yield stress, an elastic limit of two-thirds of yield, definition of yield stress as 0.2% strain offset with a slope of Young's modulus, and a nonlinear elastic strain at the elastic limit equal to a value one-third between the slopes connecting proportional limit with Young's modulus and 0.4 percent strain at yield stress. It also assumes that the nonlinear elastic strain is pulled out linearly between the elastic limit and the yield stress. This is reasonable since the relaxation stress-strain curve is linear and equal to Young's modulus from any stress level above the elastic limit.

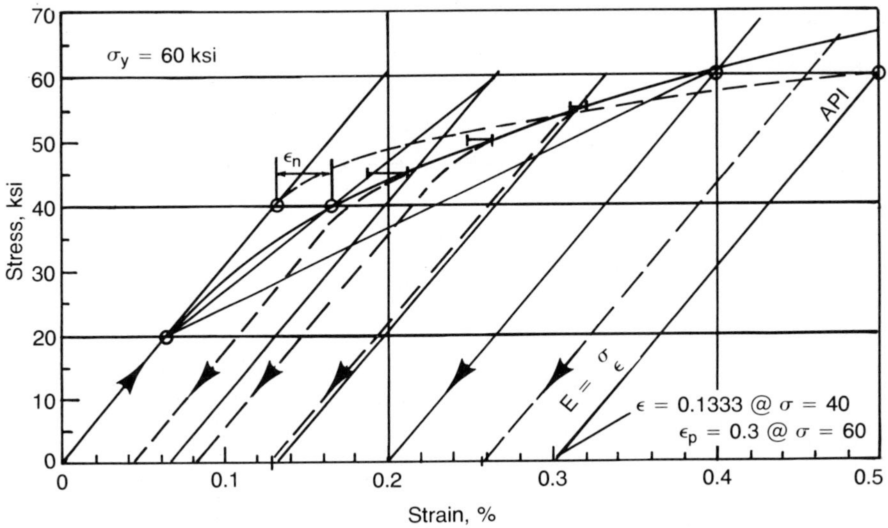

FIG. 4.77 Constructed stress-strain diagram

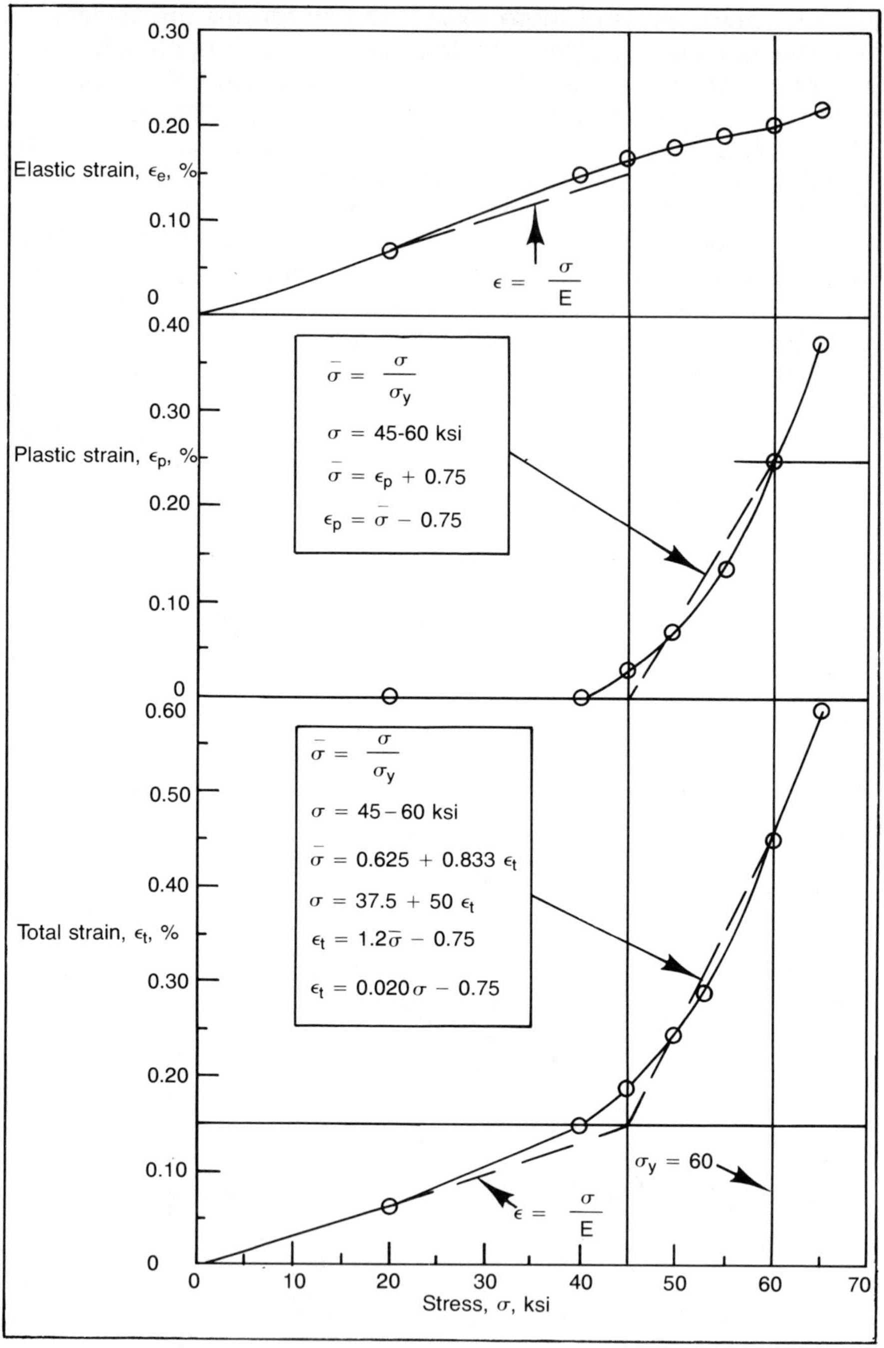

FIG. 4.78 Stress-average strain functions

The constructed stress-strain curves for both methods are shown in figure 4.77. Note that the residual plastic strain at the yield stress was 0.3% using the API procedure and 0.2% using materials science. Stepwise values from the two methods were averaged and plotted in figure 4.78 with the linearized equations to fit the average data also given in the figure.

Discussions of Procedures

The general approach used for pipeline roll analyses due to plastic overbend stress is graphically portrayed in figure 4.79. Basically, it is first assumed that the pipeline slides around a very long curved stinger having a constant curvature equal to the value of the induced plastic strain and that the suspended pipeline is free hanging, that is, no bottom force applied. The suspended pipeline is then analytically put in the normal pipe-lay position. The internal elastic and plastic strain energy and the external work for the two positions are set equal, and the resulting equations solved. The external work

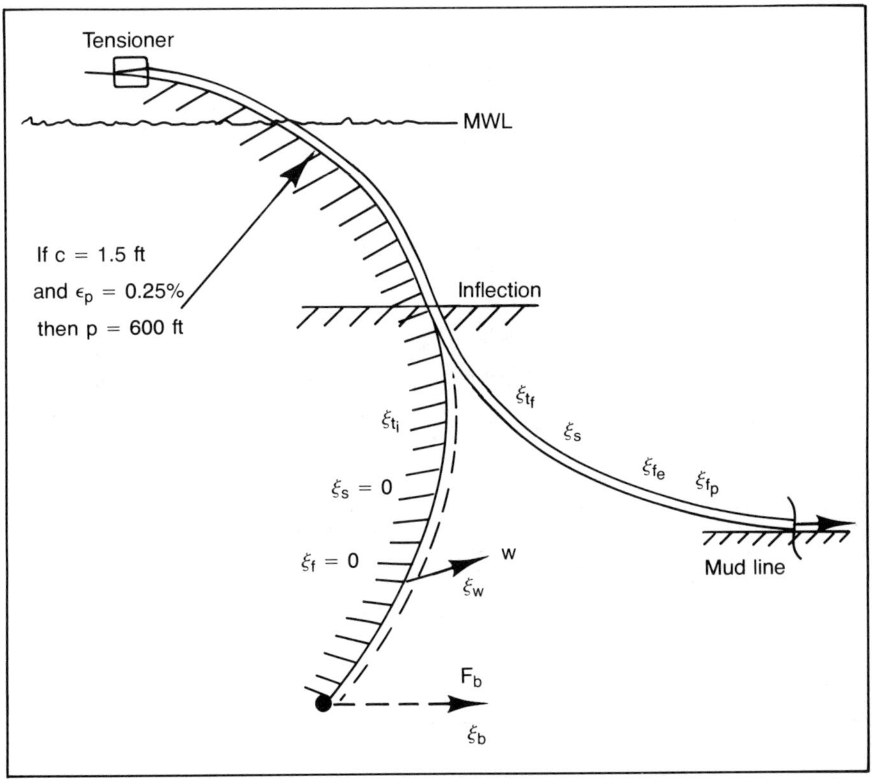

FIG. 4.79 Approach for elastoplastic roll analysis

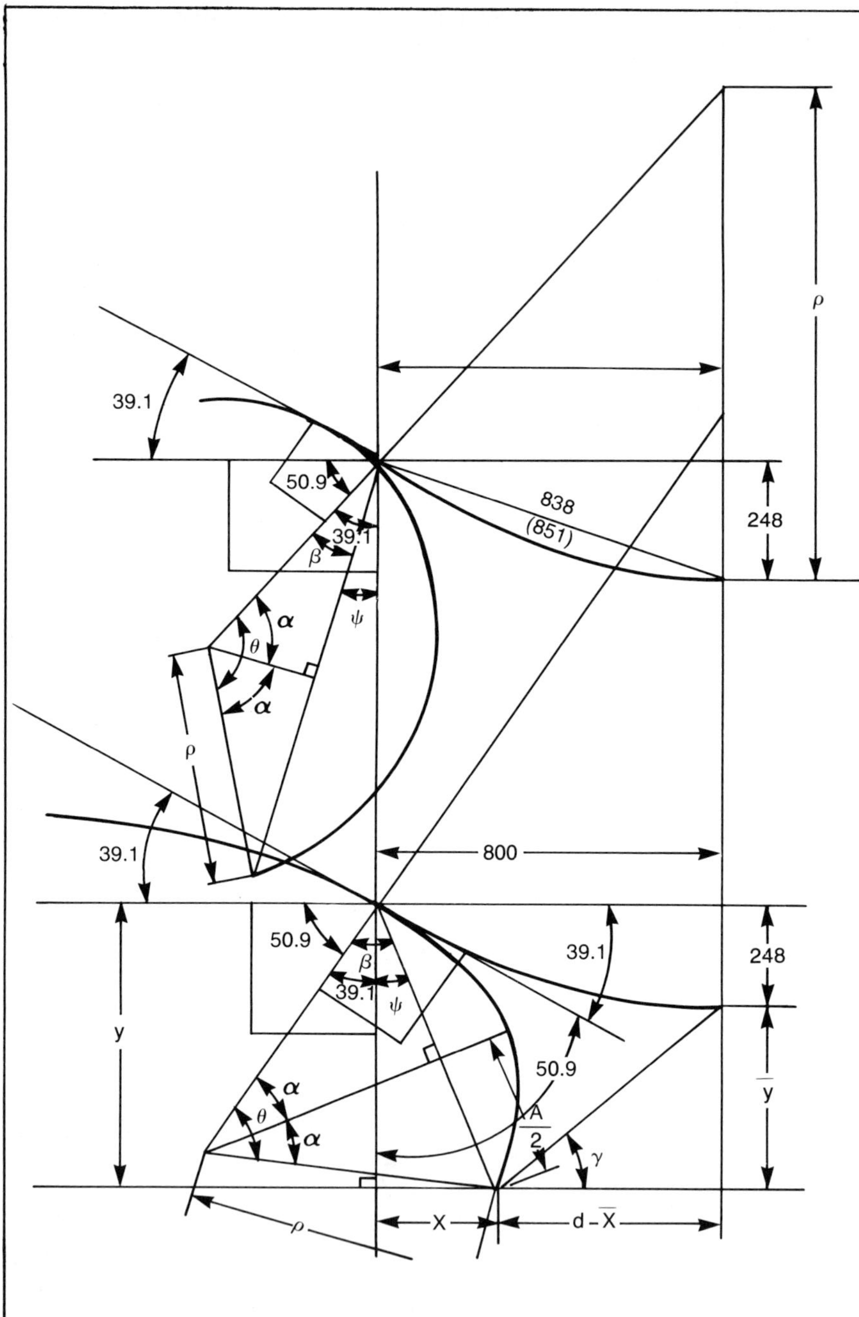

FIG. 4.80 Plastic strain geometry

consists of the bottom force applied times the horizontal distance it moves between the two positions and the gravitational work which the center of gravity of the sag bend moves vertically to the final position. The internal work is the sum of the torsional strain energy of roll, tensile strain energy of the bottom force, and the elastic and plastic flexural strain energys.

While the procedures are straight forward, the geometry (figure 4.80) gets extremely complicated. The required calculations are rather lengthy so only the final results will be given here.

Design Charts

To illustrate the magnitude of pipeline roll, the calculations were performed for one particular pipe size, grade, and concrete coating in a water depth of 410 ft. The results were plotted as a function of bottom tension and plastic strain with the corresponding total strain, stress level, and load factor also given, along with the equations in the form of a design chart (figure 4.81). The load factor is defined here as the maximum pipeline stress divided by the yield stress of the material. To see the tremendous difference a small amount of plastic strain makes, note that when the total strain increases 33% from 0.21 to 0.27%, the plastic strain level increases 100% from 0.05 to 0.10%.

Two cases are shown in the design chart. Cable lay roll means the amount of roll one suspended length incurs when the pipe-lay operation is first intitiated with the end of pipeline tied to a deadman anchor by wire rope. The second case of pipe-lay roll means the amount of roll one suspended pipeline length incurs after the barge has laid several suspended lengths, where the pipe-lay is considered fixed against torsion at each end of the suspended length. For these conditions, the amount of cable lay roll is twice the amount of pipe-lay roll.

As an example on the use of the design chart, assume that the decision has been made to continue lay operations in less than desirable weather conditions. The total strain in the overbend is calculated to be 0.27% (0.1% plastic strain) due to the pipe-lay stress with 80 k bottom tension and the induced dynamic stresses in the overbend because of barge heave and pitch. Reading directly from the chart, the pipe-lay roll would be about 56 degrees.

Using computer analyses to solve the complicated and lengthy equations, the accuracy could be expected to be within 4 to 5%. However, to simplify this presentation, the actual results were linearized, and the given design chart gives results that could be off by about 20% or so. However, the design chart expedites general analyses and is sufficiently accurate for most situations considered.

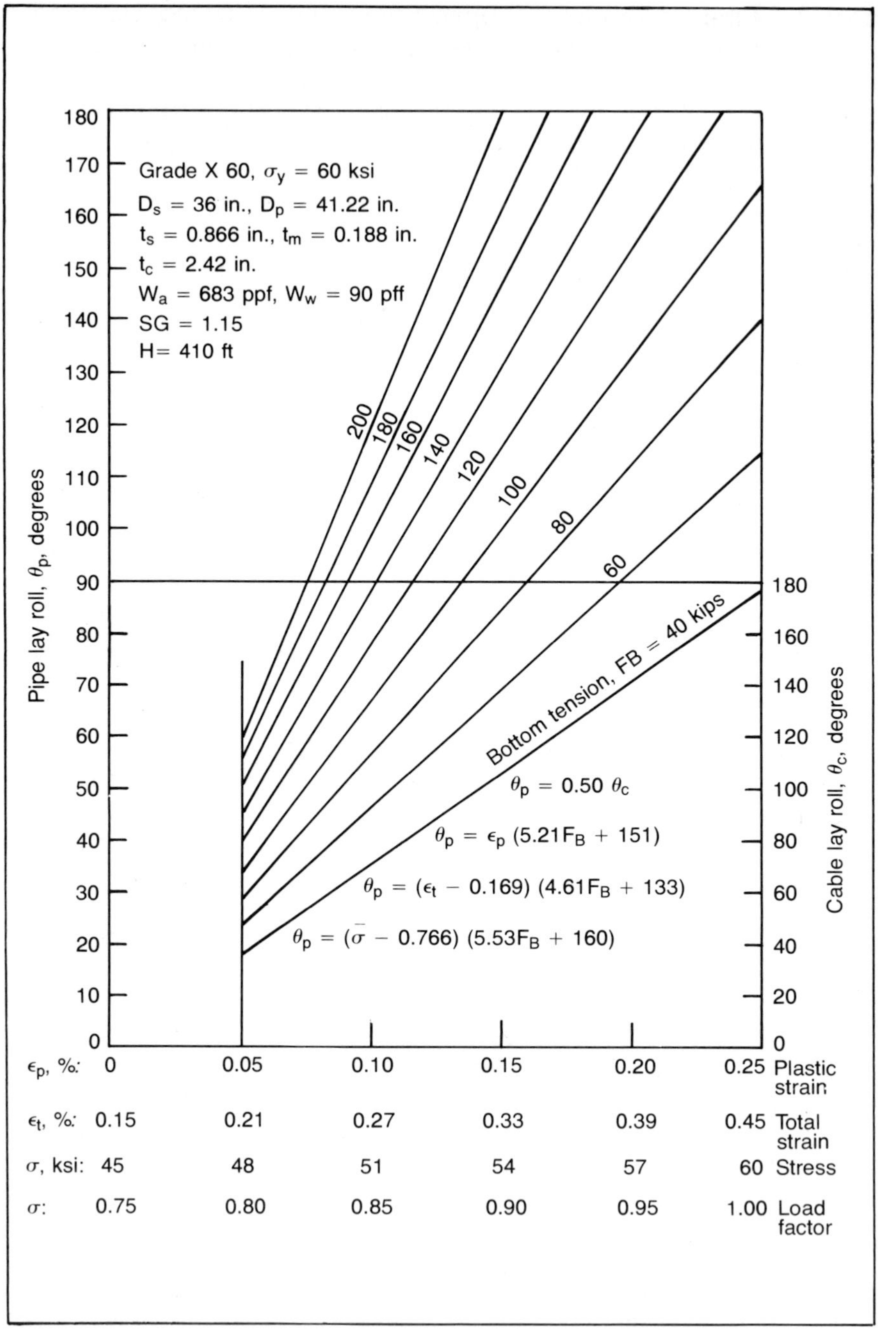

FIG. 4.81 Analysis chart for pipeline roll

REFERENCES

1. Bynum, D., Jr., "Fixed Platform Risers and the Installation of the J Tube," *Offshore*, July 76, pp. 78–81.
2. Bynum, D., Jr., "Establishing Multi-Operational Capabilities for Rigs," *Petroleum Times*, 25 July 75, pp. 33–40.
3. Bynum, D., Jr., and P. M. Lovie, "Rig Safety is Vital," *Offshore*, May 75, pp. 238–245.
4. Bynum, D., Jr., and I. H. Rapp III, "What Do You Consider Before Laying Offshore Pipelines," *Oil & Gas Journal*, Series Part 8, 5 May 75, pp. 256–261.
5. *Journal of Engineering for Industry*, ASME, February 1968, pp. 153–160.

CHAPTER

5 PRESSURE-VESSEL DESIGN

5.1 FLAT HEAD PRESSURE-VESSEL DESIGNS

In this section, we try to determine the required tensile yield strength and the thickness of a flat head welded to a cylindrical shell and thereby forming a vessel which was subjected to internal pressure. The shell-head joint had to be able to withstand at least as much pressure as the center portion of the shell where the major principal stress consisted of only the hoop membrane stress. The stress situation in and about the shell-head joint is a complex one. However, Roark has indicated a suitable approach. Our purpose was to compare the Roark procedure with the values given by Brownell's approximate relations.

The trial drum head thickness was determined by using Brownell's approximate solution (1) as follows:

The yield pressure of the shell due to hoop membrane stress is

$$P_Y = \frac{2\sigma_{Y_s} t_s}{D} = \frac{2(51.3)(0.375)}{18} = 2.14 \text{ ksi} \tag{5.1}$$

The required head thickness is given by Brownell as

$$t_h = D\sqrt{\frac{C_{1P}}{\sigma_{Y_h}}}, \text{ and } C_1 = 0.5 \text{ for a flat head.} \tag{5.2}$$

Consider a heat-treated material for the head so that $\sigma_{Y_h} = 61.5$ ksi.

Therefore

$$t_h = 18\sqrt{\frac{0.5(2.14)}{61.5}} = 2.38 \text{ in.} \tag{5.3}$$

The maximum stress in the head and in the shell was then determined using Roark's equations (2) with a trial drum head thickness of 2.50 in. The frequently used constants in the equations are

$$1 + \mu = 1.28 \qquad m = \frac{1}{\mu} = 3.57$$

$$\begin{array}{ll} 1 - \mu = 0.72 & t_h^2 = 2.50^2 = 6.25 \\ R = 8.81 & t_s^2 = 0.375^2 = 0.141 \\ R^2 = 77.6 & ID = 18 \text{ in.} \\ R^3 = 684 & t_u = 0.375 \text{ in.} \\ \sigma_y = 51.3 \text{ ksi} & \sigma_u = 72.4 \text{ ksi} \end{array}$$

To simplify the calculations, the following expressions are used:

$$F_h = \frac{Et_h^3}{12(1-\mu^2)} = \frac{30 \times 10^6(2.50^3)}{12(1-0.28^2)} = 4.24 \times 10^7 \tag{5.4}$$

$$F_s = \frac{Et_s^3}{12(1-\mu^2)} = \frac{30 \times 10^6(0.375^3)}{12(1-0.28^2)} = 1.43 \times 10^5 \tag{5.5}$$

$$C_2 = \sqrt[4]{\frac{3(1-\mu^2)}{R^2 t_s^2}} = \sqrt[4]{\frac{3(1-0.28^2)}{77.6(0.14)}} = 0.709 \tag{5.6}$$

The force and the moment acting on the head and the shell at the joint are shown in figure 5.1. The moment acting at the shell-head joint is:

$$\frac{M_o}{p} = \frac{\dfrac{R^3C_2^2F_s}{4F_h(1+\mu)} + \dfrac{2R^2C_2^3t_hF_s}{t_s(1-\mu/2)(Et_h + 2RF_s)C_2^3(1-\mu)}}{2C_2 + \dfrac{2RC_2^2F_s}{F_h(1+\mu)} - \dfrac{C_2Et_h}{Et_h + 2F_sC_2^3R(1-\mu)}} \quad (5.7) \tag{5.7}$$

$$= \frac{\dfrac{684(0.503)(1.43)(10^5)}{4(4.24)(10^7)(1.28)} + \dfrac{2(77.6)(0.356)(2.5)(1.43)(10^5)}{0.375(0.86)[30(10^6)(2.5) + 2(8.81)(1.43)(10^5)(0.356)(0.72)]}}{2(0.709) + \dfrac{2(8.81)(0.503)(1.43 \times 10^5)}{4.24(10^7)(1.28)} - \dfrac{0.709(30)(10^6)(2.5)}{30(10^6)(2.5) + 2(1.43)(10^5)(0.356)(8.81)(0.72)}}$$

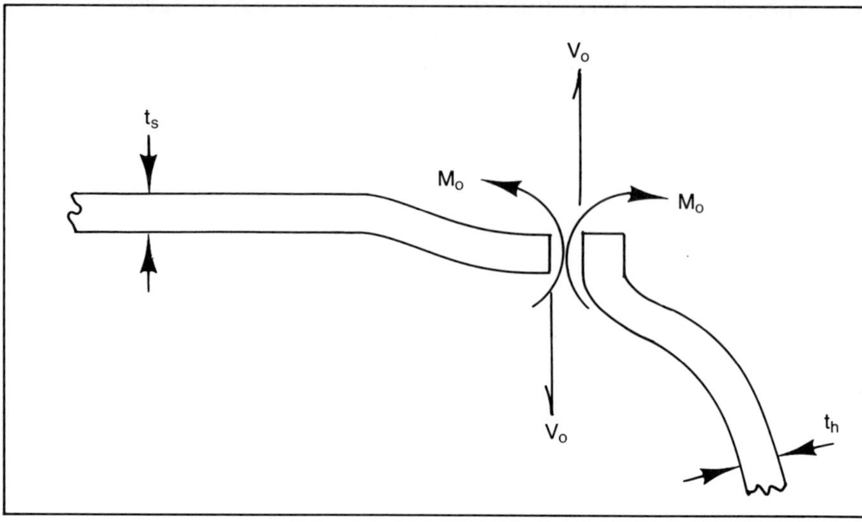

FIG. 5.1 Moment and force acting at head-shell joint

$$= 1.41 \frac{\text{in.-lb}}{\text{psi}}$$

$$= C_3$$

The shearing force acting at the head-shell joint is:

$$\frac{V_o}{p} = C_3 \left[2C_2 + \frac{2RC_2^2F_s}{F_h(1 + \mu)} \right] - \frac{R^3C_2^2F_s}{4F_h(1 + \mu)}$$

$$= 1.41 \left[2(0.709) + \frac{2(8.81)(0.503)(1.43)(10^5)}{4.24(10^7)(1.28)} \right]$$

$$- \frac{684(0.503)(1.43)(10^5)}{4(4.24)(10^7)(1.28)}$$

$$= 1.80 \frac{\text{lb}}{\text{psi}}$$

$$= C_4$$

The maximum stresses in the head occur at the head-shell joint and may be obtained by superimposing the stresses in the plate due to (a) a uniform pressure acting on a simply supported plate, (b) a pure edge moment, and (c) a shearing force distributed around the plate edge, as shown in figure 5.2.

The radial stress at the edge of a simply supported plate with uniform pressure acting on it is zero. The tangential stress, σ_1,

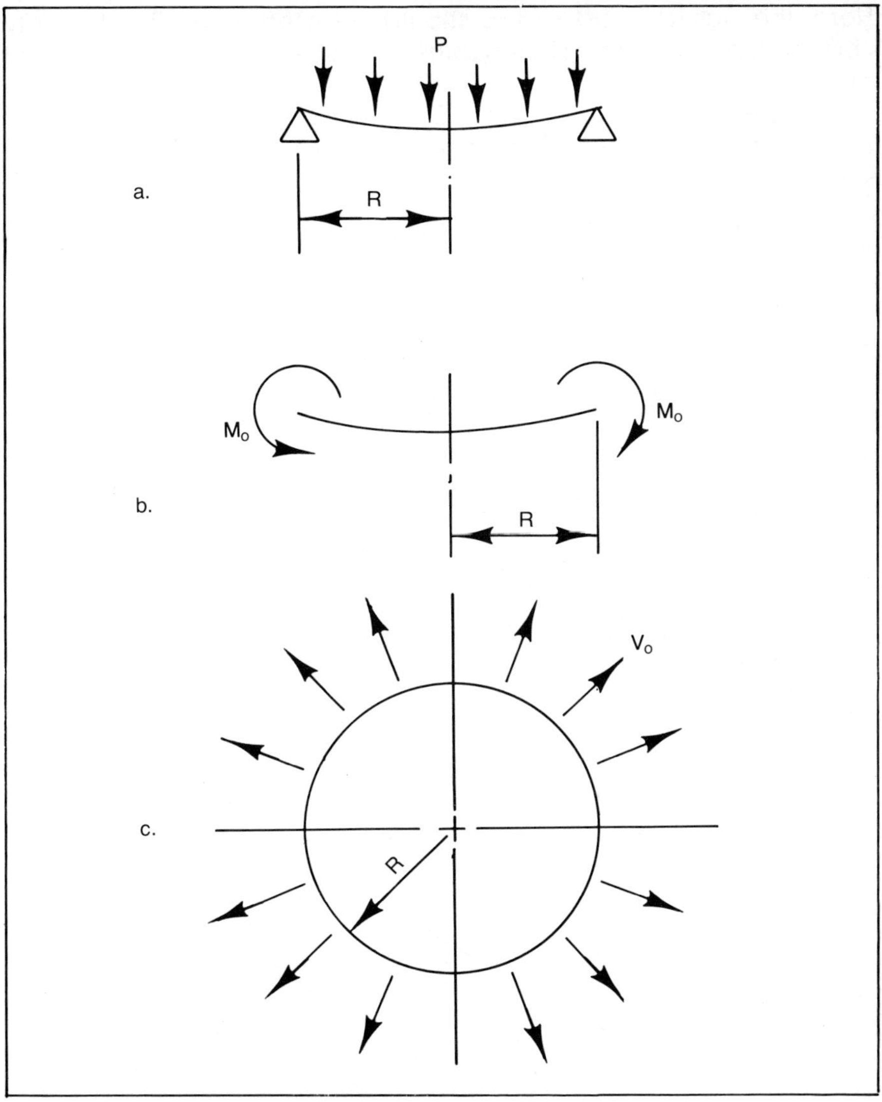

FIG. 5.2 Head loads. (a) Uniform load acting on a simply supported plate; (b) Pure edge moment acting on a plate; (c) Edge shear force

at the edge of the head, when considered as a simply supported plate with uniform pressure acting on it, is

$$\frac{\sigma_1}{p} = \frac{3R^2(m-1)}{4mt_h^2} = \frac{3(77.6)(3.57-1)}{4(3.57)(6.25)} = 6.70\,\frac{\text{psi}}{\text{psi}} \tag{5.9}$$

Here, the actual head radius should be used to be theoretically correct, but the mean shell radius, R, will suffice for all practical purposes.

The radial stress, σ_2, equals the tangential stress, σ_3, at the edge of the plate when pure edge moment is considered and these stresses are

$$\frac{\sigma_2}{p} = \frac{\sigma_3}{p} = \frac{6M_o}{t_h^2 p} = \frac{6C_3}{t_h^2} = \frac{6(1.41)}{6.25} = 1.35\,\frac{\text{psi}}{\text{psi}} \tag{5.10}$$

The stress, σ_4, at the edge of the plate due to uniform radial shear stress alone is

$$\frac{\sigma_4}{p} = \frac{V_o}{pt_h} = \frac{C_4}{t_h} = \frac{1.80}{2.5} = 0.72\,\frac{\text{psi}}{\text{psi}} \tag{5.11}$$

The stresses in the head are shown in figure 5.3. The total stress, σ_5, acting in the tangential plane of the head is

$$\frac{\sigma_5}{p} = \frac{\sigma_1}{p} - \frac{\sigma_3}{p} = 6.70 - 1.35 = 5.35\,\frac{\text{psi}}{\text{psi}} \tag{5.12}$$

The total stress, σ_6, acting in the radial plane of the head is

$$\frac{\sigma_6}{p} = \sigma_2 + \sigma_4 = 1.35 + 0.72 = 2.07\,\frac{\text{psi}}{\text{psi}} \tag{5.13}$$

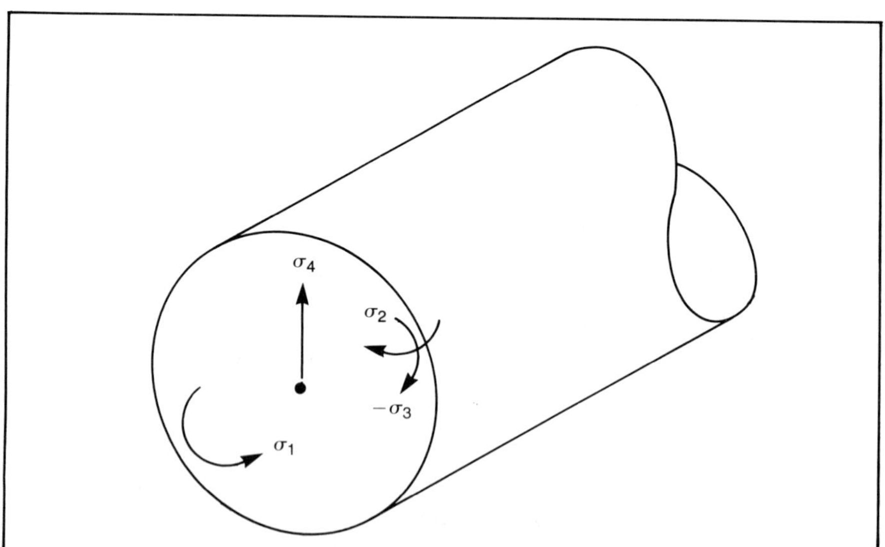

FIG. 5.3 Head stresses

The maximum stresses in the shell may be obtained by superimposing the stresses in the shell due to (a) an internal pressure acting on the cylinder, (b) a uniform radial shear force acting around the end of the cylinder, and (c) a uniform radial moment acting around the end of the cylinder as shown in figure 5.4.

The hoop membrane stress, σ_7, due to internal pressure alone is

$$\frac{\sigma_7}{p} = \frac{R}{t_s} = \frac{8.81}{0.375} = 23.5\,\frac{\text{psi}}{\text{psi}} \tag{5.14}$$

The meridional membrane stress, σ_8, due to internal pressure alone is

$$\frac{\sigma_8}{p} = \frac{R}{2t_s} = \frac{8.81}{2(0.375)} = 11.7\,\frac{\text{psi}}{\text{psi}} \tag{5.15}$$

Considering the uniform radial shear force, V_o, acting around the end of the cylinder, the maximum meridional bending stress, σ_9, resulting from this shear, occurs at

$$x = \frac{\pi}{4C_2} = \frac{\pi}{4(0.709)} = 1.11 \text{ in.} \tag{5.16}$$

and is

$$\frac{\sigma_9}{p} = \frac{1.93V_o}{C_2 t_s^2 p} = \frac{1.93C_4}{t_s^2 C_2} = \frac{1.93(1.80)}{0.141(0.709)} = 34.7\,\frac{\text{psi}}{\text{psi}} \tag{5.17}$$

In general at x,

$$\frac{\sigma_9}{p} = \frac{6C_4 e^{-C_2 x}\sin C_2 x}{t_s^2 C_2}\,\frac{\text{psi}}{\text{psi}} \tag{5.18}$$

The maximum hoop bending stress, σ_{10}, in the cylinder resulting from the edge shear force also occurs at $x = 1.11$ in. and is

$$\frac{\sigma_{10}}{p} = \mu\frac{\sigma_9}{p} = 0.28(34.7) = 9.72\,\frac{\text{psi}}{\text{psi}} \tag{5.19}$$

In general at x,

$$\frac{\sigma_{10}}{p} = \frac{6\mu C_4 e^{-C_2 x}\sin C_2 x}{t_s^2 C_2}\,\frac{\text{psi}}{\text{psi}} \tag{5.20}$$

The maximum shear stress, σ_{11}, in the cylinder resulting from the edge shear force occurs at the shell end, and is

$$\frac{\sigma_{11}}{p} = \frac{V_o}{t_s p} = \frac{C_4}{t_s} = \frac{1.80}{0.375} = 4.80\,\frac{\text{psi}}{\text{psi}} \tag{5.21}$$

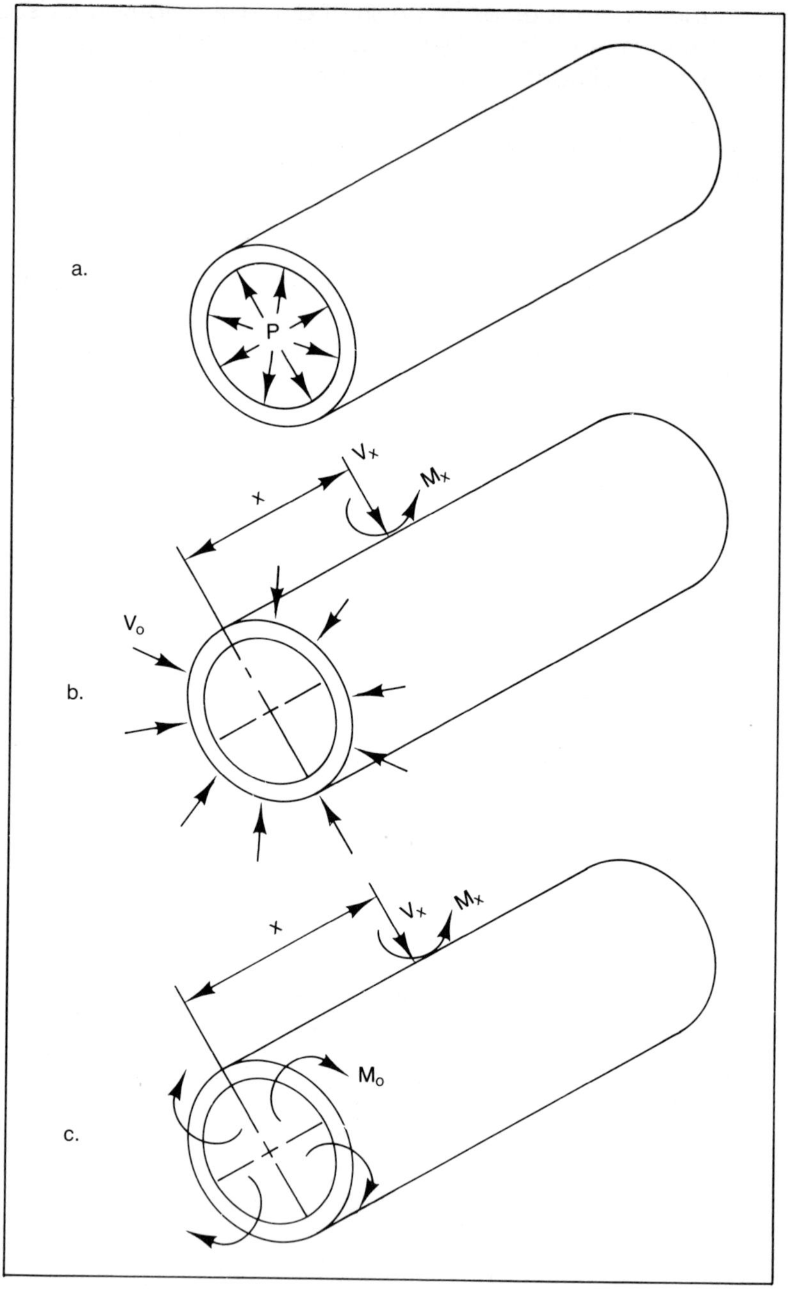

FIG. 5.4 Shell loads. (a) Internal pressure; (b) Uniform radial shear force; (c) Uniform radial moment

In general at x,

$$\frac{\sigma_{11}}{p} = \frac{C_4 e^{-C_2 x}(\cos C_2 x - \sin C_2 x)}{t_s} \frac{\text{psi}}{\text{psi}} \tag{5.22}$$

The maximum hoop membrane stress, σ_{12}, in the cylinder resulting from the edge shear force occurs at the shell end, and is

$$\frac{\sigma_{12}}{p} = \frac{2V_o C_2 R}{t_s p} = \frac{2C_4 C_2 R}{t_s} = \frac{2(1.80)(0.709)(8.81)}{0.375} = 60.0 \frac{\text{psi}}{\text{psi}} \tag{5.23}$$

In general at x,

$$\frac{\sigma_{12}}{p} = \frac{2C_4}{t_s}(C_2 R e^{-C_2 x} \cos C_2 x) \frac{\text{psi}}{\text{psi}} \tag{5.24}$$

To eliminate the trial and error type of solution to obtain the value of x where the combination of σ_9, σ_{10}, σ_{11}, and σ_{12} gives the maximum total stress, assume that the maximum values of both σ_9 and σ_{10} occur at $x = 0$ instead of at $x = \pi/4C_2$.

Considering the uniform radial moment acting around the end of the cylinder, the maximum meridional bending stress, σ_{13}, resulting from this uniform radial moment alone occurs at the shell end and is

$$\frac{\sigma_{13}}{p} = \frac{6M_o}{t_s^2 p} = \frac{6C_3}{t_s 2} = \frac{6(1.41)}{0.141} = 60.0 \frac{\text{psi}}{\text{psi}} \tag{5.25}$$

In general at x,

$$\frac{\sigma_{13}}{p} = \frac{6C_3 e^{-C_2 x}(\cos C_2 x + \sin C_2 x)}{t_s^2} \frac{\text{psi}}{\text{psi}} \tag{5.26}$$

The maximum hoop bending stress, σ_{14}, in the cylinder resulting from the radial moment alone also occurs at the shell end and is

$$\frac{\sigma_{14}}{p} = \mu \frac{\sigma_{13}}{p} = 0.28(60.0) = 16.8 \frac{\text{psi}}{\text{psi}} \tag{5.27}$$

In general at x,

$$\frac{\sigma_{14}}{p} = \frac{6C_3 \mu e^{-C_2 x}(\cos C_2 x + \sin C_2 x)}{t_s^2} \frac{\text{psi}}{\text{psi}} \tag{5.28}$$

The maximum shear stress, σ_{15}, in the cylinder resulting from the radial moment alone occurs at $x = 1.10$ in. and is

$$\frac{\sigma_{15}}{p} = \frac{0.644 C_2 C_3}{t_s} = \frac{0.644(0.709)(1.41)}{0.375} = 1.72 \frac{\text{psi}}{\text{psi}} \tag{5.29}$$

In general at x,

$$\frac{\sigma_{15}}{p} = \frac{2C_2C_3e^{-C_2x}}{t_s} \frac{\text{psi}}{\text{psi}} \tag{5.30}$$

The maximum hoop membrane stress, σ_{16}, in the cylinder resulting from the radial moment alone occurs at the shell end and is

$$\frac{\sigma_{16}}{p} = \frac{2C_3C_2^2R}{t_s} = \frac{2(1.41)(0.503)(8.81)}{0.375} = 33.3\,\frac{\text{psi}}{\text{psi}} \tag{5.31}$$

In general at x,

$$\frac{\sigma_{16}}{p} = \frac{2C_2^2RC_3e^{-C_2x}(\cos C_2x - \sin C_2x)}{t_s} \tag{5.32}$$

As before, to eliminate the trial and error type of solution to obtain the value of x where the combination of σ_{13}, σ_{14}, σ_{15}, and σ_{16} gives the maximum total stress, assume the maximum value of σ_{15} occurs at $x = 0$ instead of at $x = \pi/4C_2$. All of the stresses in the shell are shown in figure 5.5. The total meridional stress, σ_{17}, in the shell is

$$\frac{\sigma_{17}}{p} = \frac{\sigma_8}{p} + \frac{\sigma_9}{p} - \frac{\sigma_{13}}{p} = 11.7 + 34.7 - 60.0 = -13.6\,\frac{\text{psi}}{\text{psi}} \tag{5.33}$$

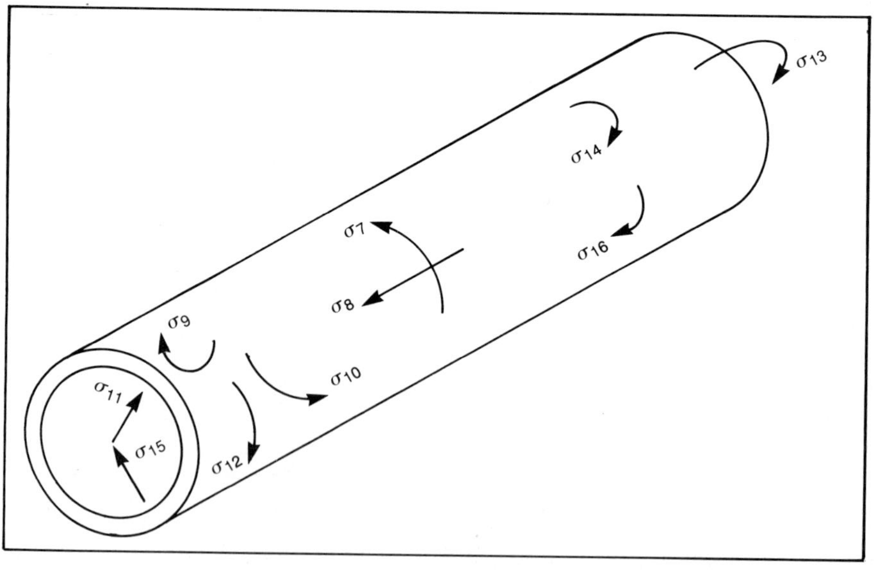

FIG. 5.5 Shell stresses

The total hoop stress, σ_{18}, in the shell is a maximum on the outside wall and is

$$\frac{\sigma_{18}}{p} = \frac{\sigma_7}{p} + \frac{\sigma_{10}}{p} - \frac{\sigma_{12}}{p} - \frac{\sigma_{14}}{p} + \frac{\sigma_{16}}{p} \tag{5.34}$$

$$= 23.5 + 9.72 - 60.0 - 16.8 + 33.3 = -10.3 \frac{\text{psi}}{\text{psi}}$$

The total radial stress, σ_{19}, in the shell is

$$\frac{\sigma_{19}}{p} = \frac{\sigma_{11}}{p} - \frac{\sigma_{15}}{p} = 4.80 - 1.72 = 3.08 \frac{\text{psi}}{\text{psi}} \tag{5.35}$$

Summarily, at the head-shell joint the maximum stress in the head is

$$\frac{\sigma_5}{p} = 5.35 \frac{\text{psi}}{\text{psi}}, \tag{5.36}$$

and the maximum stress in the shell is

$$\frac{\sigma_{17}}{p} = -13.6 \frac{\text{psi}}{\text{psi}}. \tag{5.37}$$

The hoop membrane stress in the shell

$$\frac{\sigma_7}{p} = 23.5 \frac{\text{psi}}{\text{psi}} \tag{5.38}$$

away from the shell-head joint is the largest stress in the vessel and will therefore be used to determine the yield pressure of the vessel as follows

$$P_y = \frac{\sigma_7}{23.5} = \frac{\sigma_{y_s}}{23.5} = \frac{51{,}300}{23.5} = 2{,}180 \text{ psi} \tag{5.39}$$

The required yield strength of the head is

$$\sigma_{y_h} = \sigma_5 = 5.35p_y = 5.35(2{,}180) = 11{,}700 \text{ psi} \tag{5.40}$$

The results show that the use of a mild steel head, 2.50 in. thick, will provide a safety factor of about 3,

$$\left(\text{i.e.,} \frac{\sigma_y}{\sigma_{y\,max}} = \frac{36{,}000}{11{,}700}\right) \tag{5.41}$$

on the maximum stresses at the shell-head joint when the hoop membrane stress is on the verge of yielding (an internal pressure of 2.14 ksi). Finally, the required thickness of weld is found as follows:

$$\tau = \frac{\sigma_Y}{2} = \frac{p\frac{\pi D^2}{4}}{\pi D t} = \frac{pD}{4t} \tag{5.42}$$

$$t = \frac{pD}{2\sigma_Y} = \frac{2.14(18)}{2(36)} = 0.53 \text{ in.} \tag{5.43}$$

Therefore, use a one-in. weld for a safety factor on weld shear of about two.

The head stress for a 2.5-in. plate, as given by Brownell[1] is:

$$\sigma = \frac{D^2}{t^2} C_1 p = \frac{18^2}{2.5^2}(0.5)(2{,}140) = 55{,}500 \text{ psi} \tag{5.44}$$

In the particular example the Brownell approximation is therefore conservative by a factor of almost 5, i.e.,

$$\frac{\sigma_{\text{Brownell}}}{\sigma_{\text{Roark}}} = \frac{55{,}500}{11{,}700} = 4.75 \tag{5.46}$$

5.2 CONDENSER TUBE SHEET DESIGN

Because of mathematical difficulties, most analytical investigations have been for plates perforated in triangular motif in which case a set of equivalent elastic constants could be developed because of conditions of isotropy. With these equivalent elastic constants and the classical field equations of elasticity, it is then possible to obtain a solution for the perforated plate in terms of the solution to an equivalent solid plate.

Various solutions have been developed by Horvay,[3] Malkin,[4] and Salerno and Mahoney,[5] and they have been experimentally proven to be sufficiently accurate by Deagle[6] and by Leven[7] for design purposes over a wide range of parameters. Sampson[8] has performed a rather complete three-dimensional photoelastic stress analysis of perforated strips in triangular array by utilizing the frozen-stress technique. A feasibility study of stresses in perforated strips of triangular array subjected to in-plane loading utilizing birefringent coatings, photoelasticity, electric resistance type of strain gauges, and brittle coatings has also been performed by Lemcoe.[9] Experimental values for other than slender ligaments are not equal to the analytical values because of the idealized assumptions used in the theory (namely, that the ligaments are long and slender and that uniform stress exists across the ligament width), but the accuracy of the theoretical solutions is sufficient for design purposes.

The effective elastic constants used in theoretical analysis of plates perforated in rectangular array have been determined by Daane,(10) but his solution was only for in-plane loadings. The stresses in flat plates subjected to lateral loading and containing perforations in rectangular array are anisotropic, and no proven analytical solutions exist. Mahoney, Salerno and Goldberg(11) have recently formulated a theoretical solution for this problem by making certain idealizing assumptions such as square perforations and slender ligaments. Their solution requires the aid of a high-speed digital computer to obtain numerical results; however, they have given the numerical results of stress and deflection for the plate geometry corresponding to the specimen geometries investigated by the author.

Additional papers concerning perforated-plate problems related to this study are listed in References 12 through 20.

A birefringent coating was utilized as the primary means of experimental stress analysis. This technique was first suggested by Mesnager(21) and later developed as a practical experimental tool by Zandman.(22) The principles of the birefringent-coating technique are essentially the same as in photoelasticity and are thoroughly covered in the references. Since 1953, Zandman,(27, 28) Post,(29) D'Agostino, Drucker and Liu,(30) and most recently, Day, Kobayashi and Larson,(31) Bynum,(32) as well as others, have made significant scientific contributions which have had a most pronounced effect in advancing the state of the art for applications of the birefringent-coating technique.

A limited number of electric-resistance, foil-type strain gauges were used on each specimen to enable a check to be made on the stresses calculated from the measurements of relative retardation. Deflection measurements were taken with Ames dial gauges, a mechanical displacement type of gauge.

Details of Specimens and Loading Fixture

The specimen was machined from 7075-T6, aluminum plate. The nominal stock dimensions were 5 × 32 × 32 in., and all of the test specimens were machined from this plate by successive machining operations. The allowable fabrication tolerance on the perforation diameters and the perforation pitch was ± 0.005 in. However, these parameters were measured after the machining operation for each configuration, and the actual deviation averaged only about ±0.002 in. in each case.

The solid-plate specimen was 4.872 in. thick with an overall diameter of 30 in. The effective diameter, however, was 28 in., which was the diameter of the fixture bolt-hole circle. The solid-plate test was required in order to verify that the condition of true simple support was achieved with the test fixture.

Each of the three perforated configurations tested had 261 perforations in square array with a perforation pitch of 1.250 in., and the same overall and effective diameters as the solid-plate specimen. Configuration I was 4.872 in. thick with perforation diameters of 0.812 in., thereby providing data for a ligament width to perforation pitch of S/P = 0.35 and a plate thickness to perforation pitch of H/P = 3.90. The ligament width to perforation pitch equals

TABLE 94
Nomenclature

a =	specimen radius, measured from specimen center to center of bolt holes, in.
C_b =	plastic-reinforcement correction factor for flexural loading
D =	diameter, in.
DMF =	deflection-multiplier factor, ratio of maximum deflection of perforated plate to maximum deflection of solid plate, using solid-plate test data
d =	compensator constant, div/tint of passage
E_s =	Young's modulus of specimen, psi
e =	strain, μin./in.
H =	specimen thickness, in.
K =	strain-sensitivity constant, fringe/(μin./in.) (in.)
N_n, N_o =	slope of normal and oblique incidence readings, respectively, from plot of compensator reading vs. uniform lateral load, div/psi
P =	specimen perforation pitch, in.
q =	uniform lateral pressure, psi
r =	radius, in.
S =	ligament width, in.
SCF =	stress-concentration factor, ratio of maximum principal stress in perforated plate to maximum principal stress in solid plate, using solid-plate test data
t_c =	birefringent-coating thickness on specimen, in.
w =	deflection, in.
X =	distance from boundary of hole nearest specimen center, measured along ligament center-line, in.
α =	coefficient of linear thermal expansion, μin./in. − °F
β =	direction of σ_1, deg
λ =	wave length of light source, in.
μ_s =	Poisson's ratio of specimen
σ_1, σ_2, σ_b =	maximum and minimum principal stresses, and stress at a boundary, respectively, psi
θ =	angular orientation of a specimen diameter, deg

the ratio (P − D)/P, and this term is commonly referred to as the ligament efficiency. Configuration II was 3.875 in. thick with a perforation diameter of 0.812 in., thereby providing data for S/P = 0.35 and H/P = 3.10. Configuration II warped a total of 0.012 in. during the removal of the 1 in. of thickness due to the relaxation of the residual stresses created during the manufacture of the material. This caused no error, however, since the birefringence measurements were taken in the range of linear response to load. Configuration III was 3.875 in. thick with a perforation diameter of 1.000 in., thereby providing data for S/P = 0.020 and H/P = 3.10.

Three standard ASTM 0.505-in.-diameter tensile specimens were machined from the specimen material, 7075-T6 aluminum. The tensile-test specimens corresponded to the surface, quarter-thickness and center of the plate. Both the elastic constants and mechanical properties of the specimen material were obtained from these specimens. Poisson's ratio and Young's modulus were 0.326 and 10.5×10^6 psi, respectively. The experimentally determined strengths are shown in table 95.

The loading fixture was essentially a 3-in.-thick, 30-in.-diameter steel plate. A 24-in.-diameter, 1⁄16-in.-deep cavity was machined in the specimen side of the fixture plate to permit hydraulic pressurization of the underside of the test specimen. A 7075-T6 aluminum diaphragm, 27 in. in diameter and 0.020 in. thick, was placed between the fixture plate and the specimen to seal off the specimen perforations without altering the stiffness of the specimen. A groove was provided in the fixture plate near the plate edge for a rubber O-ring seal. The specimens and the fixture each had 48 equally spaced holes, 13⁄16 in. in diameter, drilled through on a 28-in.-diameter bolt-hole circle, and three-quarter inch bolts were used to hold the specimen to the fixture. The simple-support edge condition was obtained by having the bolt holes sufficiently oversize and by using spherical washers with the bolts which attached the plate to the fixture. All of the bolts were uniformly tightened to a

TABLE 95

Plate location	Yield Strength, (0.2% offset) ksi	Ultimate Strength, ksi
Surface	69.7	76.9
Quarter-thickness	62.5	77.3
Center	47.4	65.6

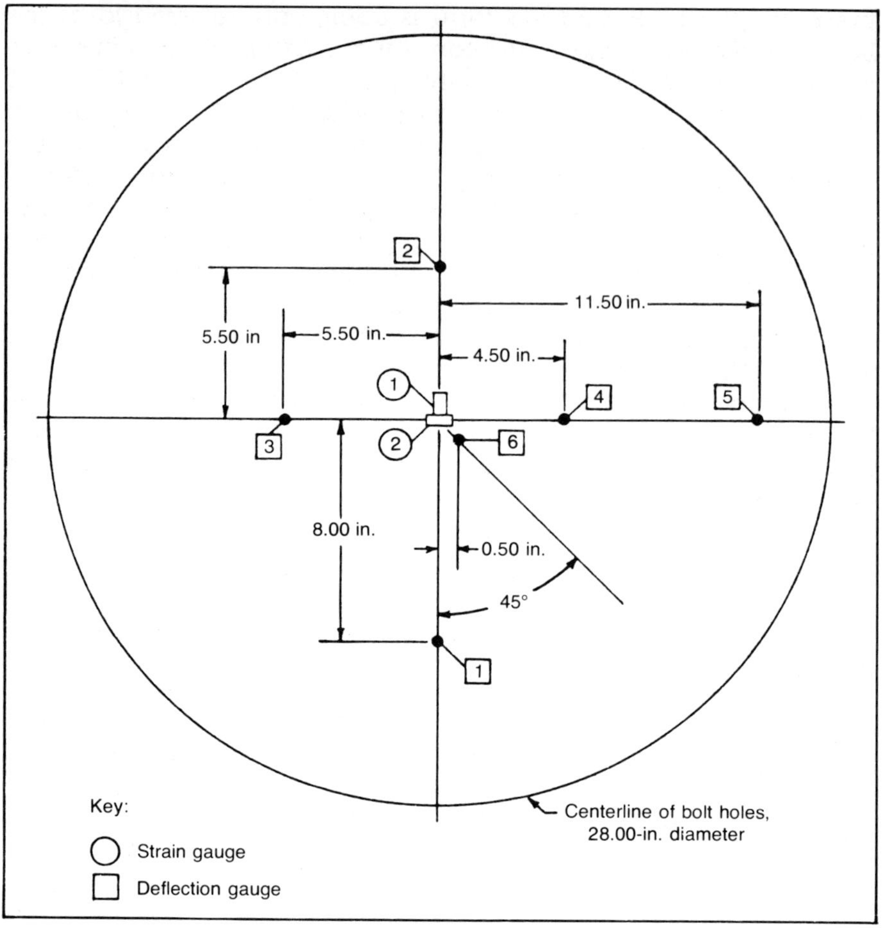

FIG. 5.6 Strain-gauge and deflection-gauge location diagram for solid-plate specimen

torque of 550 in-lb. With this simple and effective design, hydraulic pressure was applied to the diaphragm to provide uniform lateral loading of the specimen.

Experimental Procedure

A jig was designed and fabricated for mounting the Ames dial gauges, having a least count (value of smallest division) of 0.0005 in. The lateral pressure in the test fixture was indicated by the Helicoid test gauge, accuracy ±0.25%. Deflection measurements of the solid-plate specimen were taken at the locations shown on the location diagram, figure 5.6. Several deflection read-

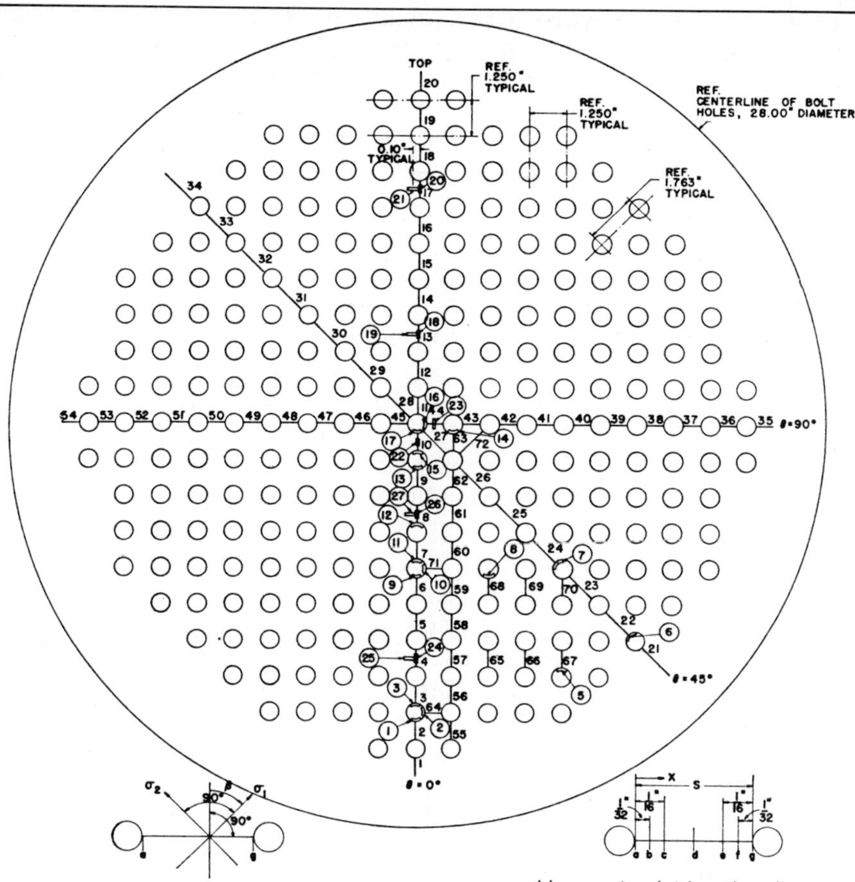

Principal stress directions

Ligament point location diagram

Notes:

1. Thickness and deflection measurements were taken at the midpoints of the ligament widths, which was ligament point d.
2. Strain gauge members are circled. All strain gauges, except those number 18 through 27, were bonded to the perforation walls near the specimen face. Strain gauges number 18 through 27 were bonded to the specimen face that was convex during loading.
3. Ligament point a was always at the hole boundary of the hole nearest to the specimen center, as shown in the ligament point location diagram. Birefringent coating readings were taken across the following numbered ligaments: 10, 9, 8, 6, 4, 2, 63, 62, 61, 59, 57, 55, 68, 65, 69, 66, 70, 67, 71, 64, 72, 28, 31, 27, 26, 25, 24, 23, 22

FIG. 5.7 Ligament and strain-gauge location diagram; specimen configurations I, II, and III

ings at each point were taken at the maximum allowable load and at zero load with each specimen configuration, and these deflection readings were averaged to obtain the reported deflection.

Deflection measurements of each of the three perforated specimens were taken at the midpoint of each ligament across three specimen diameters, corresponding to $\theta = 0$, 45 and 90 degrees on the ligament-location diagram (figure 5.7). Holes were accurately located and drilled in each side of a structural angle at a pitch of 1.250 in. on one leg and 1.763 in. on the other leg, corresponding to $\theta = 0$, 90 and $\theta = 45$ degrees, respectively, for positioning the dial gauges.

Strain gauges, having a 1⁄16-in. gauge length, were bonded to each specimen used in this study in accordance with the manufacturer's instructions.[33] Each configuration was cycled several times before taking strain-gauge readings to stabilize the strain-gauge installation.

The strain-gauge measurements were used to establish the maximum allowable load on the solid-plate specimen. Since the maximum stress occurs at the center and on the surface of a simply supported circular plate subjected to uniform lateral loading, strain gauges were placed only at the solid-plate center. Two gauges were used, as shown in figure 5.6, the second gauge used primarily as a check on symmetry. Strain-gauge readings were taken with the Baldwin strain indicator, Type K.

Strain-gauge measurements for each perforated specimen configuration were recorded by the Gilmore strain-gauge plotter, and the strain gauges on each configuration were located as shown in figure 5.7. Strain-gauge readings were taken at several different pressures with each gauge used so that the slope of the plots of strain vs. pressure could be obtained.

Type S PhotoStress liquid plastic was used as the adhesive to bond the plastic sheets to the aluminum specimen, and the coating was bonded to the specimen in accordance with the manufacturer's instructions.[34] A 7075-T6 aluminum calibration bar was also coated with the same type of plastic that was bonded to the specimen. An axial load was applied to the calibration bar, and using the plot of compensator reading vs. load, the mathematical relationships given in Ref. 35 and the applicable plastic reinforcement correction factor,[36] the K factor was computed to be 0.077 fringe/(μin./-in.)(in.).

A Budd SF/Z small field meter, which has a Babinet compensator and an oblique-incidence attachment incorporated in the design of the instrument, was used for measuring relative retardations. The compensator permitted measurements to be made to within approximately 1/57 of a fringe order. Normal incidence readings were taken first for all points studied on the ligaments, and then oblique-incidence readings were taken at all of these points except at perforation boundaries where only a normal incidence reading was required to compute the boundary stress.

Discussion of Results

The theoretical deflections for the solid-plate specimen, simply supported, were calculated on the basis of Timoshenko plate theory.[37] The deflections per unit load from bending, shear, and compression were calculated separately as a matter of interest to show the individual effect, and these individual deflections and the total deflections are shown in figure 5.8. The theoretical maximum total deflection per unit lateral load for a solid plate occurred at the specimen center and was computed to be 2.44×10^{-5} in./psi.

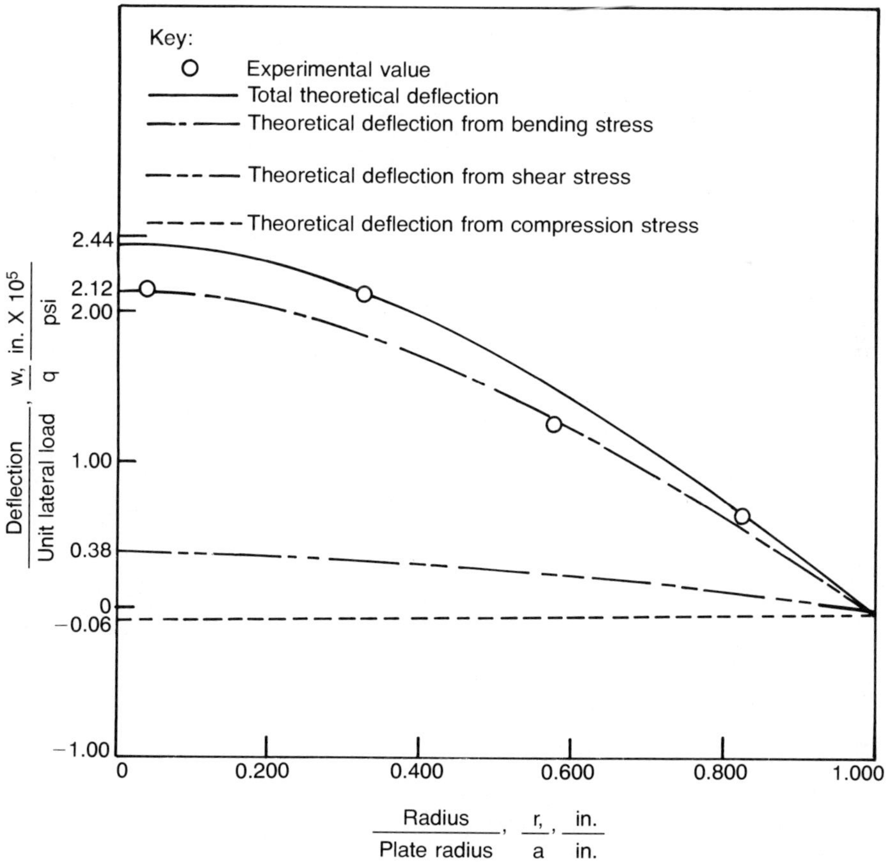

FIG. 5.8 Deflections across the solid plate

The experimentally determined deflection per unit load, w/q, for the solid plate is also shown in figure 5.8. The maximum deflection per unit load for the solid-plate specimen occurred at the specimen center and was 2.32×10^{-5} in./psi.

The maximum theoretical stress per unit load for the solid plate was 10.3 psi/psi, and the maximum experimentally determined stress per unit load was 9.76 psi/psi. For the solid plate, the maximum

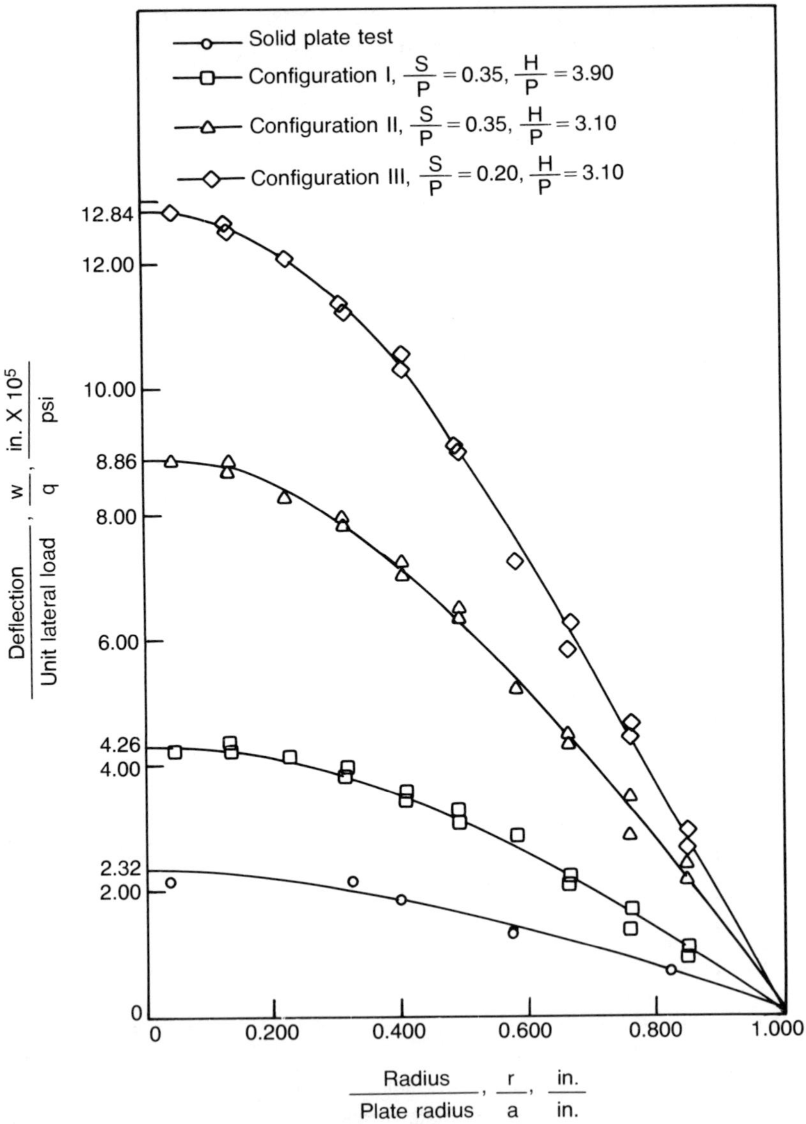

FIG. 5.9 Deflections across the specimen diameters at $\theta = 0°$

theoretical stress and deflection, for the case of simple support, were each about 5% greater than the maximum experimentally determined stress and deflection. On the basis of the excellent correlation between the results of the theoretical and experimental stress and deflection values for the solid-plate specimen, the fixture design seems to have been quite satisfactory from the standpoint of having achieved conditions of true simple support.

The deflections per unit load across the perforated specimen diameters at $\theta = 0$ degrees for each specimen configuration are plotted in figure 5.9. The maximum deflection per unit load occurred at the center of each specimen and was 4.30×10^{-5}, 8.90×10^{-5}, and 13.10×10^{-5} in./psi for configurations I, II and III, respectively. Considering the deviation between deflections measured on opposite ends of a diameter, on the diameters $\theta = 0, 90$ degrees, and the mechanical error, the total error of configuration I deflections was estimated to be within ±4%. The total error in the reported deflections of configurations II and III was estimated to be within ±2%. The deflection error of configurations II and III was less than with configuration I because of the increased stability of the test setup due to the weights placed on the deflection jig.

The maximum principal stress per unit load computed from the strain-gauge readings was 31.4, 51.3 and 71.5 psi/psi for configurations I, II and III, respectively, and occurred at ligament point 10a for each configuration. The stresses computed from the strain-gauge readings were plotted with the results from the birefringent coatings. The error due to the variance of the strain-gauge factor and the approximate orientation of the strain-gauge centerline with the principal stress direction was about 5%, and the instrument error was about 2%. Some integration effect was present, even though the gauge length was small, and this integration effect caused about 3% error. The total error was then about 10%, so the reported stresses from the strain-gauge results are accurate within ±5%.

Considering the fact that radial stresses are zero at boundaries, the tangential stresses per unit lateral load at perforation boundaries were computed from the relationship[35]

$$\frac{\sigma_b}{q} = \frac{\lambda E_s N_n}{2t_c K d(1 + \mu_s) C_b} \tag{5.47}$$

The principal stresses at interior points on the ligaments were calculated from the relationships[35]

$$\frac{\sigma_1}{q} = \frac{(0.856N_0 - N_n)1.875E_s\lambda}{dC_b t_c K(1 + \mu_s)} \tag{5.48}$$

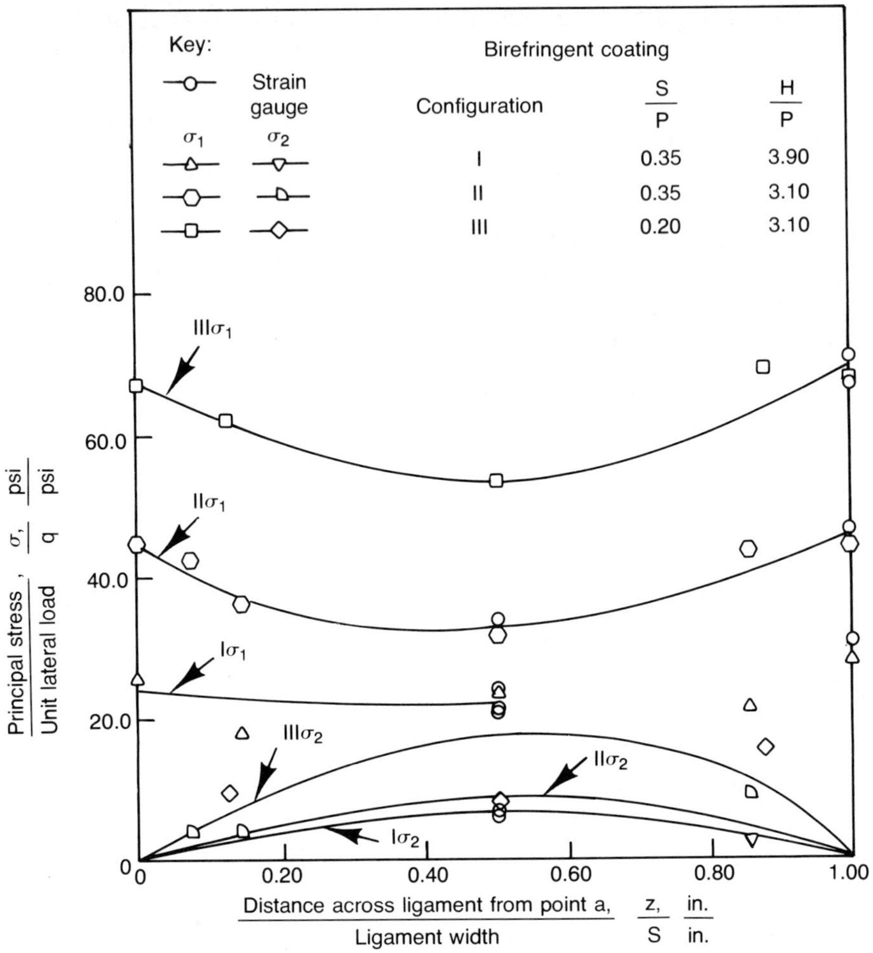

FIG. 5.10 Principal stress distribution, ligament 8

$$\frac{\sigma_2}{q} = \frac{(0.856N_0 - 0.732N_n)1.875E_s\lambda}{dC_b t_c K(1 + \mu_s)} \tag{5.49}$$

The principal stress distributions across ligament Nos. 8 and 26 are shown in figures 5.10 and 5.11, respectively, and are illustrative of all the ligaments which were studied.

The principal stress directions were oriented as shown in figure 5.7, where the direction of σ_1, the maximum principal stress, is indicated by the angle β. The angle β was less than 10 degrees for all points studied except for points on ligaments numbered 62, 67 and 72. The Budd SF/Z meter accuracy of indicating β is approxi-

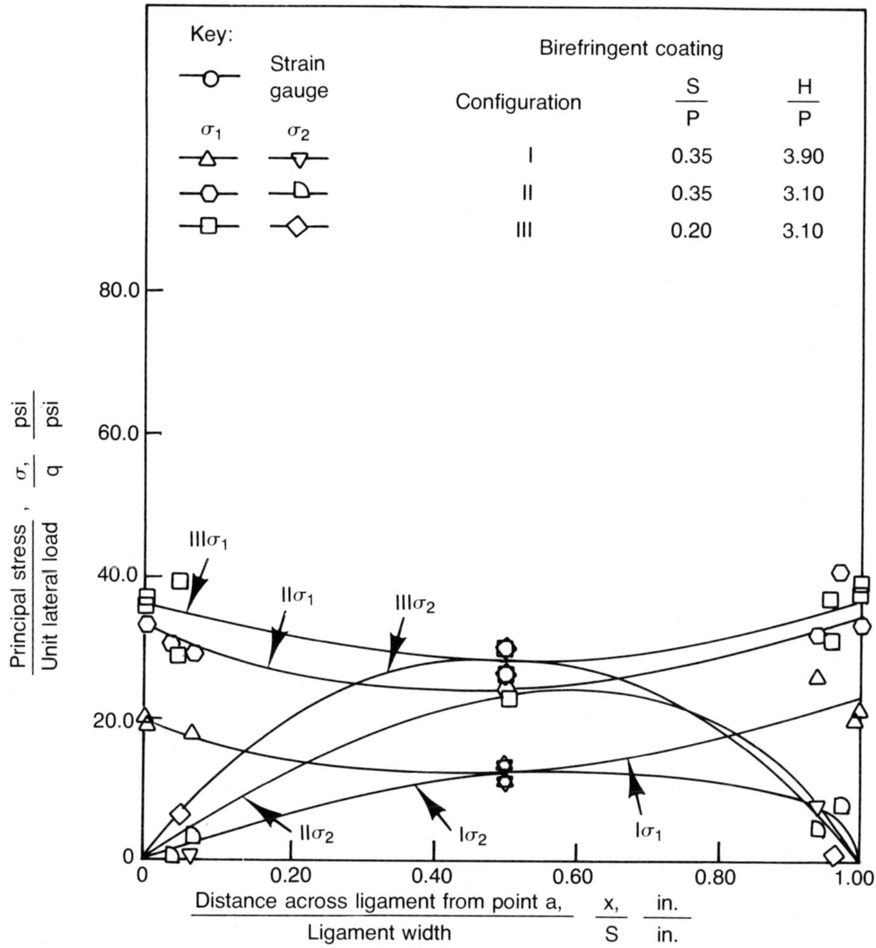

FIG. 5.11 Principal stress distribution, ligament 26

mately ±2 degrees, an error affecting the maximum principal stress less than 1%. The midpoints of ligaments on the diameter $\theta = 45$ degrees was an isotropic point for all three configurations.

The tangential stresses were considerably greater than the radial stresses, as was to be expected. The maximum principal stress occurred at the hole boundaries on ligament No. 10, and the maximum principal stress per unit load was 29.6, 49.6 and 60.0 psi/psi for configurations I, II and III, respectively.

Sometimes a small error exists in relative retardation measurements of birefringent coatings due to a thickness effect, and the magnitude of error is directly dependent upon the coating thickness to specimen thickness ratio. The exact magnitude of error from the thickness effect is presently unknown for two-dimensional strain distributions in general; however, there is no thickness-effect error for plane-stress problems. The increased sensitivity of a thicker coating was compromised in this study by using a "thin" coating so that the error due to the thickness effect was negligible.

A study(38) has been made of the thickness effect with one-dimensional variations of strain, and a more recent paper(39) suggests a procedure for determining the thickness effect on biaxial surface strains in structures having radial symmetry, if the surface curvature is known or can be determined separately. The specimens in this study were not radially symmetric, so the results from this latest paper are not applicable here. Another publication(40) provides an analytical solution for the thickness-effect correction factors in two-dimensional strain-distribution problems having a line of symmetry on the coating where secondary principal stresses are absent.

Other factors that can affect the accuracy of the birefringent-coating technique are: (*a*) Poisson effect(41) on the correlation of surface strains in the structure to relative retardations in the coating; (*b*) residual birefringence(42) caused from improper machining procedure or careless handling of the specimen; (*c*) parasitic birefringence(42) due to humidity absorption, creep, or delayed polymerization; (*d*) K factor variation with ambient temperature;(27) and (*e*) thermal stresses caused by applying the coating to a structure in thermal inequilibrium.

Also, a thermal effect(43) in a birefringent coating may exist if the ambient temperature varies appreciably, and relative retardations at the boundaries of multiple-connected bodies are then indicated with zero load since the coefficients of thermal expansion for aluminum and PhotoStress are different. This thermal effect is not indicated at interior points independent of the shear tractions at the coating-specimen interface because a hydrostatic state of plane stress is developed at these points. Since the linear coefficient of thermal expansion for plastic ranges from 20 to 110 μin./in.-°F,(44) this coefficient was experimentally determined with a Gaertner scientific microscope, using a quartz rod with a known linear coefficient of thermal expansion for instrument calibration. The linear coefficient of thermal expansion for the PhotoStress using these data was 47 μin./in.-°F, with an estimated experimental error of about 5%.

If a polariscope is oriented exactly perpendicular to a birefringent coating during normal incidence readings, the near surface of the birefringent coating will reflect a portion of the light, causing an interference in the field of view. To determine whether or not parasitic reflections were present during the experiments, the compensator was removed from the Budd SF/Z instrument, and the analyzer was rotated 90 degrees with respect to the polarizer, which made the instrument a parallel polariscope. Parasitic reflections were indicated by a glare in the changing colors with applied load. The reduction of the glare varied with the angle of cant, and the parasitic reflections were apparently eliminated with a cant of about seven degrees. Disregarding the integrating effect, a seven-degree cant caused an error of little more than 1% at the free boundaries since the relative retardation measured varies directly as the cosine of the angle of cant.

The sensitivity of the Budd SF/Z small field meter[35] for the PhotoStress used in this study was ± 38 μin./in. The normal incidence measurements were more accurate than the oblique incidence measurements because of the small integrating effect concomitant with oblique incidence.

The causes of error affecting the accuracy of the birefringent-coating measurements and the correcting procedures were thoroughly discussed in other work.[45] Most of the sources of error were eliminated by maintaining carefully controlled laboratory conditions. The accuracy of the stresses computed from the measurements of relative retardation was estimated to range from ± 2 to $\pm 8\%$ with an overall average error of $\pm 6\%$.

Conclusions

A detailed experimental analysis of thick circular plates containing circular, transverse perforations in square motif has been given. A birefringent coating was effectively utilized to enable the stress computations for the particular case of loading studied, uniformly laterally loaded plates with simply supported edges. Tube sheets, for instance, could be designed using the graphical stress results as shown, but a safety factor should still be used depending upon design criteria.

The following factors, where Roman numerals designate the particular specimen configuration, were determined from the experimental data for design purposes:

Deflection-multiplier factors:

$$DMF_I = \frac{(w_{max}/q)_I}{(w_{max}/q) \text{ solid plate}} = \frac{4.30}{2.44} = 1.76 \tag{5.50}$$

A correction factor was applied to the solid-plate results used for comparison with configurations II and III since the thickness of the solid plate was equal to the thickness of configuration I only.

$$DMF_{II} = \frac{(w_{max}/q)_I H_{II}^3}{(w_{max}/q)_{solid\ plate} H_{solid\ plate}^3} = \frac{8.90(3.875^3)}{2.44(4.872^3)} = 1.82 \quad (5.51)$$

$$DMF_{III} = \frac{(w_{max}/q)_{III} H_{III}^3}{(w_{max}/q)_{solid\ plate} H_{solid\ plate}^3} = \frac{13.10(3.875^3)}{2.44(4.872^3)} = 2.68 \quad (5.52)$$

Stress-concentration factors:

$$SCF_I = \frac{(\sigma_{max}/q)_I}{(\sigma_{max}/q)_{solid\ plate}} = \frac{29.6}{9.76} = 3.03$$

$$SCF_{II} = \frac{(\sigma_{max}/q)_{II} H_{II}^2}{(\sigma_{max}/q)_{solid\ plate} H_{solid\ plate}^2} = \frac{49.6(3.875^2)}{9.76(4.872^2)} = 3.20 \quad (5.53)$$

$$SCF_{III} = \frac{(\sigma_{max}/q)_{III} H_{III}^2}{(\sigma_{max}/q)_{solid\ plate} H_{solid\ plate}^2} = \frac{69.0(3.875^2)}{9.76(4.872^2)} = 4.45 \quad (5.54)$$

Summarily, see table 96.

TABLE 96

Configuration	$\frac{S}{P}$	$\frac{H}{P}$	DMF	SCF
I	0.35	3.90	1.76	3.03
II	0.35	3.10	1.82	3.20
III	0.20	3.10	2.68	4.45

The deflection of configurations the same as test specimens in this report, but of some material other than aluminum, would be given by

$$\frac{w}{q} = \left(\frac{w}{q}\right)_{aluminum} \left(\frac{E_{aluminum}}{E}\right) \quad (5.55)$$

The tabulated deflection data, tabulated relative retardation measurements, fixture drawings, a section on the history, principles and applications of birefringent coatings, and colored photographs of the overall isochromatics of each specimen and of a tint of passage as viewed through the Budd SF/Z meter were given in the final report.[45] The intimate details relative to this ligament study were given in a WRC bulletin[46], which is available from the offices of the Pressure Vessel Research Committee, Welding Research Council.

5.3 FILLET AND GROOVE STRESS CONCENTRATIONS

Structural discontinuities such as holes, fillets, and notches usually cause a local stress intensity at the discontinuity having a value of several times the mean stress. The value of these stress intensities must be known in order to insure adequate designs. The stress-concentration factor for many types of structural discontinuities has been given by Heywood,(47) Peterson,(48) Savin,(49) and Lipson, et al.(50), however, the factors for the particular cases covered in this section do not seem to be available from the applicable published literature.

The particular cases of structural discontinuities investigated were a single fillet or groove in specimens subjected to uniaxial and biaxial loads. The objectives of the investigation were (1) to ascertain the feasibility of using the birefringent-coating technique for future work planned, and (2) to determine several stress-concentration factors for particular specimen geometries and loading conditions.

For the studies of discontinuities with biaxial loads, a cylindrical vessel was subjected to internal hydraulic pressure to give a 2–1 biaxial stress field. The vessel had drum heads welded to the cylindrical-shell wall as shown in figure 5.12. The tensile-test specimens used in the investigations were 1 in. wide and are shown in figure 5.13.

Flat-head pressure vessels are not, of course, recommended due to the structural inefficiency of this type of design. Some sort of rounded closures are normally used so that the stresses in and about the shell-head joint are minimized. There are, however, many cases of single unit fabrication, such as the vessel used in this investigation, where the design engineer has to use flat heads due to the large expense that is involved with custom forming.

Material Properties

The 18-in.-diameter pipe used in the biaxial tests was an ASTM A106-60T steel. The Standard Certified Test Report, dated June 1, 1961, from the National Tube Division of the U.S. Steel Corp. gave the following mechanical properties of the material:

Yield strength = 51.3 ksi
Tensile strength = 72.4 ksi
Elongation in 2 in. = 39.0 %

The test report also gave the chemical analysis of the material in percent as follows:

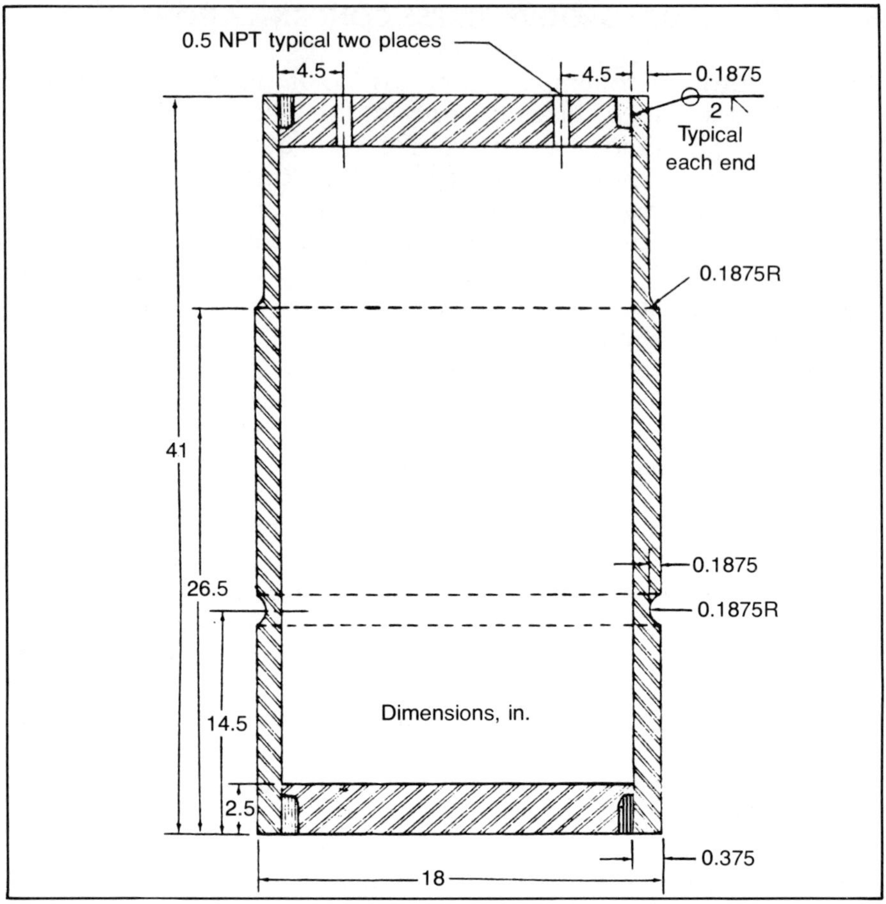

FIG. 5.12 Pressure vessel used for determining fillet and groove stress concentrations with biaxially applied load

Carbon = 0.21
Manganese = 0.65
Phosphorus = 0.008
Sulfur = 0.024
Silicon = 0.19

The pressure-vessel drum heads were fabricated from mild steel, which had a yield strength of 36 ksi and an ultimate tensile strength of 55 ksi. The test specimens which were subjected to uniaxial loads were fabricated from a heat-treated steel.

A Young's modulus of 30×10^6 psi and a Poisson's ratio of 0.28 were assumed for the pipe, drum heads, and tensile-test specimens.

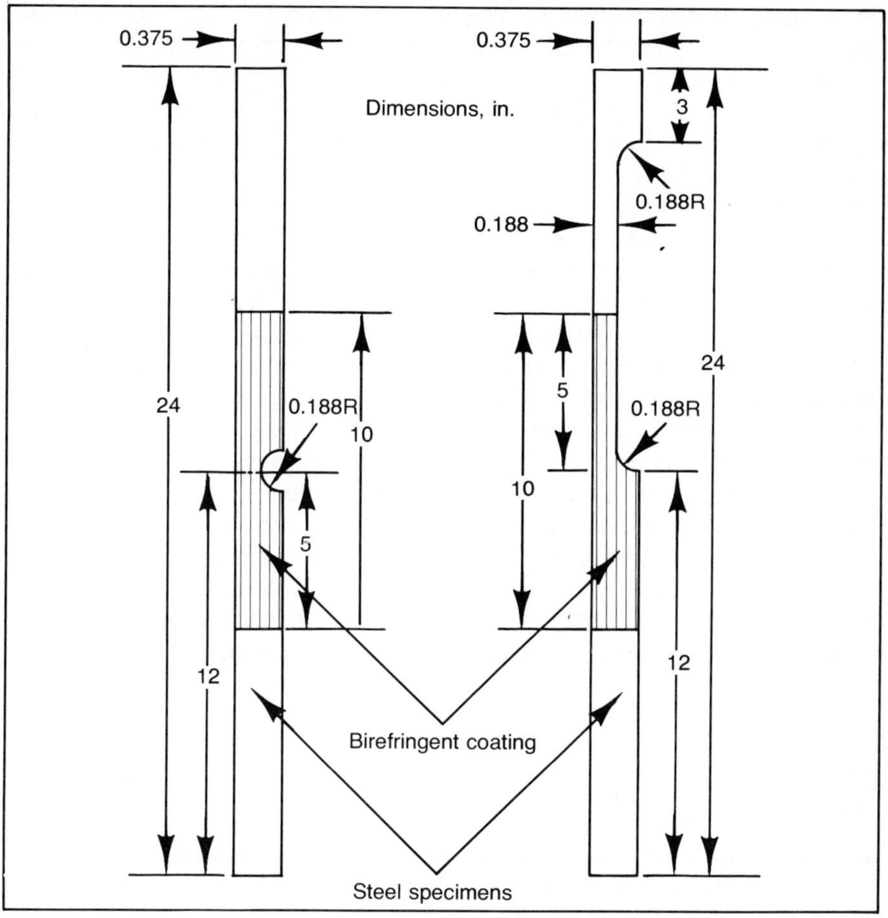

FIG. 5.13 Test specimens used for determining fillet and groove stress concentrations with uniaxially applied load

Experimental Procedure

The birefringent-coating technique was the primary means of experimental stress analysis used and was complemented with data from electrical-resistance type of strain gauges. Considerable detail concerning the birefringent-coating technique was given in WRC Bulletin No. 80[51] and will not be reiterated herein. PhotoStress was the birefringent material used for the coating, and the individual principal stresses were calculated using normal and oblique-incidence measurements of relative retardation per the instrument manufacturer's instructions.[52,53] A Budd small-field meter, Model SF/Z, was used for normal-incidence measurements, and a Budd oblique-incidence meter, Model OI/Z, was used for oblique-incidence measurements. These instruments utilize a Babinet compensator for accurately determining fractional fringe orders.

The metal foil strain gauges used were Type C6-1 × 1-32A having a 1/32-in. gauge length, 2.02 ± 1% gauge factor, and 60 ± 0.2 ohm resistance; Type C6-111 having a 1/16-in. gauge length, 1.98 ± 1% gauge factor, and 120 ± 0.5 ohm resistance; and Type C6-121 having a 1/8-in. gauge length, 2.07 ± 0.5% gauge factor, and 120 ± 0.2 ohm resistance. Measurements of relative retardation in the fillet and the groove of the pressure vessel could not be made due to the design of the oblique-incidence meter used, and the Type C6-1 × 32A gauges were used to obtain the stresses in the discontinuities for the biaxially applied loads. The other types of gauges were used at other points to serve as a basis for comparing the results from the birefringent coating. The gauge installations were made per the gauge manufacturer's instructions.[54]

The drum heads were welded to the shell, and the pressure vessel was then subjected to an internal hydraulic pressure sufficiently high so that the yield strength of the shell was exceeded in order to eliminate out-of-round effects. The actual yield pressure of the vessel was 1,950 psi even though the theoretical yield stress was 2,180. Several electrical-resistance type of strain gauges were

TABLE 97
Nomenclature

E_s =	Young's modulus of structure, psi
K =	strain sensitivity constant, fringe/(μin./in.)
N_n, N_0 =	difference of compensator divisions between load and no load, normal incidence and oblique incidence, respectively, divisions
p =	internal pressure, psi
r =	radius of fillet or groove, in.
S_1, c =	stress constants, psi/rel. retard.
SCF =	stress-concentration factor. experimental stress divided by mean stress in reduced section, numeric
W =	Babinet compensator constant, div./fringe
d =	shell inside diameter, in.
t_z, t_p =	thickness of shell, and birefringent coating, respectively, in.
t_{max}, t_{min} =	maximum and minimum thickness, respectively, in.
λ =	wave length of light, in.
μ_s =	Poisson's ratio of structure, numeric
σ =	stress, psi
x =	distance measured from intersection of fillet and shell wall having maximum thickness, or distance measured from groove centerline, in.

bonded to the inside and the outside walls of the shell to determine the shell stresses.

The fillet and groove discontinuities were then machined, and the vessel was subjected to a test pressure of 950 psi. The birefringent coating material was poured on an RTV silicone rubber platen, allowed to partially polymerize, and then contoured and bonded to the vessel using RCT reflective cement. The birefringent-coating thickness was determined by averaging several micrometer measurements of fully polymerized pieces of unbonded plastic.

Normal and oblique-incidence measurements were taken at 0 and 800 psi. Measurements could have been taken at several increments of pressure to account for any nonlinearities, but previous work[(51)] has demonstrated that this improves the accuracy by only about 3% at most, and the added expense was not considered worthwhile in this study. The vessel was hydraulically pressurized using a SwRI cycling facility.

For the case of the test specimens subjected to uniaxially applied loads, flat sheets of PhotoStress were bonded to the sides of the specimens as shown in figure 5.13. The K factor and the thickness of the plastic sheets were given by the manufacturer as 0.09 and 0.0715 ± 0.0015 in., respectively. The average thickness of the coating, consisting of the plastic sheet and the adhesive, was 0.0717 in.

The specimens were loaded using the BTE universal testing machine, 200-k capacity, and gripped so that the 0.375-in.-wide faces were adjacent to the jaws of the test machine. Strain gauges were not used on these specimens since the stress at every point studied was obtained in such a straightforward and reliable manner. The relative retardation was computed from the change of compensation divisions between 200 and 4,500 lb applied force.

The boundary stress was determined along each face of the specimen about the area of the discontinuity. The stress distribution across the minimum sections was not determined since the boundary stresses were obviously of greater magnitude.

The stress-concentration factor was probably a function of load due to the existence of bending effects. The stress distributions across various sections of the specimens and the function of the stress-concentration factor with load will be determined in planned future tests, which will be of a more comprehensive nature.

Discussion of Results

The birefringent coating used in the biaxial tests was calibrated from measurements of relative retardations in the coating at a point midway on the vessel where there was no effect from the structural discontinuities. The principal stress directions were determined by observing the isoclinics with the small field meter. Oblique-incidence measurements were then taken with the instrument oriented so that the plane containing the light beam was parallel to σ_1 or σ_2. The following mathematical relations were applicable:

Light beam parallel to σ_2:

$$\sigma_1 = S_1\left[0.707\,N_o - \frac{N_n}{2}\right] \tag{5.56}$$

$$\sigma_2 = S_1(0.707\,N_o - N_n) \tag{5.57}$$

where

$$S_1 = \frac{E_s\lambda}{t_p K_b(1+\mu_s)W} \tag{5.58}$$

Light beam parallel to σ_1:

$$\sigma_1 = S_1(0.707\,N_o + N_n) \tag{5.59}$$

$$\sigma_2 = S_1\left[0.707\,N_o + \frac{N_n}{2}\right] \tag{5.60}$$

The theoretical hoop stress for 800 psi internal pressure was

$$\sigma_1 = \frac{pd}{2t_s} = \frac{800(17.250)}{2(0.375)} = 18{,}400 \text{ psi} \tag{5.61}$$

Using the theoretical hoop stress and oblique-incidence measurements with the light beam parallel to σ_2,

$$\begin{aligned} K_{b2} &= \frac{E_s\lambda}{t_s(1+\mu_s)W\sigma_1}\left[0.707\,N_o - \frac{N_n}{2}\right] \\ &= \frac{30(10^6)(2.27)(10^{-5})}{0.0638(1.28)(56)(18400)}\left[0.707(29) - \frac{11}{2}\right] \\ &= 0.116\,\frac{\text{fringe}}{(\mu\text{in./in.})} \end{aligned} \tag{5.62}$$

and with the light beam parallel to σ_1,

$$\begin{aligned} K_{b1} &= \frac{E_s\lambda}{t_s(1+\mu_s)W\sigma_1}(0.707\,N_o + N_n) \\ &= 0.00809(0.707(4) + 11) \\ &= 0.112\,\frac{\text{fringe}}{(\mu\text{in./in.})} \end{aligned} \tag{5.63}$$

For the stress calculations with the points studied in the biaxial tests, the K factor used was:

$$K_b = \frac{K_{b2} + K_{b1}}{2} = \frac{0.116 + 0.112}{2} = 0.114 \frac{\text{fringe}}{(\mu\text{in./in.})} \tag{5.64}$$

and S_1 in this case was

$$S_1 = \frac{30(10^6)(2.27)(10^{-5})}{0.0638(0.114)(1 + 0.28)(56)}$$

$$= 1310 \text{ psi/relative retardation} \tag{5.65}$$

The stress concentration was then computed by dividing the experimentally determined stress by the appropriate theoretical hoop or meridional stress calculated at the minimum section. For both the fillet and the groove in the biaxial tests, the theoretical hoop stress at the minimum section was 46.0 psi per psi of internal pressure.

The variation of the stress-concentration factors along the wall in and about the circumferential fillet in the pressure-vessel shell is shown in figure 5.14. The meridional stress at the intersection of the fillet with the maximum thickness wall was zero, as expected, since this point was a free boundary.

The experimental hoop stress was almost equal to the theoretical hoop stress, and the experimental meridional stress was slightly

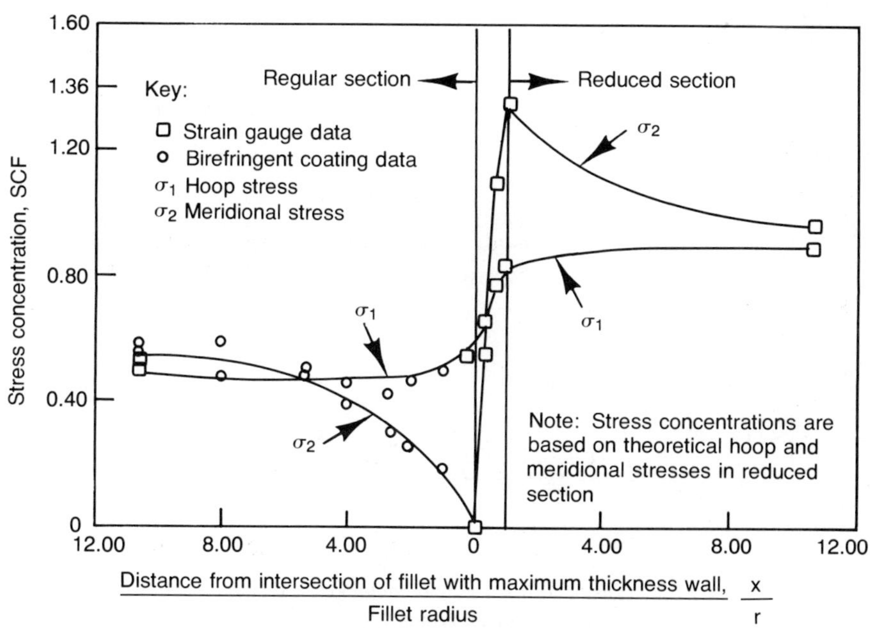

FIG. 5.14 Fillet stress-concentration factors, biaxially applied load

greater than the theoretical meridional stress in the thick-wall section at points distant from the discontinuity. No stress concentration in the thick section was equivalent to SCF = 0.5 since the theoretical stresses in the reduced section were used as a basis for the stress-concentration factors. Both the experimental hoop stress and the experimental meridional stress in the reduced section distant from the discontinuity were slightly less than the theoretical values.

The maximum stress-concentration factor of 1.36 occurred with the meridional stress and was located at the intersection of the fillet with the reduced section. As shown in figure 5.14, the stress-concentration effect from the discontinuity is most discernible for a distance of about five times the fillet radius on either side of the discontinuity.

The variation of the stress-concentration factors along the wall in and about the groove in the pressure-vessel shell is shown in figure 5.15. The meridional stress at the intersection of the groove with the outside surface of the vessel wall was zero, as expected, since this point was a free boundary.

The experimental hoop stress was almost equal to the theoretical hoop stress, and the experimental meridional stress was slightly greater than the theoretical meridional stress at points distant from

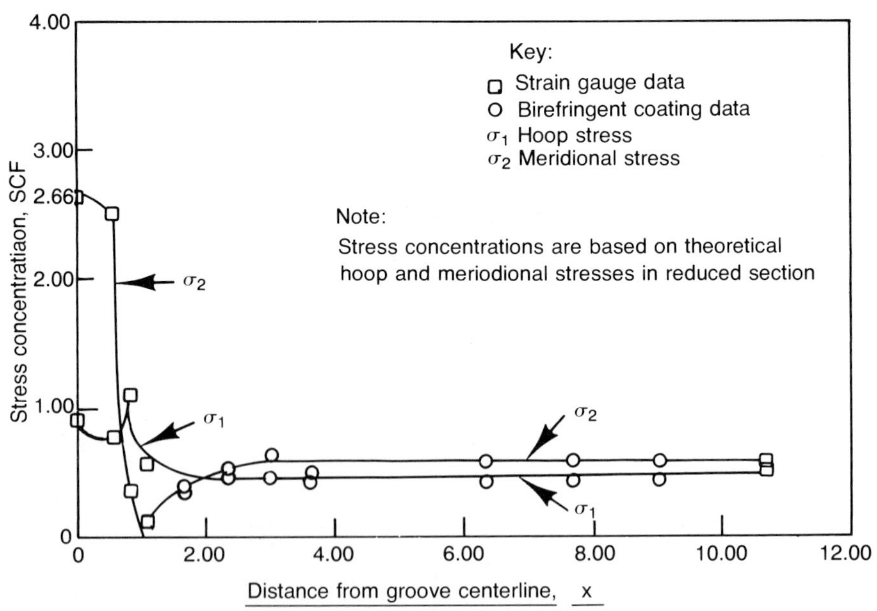

FIG. 5.15 Groove stress-concentration factors, biaxially applied load

the discontinuity. As shown by figure 5.15, the stress-concentration effect due to the discontinuity was discernible for a distance of about three times the groove radius, measured from the groove centerline.

The maximum stress-concentration factor for the hoop stress was 1.11 and occurred near the intersection of the groove with the outside shell surface. The maximum stress-concentration factor for the meridional stress was 2.66 and occurred at the groove centerline.

The difference of principal stress is related to the strain sensitivity constant, K, as follows

$$\sigma_1 - \sigma_2 = \frac{N_n \lambda E_s}{2t_p KW(1 + \mu_s)} \tag{5.66}$$

Since the stress perpendicular to a free boundary is zero, the boundary stresses obtained were determined from normal-incidence measurements of relative retardations alone, i.e., $\sigma_1 = CN_n$. Stresses obtained in this manner are quite accurate and reliable.

The stress concentration was given as the ratio of the experimentally determined stress divided by the mean theoretical stress in the reduced section. The mean theoretical stress was used as a basis for expressing the stress concentration with both the fillet and the grooved specimens, since the two specimens had equal widths, thickness, and minimum section thickness.

The variation of the stress-concentration factors along the boundary on the fillet face and the opposite face of the specimen subjected to uniaxially applied loads is shown in figure 5.16. The tangential stress at the intersection of the fillet with the regular section was zero, since this point was a free boundary.

The maximum stress-concentration factor for the fillet face was 1.88 and occurred at the intersection of the fillet radius with the reduced section. The maximum stress-concentration factor for the face of the specimen opposite the fillet face was 0.99 and occurred at $x/r = 0.66$.

The predominant effect from the discontinuity occurs within a distance, x/r, of about 12 in the reduced section for both the fillet face and the opposite face. The discontinuity is discernible for a distance of $x/r > 16$ in both faces on the regular section. A decay of the stress intensity over a distance of $x/r > 16$ is both interesting and unusual. The stress-concentration factor would be 0.5 in the regular section for no discontinuity effect, since the stress concentration was based on the mean stress in the reduced section. The stress at the intersection of the fillet with the regular section was zero, as expected, since this point was a free boundary.

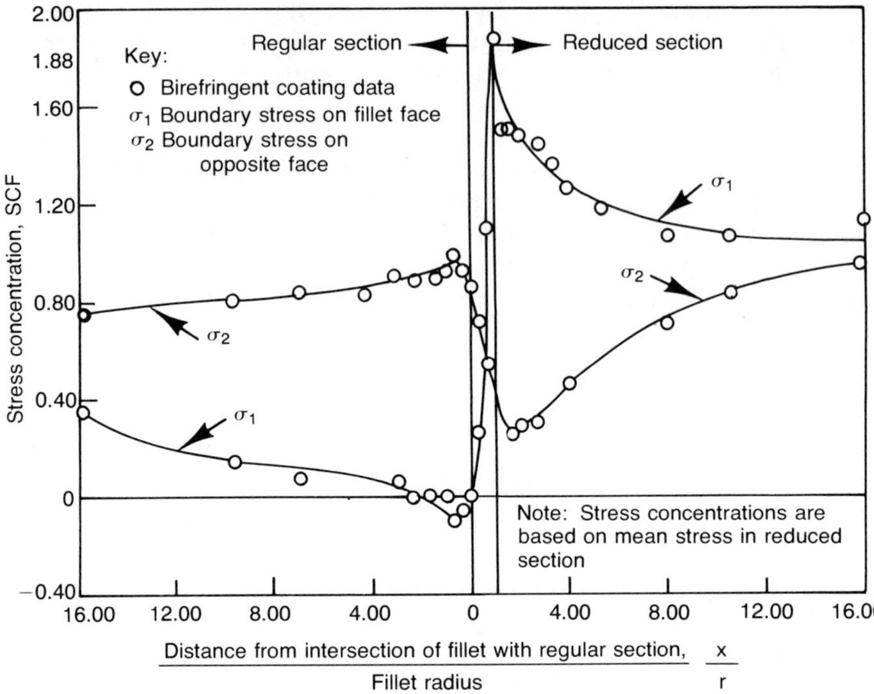

FIG. 5.16 Fillet stress-concentration factors, uniaxially applied load

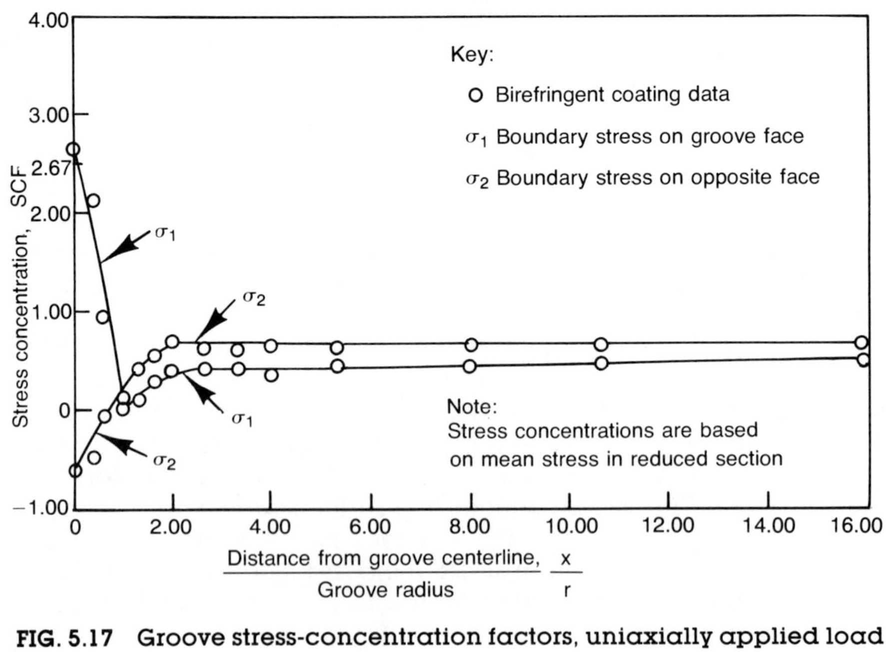

FIG. 5.17 Groove stress-concentration factors, uniaxially applied load

The variation of the stress-concentration factors along the boundary on the groove face and the opposite face of the specimen subjected to uniaxially applied loads is shown in figure 5.17. The stress at the intersection of the groove with the groove face was zero, since this point was a free boundary.

The maximum stress-concentration factor for the groove face was 2.67 and occurred at the groove centerline. The discontinuity effect in the groove face and the opposite face was discernible for a distance, x/r, of about 2. The boundary of the face opposite the groove face at and near the fillet centerline experienced a compressive stress due to the nonsymmetry of the test specimen.

The strain-gauge results were accurate within about ±3%. The difference between the true gauge stress and the real stress in the specimen could differ as much as 10% due to the integrating effect over the gauge length of the strain gauges. This effect will be greatly reduced in planned future tests by using pressure vessels of sufficient size so that the gauge length compared to the radius of the discontinuity will be small.

The estimated accuracy of the birefringent-coating results was about ±2% when only normal-incidence measurements of relative retardation were required for stress computation, and about ±6% when normal and oblique-incidence measurements were required. The error in the computed stresses was due to the usual factors which were discussed extensively in another investigation.(52)

Conclusions

In summary, the maximum stress-concentration factors (SCF), based on the nominal stress in the minimum section, determined in this investigation for the biaxially applied loads are listed below:

Discontinuity	Hoop SCF	Meridional SCF	r/t_{min}	t_{max}/t_{min}
Fillet	0.90	1.36	1	2
Groove	1.11	2.66	1	2

The fillet SCF can be compared to the results obtained by Morgan and Bizon(55), which are shown in table 98.

The maximum stress-concentration factors for the uniaxially applied loads were:

Discontinuity	SCF	r/t_{min}	t_{max}/t_{min}
Fillet	1.88	1	2
Groove	2.67	1	2

Durelli, *et al.*[(56)], found a SCF of 3.08 with a semi-infinite, photoelastic model (i.e., $t_{max}/r = 18$) having a single semicircular notch for the case of a uniaxially applied load.

The above factors are considerably higher than the comparable maximum stress-concentration factors for uniaxially applied loads on symmetrical specimens obtained by other investigators. Some of these previously determined factors for semicircular grooves and fillets are shown in table 99.

The birefringent-coating technique is ideally applicable for determining the over-all stress distribution in structures such as the specimens subjected to uniaxially applied loads which were studied

TABLE 98

Continuous surface of cylinder	D/t_{max}	t_{min}/t_{max}	$r/t_{max}-t_{min}$	Surface location	Average % variation of experimental stress compared to thin-shell theoretical stress: Meridional	Circumferential
Middle	117	0.4	1	Inner	+5.1	−5.9
Middle	117	0.4	1	Outer	+11.4	+4.3
Inner	117	0.4	1	Inner	−9.0	−11.8
Inner	117	0.4	1	Outer	−19.0	−4.0
Middle	28	0.4	~0	Inner	−3.2	−2.1
Middle	28	0.4	~0	Outer	−5.9	−10.5
Inner	28	0.4	~0	Inner	−4.0	−3.9
Inner	28	0.4	~0	Outer	−9.9	−10.9
Outer	28	0.4	~0	Inner	−5.8	−3.7
Outer	28	0.4	~0	Outer	−4.0	−7.6

TABLE 99

Discontinuity	SCF	r/t_{min}	t_{max}/t_{min}
Fillet[47]	1.43	0.5	2
Groove[47]	1.64	0.5	2
Groove[48]	1.54	0.5	2
Fillet[50]	1.28	0.5	2
Groove[50]	1.54	0.5	2
Groove[57]	1.55	0.5	2
Groove[58]	1.58	0.5	2
Fillet[59]	1.42	0.5	2
Groove[59]	1.52	0.5	2

in this investigation. In many types of problems, the stresses can be determined using the birefringent-coating technique with as good as or better than the accuracy obtained with electric-resistance type of strain gauges whose accuracy is usually taken as about $\pm 18\ \mu$in./in. The advantage of the birefringent-coating technique over photoelasticity is that an actual structure can be analyzed instead of using model analogies. Other advantages of the birefringent-coating technique are:

1. The sensitivity is a function of the plastic thickness which can be varied.
2. The stresses can be obtained exactly at the free boundaries.
3. No integrating effect exists for certain types of problems since the gauge length is a point.

5.4 STEPPED SHELL ANALYSES USING ENERGY METHODS

No theoretical procedure exists that is precisely applicable to the analysis of cylindrical shells of nonuniform thickness. Such shells are sometimes encountered in submarine design, for example where the walls are thickened in the vicinity of support fittings and/or other points of discontinuity or high loads. The usual analysis is based on classical shell theory (the shell thickness is assumed everywhere constant) or on classical ring theory (a circumferential ring of constant section modulus integral with the shell), neither of which is precisely applicable. Engineering judgement, coupled with structural tests, is then used to extrapolate the results to the actual geometries. The formulas that are derived in this section will eliminate these extrapolations and will minimize the requirements for structural tests.

The solution is an approximate one based on the theory of minimum potential energy. It makes maximum use of the high speed digital computer for the numerical computations required. As presently formulated, the loading is a single concentrated force symmetrically placed on a simply supported shell. The formulation permits the consideration of any number of thickness variations in the circumferential or meridional directions.

Air-launched solid propellant rocket motor cases are frequently equipped with support fittings or hooks by which the missile is fastened to the carrier. These fittings are designed to maintain the tie through any maneuver or acceleration which the aircraft structure is capable of surviving. So the forces and/or moments for which these fittings are designed are of a large order, and reinforcements are necessary to distribute these forces into the shell of the missile. Present practice is generally to thicken the shell wall in the vicinity of the fittings to form a ring around the entire circumference of the case.

The solutions for ring and shell combinations (85–87) indicate that the internal circumferential moments and forces in the ring rapidly attenuate away from the neighborhood of locally applied loads even when the ring thickness is everywhere constant. So to reduce weight and provide more volume for propellant (the shell is generally thickened toward the inside only), the custom is to determine the internal ring forces and moments as though the rings were of constant thickness. Then the ring is tapered in the circumferential direction to provide a decreased section in areas distant from the position of the fittings. This procedure is followed even though it is known that the variable thickness of the ring section gives a different distribution of the internal forces and moments.

In addition, the initial analysis is invariably done using techniques originally devised for rings in airplane fuselages where the skin (shell) is extremely thin compared to the stiff (high section modulus) rings. But missile motor cases are quite different since the ring is a thickened portion of the shell. Therefore, the mechanical respondence of the ring in the meridional direction in the actual motor case is not the same as that indicated by the "fuselage" type of analysis. The solutions given will provide insight into how significant the variations due to these two effects might be.

The technique uses the principle of minimum potential energy. A finite series of terms, each of which satisfies the boundary conditions and each of which contains an undetermined coefficient, is proposed as a solution for the displacements. The assumed solution is then substituted into the expression for the potential energy and the coefficients are chosen so as to minimize the potential. This procedure leads to an evaluation of the displacements at every point on the shell from which the strains and, in turn, the stresses can be determined.

The coordinate system used in this work is as shown in figure 5.18 and an example stepped cylinder for which the formulation applies is shown in figure 5.19. The given development is for the restricted conditions of simply supported ends and a concentrated force applied at the midspan. The conditions of nonsymmetrical loading and other types of loading (e.g. distributed radial-force, meridional and tangential moments or forces) are but simple extensions of the given formulation.

The theoretical development presented is valid for any number of thickness steps in the meridional and/or circumferential

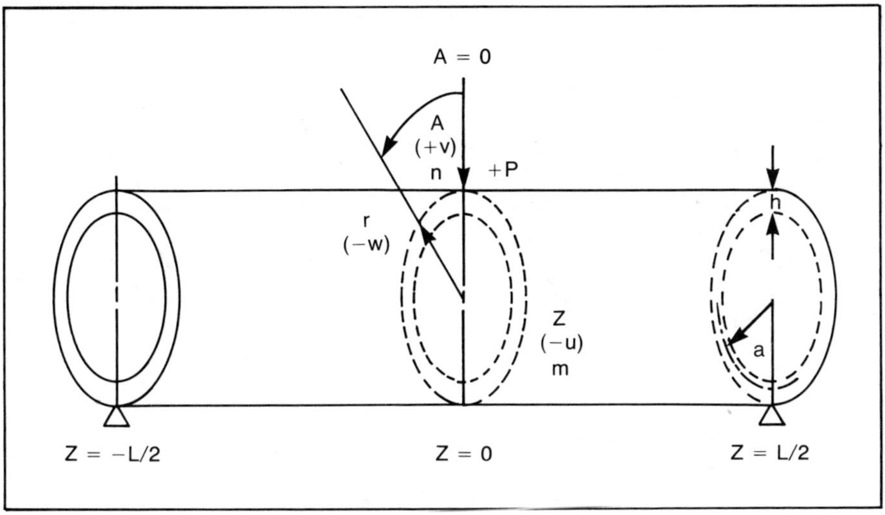

Fig. 5.18 Coordinate system

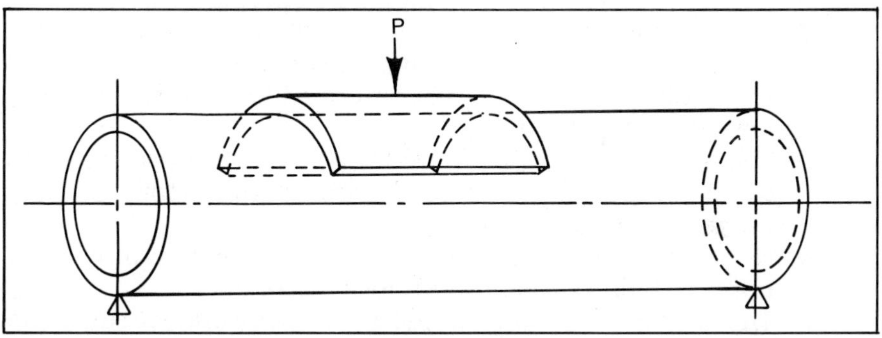

FIG. 5.19 Stepped cylinder

directions. Also, eccentric bores, such as shown in figure 5.20 can be analyzed with the given formulas by approximating the configuration with multiple circumferential steps (say 10–15). An example approximate configuration is portrayed in figure 5.20b where for clarity of illustration only three steps are shown.

Theoretical Development

The basic assumptions from which small-deflection theory for thin shells are constructed are[81]

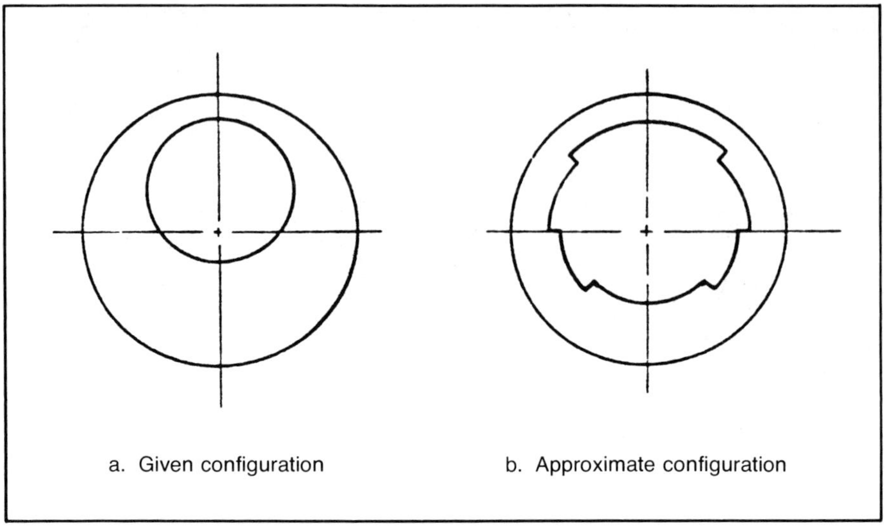

FIG. 5.20 Eccentric thickness approximation

a. The shell material is isotropic and homogeneous,
b. The displacements are small enough to neglect displacement components of order two and higher,
c. The shell thickness is small compared to the radii of curvature of its middle surface,
d. The stress components normal to the middle surface are small compared with the other stress components and may be neglected in the stress-strain relations,
e. Normals to the unstrained middle surface deform into the normals to the strained middle surface, and
f. The normals to the strained middle surface suffer no extension.

Assumption d is the condition of plane stress and assumptions e and f constitute the Kirchoff-Love hypothesis. From variational calculus[83]:

$$\mathbb{P} = \int_V U_o dV - \sum_k \int_A T_k s_k dA \tag{5.67}$$

With a single concentrated load,

$$\sum_k \int_A T_k s_k dA = P s_k \tag{5.68}$$

By definition,

$$U \equiv \int_V U_0 dV \tag{5.69}$$

Thus, equation (5.67) becomes

$$\mathbb{P} = U - P s \tag{5.70}$$

The expression for the total strain energy, using assumptions a thru f, for a circular cylinder is (equation (5.69), 74, or equation (5.81), 81):

$$U = \iint c_1 h(a^2 u_z^2 + (v_\theta - w)^2 + z a \mu u_z (v_\theta - w) + c_3 (u_\theta + a v_z)^2) dz\, d\theta + \iint c_2 h^3 (a^4 w_{zz}^2 + (w_{\theta\theta} + w)^2 + c_3 (a w_{z\theta} - u_\theta)^2 + 3_{c_3} a^2 (v_z + w_{z\theta})^2 + 2 \mu a^2 w_{zz} (w_{\theta\theta} + v_\theta) + 2 a^3 u_z w_{zz}) dz\, d\theta \tag{5.71}$$

where

$$c_1 \equiv \frac{E}{2(1 - \mu^2) a} \tag{5.71a}$$

$$c_2 \equiv \frac{E}{24(1 - \mu^2) a^3} \tag{5.71b}$$

$$c_3 \equiv \frac{1 - \mu}{2} \tag{5.71c}$$

Expanding equation (5.71)

$$U = \iint c_1 h(a^2 u_z^2 + v_\theta^2 - 2 v_\theta w + w^2 + 2 a \mu u_z v_\theta - 2 a \mu u_z w + c_3 u_\theta^2 + 2 c_3 a u_\theta v_z + c_3 a^2 v_z^2) dz\, d\theta + \iint c_2 h^3 (a^4 w_{zz}^2 + w_{\theta\theta}^2 + 2 w w_{\theta\theta} + w^2 + c_3 a^2 w_{z\theta}^2 - 2 c_3 a w_{z\theta} u_\theta + c_3 u_\theta^2 + 3 c_3 a^2 v_z^2 + 6 c_3 a^2 v_z w_{z\theta} + 3 c_3 a^2 w_{z\theta}^2 + 2 \mu a^2 w_{zz} w_{\theta\theta} + 2 \mu a^2 w_{zz} v_\theta + 2 a^3 u_z w_{zz}) d_z d_\theta \tag{5.72}$$

For a trial solution of the simply supported tube with a concentrated radial load applied at the mid-length as shown in figure 5.18, consider:

$$u = -\sum_{m,n} A_{mn} \sin \lambda \cos \omega \tag{5.73a}$$

$$v = \sum_{m,n} B_{mn} \cos \lambda \sin \omega \tag{5.73b}$$

$$w = \sum_{m,n} C_{mn} \cos \lambda \cos \omega \tag{5.73c}$$

$$m = 1, 3, 5, 7, \ldots \quad n = 0, 1, 2, 3, \ldots$$

where

$$\lambda \equiv \frac{m\pi z}{L} \tag{5.74a}$$

$$\omega \equiv n\theta \tag{5.74b}$$

Note that $m = o$ is trivial. Also, even values of m give inadmissible functions about $z = L/2$. Performing the indicated differentiations in equation (5.72) on the assumed solution equation (5.73):

$$u_z = -\sum \frac{m\pi}{L} A \cos \lambda \cos \omega \tag{5.75a}$$

$$u_\theta = \sum n A \sin \lambda \sin \omega \tag{5.75b}$$

$$v_z = -\sum \frac{m\pi}{L} B \sin \lambda \sin \omega \tag{5.75c}$$

$$v_\theta = \sum n B \cos \lambda \cos \omega \tag{5.75d}$$

$$w_{zz} = -\sum \left(\frac{m\pi}{L}\right)^2 C \cos \lambda \cos \omega \tag{5.75e}$$

$$w_{\theta\theta} = -\sum n^2 C \cos \lambda \cos \omega \tag{5.75f}$$

$$w_{z\theta} = \sum \frac{mn\pi}{L} C \sin \lambda \sin \omega \tag{5.75g}$$

where the mn subscripts have been dropped but are still understood. Substitute equations (5.73) and (5.75) in equation (5.72).

$$U = \iint c_1 h \left[\alpha^2 \left(\sum \frac{m\pi}{L} A \cos \lambda \cos \omega \right)^2 + \left(\sum n B \cos \lambda \cos \omega \right)^2 \right.$$
$$- 2 \left(\sum n B \cos \lambda \cos \omega \right) \left(\sum C \cos \lambda \cos \omega \right) + \left(\sum C \cos \lambda \cos \omega \right)^2$$
$$- 2 \alpha \mu \left(\sum \frac{m\pi}{L} A \cos \lambda \cos \omega \right) \left(\sum n B \cos \lambda \cos \omega \right)$$

$$+ 2\alpha\mu\left(\sum \frac{m\pi}{L} A\cos\lambda\cos\omega\right)\left(\sum C\cos\lambda\cos\omega\right)$$

$$+ c_3\left(\sum nA\sin\lambda\sin\omega\right)^2$$

$$- 2c_3\alpha\left(\sum nA\sin\lambda\sin\omega\right)\left(\sum \frac{m\pi}{L} B\sin\lambda\sin\omega\right)$$

$$+ c_3\alpha^2\left(\sum \frac{m\pi}{L} B\sin\lambda\sin\omega\right)^2\Bigg] dz\,d\theta$$

$$+ \iint c_2 h^3\left[\alpha^4\left(\sum\left(\frac{m\pi}{L}\right)^2 C\cos\lambda\cos\omega\right)^2\right.$$

$$+ \left(\sum n^2 C\cos\lambda\cos\omega\right)^2$$

$$- 2\left(\sum C\cos\lambda\cos\omega\right)\left(\sum n^2 C\cos\lambda\cos\omega\right)$$

$$+ \left(\sum C\cos\lambda\cos\omega\right)^2 + c_3\alpha^2\left(\sum \frac{mn\pi}{L} C\sin\lambda\sin\omega\right)^2$$

$$- 2c_3\alpha\left(\sum \frac{mn\pi}{L} C\sin\lambda\sin\omega\right)\left(\sum nA\sin\lambda\sin\omega\right)$$

$$+ c_3\left(\sum nA\sin\lambda\sin\omega\right)^2 + 3c_3\alpha^2\left(\sum \frac{m\pi}{L} B\sin\lambda\sin\omega\right)^2$$

$$- 6c_3\alpha^2\left(\sum \frac{m\pi}{L} B\sin\lambda\sin\omega\right)\left(\sum \frac{mn\pi}{L} C\sin\lambda\sin\omega\right)$$

$$+ 3c_3\alpha^2\left(\sum \frac{mn\pi}{L} C\sin\lambda\sin\omega\right)^2$$

$$+ 2\mu\alpha^2\left(\sum\left(\frac{m\pi}{L}\right)^2 C\cos\lambda\cos\omega\right)\left(\sum n^2 C\cos\lambda\cos\omega\right)$$

$$- 2\mu\alpha^2\left(\sum\left(\frac{m\pi}{L}\right)^2 C\cos\lambda\cos\omega\right)\left(\sum nB\cos\lambda\cos\omega\right)$$

$$\left. + 2\alpha^3\left(\sum \frac{m\pi}{L} A\cos\lambda\cos\omega\right)\left(\sum\left(\frac{m\pi}{L}\right)^2 C\cos\lambda\cos\omega\right)\right] dz\,d\theta \tag{5.76}$$

Consider the variation of the virtual displacements in equation (5.70), then

$$\delta\,\mathbb{P} = \delta U - P\,\delta s = 0 \tag{5.77a}$$

TABLE 100
Nomenclature

Symbol	Variable	Typical units
L	Length of shell	in.
a	Mean radius of shell	in.
h	Shell thickness	in.
E	Young's modulus of elasticity	psi
μ	Poisson's ratio	numeric
u, v, w	Longitudional (+ in + z, see fig. 5.18), tangential (+ in + θ), and radial (+ inward) displacements, respectively	in.
$\varepsilon_z, \varepsilon_\theta, \varepsilon_r$	Normal strains in z, θ, r directions, respectively	in./in.
$\gamma_{z\theta}$	Shear strain	radian
$\sigma_z, \sigma_\theta, \sigma_r$	Normal stresses in z, θ, r directions, respectively	psi
$\tau_{z\theta}$	Shear stress	psi
z, θ, r	Polar coordinates	in., rad., in.
u_z	Derivative of u with respect to z	in./in.
U_0	Strain energy density	in.-lb./in.
U	Total strain energy	in.-lb.
W	Work of external loads in virtual displacements	in.-lb.
ℙ	Potential energy	in.-lb.
T	Traction	lb.
s	Generalized displacement	in.
A	Area	in.^{2}
V	Volume	in.^{3}
A, B, C	Fourier constants	in.
ϕ	Fourier coefficients	lb./in.
m, n	z, θ, modes index, respectively	numeric
M, N	Particular values of m, n, respectively	numeric
$\hat{m}, \hat{n}$	Maximum values of m, n, respectively	numeric
ω, Ω	$n\theta$, $N\theta$, respectively	rad.
λ, Λ	$m\pi Z/L$, $M\pi Z/L$, respectively	numeric
Γ	$a\pi/L$	numeric
c_1	$E/[2(1-\mu^2)a]$	lb./in.^{3}
c_2	$E/[24(1-\mu^2)a^3]$	lb./in.^{5}
c_3	$[1-\mu]/2$	numeric
I_3^1, I_3^2	Integration numer 3 using c_1, h and c_2, h^3, respectively	—
P	Externally applied force	lb.

In terms of equation (5.73), equation (5.77a) becomes:

$$\frac{\partial U}{\partial A_{MN}} \delta A_{MN} - P \frac{\partial w}{\partial A_{MN}} \delta A_{MN} = 0 \tag{5.77b}$$

$$\frac{\partial U}{\partial B_{MN}}\,\delta B_{MN} - P\,\frac{\partial w}{\partial B_{MN}}\,\delta B_{MN} = 0 \tag{5.77c}$$

$$\frac{\partial U}{\partial C_{MN}}\,\delta C_{MN} - P\,\frac{\partial w}{\partial C_{MN}}\,\delta C_{MN} = 0 \tag{5.77d}$$

where M, N denote particular values of m, n. Note that

$$\frac{\partial w}{\partial A_{MN}} = \frac{\partial w}{\partial B_{MN}} = 0 \tag{5.77e}$$

and since the load is applied at $Z = 0$, $\theta = 0$, then

$$\frac{\partial w}{\partial C_{MN}} = [\cos(0\,M)][\cos(0\,N)] = 1 \tag{5.77f}$$

Note that the variations of A, B, C (i.e. δA, δB, δC) are arbitrary. Substituting equations (5.77e) and (5.77f) in equations (5.77b, c, and d) yields the working relations:

$$\frac{\partial U}{\partial A_{MN}} = 0 \tag{5.78a}$$

$$\frac{\partial U}{\partial B_{MN}} = 0 \tag{5.78b}$$

$$\frac{\partial U}{\partial C_{MN}} = P \tag{5.78c}$$

Now, substitute equation (5.76) into equation (5.78). To simplify this procedure, consider the following functional forms. If

$$F_1 = \left(\sum a_{mn} x_{mn}\right)^2 \tag{5.79a}$$

or

$$F_1 = a_{11}^2 x_{11}^2 + a_{21}^2 x_{21}^2 + \cdots a_{12}^2 x_{12}^2 + a_{22}^2 x_{22}^2 + \cdots + 2a_{11}x_{11}a_{21}x_{21} + 2a_{11}x_{11}a_{31}x_{31} + \cdots \tag{5.79b}$$

$$\frac{\partial F_1}{\partial a_{11}} = 2a_{11}x_{11}^2 + 2x_{11}a_{21}x_{21} + 2x_{11}a_{31}x_{31} + \cdots + 2x_{11}a_{12}x_{12} + \cdots \tag{5.79c}$$

To compact the functional form, equation (5.79c) in general can be written as:

$$\frac{\partial F_1}{\partial a_{MN}} = 2x_{MN}\sum a_{mn}x_{mn} \tag{5.79d}$$

If

$$F_2 = \left(\sum a_{mn} x_{mn}\right)\left[\sum b_{mn} y_{mn}\right] \tag{5.79e}$$

or

$$F_2 = a_{11} x_{11}[\ ] + a_{21} x_{21}[\ ] + \cdots$$
$$+ a_{12} x_{12}[\ ] + a_{22} x_{22}[\ ] + \cdots \tag{5.79f}$$

$$\frac{\partial F_2}{\partial a_{MN}} = x_{MN}[\ ] \tag{5.79g}$$

If

$$F_3 = \left(\sum n^2 a_{mn} x_{mn}\right)\left(\sum a_{mn} x_{mn}\right) \tag{5.79h}$$

$$\frac{\partial F_3}{\partial a_{MN}} = \left(N^2 x_{MN} \sum a_{mn} x_{mn}\right) + \left(x_{MN} \sum n^2 a_{mn} x_{mn}\right) \tag{5.79i}$$

To make the communication of succeeding relations more efficacious, let

$$\Lambda \equiv \frac{M\pi z}{L} \tag{5.80a}$$

$$\Omega \equiv N\theta \tag{5.80b}$$

$$\Gamma \equiv \frac{a\pi}{L} \tag{5.80c}$$

Substituting equation (5.76) in equation (5.78a), using equations (5.79) and (5.80):

$$\iint c_1 h \Big\{ 2\Gamma^2 M \cos\Lambda \cos\Omega \sum m A \cos\lambda \cos\omega$$
$$- 2\mu\Gamma M \cos\Lambda \cos\Omega \sum n B \cos\lambda \cos\omega$$
$$+ 2\mu\Gamma M \cos\Lambda \cos\Omega \sum C \cos\lambda \cos\omega$$
$$+ 2c_3 N \sin\Lambda \sin\Omega \sum n A \sin\lambda \sin\omega$$
$$-2c_3\Gamma N \sin\Lambda \sin\Omega \sum m B \sin\lambda \sin\omega \Big\} dz\, d\theta$$
$$+ \iint c_2 h^3 \Big\{ -2c_3\Gamma N \sin\Lambda \sin\Omega \sum mn C \sin\lambda \sin\omega$$
$$+2c_3 N \sin\Lambda \sin\Omega \sum n A \sin\lambda \sin\omega$$

$$+2\Gamma^3 M \cos\Lambda \cos\Omega \sum m^2 C \cos\lambda \cos\omega \Big\} dz\, d\theta = 0 \quad (5.81a)$$

Substituting equation (5.76) in equation (5.78b), using equations (5.79) and (5.80):

$$\iint c_1 h \Big\{ 2N \cos\Lambda \cos\Omega \sum nB \cos\lambda \cos\omega$$

$$-2N \cos\Lambda \cos\Omega \sum C \cos\lambda \cos\omega$$

$$-2\mu\Gamma N \cos\Lambda \cos\Omega \sum mA \cos\lambda \cos\omega$$

$$-2c_3\Gamma M \sin\Lambda \sin\Omega \sum nA \sin\lambda \sin\omega$$

$$+2c_3\Gamma^2 M \sin\Lambda \sin\Omega \sum mB \sin\lambda \sin\omega \Big\} dz\, d\theta$$

$$+ \iint c_2 h^3 \Big\{ 6c_3\Gamma^2 M \sin\Lambda \sin\Omega \sum mB \sin\lambda \sin\omega$$

$$-6c_3\Gamma^2 M \sin\Lambda \sin\Omega \sum mnC \sin\lambda \sin\omega$$

$$-2\mu\Gamma^2 N \cos\Lambda \cos\Omega \sum m^2 C \cos\lambda \cos\omega \Big\} dz\, d\theta = 0 \quad (5.81b)$$

Substituting equation (5.76) in equation (5.78c), using equations (5.79) and (5.80):

$$\iint c_1 h \Big\{ -2 \cos\Lambda \cos\Omega \sum nB \cos\lambda \cos\omega$$

$$+ 2 \cos\Lambda \cos\Omega \sum C \cos\lambda \cos\omega$$

$$+ 2\mu\Gamma \cos\Lambda \cos\Omega \sum mA \cos\lambda \cos\omega \Big\} dz\, d\theta$$

$$+ \iint c_2 h^3 \Big\{ 2\Gamma^4 M^2 \cos\Lambda \cos\Omega \sum m^2 C \cos\lambda \cos\omega$$

$$+ 2N^2 \cos\Lambda \cos\Omega \sum n^2 C \cos\lambda \cos\omega$$

$$- 2 \cos\Lambda \cos\Omega \sum n^2 C \cos\lambda \cos\omega$$

$$- 2N^2 \cos\Lambda \cos\Omega \sum C \cos\lambda \cos\omega$$

$$+ 2 \cos\Lambda \cos\Omega \sum C \cos\lambda \cos\omega$$

$$+ 2c_3\Gamma^2 MN \sin\Lambda \sin\Omega \sum mnC \sin\lambda \sin\omega$$

$$- 2c_3\Gamma MN \sin\Lambda \sin\Omega \sum nA \sin\lambda \sin\omega$$

$$- 6c_3\Gamma^2 MN \sin\Lambda \sin\Omega \sum mB \sin\lambda \sin\omega$$

$$+ 6c_3\Gamma^2 MN \sin\Lambda \sin\Omega \sum mnC \sin\lambda \sin\omega$$

$$+ 2\mu\Gamma^2 M^2 \cos\Lambda \cos\Omega \sum n^2 C \cos\lambda \cos\omega$$

$$+ 2\mu\Gamma^2 N^2 \cos\Lambda \cos\Omega \sum m^2 C \cos\lambda \cos\omega$$

$$- 2\mu\Gamma^2 M^2 \cos\Lambda \cos\Omega \sum nB \cos\lambda \cos\omega$$

$$\left. + 2\Gamma^3 M^2 \cos\Lambda \cos\Omega \sum mA \cos\lambda \cos\omega \right\} dz\, d\theta = P \quad (5.81c)$$

To simplify the integration of equation (5.81), use the following trigonometric identities

$$\sin\Lambda \sin\Omega \sin\lambda \sin\omega = \tfrac{1}{4}\,[\cos(\Lambda - \lambda) - \cos(\Lambda + \lambda)][\cos(\Omega - \omega) - \cos(\Omega + \omega)] \quad (5.82a)$$

$$\cos\Lambda \cos\Omega \cos\lambda \cos\omega = \tfrac{1}{4}\,[\cos(\Lambda - \lambda) + \cos(\Lambda + \lambda)][\cos(\Omega - \omega) + \cos(\Omega + \omega)] \quad (5.82b)$$

since h varies with Z and θ as indicated in equations (5.81), alternately apply the faciends, $c_1 h$ and $c_2 h^3$, to equations (5.82) and integrate (using equations 5.74 and 5.80):

$$\iint c_1 h(\text{EQ. } 5.82a)\, dz\, d\theta = \tfrac{1}{4} \sum_{ijk} c_1 h_k \left[\frac{\sin(\Lambda - \lambda)}{(M - m)\frac{\pi}{L}} - \frac{\sin(\Lambda + \lambda)}{(M + m)\frac{\pi}{L}} \right]_{z_i}^{z_{i+1}} \left[\frac{\sin(\Omega - \omega)}{N - n} - \frac{\sin(\Omega + \omega)}{N + n} \right]_{\theta_j}^{\theta_{j+1}} \quad (5.82c)$$

Equation (5.82c) will be written in functional form, denoting particular faciend by a superscript on I, as:

$$I_1^1 \equiv \tfrac{1}{4} \sum c_1 h[Q_1 - Q_2][Q_3 - Q_4] \quad (5.82d)$$

Integration of equation (5.82a) using the second faciend is, in functional form,

$$I_1^2 \equiv 1/4 \sum c_2 h^3 [Q_1 - Q_2][Q_3 - Q_4] \quad (5.82e)$$

Integration of equation (5.82b), using the two faciends alternately, yields in functional form:

$$I_2^1 \equiv 1/4 \sum c_1 h [Q_1 + Q_2][Q_3 + Q_4] \quad (5.82f)$$

$$I_2^2 \equiv 1/4 \sum c_2 h^3 [Q_1 + Q_2][Q_3 + Q_4] \quad (5.82g)$$

A more rigorous exposition would account for the variation of the mean radius, a, as h varies, and this could be interjected in the summations in the above equations. Within the applicable range of the parameters, however, this effect would be imperceptible for all practical purposes.

Note that for certain values of the indices, some of the terms in equation (5.82c) can become unbounded. Observing the forms in equations (5.82a and b), the conditions on the terms in equations (5.82d–g) are seen to be as follows:

When

$$M = m,\ Q_1 = z \quad (5.82h)$$

When

$$N = n,\ Q_3 = \theta \quad (5.82i)$$

When

$$N = n = 0,\ Q_3 = Q_4 = \theta \quad (5.82j)$$

Integrating equations (5.81) using equations (5.82) and rearranging so that the Fourier constants are ordered:

$$\sum \Big\{ A_{mn} \left[\Gamma Mm\, I_2^1 + \frac{c_3 Nn}{\Gamma} (I_1^1 + I_1^2) \right] + B_{mn} [-\mu Mn\, I_2^1 - c_3 Nm\, I_1^1] + C_{mn} [\mu M\, I_2^1 - c_3 N mn\, I_1^2 + \Gamma^2 Mm^2 I_2^2] \Big\} = 0 \quad (5.83a)$$

$$\sum \Big\{ A_{mn} [-\mu Nm\, I_2^1 - c_3 Mn\, I_1^1] + B_{mn} \left[\frac{Nn}{\Gamma} I_2^1 + c_3 \Gamma Mm (I_1^1 + 3 I_1^2) \right] + C_{mn} \left[-\frac{N}{\Gamma} I_2^1 - \mu \Gamma Nm^2 I_2^2 - 3 c_3 \Gamma Mmn\, I_1^2 \right] \Big\} = 0 \quad (5.83b)$$

$$
\begin{aligned}
\sum\Bigg\{&A_{mn}[\mu m\, I_2^1 + \Gamma^2 M^2 m\, I_2^2 - c_3 MNn\, I_1^2] \\
&+ B_{mn}\left[-\frac{n}{\Gamma} I_2^1 - \mu\Gamma M^2 n\, I_2^2 - 3c_3\Gamma MNm\, I_1^2\right] \\
&+ C_{mn}\Bigg[\frac{I_2^1}{\Gamma} + 4c_3\Gamma MN\, mn\, I_1^2 \\
&+ \Bigg(\Gamma^3 M^2 m^2 + \frac{N^2 n^2 - n^2 - N^2 + 1}{\Gamma} \\
&+ \mu\Gamma\, M^2 n^2 + \mu\Gamma\, N^2 m^2\Bigg) I_2^2\Bigg]\Bigg\} = \frac{P}{2\Gamma}
\end{aligned}
\qquad (5.83c)
$$

Since n = 0, 1, 2 . . . and m = 1, 3, 5 . . ., then for an equal number of incremental advances in n and m,

$$\hat{m} = 2\hat{n} + 1 \qquad (5.84)$$

Equations (5.83) can be expressed in matrix format as:

$$
\begin{bmatrix}
\phi_{11}^A & \phi_{21}^A \cdots & \phi_{12}^A & \phi_{22}^A \cdots & \phi_{11}^B \cdots & \phi_{11}^C \cdots \\
\phi_{11}^A & \phi_{21}^A \cdots & & mn & & \\
\phi_{11}^A & \phi_{21} \cdots & & \xrightarrow{\quad J \quad} & & \\
 & & \downarrow I & \text{FORTRAN PROGRAM} & & \\
 & & & & & \\
\end{bmatrix}
\begin{Bmatrix}
A_{11} \\ A_{21} \\ \vdots \\ A_{12} \\ \vdots \\ B_{11} \\ \vdots \\ C_{11} \\ \vdots
\end{Bmatrix}
=
\begin{Bmatrix}
O \\ O \\ \vdots \\ O \\ \vdots \\ O \\ \vdots \\ P \\ \vdots
\end{Bmatrix}
\qquad (5.85)
$$

With the assumption of equation (5.84) and observing the pattern in equation (5.85), the size of the complete matrix is seen to be $3(\hat{n} + 1)^2$ by $3(\hat{n} + 1)^2$ Therefore, each of equations (5.83) are used $(\hat{n} + 1)^2$ times. Observing the column of Fourier constants, note that in the first equation, the particular values of m, n (i. e., M, N) in all of the Fourier coefficients, ϕ, in the first row are 1, 1; in the second equation, M, N = 1, 2 etc. Also, in the $(\hat{n} + 1)^2 + 1$ equation, M, N = 1, 1 etc. With equation (5.85) then set up explicitly, the equations are solved simultaneously for the Fourier constants. The displacements are then determined from equations (5.73), the strains are determined from equation (5.67–5.69), and the stresses are determined from equations (5.71–5.73).

5.5 DESIGN CRITERIA FOR CORRUGATED PLATE WATER-OIL SEPARATOR

General Criteria of Gravity-type Separators (per Pielkenroad)

Under ideal operating conditions, gravity-type separators will remove from the entering fluid flow all particles that have a rising velocity (or sedimentation velocity) which is equal or larger than the overflow rate of the fluid flow through the separator. Hence:

$$V_s \geq V_o \tag{5.86}$$

in which:

V_s = Particle rising or sedimentation velocity in M/Hr.

V_o = Fluid overflow velocity in M/Hr.

Particles with a rising (or sedimentation) velocity equal to or greater than the overflow rate will be removed completely. Particles with a lower rising velocity will be separated according to the ratio:

$$\frac{V_s}{V_o} \tag{5.87}$$

The design of a gravity type separator is further based on the formula:

$$A_{eh} = \frac{Q}{V_o} \tag{5.88}$$

in which:

A_{eh} = The effective horizontal separation area in M^2.

Q = Separator capacity in M^3/Hr.

V_o = Fluid overflow velocity in M/Hr.

Determination of Effective Separation Area of a Corrugated Plate Separator Pack

Pielkenroad corrugated plate separator packs contain 47 corrugated plates. The true area of each plate is 1.8 M^2 or 19.5 sq. ft. Thus the total plate area (A_p) in a plate pack is:

$$47 \times 1.8\ M^2 = 86\ M^2 = 930 \text{ sq. ft.}$$

For the purpose of calculating the separation capacity of a corrugated plate pack, the total plate area is only one factor among a number of others that need to be considered.

Between the 47 corrugated plates in the plate pack, there are 46 separator compartments. The total separation area (A_s) provided in these compartments is:

$$46 \times 1.5\ M^2 = 69\ M^2$$

since the projected area of each separation compartment is 1.5 M^2.

When the pack is mounted in an inclined position, the horizontally projected area (A_h) provided by each plate pack is:

$$A_h = A_s \text{ x cosine } \alpha \tag{5.89}$$

in which:

α = Angle of inclination

In the case of a 45° angle of inclination, the horizontal projection of the separation area is:

$$0.708 \times 69\ M^2 = 48.8\ M^2$$

This area is the *theoretically* effective horizontal separation area, called A_{th}.

In practice, the actual effective horizontal separation area (A_{eh}) is smaller than the theoretically effective area because of disturbing effects in the separator's operation such as short circuiting and turbulence. Mathematically, the relationship between A_{eh} and A_{th} can be expressed as:

$$A_{eh} = \frac{A_{th}}{F} \tag{5.90}$$

in which:

A_{eh} = Actual effective separation area.

A_{th} = Theoretically effective area.

F = A safety factor to allow for short circuiting and turbulence.

In a Pielkenroad corrugated plate separator pack, little if any short circuiting or turbulence takes place. Experience in a number of industrial installations indictates that an overall safety factor of 5% provides a conservative estimate for the actual separation area to be provided in an installation. Hence, the actual effective area of a corrugated pack with 47 plates, mounted under a 45° incline is:

$$\frac{48.8}{1.05} = 46.5\ M^2 = 500 \text{ sq. ft.}$$

For comparison, the American Petroleum Institute manual on the disposal of refinery wastes recommends two safety factors F_1 and F_2 for API basins without depth subdivisions:

F_1 to allow for short circuiting of flow through the basin. An average value for F_1 is 1.2.

F_2 to allow for turbulence occurring in the basin. F_2 is a function of the flow velocity and the rising velocity of the oil droplets. An average value for F_2 is 1.37.

The product of F_1 and F_2 yields the overall safety factor of 1.65 for the API basin.

Example of the Calculation of the Required Number of Plate Packs

Assume a waste water stream is contaminated with dispersed oil. The separator design to be based upon the complete removal of oil droplets 60 microns in size and larger, the following basic design data are available:

Maximum flow of waste water:	130 M^3/Hr. = 575 gpm
Temperature of water:	20°C
Density of oil:	0.900

The rising rate of an oil droplet is defined by Stoke's Law, which can be expressed as follows:

$$V_s = \frac{g}{18\mu} (\rho_w - \rho_o) D^2 \tag{5.91}$$

In which:

V_s = Velocity of rise of the particles in cm/sec.

g = Gravity constant in cm/sec.^{2}

μ = Viscosity of water in poises.

ρ_w and ρ_o

= Density of water and oil respectively.

D = Diameter of the particles in cm.

At a temperature of 20°C, the density of water is 0.998 and the viscosity 1.00×10^{-2} poise. Hence, the rising rate (V_s) of an oil droplet with a diameter of 60 microns is:

$$V_s = \frac{981}{18 \times 10^{-2}} \times (0.998 - 0.900) \times 36 \times 10^{-6}$$

$$= 0.0192 \text{ cm/sec.} = 0.7 \text{ M/Hr.}$$

Since all oil droplets 60 microns in size must be removed, the overflow velocity of the separator V_o must be equal to the rising velocity V_s of the 60 micron oil droplets.

Hence:

$$V_o = V_s = 0.7 \text{ M/Hr.}$$

According to formula 3:

$$A_{eh} = \frac{Q}{V_x}$$

or in this application:

$$A_{eh} = \frac{130 \text{ M}^3\text{/Hr.}}{0.7 \text{ M/Hr.}} = 186 \text{ M}^2 = 2{,}000 \text{ sq. ft.}$$

Since the effective separation area in each Pielkenroad corrugated plate pack is 46.5 M^2 or 500 sq. ft., the total number of CPS packs needed in this application is:

$$\frac{186}{46.5} = 4$$

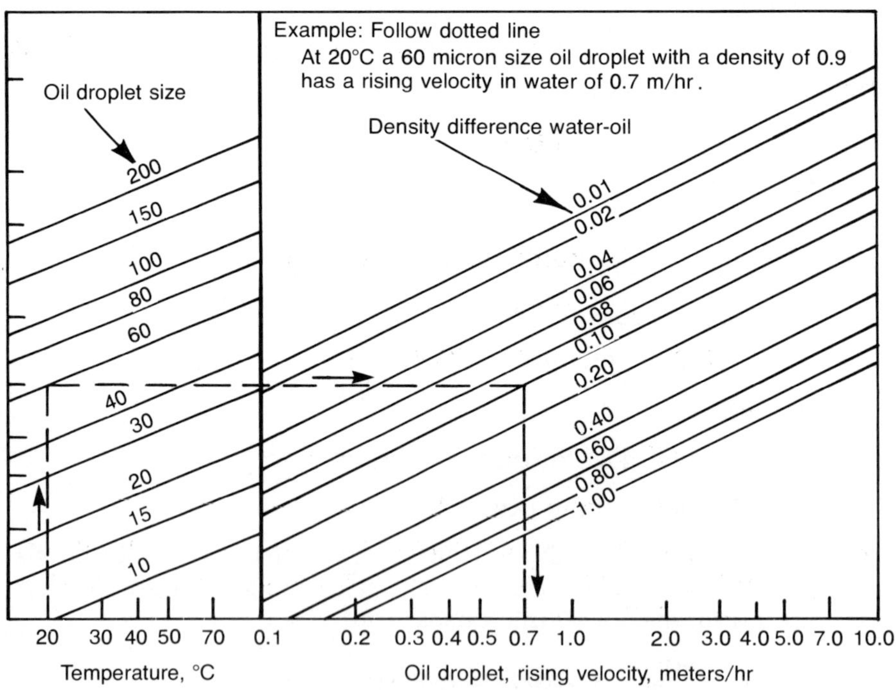

FIG. 5.21 Corrugated plate water-oil separator design chart

To eliminate the tedious calculations, the correlation between:

- Oil droplet size
- Oil/water density difference
- Water viscosity (correlated with temperature)
- Droplet rising rate

is given in figure 5.21.

The dotted line shown on figure 5.21 represents the parameters of the example.

5.6 EXPERIMENTAL STRESS ANALYSES OF A SUBMARINE

The work described covers the experimental stress analysis of a 0.286 scale, hemispherical model which was subjected to loads of uniform external pressures.(88) The model simulated the hull of an oceanographic vessel designed and fabricated by the General Mills Company for the Woods Hole Oceanographic Institution, under the auspices of the Office of Naval Research. The full scale design has been referred to as Alvin by the organizations.

The objective of the experimental stress analysis of the scaled model was a verification of the adequacy of the full scale design, considering the laws of structural similitude. The author was unable to locate in the literature any previous theoretical developments or experimental results for the analysis of structures having a similar geometry. Therefore, even though the Alvin submarine is now operational, the data presented here remains timely.

Tests were conducted with two model configurations, the cross sections of which are given in figure 5.22. Both model configurations had a plexiglass window. Dimensions A and B were 1.075 in. and 3.082 in., respectively, in Configuration 1, and 1.434 in. and 3.440 in., in Configuration 2. The thickness variation of Configuration 1 was checked by utilizing a granite flat (flat within 0.0001 in.) and a mechanical displacement type of gauge. All dimensions were found to be well within the specified tolerances of ±0.015 in. and ±15 min on linear and angular dimensions.

After the test with Configuration 1 was completed, the port was machined to the larger diameter, and a larger window was fabricated for the test of Configuration 2. Through holes similating cable penetrations and half-through holes, drilled in the convex surface for simulation of bolt holes, were also machined in Configuration 2 as shown in figure 5.23. The cable penetrations and the half-through holes were drilled parallel, with the through holes directed to the center of the 11.373 in. radius of the concave model surface.

The cable penetrations were sealed with plugs having 0.005-in. clearance with the hole walls and specially made O rings. Since strain gauges were installed at the edge of the holes, the plugs were welded to brackets which overspanned the holes. The ends of the bracket legs were ground to fit the contour of the model, and the O rings were located at the half-thickness of the shell wall with the plugs in place.

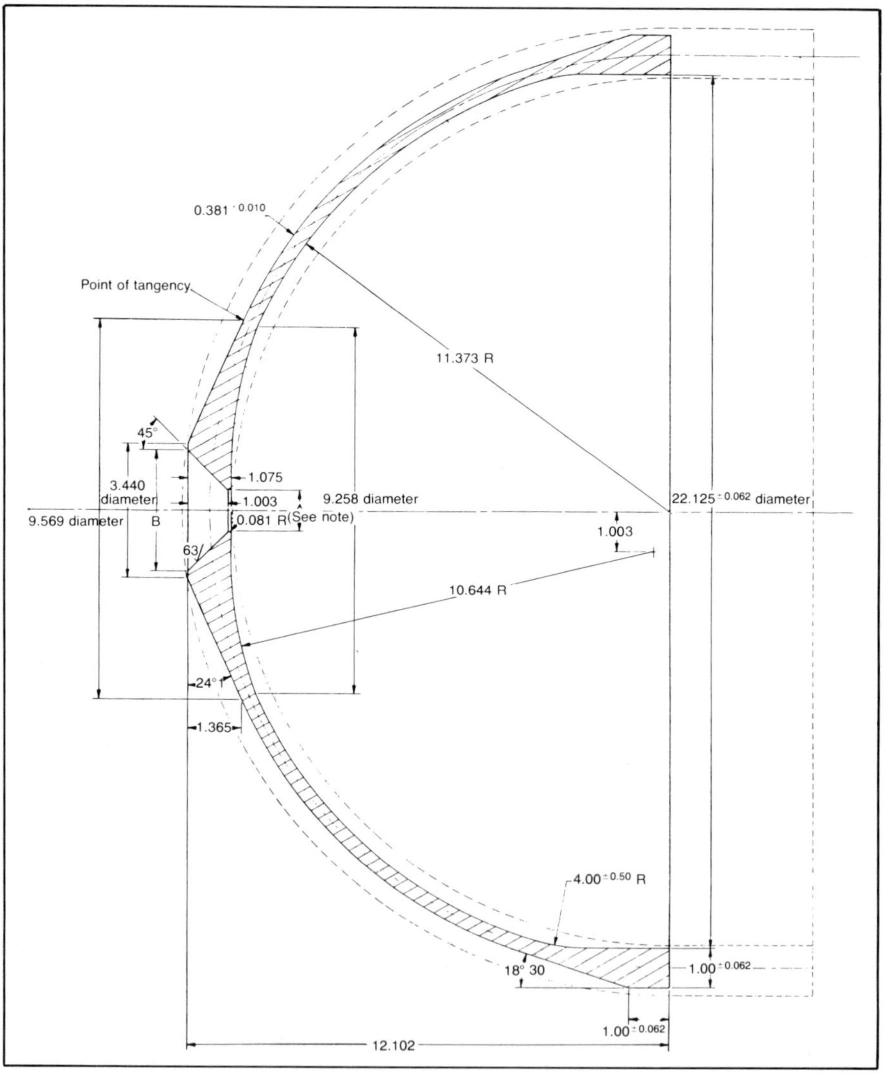

FIG. 5.22 Cross section of models

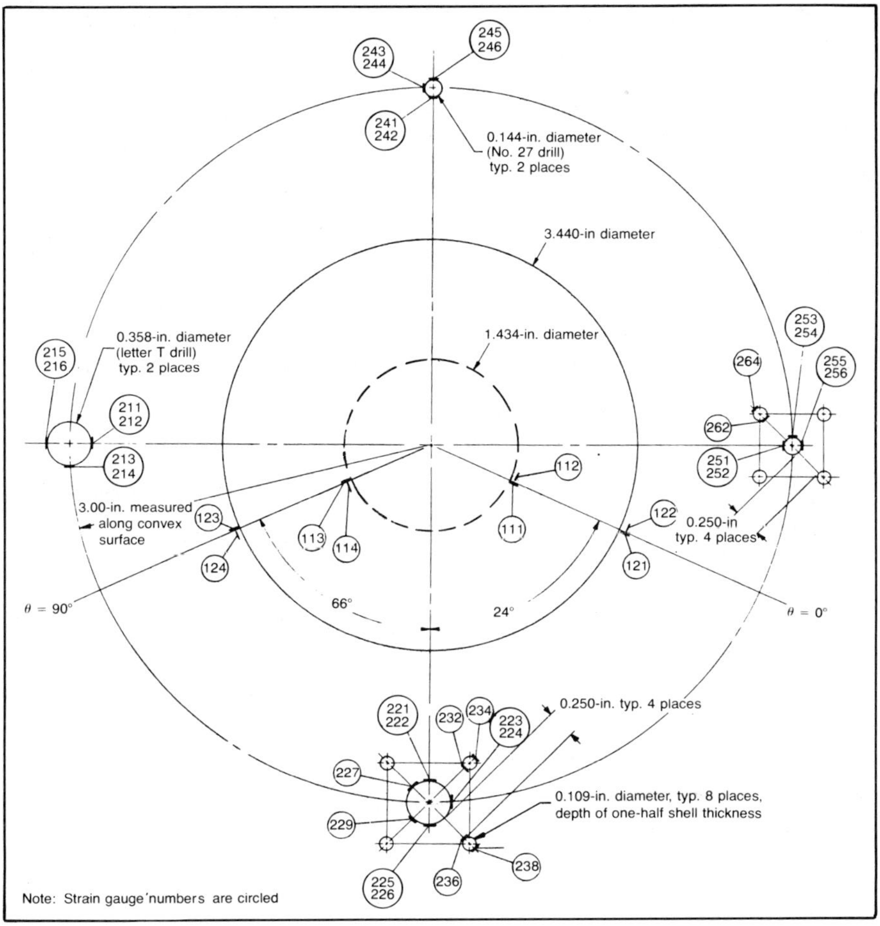

FIG. 5.23 Details of configuration 2

Material Properties

The formed hemispherical head was made from 7079-T6 aluminum, and was machined to the final dimensions. Typical mechanical properties of the material were as follows:[89]

$\sigma_u = 77$ ksi

$\sigma_y = 66$ ksi

Elongation in 2 in. = 4% minimum

Brinell hardness number = 150, (500 kg load, 10 mm ball)

$\tau_u = 48$ ksi

$$\sigma_e = 22 \text{ ksi, (Based on } 5 \times 10^8 \text{ cycles of reversed stress)}$$

$$E = 10.4 \times 10^6 \text{ psi}$$

A Poisson's ratio of 0.33 for the aluminum was used in the stress computations.

The plexiglass port had the following mechanical properties:[90]

Tensile strength	ASTM D-638-46T, ¼″ specimen (0.2″/min)	7 ksi
Flexural strength	ASTM D-790-45T, Span-depth ratio of 16(0.1″/min)	12 ksi
Compressive strength	ASTM D-695-44T (0.2″/min)	12 ksi

A Young's modulus of elasticity of 400 ksi and a Poisson's ratio of 0.5 for the plastic were used in the stress computations.

Theoretical Stresses

The Lame equations[91] for the normalized stresses in thick-walled spheres or in hemispheres subjected to uniform external pressure, zero internal pressure, are:

$$\frac{\sigma_m}{P_o} = \frac{b^3(2R^3 + a^3)}{2R^3(a^3 - b^3)} \tag{5.92}$$

$$\frac{\sigma_r}{P_o} = \frac{b^3(R^3 - a^3)}{R^3(a^3 - b^3)} \tag{5.93}$$

Considering the dimensions given in figure 5.22, the calculated normalized surface stresses on the concave side of the shell were:

$$\frac{\sigma_m}{P_o} = 15.9 \text{ psi/psi} \tag{5.94}$$

$$\frac{\sigma_r}{P_o} = 0 \tag{5.95}$$

The calculated normalized surface stresses on the convex side of the shell were:

$$\frac{\sigma_m}{P_o} = 15.4 \text{ psi/psi} \tag{5.96}$$

$$\frac{\sigma_r}{P_o} = -1.0 \text{ psi/psi} \tag{5.97}$$

The small difference in the magnitudes of the stresses on the concave and convex surfaces of the head indicated that the head could be considered as a thin shell. Using membrane theory, the calculated stress per unit external pressure was

$$\frac{\sigma}{P_o} = \frac{r}{2t} = \frac{11.373}{2(0.381)} = 14.9 \text{ psi/psi} \tag{5.98}$$

Considering the allowable shear shrength to be one-half of the tensile yield strength, the allowable pressure on the plexiglass window from the standpoint of shear stress was:

$$P_o = \frac{2t_\tau}{r} = \frac{t_\sigma}{r} = \frac{1.00(7000)}{0.537} = 13.0 \text{ ksi} \tag{5.99}$$

Hence, the shear stress in the window was seen to be only a minor consideration.

Test Procedure

Electrical resistance type of strain gauges were used to measure the strain produced by the external pressure applied to the hemispherical model. Detailed information on the various types of gauges used are given in table 101. Rectangular rosettes of the gauges were installed on the concave and convex side of the head in a manner such that for any particular rosette, a corresponding rosette was located on the opposite surface (on a radial line drawn from the center of curvature of the concave surface). A row of rosettes was installed across the meridian, $\theta = 0°$, with the orientation as given in figure 5.23, and at $\theta = 90°$ in order to ascertain the degree of symmetrical response of the vessel.

In the test setup, a paper gasket with sealing compound was placed between the lid of the pressure facility and the model to effect a seal. The model was held to the lid of the pressure facility by means of a turnbuckle bar until the initial seal between the lid and model was formed. The bar was then loosened before beginning the tests. End blocks on the bar were bonded to the shell with an epoxy plastic for the test setup of Configuration 1. This arrangement proved unsatisfactory so small indentions were made in the model to position the bar for the test of Configuration 2.

The edge of the port window was coated with sealing compound and the window was taped in place for the tests of Configuration 1. A small amount of oil seeped between the window and the head before the initial seal was obtained. In order to eliminate this leakage, the window was bonded to the shell with Duco cement for the tests of Configuration 2.

TABLE 101
Characteristics of strain gauges used

Strain gauge type number	Manufacturer's strain gauge number	Gauge length, inches	Electrical resistance, R, ohms	Gauge factor, GF	Correction factor, CF
1	C12-1X1-64*	1/64	120 ± 1	1.83 ± 3%	1.114
2	C12-111*	1/16	120 ± 0.5	2.04 ± 1%	1
3	C6-111*	1/16	120 ± 0.5	2.01 ± 1%	1.015
4	C6-121*	1/8	120 ± 0.2	2.07 ± 0.5%	0.985
5	C6-121*	1/8	120 ± 0.2	2.02 ± 0.5%	1.010
6	EA-13-031DE-120**	1/32	120 ± 0.4	2.07 ± 1%	0.985
7	EA-13-031EC-120**	1/32	120 ± 0.4	2.01 ± 1%	1.015

* Available from Budd Co., P.O. Box 245, Phoenixville, Pa.
** Available from Micro-Measurements, Inc., 38905 Chase Road, Romulus, Michigan.

A Gilmore Strain Indicator was used to record the strains in the model. The model was cycled between 100 and 1,600 psi external pressure several times in order to stabilize the strain gauge installations. Then the strains were recorded at each increment of 300 psi from 100 to 1,600 psi and finally again at 100 psi. For convenience during the tests, all channels in the Gilmore were set for a gauge factor of 2.04. The corrected strain was found by applying a factor to the indicated strain since

$$\varepsilon_{\text{true}} = \frac{\Delta R}{R(GF)_{\text{true}}} \tag{5.100}$$

$$\varepsilon_{\text{indicated}} = \frac{\Delta R}{R(GF)_{\text{set}}} \tag{5.101}$$

$$\varepsilon_{\text{corrected}} = \varepsilon_{\text{indicated}}\left(\frac{GF_{\text{set}}}{GF_{\text{true}}}\right) = (\varepsilon_{\text{indicated}})(CF) \tag{5.102}$$

The stresses were computed using the plane stress relations. Since the radial stress and strain are zero at a free boundary, the stresses at the boundaries of the cable penetrations were computed directly from the single strain gauge readings.

The pressure effects on the type of gauges used in this study have been found to be negligible at the given low pressures[(92)] To insure that the applied pressure would not affect the indicated strains, compensation gauges were placed on blocks of the same material as the shell and the port window. The gauges on these blocks were subjected to atmospheric pressure and to the applied external pressure for compensation of the gauges on the concave and convex surfaces of the shell, respectively, during the tests.

The usual three-wire lead hookup for self-temperature compensating gauges could not be used since the Gilmore Strain Gauge Plotter requires a dummy gauge in the circuit. The urgent need for the results necessitated the use of stock strain gauges, and since insufficient aluminum compensated gauges were on hand, some steel compensated gauges were used (table 101). Consequently, one of each of these two types of gauges was installed on the aluminum dummy blocks inside and outside of the pressure facility and appropriately wired in the circuitry.

Discussion of Results

Results for Configuration 1. The normalized stress distribution and variation of shell thickness are plotted in figure 5.24. This graphical representation of the stress distribution is self-explanatory, and the membrane stresses are shown to be the largest stresses in the structure. The average experimentally de-

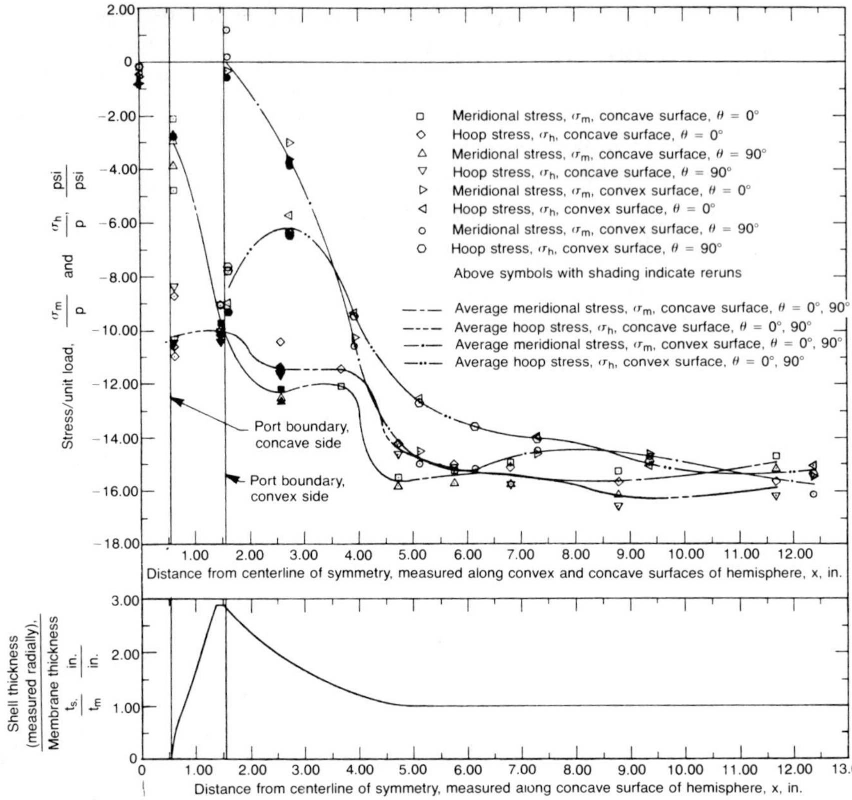

FIG. 5.24 Normalized stress distribution and variation of shell thickness, configuration 1

termined membrane stress per unit load of 15.5 psi/psi can be compared to the theoretical membrane stress of 14.9 psi/psi.

Results for Configuration 2. The normalized stress distribution across meridians (not coincident with the cable penetrations) is presented in figure 5.25. The membrane stresses on the meridians were again of larger magnitude than in the thickened section containing the port, and the average membrane stress of about the same magnitude as for Configuration 2.

The boundary stresses about the simulated cable penetrations and bolt holes are given in tables 102 and 103, with strain gauge locations as given in figure 5.23. Insufficient data was obtained around the circumference of these discontinuities to obtain graphical representations; however, the gauges were located such that the given results provided the significant data necessary for design purposes.

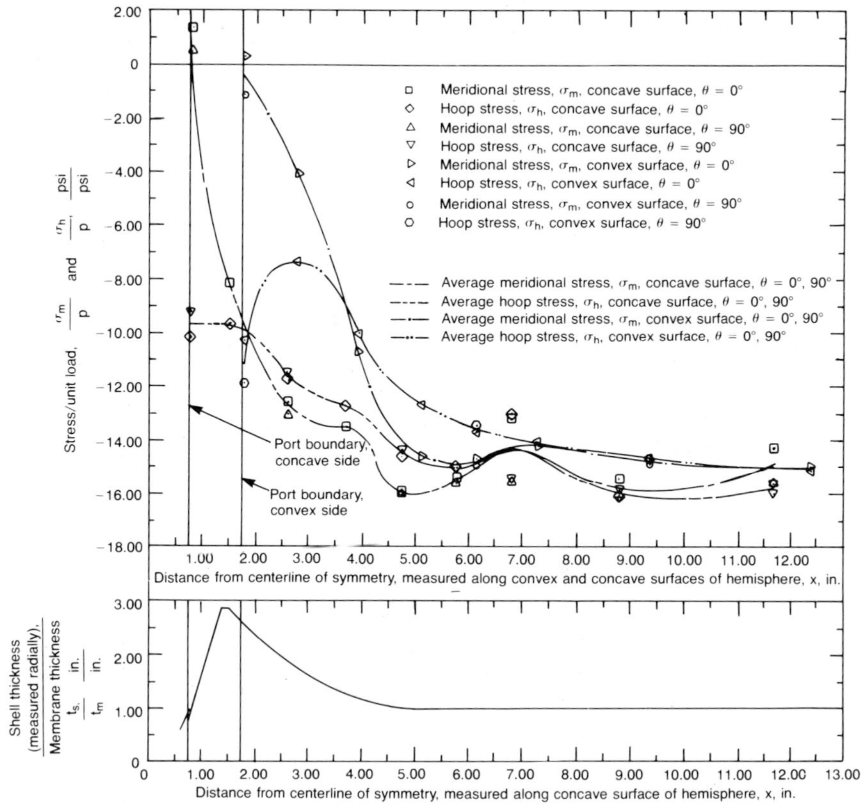

FIG. 5.25 Normalized stress distribution and variation of shell thickness, configuration 2

The stress concentration factors for the two sizes of cable penetrations without bolt holes, based on the theoretical membrane stress, were 1.28 for the 0.358-in. diameter hole, and 0.97 for the 0.144-in. diameter hole. The location of the maximum stress for each of these two sizes of holes was on the concave surfaces at the hole boundary on the penetration centerline circle. These factors are comparable to the theoretical factor of two for an infinite plate of uniform thickness having a circular hole, and subjected to a biaxial stress field.[(93)]

The stress concentration factor for the two sizes of cable penetrations with bolt holes, again based on the theoretical membrane stress, were 1.21 for the 0.358-in. diameter hole and 1.03 for the 0.144-in. diameter hole. The location of these maximum stresses corresponded to the locations of maximum stress for the cable penetrations without bolt holes.

TABLE 102

Strain gauge data and results for gauges at boundaries of cable penetrations without bolt holes, configuration 2

Strain gauge number*	Strain gauge type	Correction factor, CF	Corrected strain, ε, μ-in./in. (at p = 1500 psi)	Boundary stress per unit load, $\frac{\sigma_b}{p}$, $\frac{psi}{psi}$	Distance from edge of hole, in.
Gauges around 0.358″ Diameter Cable Penetration without Bolt Holes					
211	6	0.985	−2380	−16.5	0.05
212	6	0.985	−1360	9.39	0.06
213	6	0.985	−2740	−19.0	0.04
214	6	0.985	−1090	−7.54	0.06
215	6	0.985	−2470	−17.2	0.05
216	6	0.985	−1540	−10.7	0.05
Gauges around 0.144″ Diameter Cable Penetration without Bolt Holes					
241	6	0.985	−1420	−9.83	0.12
242	6	0.985	−1090	−7.54	0.06
243	6	0.985	−2080	−14.4	0.04
245	6	0.985	−1620	−11.2	0.10
246	6	0.985	−1190	−8.,25	0.05

*Odd-numbered gauges were located on concave surface, and even-numbered gauges were located on convex surface.

The stresses around the bolt holes near the 0.358-in. diameter cable penetration were greater than those near the 0.144-in. diameter, and the stresses around the bolt holes for either case were less than that at any of the cable penetrations. The bolt hole boundaries were at the same distance from the boundaries of the cable penetrations for both sizes of the penetrations. The stress concentration factors for the cable penetrations decreased directly as a function of the ligament efficiency, i.e., the stress intensity decreased as the distance between the penetration and the bolt hole centerlines decreased, as would be expected.[94]

General Comments. Some question may be raised as to the effect of the plug arrangement for sealing the cable penetrations. The pressure acting on the hole walls between the convex surface and the O-ring seal can easily be shown to have a negligible effect on the overall or local stress distribution, intensity, or direction. The 0.005" clearance provided between the plug and the hole wall was demonstrated in the Gilmore strain plot to be adequate, since the strain-pressure plots for the gauges around each hole were linear. The plots would have been nonlinear with the occurrence of binding between the plugs and the perforation walls.

Note that since the reported stresses were computed using measurements from strain gauges which have a definite gauge length, the stresses as given are the integrated stresses over that gauge length, and not the stress at a point. Also, to have installed the gauges with the gauge centerline at the boundaries of the discontinuities was physically impossible and the actual locations are given in the tables. Comparing the reported results with experience in this area of experimental stress analysis, the true stress at a point on the boundaries of the discontinuities could be as much as 35% higher than the stresses reported herein.

In the absence of flexural stresses, the theoretical stress per unit load at the penetration boundary would be 15.8 psi/psi, based on a stress concentration factor of 2 applied to the theoretical membrane stress, considering the shell thickness at the hole boundary. The experimental stresses given in tables 102 and 103 vary between 20% greater and 66% less than the theoretical value.

If a more accurate determination of stress at the boundaries of the discontinuities had have been required, the birefringent coating technique could have been used most satisfactorily since the stresses could then be obtained exactly at the hole boundary.[94] Also, the individual principal stresses could be obtained from measurements of relative retardation with normal and oblique incidence without any integration effect since the specimen was symmetrical.

TABLE 103
Strain gauge data and results for gauges at boundaries of cable penetrations and bolt holes, configuration 2

Strain gauge number*	Strain gauge type	Correction factor, CF	Corrected strain, ε, μ-in./in. (at p = 1500 psi)	Boundary stress per unit load, $\frac{\sigma_b}{p}, \frac{psi}{psi}$	Distance from edge of hole, in.
Gauges around 0.358″ Diameter Cable Penetration with Bolt Holes					
221	6	0.985	−2310	−16.0	0.07
222	6	0.985	−1250	−8.63	0.08
223	6	0.985	−2600	−18.0	0.04
224	6	0.985	−1020	−7.10	0.04
225	6	0.985	−2310	−16.00	0.05
226	6	0.985	−1830	−12.7	0.10
227	6	0.985	−2300	−16.0	0.05
Gauges around 0.109″ Diameter Bolt Holes near 0.358″ Diameter Cable Penetration					
232	6	0.985	−827	−5.73	0.04
234	6	0.985	−732	−5.08	0.03
236	6	0.985	−1470	−10.2	0.04
238	6	0.985	−1430	−9.94	0.04

Gauges around 0.144″ Diameter Cable Penetration with Bolt Holes					
251	6	0.985	−1770	−12.2	0.05
252	6	0.985	−1100	−7.65	0.05
253	6	0.985	−2220	−15.4	0.06
254	6	0.985	−772	−5.35	0.05
255	6	0.985	−1650	−11.5	0.06
256	6	0.985	−1300	−9.01	0.06
Gauges around 0.109″ Diameter Bolt Holes near 0.144″ Diameter Cable Penetration					
262	6	0.985	−827	−5.73	0.03
264	6	0.985	−749	−5.19	0.04

*Odd-numbered gauges were located on concave surface, and even-numbered gauges were located on convex surface.

Note the manner in which the curves in figures 5.24 and 5.25 were drawn through the data points. Observation of this data over the specimen meridians, $\theta = 0°$, and $\theta = 90°$, indicates that the stress response of the structure was slightly nonsymmetric. Somewhat smoother curves would have resulted if separate curves had been drawn through the data points over these two meridians. Since the degree of nonsymmetric response was so slight at most values of t_s/t_m, the curves were drawn through the average value of the stresses obtained from the two meridians.

Accuracy of Results. The maximum deviation between the results reported in this investigation and the true gauge stresses was estimated to be ±4% maximum, neglecting the integration effect over the gauge length and the small distance between the gauge centerline and the boundaries of the discontinuities, (i.e., using the same technique the experimental results could be duplicated within an accuracy of ±4%).

Conclusions

The significant data obtained from the study were as follows:

(1) The maximum stress in the Configuration 1 model was the membrane stress, and this stress was almost equal to the theoretical membrane stress.

(2) The maximum stress in the Configuration 2 model, except at the cable penetrations, was the membrane stress, and this stress was of about the same intensity as that with Configuration 1.

(3) The cable penetrations caused a stress of higher magnitude than the membrane stress, with the higher stress concentration factor for the larger size of hole.

(4) The stress concentration factor at the cable penetration was decreased when bolt holes were used, and decreased further for the case of the bolt holes located closer to the cable penetration.

(5) The magnitude of the stresses at the boundaries of the bolt holes was less than the membrane stress.

REFERENCES

1. Brownell, L. E., and Young, E. H., *Process Equipment Design*, 1959, John Wiley.
2. Roark, R. J., *Formulas for Stress and Strain*, 2nd Ed., McGraw-Hill Book Company, 1943.

3. Horvay, G., "The Plane-Stress Problem of Perforated Plates," Jnl. Appl. Mech., 19, No. 3, 355–360 (Sept. 1952).
4. Malkin, I., "Notes on a Theoretical Basis for Design of Tube Sheets of Triangular Layout," Trans. ASME, 74, 387 (1952).
5. Salerno, V. L., and Mahoney, J. B., "A Review, Comparison and Modification of Present Deflection Theory for Flat Perforated Plates," Welding Research Council Bulletin No. 52, 345 E. 47th Street, New York 17, N.Y. (July 1959).
6. Deagle, L., "Correlation of Experimental Data with Theory for Perforated Plates with a Triangular Hole Array," Welding Research Council Bulletin No. 52 (July 1959).
7. Leven, M. M., "Photoelastic Determination of Stresses in Tube Sheets and Comparison with Calculated Values," Research Report 100-FF996-R9-X,* Westinghouse Research Laboratories (Dec. 1959).
8. Sampson, R. C., "Photoelastic Analysis of Stresses in Perforated Material Subject to Tension or Bending," Research Report 100FF996-R8-X,* Westinghouse Research Laboratories (Oct. 1959).
9. Lemcoe, M. M., "Feasibility Studies of Stresses in Ligaments," Welding Research Council Bulletin No. 65 (Nov. 1960).
10. Daane, Robert A., "The Effective Elastic Constants of Perforated Plates and Shells," Ph.D. thesis, University of Wisconsin (1959).
11. Mahoney, J. B., Salerno, V. L. and Goldberg, M. A., "Analysis of a Perforated Circular Plate Containing a Rectangular Array of Holes," Welding Research Council Bulletin No. 80 (Aug. 1962).
12. Duncan, J. P., "Rational Design of Stayed Tube Plates," Engineering, London, (12 Oct. 1956).
13. Gardner, K. A., "Heat-Exchanger Tube-Sheet Design," Jnl. Appl. Mech., 15, 377 (1948).
14. Gardner, K. A., "Heat-Exchanger Tube-Sheet Design-2, Fixed Tube Sheets," Ibid. 19, 159–166 (1952).
15. Discussion of Ref. No. 12, Ibid., 19, 573 (1952).
16. Horvay, G., "Thermal Stresses in Perforated Plates," Proc. First U.S. Nat'l Cong. Appl. Mech., ASME, New York, 247 (1951).
17. Horvay, G., "Bending of Honeycombs and of Perforated Plates," Jnl. Appl. Mech., 19, 122 (1952).
18. Discussion of Ref. No. 17, Ibid. 19, 406 (1952).

* These reports were published in the April 1960 Bettis Technical Review and may be obtained from the Office of Technical Services, Department of Commerce, Washington 25, D.C.

19. Miller, K. A. G., "The Design of Tube-Plates in Heat-Exchangers," Proceed. (B) Inst. Mech. Engrs., 1B, 215 (1952).
20. Radkowski, P. O., "Stresses in a Plate Containing a Ring of Circular Holes and a Central Circular Hole," Proc. Second U.S. Nat'l Cong. Appl. Mech., ASME, New York, 277 (1954).
21. Mesnager, M., "Sur la Détermination Optique des Tensions Intérieures dans les Solides à Trois Dimensions," Comptes Rendus, Paris, 190, 1249 (1930).
22. Zandman, F., "Analyse des Contraintes par Vernis Photoélastiques," G.A.M.A.C., 15 Rue Vauquelin, Paris 5, France (1953).
23. Hetényi, M., Editor, *Handbook of Experimental Stress Analysis*, John Wiley and Sons, Inc., New York, N.Y. (1950).
24. Coker, E. G., Filon, L. N. G. and Jessop, H. T., *A Treatise on Photo-Elasticity*, 2nd Ed., Cambridge Univ. Press, 32 East 57th St., New York 22, N.Y. (1957).
25. Frocht, M. M., *Photoelasticity*, John Wiley and Sons, Inc., New York, N.Y., 1 (1941).
26. Flügge, S., Editor, *Elasticity and Plasticity, Encyclopedia of Physics*, 6, Springer-Verlag, Berlin (1958).
27. Zandman, F., *PhotoStress, Principles and Applications*, Budd Co., P.O. Box 245, Phoenixville, Pa.
28. Zandman, F., and Maier, H. N., "6 New Techniques for Photoelastic Coatings," Prod. Eng. 45 (July 24, 1961).
29. Post, D., and Zandman, F., "Visible Stress," Sci. Technol., 1, 24–33 (Jan. 1962).
30. D'Agostino, J., Drucker, D. C., Liu, C. K. and Mylonas, C., "An Analysis of Plastic Behavior of Metals with Bonded Birefringent Plastics," Proc. SESA, XII, No. 2, 115–122 (1955).
31. Day, E. E., Kobayashi, A. S. and Larson, C. N., "Fringe Multiplication and Thickness Effects in Birefringent Coating," *Experimental Mechanics*, 2, No. 4, 115–122 (Apr. 1962).
32. Bynum, D. J., "Birefringent Coating Analysis of a Tensile Plate Containing a Heterogeneous Inclusion, Utilizing Fringe Multiplication," unpublished M.S. thesis, University of Washington (1961).
33. "Installation Instructions for Tatnall 101 Series Epoxy Back Strain Gauges Using Epoxy Adhesives," Bulletin IB-6102, Budd Company, P.O. Box 245, Phoenixville, Pa.
34. "Instructions for Applying PhotoStress Sheet Plastic," Bulletin IB-8004, Budd Company, P.O. Box 245, Phoenixville, Pa. (Feb. 18, 1959).
35. "Instructions for the Use of PhotoStress Small Field Universal Meter," Bulletin PS-3044, Budd Company, P.O. Box 245, Phoenixville, Pa. (Mar. 21, 1960).

36. Zandman, F., Redner, S. S. and Riegner, E. I., "Reinforcing Effect of Birefringent Coatings," *Experimental Mechanics*, 55–64 (Feb. 1961).
37. Timoshenko, S., and Woinowsky-Krieger, S., *Theory of Plates and Shells*, 2nd Ed., McGraw-Hill Book Co., Inc., New York, N.Y. (1959).
38. Duffy, J., "Effects of the Thickness of Birefringent Coatings," Technical Report No. 34, Brown University, Dept. of the Army Project 503–06–005, (Apr. 1958).
39. Lee, T. C., Mylonas, C. and Duffy, J., "Thickness Effects in Birefringent Coatings and Radial Symmetry," *Experimental Mechanics*, 134–142 (Oct. 1961).
40. Chen, W. T., and Duffy, J., "Effects of the Thickness of Birefringent-coatings for Two-Dimensional Variations in Surface Strain," U.S. Army Research Office (Durham), DA Proj. No. 59901004, Technical Report No. 2 (Feb. 1962).
41. Post, D., and Zandman, F., "The Accuracy of the Birefringent-coating Method for Coatings of Arbitrary Thickness," *Experimental Mechanics*, 21–32 (Jan. 1961).
42. "Method of Correction for Parasitic Birefringence during Photo-Stress Tests," Bulletin PS-5051, Budd Company, P.O. Box 245, Phoenixville, Pa. (Mar. 8, 1960).
43. Zandman, F., Redner, S. S. and Post, D., "Photoelastic Coating Analysis in Thermal Fields," Instruments Division, Budd Company, Phoenixville, Pa.
44. Miner, D. F., and Seastone, J. B., Editors, *Handbook of Engineering Materials*, 1st Ed., John Wiley & Sons, Inc., New York, N.Y. (1958).
45. Bynum, D. J., and Lemcoe, M. M., "PhotoStress Study of Stresses in Thick Perforated Circular Plates under Lateral Loading," Final Report (Dec. 15, 1961), Southwest Research Institute Project Number 1023–3, Prepared for Pressure Vessel Research Committee, Welding Research Council, 345 E. 47th St., New York, N.Y.
46. Bynum, D. J., and Lemcoe, M. M., "Stresses and Deflections in Laterally Loaded Perforated Plates," Welding Research Council Bulletin No. 80 (Aug. 1962).
47. Heywood, R. B., *Designing by Photoelasticity*, Chapman and Hall, Ltd., London, 162–177 (1952).
48. Peterson, R. C., *Stress Concentration Design Factors*, John Wiley and Sons, Inc., New York, 27 (1953).
49. Savin, G. N., "Stress Concentration Around Holes," Translated from Russian, Pergamon Press, 1961.
50. Lipson, C., Noll, G. C., and Clock, L. S., *Stress and*

Strength of Manufactured Parts, 1st Ed., McGraw-Hill Book Co., New York, 80–85 (1950).

51. Bynum, D. J., and Lemcoe, M. M., "Stresses and Deflections in Laterally Loaded Perforated Plates," Bulletin No. 80, August 1962, Welding Research Council, 345 E. 47th St., New York, N.Y. 10017
52. "Operating Instructions, SF/Z Meter," BN 8002, July 30, 1957, Budd Co., P.O. Box 245, Phoenixville, Pa.
53. "Operating Instructions, OI/Z Meter," BN 8003, June 20, 1958, Budd Co., P.O. Box 245, Phoenixville, Pa.
54. "Installation Instructions for Tatnall 101 Series Epoxy Back Strain Gauges Using Epoxy Adhesives," IB-6012, Budd Co., P.O. Box 245, Phoenixville, Pa.
55. Morgan, W. C., and Bizon, P. T., "Experimental Investigation of Stress Distributions Near Abrupt Change in Wall Thickness in Thin-Walled Pressurized Cylinders," Tech. Note D-1200, NASA, 20, June 1962.
56. Durelli, A. J., Lake, R. L., and Phillips, E., "Stress Distributions in Plates Under a Uniaxial State of Stress, with Multiple Semicircular and Flat-Bottom Notches," Proc. First Natl. Congr. Appl. Mech., ASME, 309–315 (1951).
57. Neuber, H., "Theory of Notch Stress: Principles for Exact Stress Calculation," Translation 74, David Taylor Model Basin, 1945.
58. Frocht, M. M., Guernsey, R., Jr., and Landsberg, D., "A Photoelastic Re-Examination of Notched Tension Bars," Jnl. Appl. Mech., 19 (1), 124, March 1952.
59. Frocht, M. M., "Photoelastic Studies in Stress Concentration," Mech. Engr., 58 (8), 485–490 (1936).
60. Love, A. E. H., *Mathematical Theory of Elasticity*, Cambridge Univ. Press, London, England, 4th Ed., 1934, p. 519–31.
61. Byrne, R., Jr., "Theory of Small Deformations of a Thin Elastic Shell," Univ. Cal., Pub. in Math., new series, V 2, 1944, p. 103–52.
62. Green, A. E., and Zerna, W., "The Equilibrium of Thin Elastic Shells," *Qrtly J. Mech. and Appl. Math.*, V 3, part 1, Mar 1950, p. 9–22.
63. Reissner, E., "Note on the Expressions for the Strains in a Bent Thin Shell," *American J. of Math.*, V 64, 1942, p. 768–72.
64. Reissner, E., "Stress-Strain Relations in the Theory of Thin Elastic Shells," *J. Math. and Phys.*, V 31, Jul 1952, p. 109–19.
65. Langhaar, H. L., "A Strain-Energy Expression for Thin Elastic Shells," *J. Appl. Mech.*, June 1949, p. 183–9.
66. Basset, A. B., "On the Extension and Flexure of Cylindrical and Spherical Thin Elastic Shells," *Phil. Trans. Royal Society*

(London), series A, V 181, 1890, p. 443–80.
67. Flugge, W., "Die Stabilitat der Kreiszylinderschale," *Ingenieur Archiv*, V 3, 1932, p. 463–506.
68. Kennard, E. H., "The New Approach to Shell Theory: Circular Cylinders," *J. Appl. Mech.*, V 20, Mar 1953, p. 33–40.
69. Kennard, E. H., "Cylindrical Shells: Energy, Equilibrium, Addenda and Erratum," *J. Appl. Mech.*, V 22, Mar 1955, p. 111–16.
70. Kennard, E. H., "Approximate Energy and Equilibrium Equations for Cylindrical Shells," *J. Appl. Mech.*, V 23, Dec 1956, p. 645–6.
71. Salerno, V. L., and B. Levine, "Buckling of Circular Cylindrical Shells With Evenly Spaced Equal Strength Circular Ring Frames, Part I," Polytechnic Inst. of Brooklyn, PIBAL Report no. 167, Apr 1950.
72. Hildebrand, F. B., E. Reissner, and G. B. Thomas, "Notes on the Foundations of the Theory of Small Displacements of Orthotropic Shells," NACA TN 1833, Mar 1949.
73. Epstein, P. S., "On the Theory of Elastic Vibrations in Plates and Shells," *J. Math. and Phys.*, V 21, 1942, p. 198–209.
74. Bleich, H. H., and F. DiMaggio, "A Strain-Energy Expression for Thin Cylindrical Shells," *J. Appl. Mech.*, Sep 1953, p. 448–9.
75. Langhaar, H. L., and D. R. Carver, "On the Strain Energy of Shells," *J. Appl. Mech.*, Mar 1954, p. 81–2.
76. Vlasov, V. S., "Basic Differential Equations in General Theory of Elastic Shells," NACA TM 1241, Feb 1951.
77. Osgood, W. R., and J. A. Joseph, "On the General Theory of Thin Shells," *J. Appl. Mech.*, Dec. 1950, p. 396–8.
78. Flugge, W., *Statik und Dynamik der Schalen*, Edwards Brothers, Inc., Ann Arbor, Mich., 1943.
79. Koiter, W. T., "A Strain-Energy Expression for Thin Elastic Shells—Discussion," *J. Appl. Mech.*, V 72, 1950, p. 107.
80. Pfluger, A., *Stabilitatsprobleme der Elastoatatik*, J. Springer, Berlin, Germany, 1950.
81. Haywood, J. H., and L. B. Wilson, "The Strain-Energy Expression for Thin Elastic Shells," *J. Appl. Mech.*, Dec 1958, p. 546–52.
82. Boley, B. A., and Weiner, J. H., *Theory of Thermal Stresses*, 1960, John Wiley, p. 262–9.
83. Sokolnikoff, I. S., *Mathematical Theory of Elasticity*, 1956, McGraw-Hill, p. 377–465.
84. Bijlaard, P. P., "Stresses from Local Loading in Cylindrical Pressure Vessels," *Trans. ASME*, Aug 1955, p. 805–16.
85. Cowper, G. R., and H. A. Tulloch, "Tables for the Stress Analysis

of Uniform Circular Rings," Mech. Engr. Rep. MS-110, Aug 1964, Natl. Aero. Estab., Natl. Res. Council of Canada, Ottawa, Canada.

86. Wignot, J. E., H. Combs, and A. F. Ensrud, "Analysis of Circular Shell-Supported Frames," NACA TN 929.

87. Kempner, J., and J. E. Duberg, "Charts for Stress Analysis of Reinforced Circular Cylinders under Lateral Loads," NACA TN 1310.

88. Bynum, D., Jr., and R. C. DeHart, "Experimental Stress Analysis of a Model of the Alvin Hull," *Final Report*, 3 Apr. 1963, Proj. No. 1206-3, Southwest Research Institute, 8500 Culebra Road, San Antonio, Texas.

89. *Alcoa Aluminum Handbook*, 1959, Aluminum Co. of America, Pittsburgh, Pa., p. 72.

90. *Modern Plastics Encyclopedia*, Vol. 35, No. 1A, Sept 1957, 575 Madison Ave., New York 22, N.Y., p. 83, 315.

91. Timoshenko, S., and J. N. Goodier, *Theory of Elasticity*, 2nd Ed., McGraw-Hill Book Co., 1951, p. 359.

92. Gerdeen, J. C., "Effects of Pressure on Small Foil Strain Gauges," *Exp. Mech.*, Vol. 3, No. 3, Mar. 1963, p. 73–80.

93. Peterson, R. E., *Stress Concentration Design Factors*, 1953, John Wiley & Sons, Inc., New York, p. 93.

94. Bynum, D., Jr., and M. M. Lemcoe, "Birefringent Coating Analysis of Laterally Loaded Perforated Plates," *Exp. Mech.*, June 1963, p. 140–147.

CHAPTER

6 STATIC STRUCTURAL DESIGN

6.1 DESIGN OF SKEWED PLATES

The use of skewed plates is highly desirable in certain types of problems where the possibility of exciting the resonant frequency of the plates with dynamic loading must be minimized.

Certain aspects of the theoretical behavior of skewed plates have been analyzed. Williams[1], Lardy[2], Tungl[3] and, more recently, Morley[4] have derived the relationships governing the theoretical behavior of plates having particular aspect ratios of plate angle to plate width. The only numerical data published, aside from Morley's results, have been obtained by Favre[5] and Jensen[6], and presented by Timoshenko.[7] The design curves for the moment and rigidity coefficients presented are based on a compilation of the numerical results given by Morley and Timoshenko.

Sketches of the plate edge conditions and the relation for plate weight are given on the nomograms. The maximum moment on the plate is given by:

$$M_{max} = \beta q a^2$$

TABLE 104
Nomenclature:

a =	plate edge length, or height of plate, inches
E =	modulus of elasticity, psi
F =	safety factor
q =	uniform lateral load per unit area, psi
t =	thickness of plate, inches
W =	weight, lb
Z =	section modulus, cu in
α =	rigidity coefficient
β =	moment coefficient
δ =	deflection, inches
ν =	Poisson's ratio
ρ =	density, lb/cu in
σ =	maximum principal stress, psi
ψ =	acute angle of plate, deg

Combining the relation for the section modulus $Z = t^2/6$ yields the expression for the maximum principal stress given on the stress monogram. The deflection is given by the relation

$$\delta = \alpha q a^4/D$$

where

$$D = Et^3/[12(1 - v^2)] \tag{6.1}$$

Substituting a value of 0.3 for Poisson's ratio yields the expression for the deflection given on the deflection nomogram. This simplification causes less than 1% error when plates of steel or aluminum are designed using this expression. Considering this error and the error resulting from simplifications in the numerical analysis for the moment and rigidity coefficients, the solution obtained from the nomogram relations deviates from the true theoretical values by less than 4%.

The limits for the aspect ratio of plate width to thickness given on the nomograms were necessitated by assumptions used in the theoretical derivations. The theoretical relations are not valid for very small t/a ratios since the plate would act as a membrane. Also, shearing stresses are significant for plates having very large t/a ratios, and the shear stress was not considered in the theoretical derivations.

Example: Find the required thickness and yield strength of a rhombic, aluminum plate having an edge length of 50 in., an acute angle of 50 degrees, simple supports at each of the four edges and a maximum deflection of 0.1 in. when the plate is uniformly loaded with a lateral pressure of 1000 psi. A safety factor of 1.5 on the yield strength is used to specify the plate material.

Solution: The problem corresponds to Case 1. Using Nomogram 2, enter at $\psi = 50$, projecting upward to intersect the curve for Case 1 and projecting horizontally to the α scale. Align this intersection with $E = 10.5 \times 10^6$, intersecting reference line 1. Align $\delta = 0.1$ with $q = 1000$, intersecting reference line 2. Align the intersections on reference lines 1 and 2, intersecting reference line 3. Align this intersection with $a = 50$, and read $t = 4.8$ in. Using Nomogram 1, enter at $\psi = 50$, projecting to Case 1, thence horizontally to the β scale. Align this intersection with $t = 4.8$, intersecting reference line 1. Align $a = 50$ with $q = 1000$, intersecting reference line 2.

Align the intersections on reference lines 1 and 2, and read $\sigma = 24{,}000$ psi (figures 6.1, 6.2). Thus, with a safety factor of 1.5, the material must have a minimum yield strength of 36,000 psi.

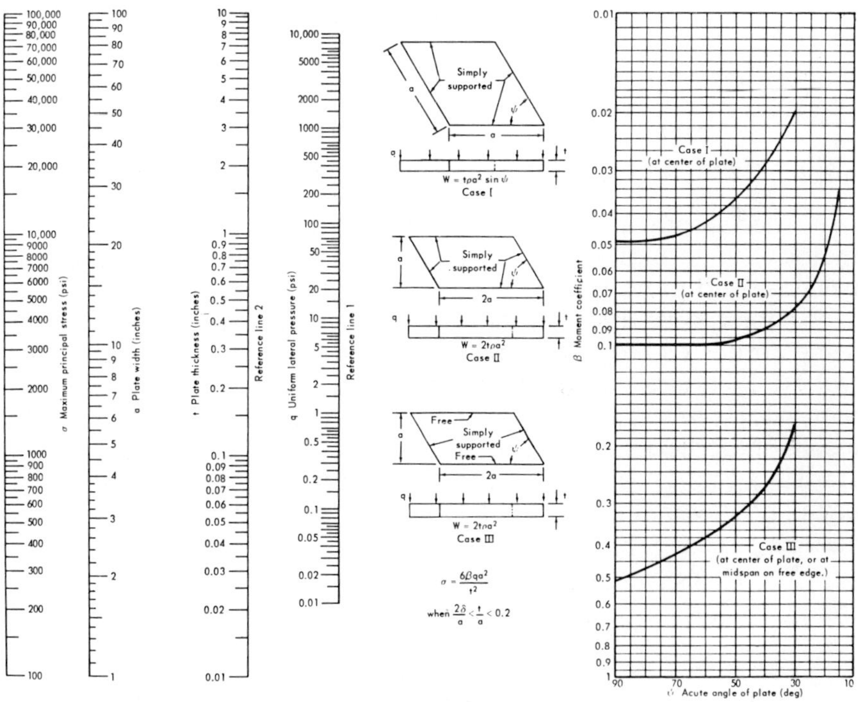

FIG. 6.1 Nomogram 1—stress

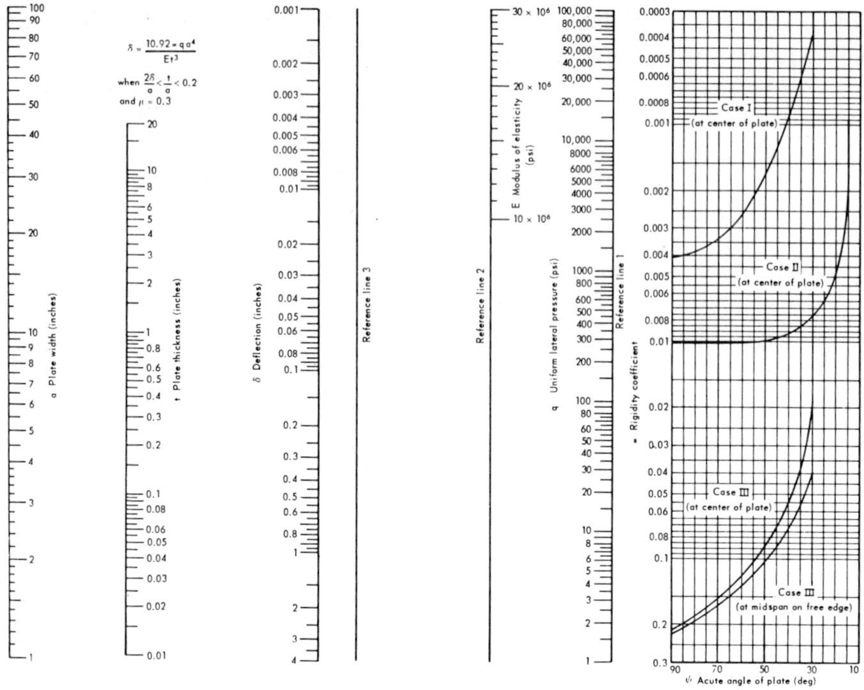

FIG. 6.2 Nomogram 2—deflection

6.2 DESIGN OF FINITE PLATES WITH FORCE-FITTED CYLINDERS

The accompanying nomogram applies to the design of symmetrical, finite plates with force-fitted solid, circular cylinders, when cylinder and plate have the same elastic constants.

Photoelastic investigations using "Araldite" plastic give the same results as the Lame solution for stresses in concentric, thick-walled cylinders having a proportional interference at their common radius when the aspect ratio is between 0.2 and 0.6.[10] The results indicated maximum stress decreased nonlinearly with increasing aspect ratios between 0.6 and 0.7. To simplify the nomogram, however, it is constructed to give constant values of stress for this range (0.6–0.7).

The nomogram solves the equation

$$\sigma_M = iE[(D/W)^2 + 1]/D \tag{6.2}$$

For an infinite plate (D/W = 0), with the cylinder and the plate having the same elastic properties, the theoretical solution for the maximum stress in the plate reduces to:[11]

TABLE 105
Nomenclature:

i =	radial interference, inches
D =	diameter of hole, inches
W =	width of plate, inches
μ =	Poisson's ratio
E =	modulus of elasticity, psi
σ_M =	maximum stress, psi
D/W =	aspect ratio

$$\sigma_M = 2Ei/D(1 + \mu)$$

Example: Find the maximum stress in a 2-in.-wide steel plate when the hole diameter is 0.90 in. and the radial interference between the hole and a forced-fit cylinder is 0.001 in.

Solution: Align D = 0.90 (lower right) with W = 2.0, intersecting D/W = 0.45. Project vertically upward to the curve, thence horizontally to reference line 1 (figure 6.3). Align this intersection with D = 0.9 (left), intersecting reference line 2. Align this intersection with E = 30 × 10^6, intersecting reference line 3. Align this intersection with i = 0.001, intersecting σ_M = 40,000 psi.

Conversely, the radial interference could be found, if the geometry and stress level were given, by connecting the given stress level point with the intercept on reference line 3.

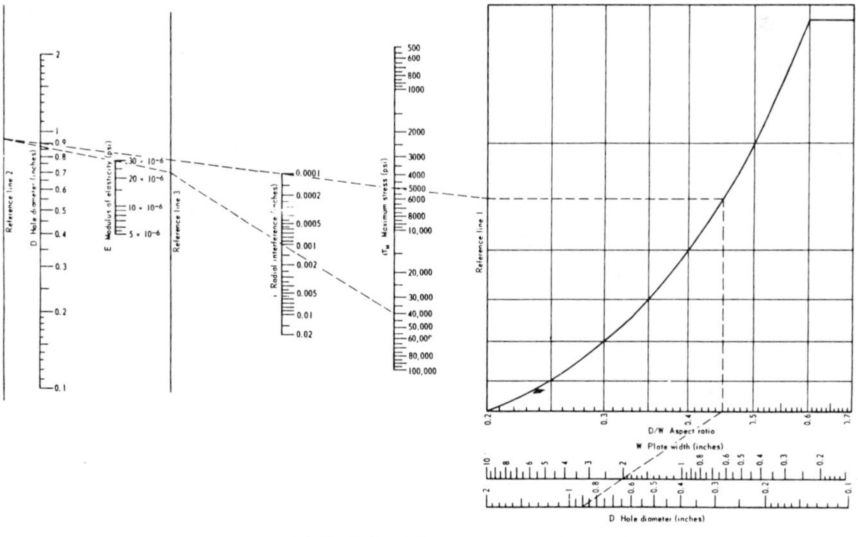

FIG. 6.3 Nomogram 1

6.3 FILLET AND GROOVE DESIGN FACTORS

Structural discontinuities such as holes, fillets, and notches usually cause a local stress intensity at the discontinuity with values several times the mean stress. These stress intensities must be known to insure adequate design.

The stress concentration factors for many types of structural discontinuities have been given.[(12–15)] The factors for the particular cases covered in this article are not available, however, from the applicable published literature.

The particular cases of structural discontinuities recently investigated[(16)] were fillets and grooves in specimens subjected to uniaxial and biaxial loads. The test specimens subjected to uniaxial loads were nonsymmetrical. For the studies of discontinuities with biaxial loads, a cylindrical vessel was subjected to internal hydraulic pressure to give a 2 to 1 biaxial stress field. The vessel is shown in figure 5.12. The 1-in.-wide test specimens that were subjected to uniaxial loads are shown in figure 5.13.

The birefringent coating technique was the primary means of experimental stress analysis used in the studies and was complemented with data from electrical resistance type of strain gauges. See Reference (17) for details concerning the birefringent coating technique.

The experimentally determined maximum stress concentration factors, based on the nominal stress in the minimum section, for the case of biaxially applied loads are listed in table 107.

TABLE 106
Nomenclature:

p	=	internal pressure, psi
r	=	radius of fillet or groove, inches
SCF	=	stress concentration factor
D	=	inside shell diameter, inches
t_{max}	=	maximum thickness, inches
t_{min}	=	minimum thickness, inches
σ	=	principal stress, psi
σ_H	=	hoop stress, psi
σ_M	=	meridional stress, psi
σ_y	=	yield stress, psi
F	=	force, lb
A	=	area, sq in
w	=	width, inches
SF	=	safety factor

TABLE 107

Discontinuity	SCF Hoop	SCF Meridional	$\frac{r}{t_{min}}$	$\frac{t_{max}}{t_{min}}$
Fillet	0.90	1.36	1	2
Groove	1.11	2.66	1	2

TABLE 108

Continuous Surface of Cylinder	$\frac{D}{t_{max}}$	$\frac{t_{min}}{t_{max}}$	$\frac{r}{t_{max} - t_{min}}$	Variation
Middle	117	0.4	1	5.9
Inner	117	0.4	1	11.8
Middle	28	0.4	=0	10.5
Inner	28	0.4	=0	10.9
Outer	28	0.4	=0	7.6

In recent tests of cylinders with reduced sections of various geometry, Morgan and Bizon(18) found the percent variations of experimental stress compared to the hoop stress given by thin shell theory as listed in table 108.

Now, since the ratio of the stresses in the walls of a cylindrical pressure vessel obtained by thick and thin shell theory is

$$\sigma_{thick}/\sigma_{thin} = 1.10 \tag{6.3}$$

when the ratio of the shell diameter is

$$D_{outside}/D_{inside} = 1.2 \tag{6.4}$$

the experimental data were used to formulate the following equation for general use of cylindrical shells having a reduced section along a meridian:

$$\begin{aligned}\sigma_y &= (1.2)(1.11)[pd(SF)/2t_{min}] \\ &= 0.62[pd(SF)/t_{min}]\end{aligned} \tag{6.5}$$

when $D_o/D_i \gtreqless 1.2$

$$2 \gtreqless t_{max}/t_{min}$$
$$1 \gtreqless r/t_{min}$$

Inner, middle or outer surface is continuous.

For conservative, general design of pressure vessels having a circumferential groove, the following equation can be used pending the development of applicable theory or the generation of additional experimental data:

$$\sigma_Y = (1.1)(2.66)[pd(SF)/4t_{min}]$$
$$= 0.73[pd(SF)/t_{min}] \tag{6.6}$$

when $D_o/D_i \lesseqgtr 1.2$

$$2 \lesseqgtr t_{max}/t_{min}$$
$$1 \lesseqgtr r/t_{min}$$

Inner surface is continuous.

For the case of uniaxially applied loads, the experimentally determined stress concentration factors[16], based on the nominal stress in the minimum section, are those shown in table 109. These stress concentration factors can be used for general, conservative design, pending additional available data, when the geometry parameters are:

$$1 \lesseqgtr (r/t_{min}) \text{ and } (t_{max}/t_{min}) \gtreqless 2$$

These stress concentration factors are considerably higher than the comparable maximum stress concentration factors for uniaxially applied loads on symmetrical specimens having two discontinuities, given by Heywood[12] (table 110).

Example 1: Given a cylindrical pressure vessel subjected to internal pressure, with a circumferential groove in the shell and the geometry D = 50 in., t_{max} = 1 in., t_{min} = 0.5 in. and r = 0.5 in. If the

TABLE 109

Discontinuity	SCF	$\frac{r}{t_{min}}$	$\frac{t_{max}}{t_{min}}$
Fillet	1.88	1	2
Groove	2.67	1	2

TABLE 110

Discontinuity	SCF	$\frac{r}{t_{min}}$	$\frac{t_{max}}{t_{min}}$
Fillet	1.4	0.5	2
Groove	1.7	0.5	2

yield strength of the material is 50,000 psi and if a safety factor of 1.5 is used based on the yield strength, determine the allowable pressure.

Solution: Using Eq. 2:

$$p = \frac{(50{,}000)(0.5)}{(0.73)(50)(1.5)} = 457 \text{ psi} \tag{6.7}$$

Example 2: Given a nonsymmetrical bar, subjected to a uniaxially applied load, with a groove and having the geometry $r = 0.5$ in., $t_{max} = 1$ in., $t_{min} = 0.5$ in. and $w = 1$ in.

If the yield strength and the safety factor are as given in Example 1, determine the allowable applied force.

Solution:

$$F = \frac{(\sigma_Y)(w)(t_{min})}{(SF)(SCF)}$$

$$= \frac{(50{,}000)(1)(0.5)}{(1.5)(2.67)} = 6250 \text{ lb} \tag{6.8}$$

6.4 PLASTIC STRESS ANALYSES

Assuming the principal strains in a structure are known, this section gives a simplified procedure for calculating the plastic state of stresses. The strains could be determined from strain gauges. The procedure is applicable only for the first application of load beyond the yield strength.

Observe from figure 6.4 that with increasing load along path OA into the plastic regime of a pristine structure, a strain ε_1, is indicated. However, as the load is relieved, the stress-strain path will be AB with a resultant plastic deformation of OB. The second time that load is applied, the stress-strain path will be along BA. Hence, it is obvious that if plastic analyses from future structural tests are to be meaningful, all tests must be within the elastic regime except for the last test, which would logically be for the most critical loading conditions. The last test of a particular structure would then be closely monitored and guided through the yield region to ultimate loads so that true, varying margins of safety could be determined for the plastic response.

The true margin of safety based upon ultimate loads for a complex structure vary *nonlinearly* and could *oscillate*, even though the margin, in general, decreases with increasing load (figure 6.5). The nonlinearity is due to the intrinsic material character (property). The oscillation results from strain hardening and *changing* points

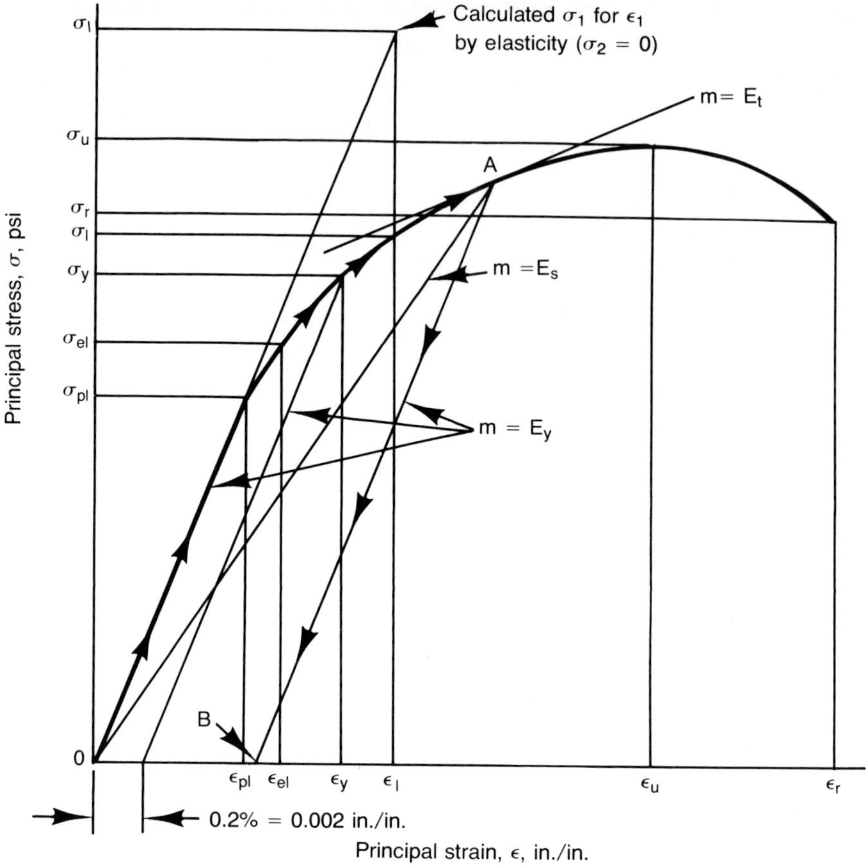

FIG. 6.4 Intrinsic topography

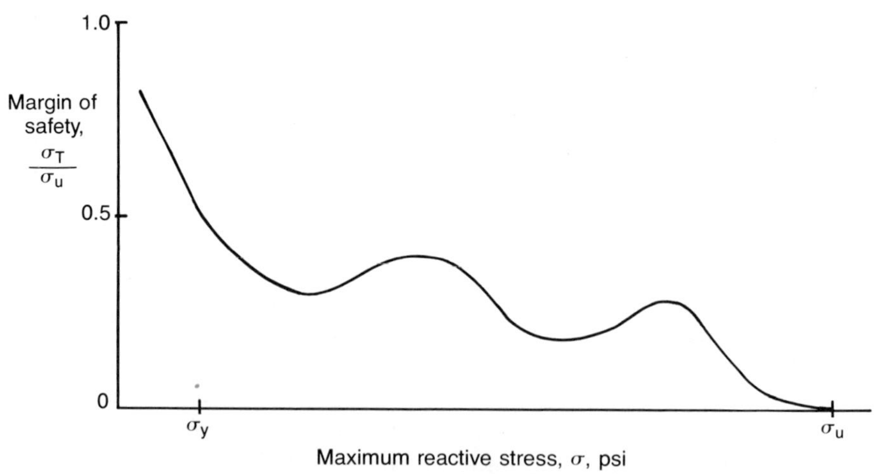

Fig. 6.5 Variation in margin of safety

of maximum stress. Even though the structure will initially yield at the point of maximum stress, a redistribution of the stress state because of this yielding could under certain conditions cause the structure to yield at a different point (or area) with an increase in load.

Published literature on procedures for conducting plastic analysis with given strains should be available. However, the only applicable document known to the author is a technical note by Ades.* The generally available textbooks on plasticity seem to be restricted

TABLE 111
Nomenclature

Symbol	Variable	Typical units
E_Y	Young's modulus, σ/ε for $\sigma \leq \sigma_{PL}$	psi
E_t	Tangent modulus, slope of the tangent to the σ vs ε curve at some particular ε value	psi
E_s	Secant modulus, slope of the line originating at the origin and intersecting the σ vs ε strain curve at some particular ε value	psi
σ_{el}	Elastic limit, maximum value of stress on the σ vs ε curve without plastic deformation	psi
σ_{pl}	Proportional limit of linearity on σ vs ε curve	psi
σ_Y	Yield stress, stress value corresponding to the intersection of the σ vs ε curve with a line whose slope is E_Y and which originates at $\varepsilon = 0.002$ in./in.	psi
σ_{rs}	Rupture strength, stress value corresponding to maximum strain on the σ vs ε curve	psi
σ_u	Ultimate strength, maximum value of stress on the σ vs ε curve	psi
ε	Principal strain	in./in.
μ	Poisson's ratio	numeric
m	Slope	—
V	Functional	—
c, $\bar{c}$, k	Constants	—
B	Biaxial stress ratio, σ_2/σ_1	numeric

Subscripts

1, 2	principal directions
E	elastic
P	plastic
T	total

* Ades, C. S., "Reduction of Strain Results in the Plastic Range," J. Aero/Space Sc., June 1959.

to purely theoretical matters, such as limit analysis based on the St. Venant model.

A part of the basis for Ades' philosophy (i.e., assumptions) is not clear. Further, he utilizes the secant modulus, whereas it appears the variation of the tangent modulus would be more appropriate. Also, Ades' procedure makes use of an iterative process, while the procedure given here is direct and requires less computer time for solution. This procedure may or may not give a solution more accurate than the Ades' approach and is not precise, but it is being offered as a simple approximate solution to the problem.

Before proceeding, the basic definitions from the theory of strength of materials, given in table 111 and in the intrinsic topography (figure 6.4), might be reviewed for advantage.

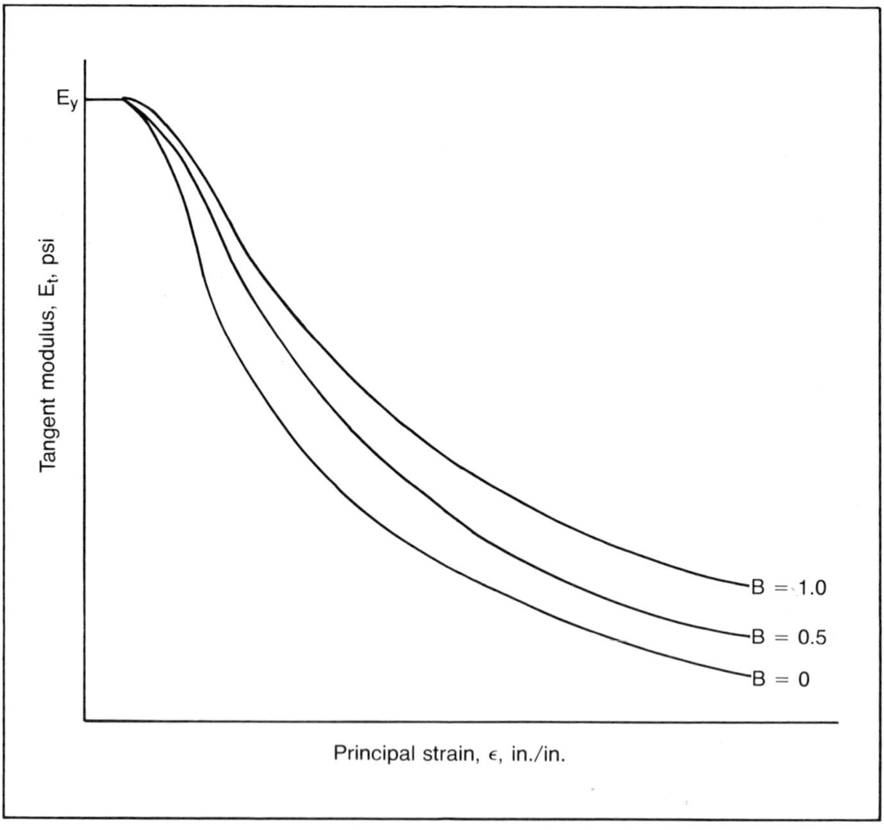

FIG. 6.6 Derived modulus

Development

Assumptions:

1. Consider a state of plane stress.
2. The material is homogeneous.
3. The material may be anisotropic.
4. Hooke's law is applicable for the elastic regime.
5. The σ vs ε curve is the same for tension and compression.
6. Poisson's ratio is 0.5 at the ultimate strength of the material.

Hooke's law is:

$$\varepsilon_1 = \frac{\sigma_1}{E_1} - \mu_1 \frac{\sigma_2}{E_2} \tag{6.9a}$$

$$\varepsilon_2 = \frac{\sigma_2}{E_2} - \mu_2 \frac{\sigma_1}{E_1} \tag{6.9b}$$

Solving equations (6.9) simultaneously for the stresses gives:

$$\sigma_1 = \frac{E_1 \varepsilon_1}{1 - \mu_1 \mu_2} + \frac{\mu_1 E_1 \varepsilon_2}{1 - \mu_1 \mu_2} \tag{6.10a}$$

$$\sigma_2 = \frac{E_2 \varepsilon_2}{1 - \mu_1 \mu_2} + \frac{\mu_2 E_2 \varepsilon_1}{1 - \mu_1 \mu_2} \tag{6.10b}$$

If the material is isotropic and completely elastic, equations (6.10) reduce to the more familiar form:

$$\sigma_1 = \frac{E}{1 - \mu^2}(\varepsilon_1 + \mu \varepsilon_2) \tag{6.11a}$$

$$\sigma_2 = \frac{E}{1 - \mu^2}(\varepsilon_2 + \mu \varepsilon_1) \tag{6.11b}$$

It will be convenient to write the total stress for the plastic state as

$$\sigma_{1T} = \sigma_{1E} + \sigma_{1P} \tag{6.12a}$$

$$\sigma_{2T} = \sigma_{2E} + \sigma_{2P} \tag{6.12b}$$

where σ_{1E} and σ_{2E} are given by equations (6.11), using

$$E = E_y \tag{6.13a}$$

$$\mu = \mu_E \tag{6.13b}$$

Using the stress-strain curve determined from a uniaxial tensile test ($B = 0$), the tangent modulus can be plotted (figure 6.6). Biaxial material test data are available less frequently, but the corresponding tangent modulus should also be plotted when such data is given. An appropriate equation for the family of curves in figure 6.6 can be determined using a polynominal representation,

$$E_t = (\bar{c}_0 + \bar{c}_1 \varepsilon + \bar{c}_2 \varepsilon^2 + \cdots)(K_0 + K_1 B + K_2 B^2 + \cdots) \tag{6.14}$$

where the number of constants chosen depends upon the desired accuracy of the curve fit. In order for E_t to degenerate to E_y for the uniaxial tensile, elastic state

$$K_0 = 1 \tag{6.15a}$$

$$\bar{c}_0 = E_y \tag{6.15b}$$

The rest of the constants in equation (6.14) can be readily determined by substituting various values of E_t with corresponding values of ε and solving the resultant algebraic equations simultaneously.

Since the variation of Poisson's ratio in the plastic regime is hardly ever available, let us derive an approximate relation. We know that Poisson's ratio is 0.5 for large plastic strains; therefore, a parabolic plastic variation is certainly reasonable, i.e., functionally,

$$y = a + b\sqrt{x} \tag{6.16a}$$

Assuming $\mu = \mu_E$ when $\varepsilon \leqq \varepsilon_y$ (6.16b)

equation (6.16a) becomes:

$$\mu_{P1} = \mu_E + (0.5 - \mu_E)\sqrt{\left|\frac{\varepsilon_1 - \varepsilon_y}{\varepsilon_u - \varepsilon_y}\right|} \tag{6.17a}$$

$$\mu_{P2} = \mu_E + (0.5 - \mu_E)\sqrt{\left|\frac{\varepsilon_2 - \varepsilon_y}{\varepsilon_u - \varepsilon_y}\right|} \tag{6.17b}$$

By definition, let

$$V_1 \equiv \int_{\varepsilon_y}^{\varepsilon_1} E_t d\varepsilon, \qquad V_2 \equiv \int_{\varepsilon_y}^{\varepsilon_2} E_t d\varepsilon \tag{6.18a, b}$$

where εy is the yield strain, which must have the same sign as the particular reactive strain. Using equations (6.14 and 6.15) in equation (6.18):

$$V_1 = E_y\left[\varepsilon + \frac{c_1 \varepsilon^2}{2} + \frac{c_2 \varepsilon^3}{3} + \cdots\right][1 + K_1 B + K_2 B^2 + \cdots] \tag{6.19a}$$

$$V_2 = E_y\left[\varepsilon + \frac{c_1 \varepsilon^2}{2} + \frac{c_2 \varepsilon^3}{3} + \cdots\right][1 + K_1 B + K_2 B^2 + \cdots] \tag{6.19b}$$

where $c_1 = \dfrac{\bar{c}_1}{E_y}$, $c_2 = \dfrac{\bar{c}_2}{E_y}$, etc. (6.20)

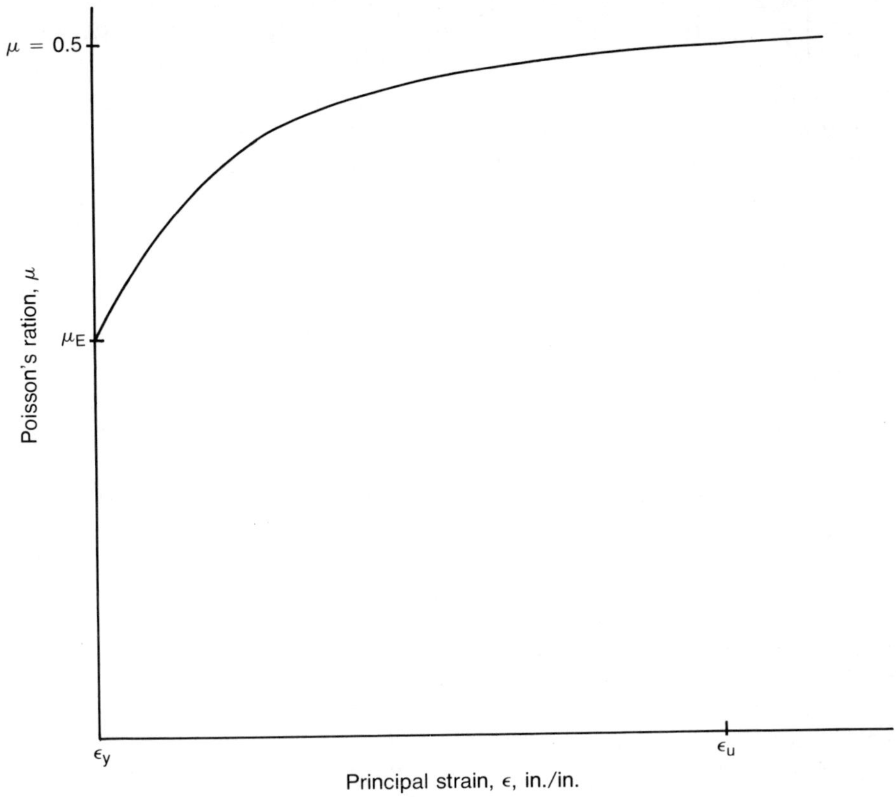

FIG. 6.7 Variation of Poisson's ratio

If the biaxial tangent modulus is not readily available, setting all the K constants to zero should still give a reasonable solution. Now using the V definitions, equations (6.10) for the plastic regime are

$$\sigma_{1P} = \left[\frac{1}{1 - \mu_{1P}\mu_{2P}}\right][V_1 + \mu_{1P}V_2] \tag{6.21a}$$

$$\sigma_{2P} = \left[\frac{1}{1 - \mu_{1P}\mu_{2P}}\right][V_2 + \mu_{2P}V_1] \tag{6.21b}$$

The total stress given by equation (6.12) can now be written explicitly with the conditional equations for use with digital computer operations so that the same base equations can be used for elastic and plastic strains. The base equations are:

$$\sigma_{1T} = \left[\frac{E_y}{1 - \mu_E^2}\right][\varepsilon_{1y} + \mu_E\varepsilon_{2y}] + \left[\frac{1}{1 - \mu_{1P}\mu_{2P}}\right][V_1 + \mu_{1P}V_2] \tag{6.22a}$$

$$\sigma_{2T} = \left[\frac{E_y}{1 - \mu_E^2}\right][\varepsilon_{2y} + \mu_E \varepsilon_{1y}] + \left[\frac{1}{1 - \mu_{1P}\mu_{2P}}\right][V_2 + \mu_{2P}V_1] \quad (6.22b)$$

where μ_P and V are given by equations (6.17) and (6.19), respectively. When ε_1 and ε_2 are known principal reactive strains (say, determined from the strain gauge rosettes), the conditions on equation 6.22 are as follows.

1. If $\varepsilon_1 < \varepsilon_y$, $\varepsilon_2 < \varepsilon_y$, then let
   $\varepsilon_{1y} = \varepsilon_1$, $\varepsilon_{2y} = \varepsilon_2$, $V_1 = V_2 = 0$ (6.23a)
2. If $\varepsilon_1 > \varepsilon_y$, $\varepsilon_2 > \varepsilon_y$, then use eqs. 6.22 as they are (6.23b)
3. If $\varepsilon_1 < \varepsilon_y$, $\varepsilon_2 > \varepsilon_y$, then let $\varepsilon_{1y} = \varepsilon_1$, $V_1 = 0$ (6.23c)
4. If $\varepsilon_1 > \varepsilon_y$, $\varepsilon_2 < \varepsilon_y$, then let $\varepsilon_{2y} = \varepsilon_2$, $V_2 = 0$ (6.23d)

Remember that ε_y must have the same sign as the associated reactive strain, ε_1 (or ε_2), in equations (6.19), (6.22), and (6.23).

6.5 PHOTOELASTIC OBSERVATIONS WITH POLYURETHANE

This work was originally done for designing the grain in solid-propellant types of missiles. The procedures can be used for improved design of blowout-preventer bags or other rubber products in the petroleum industry today. In developing the embedded-particle technique for obtaining measurements of internal displacements (and thus internal strains) in solid propellant grains,[(19–23)] an investigation[(24)] was undertaken to determine if the grain's deformational behavior was disturbed or distorted because of the very presence of the inclusions.

Since the particles were to be placed while the grain was being cast, it was fairly certain that residual deformations would surround each inclusion as the result of shrinkage during curing. The intensity and distribution of such residuals would be dependent on the particle material, size, and geometry. Spacing of the particles was also considered important. If the particles were too close to one another, the volumes of disturbed material around each inclusion would merge and tend to reinforce each other.

Finally, there was the question of whether or not the particles and the accompanying residuals would significantly interfere with, or distort, the load-induced displacement patterns.

Many of the cylindrical grains containing particles in transverse planes were formed by segmentally casting individual layers. Although the casting and curing procedures were carefully

TABLE 112
Nomenclature

C	= stress optical constant, psi-in./order
F	= force, lb
f	= fringe order, numeric
G_1	= instantaneous shear modulus, psi
G_2	= retarded shear modulus, psi
G_3	$= \dfrac{G_1 G_2}{G_1 + G_2}$, psi
h	= specimen thickness, in.
t	= time, sec
w	= specimen width, in.
α	$= \dfrac{G_2}{\eta_2}$
γ	= shear strain, μin./in.
η_2, η_3	= viscosities
σ_1, σ_2	= major and minor principal stresses, respectively, psi
τ	= maximum shear stress, psi

controlled to minimize the layer-to-layer interface, a discontinuity did exist. The extent to which the presence of these interfaces distorted the cylindrical grain's load-induced deformation behavior was also investigated.

The initial investigations were of a qualitative nature. Use was made of the fact that the transparent, unfilled binder used for several solid-propellant candidates exhibited birefringence (when viewed with polarized light) which was directly related to the principal strains. Specimens of Adiprene L were cast, and these specimens contained various types of inclusions; other cylindrical models contained segmentally cast interfaces and combinations of interfaces and particles. Subsequent studies of a preliminary nature were made to ascertain whether or not this material could be used for quantitative photoelastic investigations.

Previous Investigations of Photoelastic Materials Useful for Grain Design

The first published work concerned with determining the stresses in the grains of rocket engines seems to have been by Durelli, et al.[(25)] Later, Ordahl and Williams[(26)] determined certain stress-concentration factors for design purposes by observing the isochromatics in photoelastic specimens which were fabricated from CR-39 and loaded by means of a jig similar to the one used in previous work.[(25)] Table 113[(27)] gives some of the characteristics of some commonly used photoelastic materials, including the material studied in this investigation.

TABLE 113

Properties of selected photoelastic materials

Material	Stress-optical coefficient, psi-in./fringe	Elastic behavior	Machinability	Freedom from aging effects	Effect of temperature on physical and optical properties	Cost
Glass	Mean value 1500	Excellent	Poor	Yes	Little	Low
Celluloid	Variable 120–160	Poor	Fair	Yes	Pronounced	Low
Bakelite (BT-61-893)	89	Good	Fair	No	Little	High
Marblette						
As received	50	Poor	Fair	Almost	Pronounced	Moderate
Annealed	75	Poor	Fair	Almost	Pronounced	Moderate
CR-39	97–100	Good	Fair	No	Little	Moderate
Polyurethane	1.09	Good	Excellent	Yes	—	Low

The Thiokol Chemical Corp. has contributed much valuable information useful in the field of rocket engine design. Under a past contract with the Air Force, the investigators at Thiokol performed their initial tests[(28)] using photoelasticity techniques with specimens fabricated from CR-39. The investigators at Thiokol later performed a feasibility study[(29)] for determining the compatibility of various materials that could be used as a birefringent coating. The materials which they evaluated were:

1. Three Thiokol solithane polymer-castor oil formulations whose moduli of elasticity ranged from 160 to 500 psi
2. PhotoStress, having a modulus of elasticity of 350 psi and a Poisson's ratio of approximately 0.4
3. Hysol formulations, type 4485, having a modulus of elasticity of 450 psi and a Poisson's ratio of 0.465
4. Two Aero-jet-General formulations[(30)] designated LM-2 and IM-3, having a modulus of elasticity of 228 and 2680 psi, respectively, at 80°F.

The relative strain sensitivities of the materials tested are listed in table 114.

Upon testing the coatings on propellant bars at 0°F and then at 150°F, it was noted that a birefringence due to the difference of the coefficients of thermal expansion between the coating and the propellant was indicated with LM-2 and IM-3. Further tests[(31)] showed that Hysol 4485 has a coefficient of thermal expansion nearest in value to that of the propellant.

Considerable work[(32 and 33)] in evaluating the physical properties of Adiprene L has already been conducted, and the essence of these studies is presented here since the formulation used in this study so nearly approximated the formulation used in the previous tests. The mechanical properties of Adiprene L at room and elevated temperatures are given in table 115. The curing

TABLE 114

Material	Relative strain sensitivity, μin./in.-lb
Hysol 4485	62.4
Thiokol (53% polymer)	58.2
Aerojet IM-3	48.1
Thiokol (49% polymer)	35.3
Photo Stress	25.0
Aerojet LM-2	9.8

TABLE 115

Mechanical properties of adiprene L at elevated temperatures

Adiprene L	100 parts by weight
MOCA	11 parts by weight
Mix temp	212°F
Cure, hr at temp	3 hr at 212°F

Test temp, °F	75	158	212	250	302
Stress at 100% elongation, psi	950	825	800	800	550
Stress at 300% elongation, psi	2100	1650	—	—	—
Tensile strength, psi	4225	1925	1250	1200	675
Elongation at break, %	440	330	230	130	210
Hardness, Shore A	90	90	90	90	89
Graves tear, pli	515	485	390	330	303
Compression set, ASTM A, % 22 hr	0.6	1.6	3.7	19.5	—
Compression set, ASTM B, % 22 hr	—	20	72	100	100
Deflection at 400 psi					
1 hr, %	9	6.2	5.9	4.6	—
22 hr, %	9	5.4	7.6	17.1	—

agent used with the Adiprene L maintains its hardness and exhibits almost no creep with temperatures as high as 250°F. As is shown in table 115, the compressive set of Adiprene using ASTM Procedure A is 0.6 percent for a room temperature of 75°F and a test time of 22 hours.

Figure 6.8 shows the variation in tensile stress with elongation due to various hardnesses of vulcanate. The tensile stress-elongation curves for Adiprene and rubber are shown in figure 6.9. Natural rubber is, of course, a convenient reference material for comparing physical properties. Even though fillers can be used to increase the tensile stress of rubber for a particular elongation, the tensile stress of the modified rubber is still considerably lower than that of the Adiprene.

Figure 6.10 shows the variation of tensile stress at 100 percent elongation and the ultimate elongation of Adiprene as a function of hardness. Figure 6.11 gives the compressive stress at 10% strain for rubber and Adiprene as a function of hardness, and figure 6.12 gives the compressive stress of Adiprene and rubber as a function of compressive strain.

The condition of the load surfaces has a pronounced effect on the compressive stress–strain relationships for short elastomeric specimens. The highest compressive stress per unit compressive

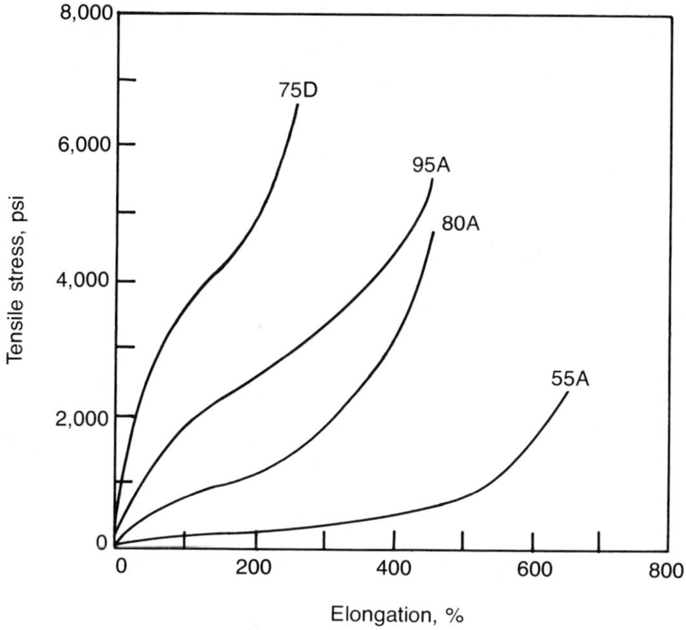

FIG. 6.8 Tensile stress-elongation function of Adiprene hardness

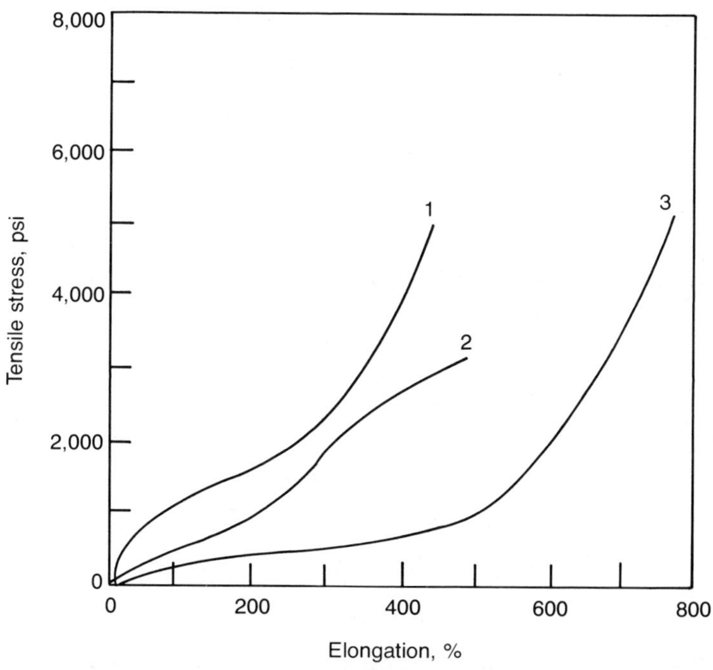

FIG. 6.9 Tensile stress-elongation of Adiprene and rubber. (1) Adiprene L (with no carbon black); (2) Natural rubber with 40 parts Haf Black; (3) Natural rubber with no carbon black

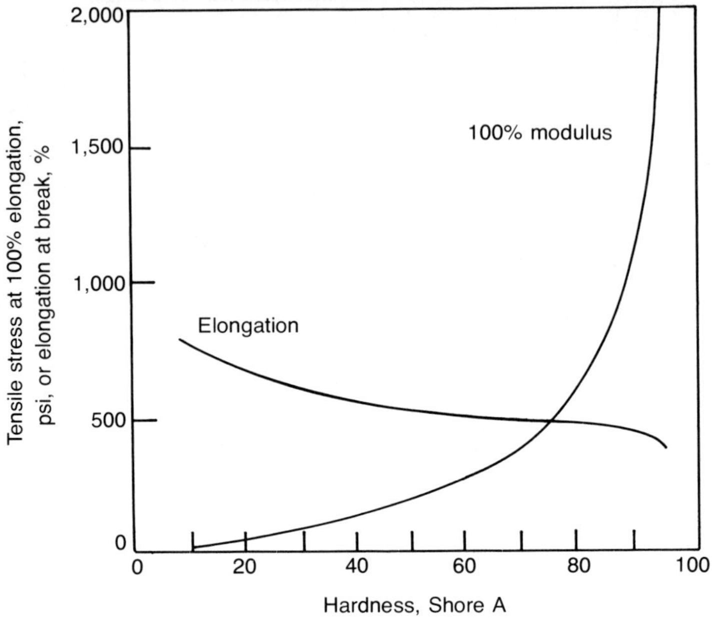

FIG. 6.10 Ultimate elongation function of Adiprene hardness

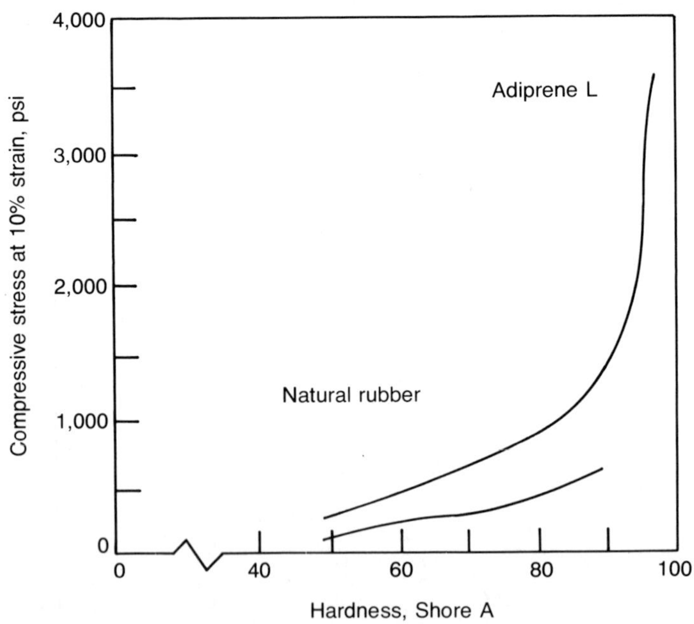

FIG. 6.11 Compressive-stress function of Adiprene hardness

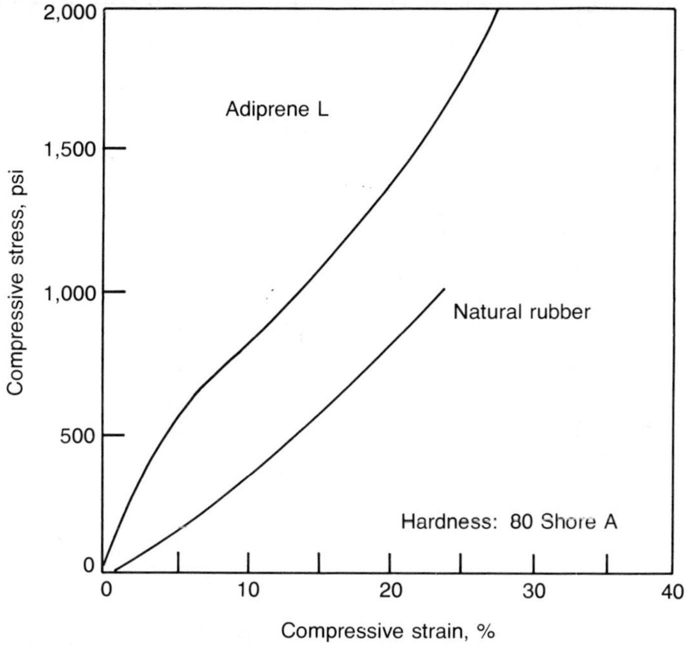

FIG. 6.12 Compressive stress-strain of Adiprene and rubber

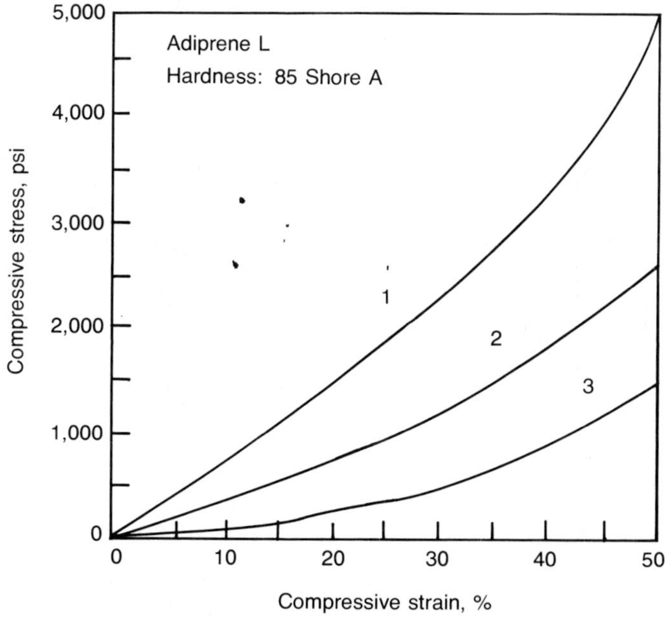

FIG. 6.13 Compressive stress-strain function of load surface. (1) Load surface bonded to metal plates; (2) Load surface unbonded, clean and dry; (3) Load surface unbonded, lubricated

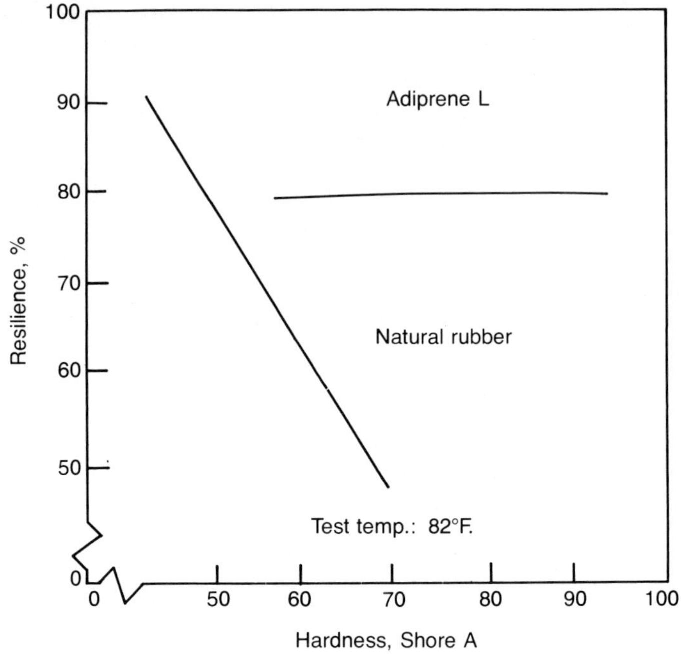

FIG. 6.14 Resilience function of hardness for Adiprene and rubber

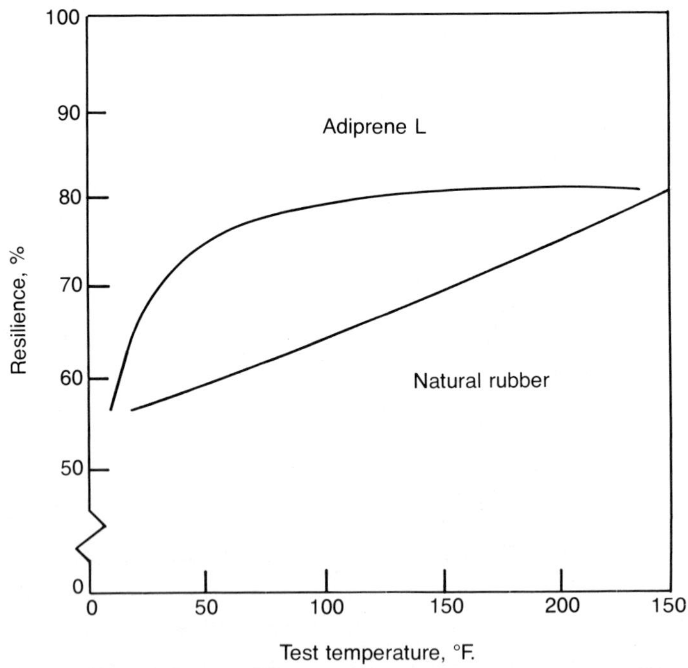

FIG. 6.15 Resilience function of temperature for Adiprene and rubber

strain is obtained when the elastomer is bonded to the load areas (figure 6.13).

The resilence of Adiprene and rubber is shown plotted as a function of hardness and temperature in figures 6.14 and 6.15, respectively. The resilience of Adiprene is approximately constant over the hardness range of 60–95 Shore A and over the temperature range of 85–230°F.

Experimental Procedure

The formulation of the material used in the test specimens consisted of 21 g of castor oil and 1.7 g of 1.4 butane diol per 100 g of Adiprene L. Each component was heated to a temperature of 180°F before mixing. The mixture was then degassed in a vacuum oven at 180°F. After pouring the polyurethane in the cylindrical, aluminum mold, the mix was again degassed at 180°F. The pot life was usually about 2 hours. The mixture was then cured in an oven at 285°F for 2 hours. Considerable details of the various casting techniques have been given by Miguel and Duran[34] and will not be reiterated here.

All of the test model grains had a 3-in. nominal outside diameter. Various types of particles were embedded in the Adiprene models (0.001-in. diameter wires, and 1/32-in. nominal diameter glass and lead pellets) so that the effect of the inclusions could be studied.

Several specimens were segmentally cast with oriented inclusions located over the segment interfaces. A segment was poured and allowed to partially polymerize so that the Adiprene would support the weight of the inclusion. Then, long before this segment became fully polymerized, the next segment was poured.

Discussion

Photoelastic Properties of Tolylene Di-isocyanate. The stress–strain phenomenological behavior of a linear viscoelastic material can be represented by mechanical models. One finite element model which appears to be of interest consists of combining in series a Kelvin model[35] to represent the delayed elasticity, a Hookean model to represent the instantaneous elasticity, and a Newtonian model to represent the flow mechanism (figure 6.16).[36] The resultant creep stress–strain–time relationship is then

$$\gamma = \frac{\tau}{G_1} + \frac{\tau}{G_2}(1 - e^{-\alpha t}) + \frac{\tau t}{\eta_3} \tag{6.24}$$

If a sufficiently long period of time is considered, the delayed elastic response in the Kelvin model can be neglected. Moreover,

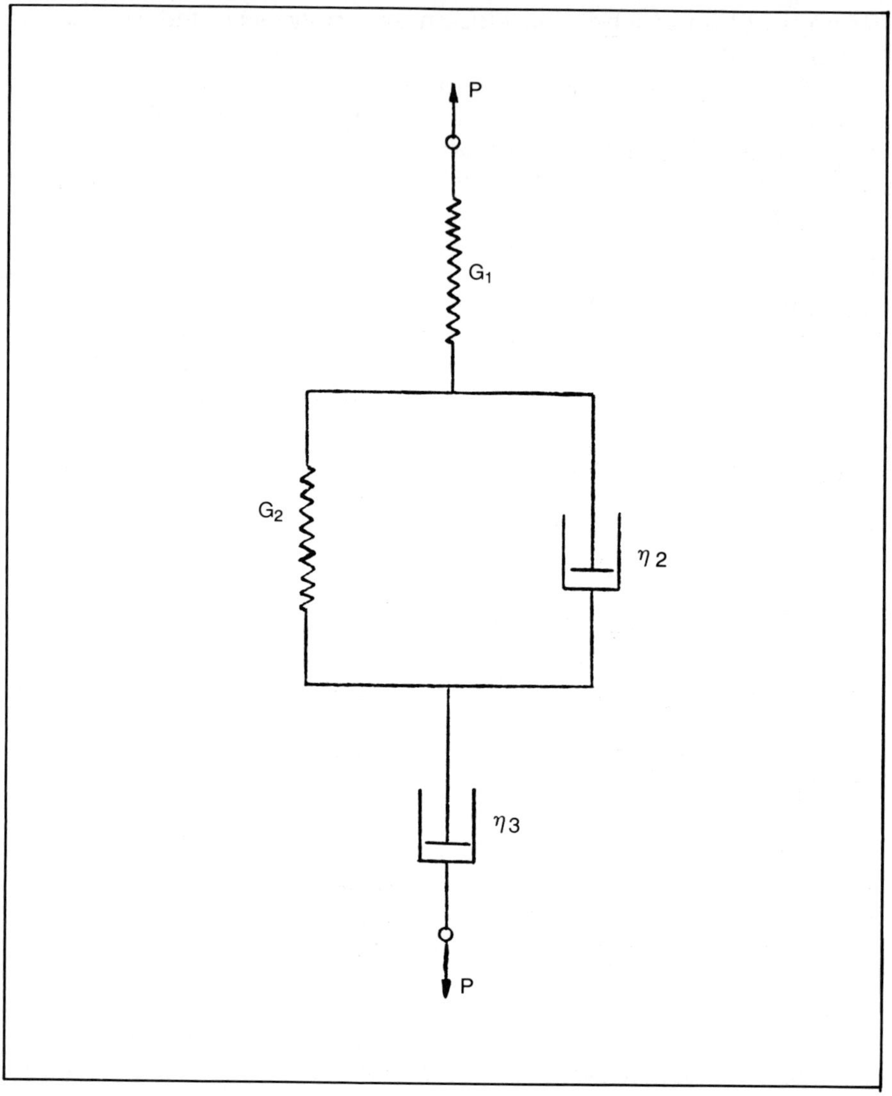

FIG. 6.16 Mechanical model representing an idealized viscoelastic material

if η_3 is very large at low temperatures (i.e., room temperature) in comparison to the product of stress and time, then the equilibrium stress–strain relation is simply

$$\gamma = \frac{\tau}{G_3}$$

The stress-optical constant of the polyurethane was determined by applying the stress-optic law

$$\sigma_1 - \sigma_2 = \frac{cf}{h}$$

to a square specimen uniformly loaded in compression. The measurements were taken after the specimen material reached its equilibrium state. Since $\sigma_2 = 0$ for a uniaxial compression load,

$$\sigma_1 = \frac{F}{hw}$$

$$c = \frac{hF}{fhw} = \frac{F}{fw}$$

The value of the stress-optical constant determined was 1.09 psi-in./order, so

$$\tau_{max} = \frac{\sigma_1 - \sigma_2}{2} = \frac{0.54f}{h} \text{ psi} \qquad (6.25)$$

In addition to the problems that would be involved if equilibrium stress–strain states were not considered, it appears that the only other significant limitation in using tolylene di-isocyanate as a photoelastic material is that thick sections could not be used in dynamic analysis. The thermal conductivity of urethane elastomers varies from 1.181 to 0.754 Btu/(hr)(sq ft)(°F/in.) for Shore Durometer hardness of 55A to 75D. Such low conductivities prevent adequate internal-heat dissipation, resulting from internal friction during high-frequency flexing, and premature failures of the specimens would most likely occur. This particular problem could be circumvented, however, by using specimens of thin sections for dynamic analysis.

The polyurethane used seemed to be free from any time-aging effects. Over 30 specimens were cast and machined, and no residual birefringence could be detected after a period of 6 months. The polyurethane was found to be easily machinable.

Conclusions

The investigations show that tolylene di-isocyanate polyurethane appears to be an excellent photoelastic material. Due to the similarity in the physical properties of the material compared to hard rubber, the material is ideally suited for photoelastic investigations of rubber products.

Further studies should be conducted with tolylene di-isocyanate to determine the variation of the stress-optical constant as a function of the time period of cure, cure temperature, strain rate, strain relaxation, and ambient test temperature. The stress-optical-constant function of strain rate and strain relaxation for polyurethane could best be determined experimentally in a fashion after Stein, et al.,[37] who ascertained the applicable parameters for low and high-density polyethylenes. These parameters cannot be directly correlated to polyurethane since this material is amorphous, whereas polyethylene has a crystalline structure. With the recent addition of specialized testing machines, including the availability of an Instron testing machine, the author tentatively plans to conduct such tests with polyurethane.

The shrinkage stresses developed during the cast appear to be of small magnitude. Accurate values of residual birefringence could be determined by utilizing photoelastic compensation, i.e., with a Soleil-Babinet compensator if a white-light source were used, with a Coker compensator, or by goniometric compensation (Tardy method) in conjunction with a full-wave plate. Very thin tensile-test specimens containing inclusions should be used to ascertain the stress-concentration effect due to inclusions of various geometries, material and location within the specimen.

6.6 HISTORY OF THE BIREFRINGENT COATING TECHNIQUE

The birefringent coating technique of experimental stress analysis was originated by Mesnager[38] in France in 1930. He was the first to suggest bonding a photoelastic material to a polished surface, and then relating the strains in the coating to the surface strains in the structure.

Mabboux[39] in France (1932) suggested using heterogeneous inclusions of photoelastic material in concrete structures and using a Norrenberg doubler to measure the relative retardation. This arrangement has limited applications since the inclusion merely acts as a single strain gauge. Oppel[40] in Germany (1937) determined notch stresses with a birefringent coating. The plastic which he used, however, was not sufficiently optically sensitive for general applications. D'Agostino[41] and his colleagues reported a birefringent coating technique in 1954, but their results were applicable only for flat parts.

Zandman, while in France, developed an epoxy formula and reported in 1953 that the plastic could be used satisfactorily on a practical basis for elastic or plastic ranges and for any size or shape. The Zandman method[42] has been used widely in Europe and was introduced in the United States in 1956. The plastic which Zandman discovered is now referred to as Photostress and is commercially available from the Budd Company.

Principles of the Birefringent Coating Analysis

Birefringent coatings complement the functions of conventional photoelasticity, brittle lacquer, and electric resistance strain gauges. The birefringent coating is, in a sense, analogous to an infinite number of strain gauges, since the whole field strains are immediately identifiable by the contrast of colors.

Basically, birefringent coating principles are the same as in photoelasticity. The mathematical formulas expressing the fundamental relationships used in photoelastic analysis have been thoroughly developed in the literature[43–47] and will not be reiterated here.

The primary difference between photoelasticity and the birefringent coating analysis is that a transparent model, such as glass or plastic, is utilized in photoelasticity, whereas the actual structure is tested in a birefringent coating analysis. When a transparent coating is applied to a structure, the coating follows the surface strains of the structure. With applied load, the coating exhibits a relative retardation of polarized light proportional to the strain because of the doubly refractive, or birefringent, characteristic of the coating Since the light waves are polarized at 90° to each other, corresponding to the principal stress directions, only the difference of principal stresses is obtained with normal incidence. Fortunately, there are several means of determining the individual principal stresses.

A common procedure for determining the individual principal stresses is by first determining the isopachics, the sum of the principal stresses. Knowing at a point, the isopachic and the isochromatic, the principal stress difference, the individual principal stresses can be readily calculated.

Another method for determining the individual principal stresses is by integrating Filon's transformations of the Lame'-Maxwell equations, as illustrated in Reference (44). The principal stress at a boundary equals the isochromatic, since the minor principal stress, perpendicular to and at the boundary, is zero in a plane stress problem. From a consideration of the boundary stresses, the isoclinics, and the isochromatics, a satisfactory solution may be obtained by graphical integration. Other means sometimes used for determining the individual principal stresses is the shear difference method and the slope equilibrium method, as illustrated in Reference (44).

Another common way to determine the individual principal stresses is by the consideration of oblique incidence. Mathematical relationships exist for finding the principal stresses from normal and oblique incidence measurements of relative retardation.

The shear stresses in a structure can be fairly accurately determined by visual observation of the birefringent coating and by utilizing a strain-stress color chart[48]. A method commonly used to more accurately determine whole field shear stresses is the Tardy Method[49] (goniometric compensation), which utilizes fringe shift from analyzer rotation. Beginning with a dark field circular polariscope, which has crossed polarizers and crossed quarter wave plates, rotate the analyzer 90° This causes the fringes to shift to a light field. An additional 90° rotation in the same direction causes the fringes to shift further to a dark field again. Knowing the fringe constant, f, the stress at some point between adjacent fringes can be determined within an accuracy of about f/50.

Fringe multiplication[50–52] is another method useful in determining whole field shear stresses. This technique has been successfully used in experimental work with birefringent coatings, but the limitations are inplane loading and studies of flat surfaces. The surface area being studied has to have a specular finish and be optically flat. This can be achieved by various electropolish or electroplating procedures. With a system of mirrors, the number of fringes can be increased as many as five times, dependent upon the quality of the specimen surface.

The Babinet Compensator is a means of adding a known birefringence, since the compensator is a calibrated quartz wedge which is optically flat on each side. The compensator constant is the number of divisions between successive tints of passage, which is usually considered the sharp distinction between the erubescent and blue colors.

The difference in wedge thickness for the number of divisions corresponding to the compensator constant is one wavelength of white light, 22.7×10^{-6} in. For an incremental change of load on the specimen, the tint of passage moves along the compensator. The number of divisions that the tint of passage moves on the compensator per unit load is used in the stress computations.

The K factor (quasi-stress optical constant) is ordinarily calibrated using readings of relative retardation in a birefringent coating, applied to a simple uniaxial tensile specimen or to a cantilever. Relative retardations in the prototype coating are related to the prototype stress using this constant and the prototype coating thickness. The coating thickness on the prototype is measured after contouring but before cementing to the prototype.

Critical Aspects of Birefringent Coating Analysis

Quality of Bond. The quality of the bond of the birefringent coating to the specimen is sometimes important. If the plastic and the specimens have the same geometry and the same end displacements, however, the relative retardation of the plastic is unaffected[53] by the quality of bond.

The manufacturer recommends that Photostress be heated in order to remove moisture and, of course, be allowed to cool before using. If moisture-free Photostress liquid turns a milky color when the hardener is added, the shelf life has been exceeded. This discoloration of the plastic makes readings of relative retardation impossible.

If a PhotoStress coating needs to be removed, the plastic can be pried from the specimen after being subjected to a thermal shock. Also, pure dimethyl formamide or nitromethane will buckle the plastic somewhat, enabling the plastic to be peeled from the specimen.

Thickness Effect. A small error exists in relative retardation measurements of birefringent coatings due to a thickness effect. The magnitude of error is directly dependent upon the coating thickness to specimen thickness ratio, but the exact magnitude of error from the thickness effect is presently unknown for two-dimensional strain distributions in general.

A study[54] has been made of the thickness effect with one-dimensional variations of strain, which obviously has negligible practical significance. A more recent paper[55] suggests a procedure for determining the thickness effect on biaxial surface strains in structures having radial symmetry, if the surface curvature is known or can be determined separately.

Reinforcement Correction Factor. The birefringent coating applied to a structure carries a small part of the applied load, and the surface strain of the structure is therefore reduced. The correction factors for this reinforcing effect have been derived and plotted in the literature[56] for several important types of loading.

An optimum birefringent coating thickness can be selected for combined loading of certain structures. The correction factor curves for primary and flexural stress vs. thickness ratio[56] intersect. So, for any particular structural thickness, a birefringent coating thickness can be selected so that the two types of correction factors are equal.

Poisson Contraction. The assumption that a birefringent coating has the same strain distribution as the surface strains of the structures is not exactly true, especially in the propinquity of discontinuities such as holes or notches or at a boundary.[53] If the coating is well bonded and the study area is remote from a discontinuity, the assumption is valid for thin coatings. Only a small error exists for thick coatings.

If the study area is near a discontinuity and the coating is thick, the deviation of relative retardation, $\Delta\delta$, due to the Poisson contraction for uniaxial tensile loading, for instance, is given by

$$\Delta\delta = \frac{\mu_S - \mu_C}{1 + \mu_S} \tag{6.26}$$

This deviation is small for most metals, even with thick coatings.

Residual Birefringence. Residual birefringence in a coating is usually of a permanent nature and is caused from improper machining procedure or careless handling of the specimen. If the point of interest is at a boundary, the residual stress can be readily calculated since the stress direction is known.

If the point of interest is not at a boundary and the residual stress is not excessive, compensator readings vs. load will give a curve which is initially nonlinear. With increased load, the curve becomes a straight line. The nonlinearity exists because the residual stress and the induced stress have different direction. The stress can be calculated from the slope of the straight portion of the curve.

A procedure exists for determining the stress at an interior point from a load, even when the residual birefringence approaches the value of relative retardation at some load which causes a stress just under the yield strength. Here the plot of relative retardation versus load would be completely nonlinear. The procedure is rather time consuming, and applying a new coating would be more expedient if many points of interest were involved.

Parasitic Birefringence. Parasitic birefringence is due to humidity absorption, creep, or delayed polymerization. A layer of adhesive is usually applied to the machined surfaces of the birefringent coating so that the humidity absorption will be negligible. If the additional adhesive is not applied, humidity absorption would cause the plastic to swell, causing a birefringence.

The plastics will creep if a load is steadily maintained for a prolonged period of time. A temporary parasitic birefringence would be exhibited with the load relaxation. Creep effects can, however, be accounted for, and the relative retardations adjusted accordingly.

Photostress polymerizes most in the first few hours after application; however, the manufacturer suggests that the adhesive be given at least 24 hours to set before using. During a month or so after application, the adhesive continues to polymerize slightly which causes a contraction of the coating. The parasitic birefringence from this delayed polymerization, as well as from humidity absorption, can be handled in the same manner as residual birefringence.

K Factor Temperature Variation. The K factor of Photostress varies in a nonlinear fashion with temperature, as shown in Reference (42). This effect must, therefore, be considered in problems concerning thermal stress analysis.

Thermal Effect. A thermal effect in a birefringent coating exists when the temperature varies. A relative retardation is indicated if the thermal coefficients of expansion for the prototype and the birefringent coating are different.

The coefficient of thermal expansion has been given by Zandman[(58)] as 20 (micro-in.)/(in.-°F) for 0.070-in. thick Photostress. Since the coefficient of thermal expansion for plastics ranges from 20 to 110 (micro-in.)/(in.-°F)[(59)], the figure given by Zandman appears low.

The coefficient of thermal expansion for Photostress has been experimentally determined[(60)] with a Gaertner Scientific Microscope, using a quartz rod with a known coefficient for instrument calibration. The gauge length was the distance between diamond scratch marks on the quartz and between scribed marks on the Photostress. The quartz and Photostress gauge lengths were measured at room temperature. Then the gauge lengths were measured in a refrigerated storage room. Of course, the microscope was allowed to reach thermal equilibrium before measurements were taken. Measurements were as follows:

Temperature, °F	76	79	42
	Gauge Length, L, in.		
Photostress	3.8270		3.8217
Quartz		3.6166	3.6174

The coefficient of thermal expansion for the Photostress from this data was 47 (micro-in.)/(in.-°F). The scribed marks were about 0.002 in. wide on the Photostress, and about 0.0005 in. wide on the quartz, but the point of observation was the midpoint of the scribed mark. The experimental error was estimated to be less than 5%.

A second kind of thermal stress can also be caused by applying the birefringent coating to a massive specimen not in thermal equilibrium. The thermal stresses of the second kind are of the same order and are governed by the same parameters as the first kind mentioned.

Parasitic Reflections. If a polariscope is oriented perpendicular to a birefringent coating during normal incidence readings, the near surface of the birefringent coating will reflect a portion of the light. This sometimes causes an interference in the field of view. To determine whether or not parasitic reflections are present during experiments, the compensator is removed from the field of view and the reticle is rotated 90° with respect to the analyzer, with a parallel polariscope resulting.[(61)] Parasitic reflections are indicated by a glare in the changing colors with applied load. The reduction of the glare varies with the angle of cant, and the parasitic reflections are apparently eliminated with a cant of about 7°. Disregarding the small integrating effect, a 7° cant causes an error of little more than 1% at the free boundaries, since the relative retardation measured varies directly as the cosine of the angle of cant.

Double Refraction Effect from Oblique Incidence. Plane polarized light can be obtained by reflection if the light source is obliquely oriented with the reflective surface.[(45)] Light is almost completely plane polarized if the angle of incidence is $\tan^{-1} n$, where n is the refractive index of the material. This critical angle of oblique incidence must be avoided as the optical path traverses the polariscope.

Conclusions

The technique of birefringent coating analysis is useful in determining:

a. The elastic or plastic strain distribution from static or dynamic loading.
b. Instantaneous detection of stress concentration areas.
c. Stress concentration factors around cracks, small holes, fillets, and notches.[(62)]
d. Residual strains or stress from plastic deformations.
e. Stresses in composite materials such as reinforced concrete.
f. Stresses in anisotropic materials.[(63)]
g. Stresses in and around the weld of dissimilar metals.
h. Stresses in and around heterogeneous inclusions.[(50)]

Some of the special advantages of birefringent coating analysis are as follows:

a. The size of the specimen can vary from a small crystalline grain to very large structures.
b. The tests can be telemetered.
c. Measurements can be automated by means of photocells and servomechanisms.
d. An analysis can be made in almost any atmospheric condition.
e. An analysis can be made with the specimen completely submersed in any transparent liquid such as water or oil.
f. With normal incidence readings, the strain (or stress) is obtained at a point rather than over a finite gauge length.

g. The individual principal stresses at a point can be obtained with great accuracy from normal and oblique incidence readings if the specimen is symmetrical.
h. If a specimen is not symmetrical, the integrating effect due to gauge length can be varied when the individual principal stresses are to be obtained.
i. The strains (or stresses) can be obtained exactly at the edge of free boundaries.

Disadvantages of the birefringent coating analysis are that extreme care must be exercised in order to obtain accurate results and that the experimentalist must have a working knowledge of the general principles of conventional photoelasticity to utilize the technique. Also, the correlation of prototype stress and indicated relative retardations sometimes becomes involved when several correction factors have to be applied.

6.7 DESIGN BY FRINGE MULTIPLICATION WITH MIRRORED BIREFRINGENT COATINGS

When performing photoelastic studies utilizing fringe multiplication,[66] sources of error usually encountered include the integrating effect and the scatter effects. The integrating effect due to strain averaging through thick coatings can be minimized by having the oblique angle of light incidence correspond with the minimum strain gradient. Light scatter due to the diffuse, mirrored birefringent coatings usually used can be minimized with a technique whereby the structure surface is electroplated to obtain a specular finish for use of unmirrored birefringent coatings.[67]

To experimentally determine the order of worst-case error, a curved beam was chosen since this structure has a large strain gradient and could be conveniently loaded in the given reflective polariscope. The reflective polariscope used in this study, and specially designed for use with fringe multiplication, has been described in Ref. (68). A thick coating (thick relative to the structure) was used, and the plane of oblique incidence was set along the maximum strain gradient. Fringe-multiplication data from the birefringent coating and photoelastic-model data were then compared to the approximate Winkler-Bach theory[69] for an estimate of the total error.

The curved beams were of constant cross section and 1.5 in. in height. The radius of curvature measured to the centroid of the beam cross section was 3.75 in., and the eccentricity measured from the load azimuth to the centroid was 4.75 in. The

CR-39 photoelastic model was 0.373 in. thick and the calibrated stress optical constant was 92 psi-in./fringe.

The 0.161-in.-thick aluminum model had a 0.122-in.-thick, mirrored birefringent coating with a K factor of 0.09. Two passages (conventional coating isochromatics, i.e., light traversed coating two times) were about the convex and concave boundaries.

The photoelastic-model data were permutated for thickness and then normalized and plotted with the normalized-coating data and Winkler-Bach results in figure 6.17. As shown by the plot, the photoelastic-model data are in close agreement with theory, but the concave boundary stresses obtained using two and four coating passages were about 16% and 29%, respectively, in error compared to theory. The analogous errors for the convex boundary with two and four passages were about 8% and 19%,

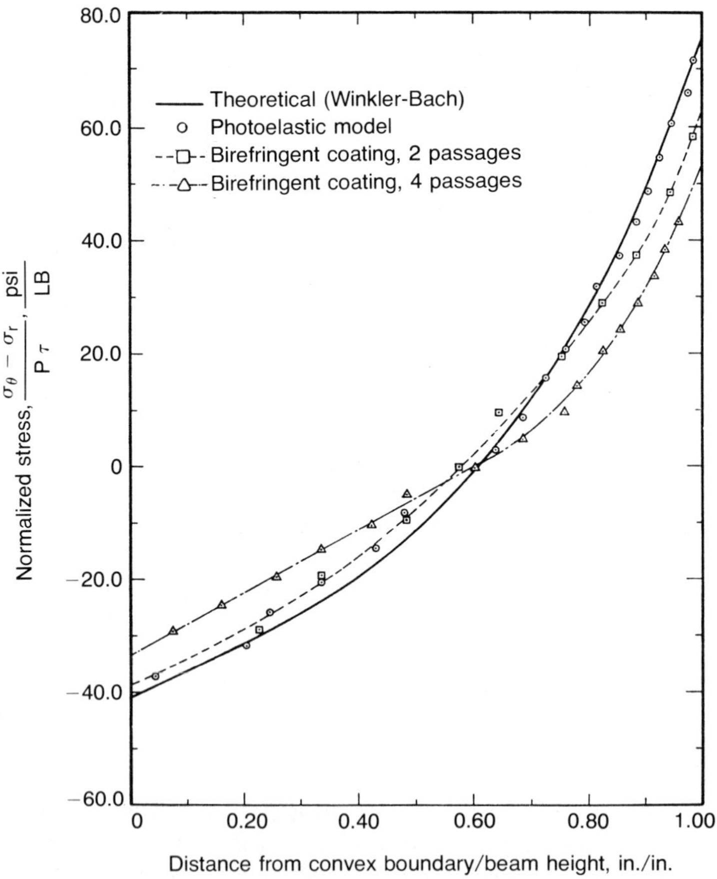

FIG. 6.17 Normalized stress variation across line of symmetry

respectively. The lateral displacement of the experimentally determined point of zero shear, relative to the location of the theoretical value and due to the integrating effect, can also be observed in figure 6.17.

Analysis of flat surfaces using fringe multiplication can be more efficacious than point-by-point methods, such as goniometric compensation, and the results of the experiment should, therefore, not be construed as an adverse implication. The results, however, vividly demonstrate the importance of the preferential plane of oblique-light incidence which is along the minimum strain gradient.

6.8 DESIGN BY FRINGE MULTIPLICATION WITH UNMIRRORED COATINGS

The primary objective of the investigation was the development of a technique for utilizing fringe multiplication with unmirrored birefringent coatings. Mirrored birefringent plastic sheets were successfully used in Larson's investigations with fringe multiplication.(70) However, the quality of the bond between the structure and the coating is sometimes uncertain when using mirrored coatings. The coating and the structure must have the same edge displacements in order to correlate the relative retardations with the strains or stresses in the structure.(71) Whether or not the coating is properly bonded can be clearly ascertained with unmirrored coatings, but then the structure must have a specular finish. A second objective of the investigation was the application of the technique for a cursory analysis of a finite tensile plate containing a heterogeneous inclusion and having lift (i.e., separation between plate and inclusion).

The Zandman method(72) is ideally suited for the study of problems such as the welds between two dissimilar metals(73) or for the analysis of structures containing heterogeneous inclusions. The use of fringe multiplication for increasing the photoelastic effect is *sometimes* desirable since whole field retardations can be observed, as opposed to point-by-point methods such as goniometric compensation.

Previous Investigations

Fringe Multiplication. Fringe multiplications are obtained by placing a partially silvered mirror in a reflective polariscope so that the light ray traverses the birefringent material several times. Mathematical relationships for fringe multiplication have been thoroughly covered in the literature(74–77) and will not be reiterated here.

TABLE 116
Nomenclature

Symbol	Variable	Typical units
a	Radius of inclusion	inch
C	Light intensity reduction coefficient	percent
D	Inclusion diameter	inch
E	Young's modulus of elasticity	psi
F	Force	lb_F
G	Modulus of rigidity, $\frac{E}{2(1 + \mu)}$	psi
K	Optical strain sensitivity constant	numeric
i	Radial interference	inch
M	Transmittance	percent
N	Number of fringe multiplications	numeric
P	Forced-fit contact pressure	psi
R	Reflectivity	percent
r	Radius	inch
T	Average stress, $\frac{F}{t_p(W - D)}$	psi
t	Thickness	inch
W	Plate width	inch
W/D	Aspect ratio	numeric
μ	Poisson's ratio	numeric
θ	Angle from direction of load	degrees
τ	Maximum shear stress, $\frac{\sigma_1 - \sigma_2}{2}$	ksi
σ_1, σ_2	Major and minor principal stresses, respectively	psi
σ_θ, σ_r	Tangential and radial stresses, respectively	psi
σ_y, σ_u	Tensile yield strength, and ultimate tensile strength, respectively	psi
γ, β, δ	Constants	numeric
α	Relative retardation	inch

Subscripts

d	Inclusion
p	Plate
c	Birefringent coating

Optimum mirror properties for fringe multiplication, assuming zero mirror absorption, have been derived by Post.[74] The light intensity exponentially decreases with increasing fringe multiplications, which necessitates a high-intensity light source for fringe multiplication polariscopes. Also, a different mirror transmittance should be used for each particular number of fringe multiplications (figure 6.18).

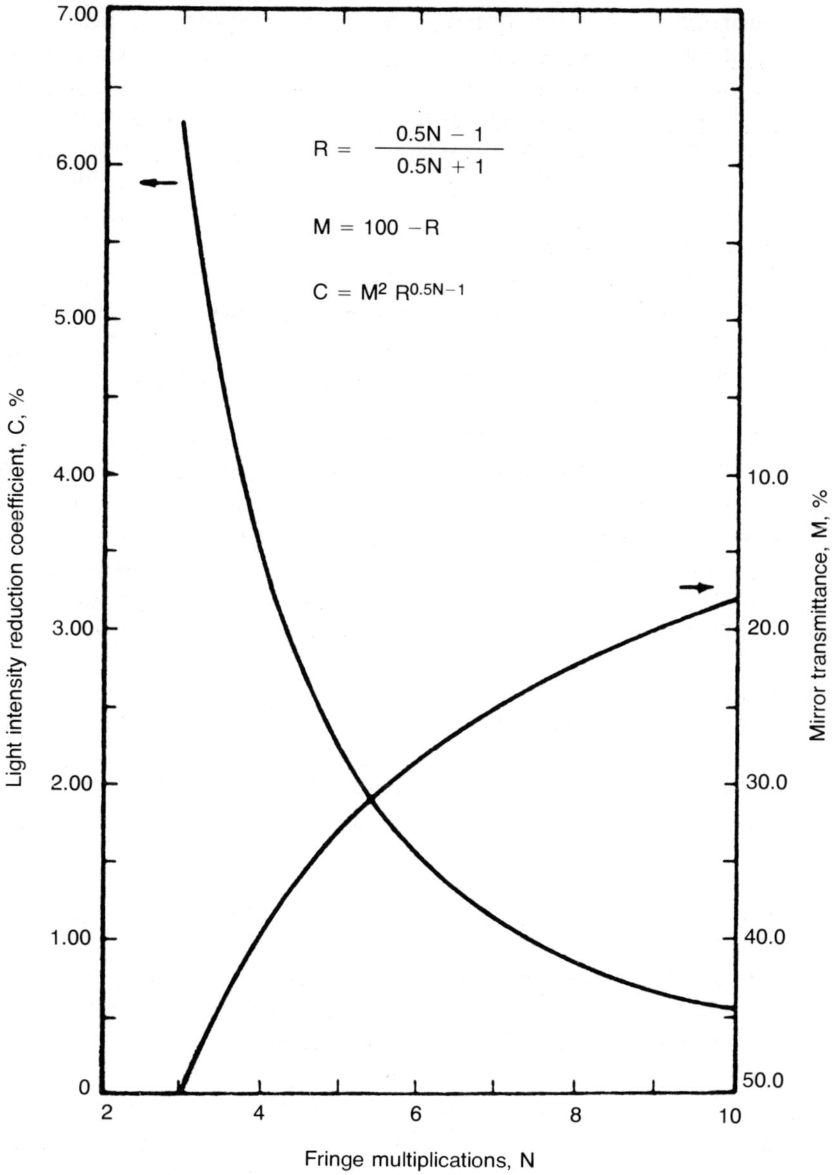

FIG. 6.18 Optimum mirror properties for minimum reduction of light intensity

Experimental Problem Solutions. The problem of a tensile plate containing a heterogeneous inclusion was the subject of a photoelastic study conducted by Jessop, Snell, and Holister.[78] In their tests, they used Araldite plates 1-in. wide and 0.125-in. thick with four hole sizes of 0.125, 0.250, 0.375 and 0.500-in. diameters.

Two degrees of pin interference were examined for each hole size. Most of their tests were performed using brass or Araldite pins, but a few of the pins were made of Bakelite. In their analysis, Jessop et al. gave the plate shear stress and the stress concentration factors across the minimum section and at the hole boundaries, 0 and 90° from the direction of uniaxially applied plate load. However, the variation of the plate shear stress around the hole boundary was not given.

Theoretical Problem Solutions. Several theoretical solutions[79–86] exist for the stresses in a tensile plate containing heterogeneous inclusions. Each of these solutions, however, includes several of the following rather stringent assumptions: (a) body forces are negligible; (b) the plate and inclusion materials are each to be isotropic and homogeneous; (c) the plate is infinite and in a generalized plane state of stress; (d) the inclusion and the plate thicknesses are equal and friction between them is neglected; (e) zero lift occurs between inclusion and plate, i.e., all-around contact is maintained for all loads; (f) inclusion diameter and plate hole diameter are equal; and (g) the plate with the inclusion is considered continuous, i.e., singly connected.

Experimental Apparatus

The reflective type polariscope (designed to utilize fringe multiplication) is used in this investigation. Intimate details of this particular polariscope (including schematics of the optics and the optical path and a detailed description of each component) can be found elsewhere[70, 77, 87] and will not be repeated here.

The specimen was loaded by means of a fine pitched screw, and the load was read from a Dillon Dynometer which had been previously calibrated against a Tinus-Olsen testing machine. The clevis bolt connected to the lower end of the specimen was mounted in a ball bearing housing. This prevented any appreciable, induced torsional stresses in the specimen while applying a uniaxial tensile load. The specimen was firmly bolted between two pairs of reinforcing plates, and the bolt holes in the specimen were made larger than the bolt holes in the reinforcing plates. The bolts were sufficiently tight so that the applied load on the pin was transferred to uniform loading on the specimen.

Experimental Procedures

Surface Preparations. During the investigations, aluminum specimens with the following surface treatments were tested for optimum reflectivity and fringe multiplication: (a) hard, chromium electroplating; (b) nickel electroplating; (c) silver electroplating; (d) chemical polish; (e) wheel polish; (f) buff, electropolish, buff; (g) buff, electropolish; and (h) electropolish. The hard, chromium electroplated finish was more compatible with the light wavelength of the mercury vapor lamp than with any of the other surface treatments investigated, with the exception of silver electroplating which gave the same number of multiplications. These multiplications were not, however, as distinct as with the chromium electroplating.

The specimens with the chromium electroplating were nickel electroplated prior to plating the final hard chromium layer. The chromium, nickel, and silver electroplated specimens were prepared for plating using the following processes: (a) vapor degreasing, (b) caustic etch, (c) nitric hydrofluoric dip, (d) sodium zinc immersion, (e) nitric acid dip, (f) sodium zincate dip, and (g) copper electroplating. The chemical polish was achieved by using the following dips: (a) acidic, (b) phosphoric, (c) sulfuric, (d) nitric, and (e) chromic. The specimens that were wheel polished were followed by a hand polish on plate glass with a silicon carbide-water solution and an alumina-water solution. The Alzac electropolish process[(88)] was used with the following dips: (a) 2.5% fluroboric, (b) phosphoric, (c) water rinse, and (d) nitric.

A great deal of difficulty was encountered in bonding the birefringent coating to the specimens. Ideally, the surface should be slightly roughened before applying the birefringent epoxy coatings. This was not possible in these experiments because a rough surface would scatter the light rays. A specular finish is required for fringe multiplication.

The chromium plating was harder than the tool steel milling cutters used, and frequent sharpening of the cutters during finish machining operations was required. The chromium chipped in flakes due to its brittleness, causing several coatings to peel during the machining. When forcing the inclusion into the plate, several coatings became unbonded. A satisfactory specimen was finally produced by most careful machining, and by making the hole diameter in the birefringent coating on the plate 0.004 in. larger than the inclusion diameter.

Details of Specimens. The tensile plate was 0.160 in. thick, 7075-T6 aluminum, and the inclusion was 0.160 in. thick, C 1020 steel. The inclusion diameter was 0.3925 in. and the plate hole diameter was 0.3920 in. The plate widths and applied loads are given in table 117, and the mechanical properties are given in table 118. The inclusion diameter and the diametrical interference of the inclusion with the hole in the plate were held constant. The various aspect ratios, W/D, were obtained by successive machining operations on the original configuration.

Birefringent Coating. The birefringent coating, consisting of the plastic sheet and the thin layer of adhesive, was 0.052 ± 0.002 in. thick and had a K factor of 0.08. The plastic was bonded to the specimen per the manufacturer's instructions.[(89)] Lines were scribed 0.25 in. apart on the coating, beginning at the specimen center, for ease in determining the scale factor in photographic enlargements of the isochromatics.

Discussion of Results

As many as 5 fringe multiplications (i.e., light traversing through the coating 10 times) could be visually observed, but only three multiplications could be satisfactorily photographed due to the length of time required for exposure. Exposure times for the isochromatic photographs were 45 seconds without fringe multiplication, 2 minutes and 30 seconds for 1 multiplication, and 9 minutes for 2 multiplication. The maximum shear stress was computed from the experimental data using the familiar expression

$$\tau = \frac{\alpha E_s N}{2t_c K(1 + \mu_s)} \tag{6.27}$$

As the ratio of the induced stress divided by the contact pressure was increased, the two points of zero shear stress, occurring above and below the inclusion boundary and on the axis of the uniaxially applied load, moved closer to the hole boundary. This was because the applied tension had the effect of adding a compressive tangential stress and a tensile radial stress to the tensile tangential stress and compressive radial stress initially existing because of the forced fit inclusion.

Muskhelishvili's[(79)] theoretical solution (infinite plate; zero lift) for the problem was used for a qualitative comparison with the experimental results. Using conventional nomenclature, the force-fit contact pressure and the resultant stresses in the plate are given by the relations:

$$P = \frac{2iG_p G_d}{a[G_d + G_p(1 - 2\mu_d)]} \tag{6.28}$$

$$\sigma_r = -\frac{Pa^2}{r^2} \tag{6.29}$$

$$\sigma_\theta = \frac{Pa^2}{r^2} \tag{6.30}$$

$$\tau = \frac{\sigma_\theta - \sigma_r}{2} = \frac{Pa^2}{r^2} = \frac{2iaG_p G_d}{[G_d + G_p(1 - 2\mu_d)]r^2} \tag{6.31}$$

If the inclusion is not force-fitted and a uniaxial tensile load is applied to the plate, the resultant plate stresses are:

$$\sigma_r = \frac{T}{2}\left[1 - \frac{\gamma a^2}{r^2} + \left(1 - \frac{2\beta a^2}{r^2} - \frac{3\delta a^4}{r^4}\right)\cos 2\theta\right] \tag{6.32}$$

$$\sigma_\theta = \frac{T}{2}\left[1 + \frac{\gamma a^2}{r^2} - \left(1 - \frac{3\delta a^4}{r^4}\right)\cos 2\theta\right] \tag{6.33}$$

where

$$\gamma = \frac{G_p(1 - 2\mu_d) - G_d(1 - 2\mu_p)}{G_d + G_p(1 - 2\mu_d)} \tag{6.34}$$

$$\beta = \frac{G_p(6 - 4\mu_d) - 2G_d}{G_p(3 - 2\mu_d) + G_d(5 - 6\mu_p)} \tag{6.35}$$

$$\delta = -\frac{G_p(3 - 2\mu_d) + G_d(1 - 2\mu_p)}{G_p(3 - 2\mu_d) + G_d(5 - 2\mu_p)} \tag{6.36}$$

But this latter solution is based on the premise of zero lift at $\theta = 0$ and $r = a$ (see insert, figure 6.19). A permissible solution is obtained by superimposing this latter solution upon the stresses resulting from the force-fit so that $\sigma_r < 0$ anywhere along the boundary, $r = a$. Using the material properties listed in table 118, then

$$P = 8.68 \text{ ksi}, \gamma = -0.187, \beta = 0.0982, \text{ and } \delta = -0.224. \tag{6.37}$$

For zero lift,

$$T[1 - 0.5\gamma - \beta - 1.5\delta] - P \leqq 0 \tag{6.38}$$

and for the given material properties, this relation is

$$2.37T - 8.68 < 0 \tag{6.39}$$

TABLE 117
Experiment schedule

Model Number	Plate width, W, inches	Aspect ratio, $\frac{W}{D}, \frac{\text{inch}}{\text{inch}}$	Load, F, kips	Average stress, T, ksi
1	4	10.02	8.50	14.7
2	2.376	6.04	8.50	26.8
3	1.564	3.98	6.50	34.6
4	0.750	1.91	3.50	61.1

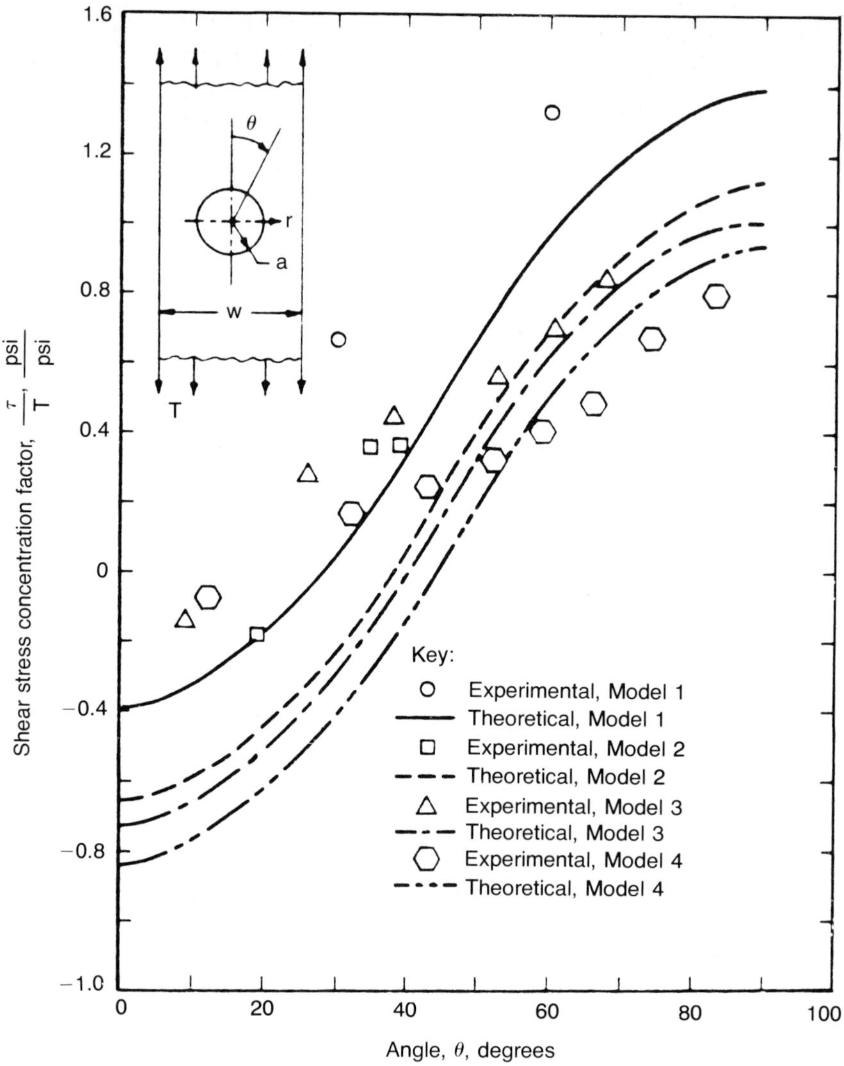

FIG. 6.19 Distribution of shear stress intensity around inclusion at r = a

Substituting the average stresses (table 117) into this last equation indicated that lift occurred in each of the tests. Since the relative retardations would have been of such small magnitude for the condition of zero lift and since only infinite plate theory was available, it was deemed more desirable to use the loads of higher magnitude and then to use the theoretical results only for qualitative purposes of comparison.

TABLE 118
Mechanical properties of specimen

	C1020 Inclusion	7075 T-6 Plate
Young's modulus, E, psi	30×10^6	10.4×10^6
Poisson's ratio, μ, numeric	0.28	0.33
Modulus of rigidity, G, psi	11.7×10^6	3.90×10^6
Tensile yield stress, σ_y, ksi	42	66
Ultimate tensile strength σ_u, ksi	57	77

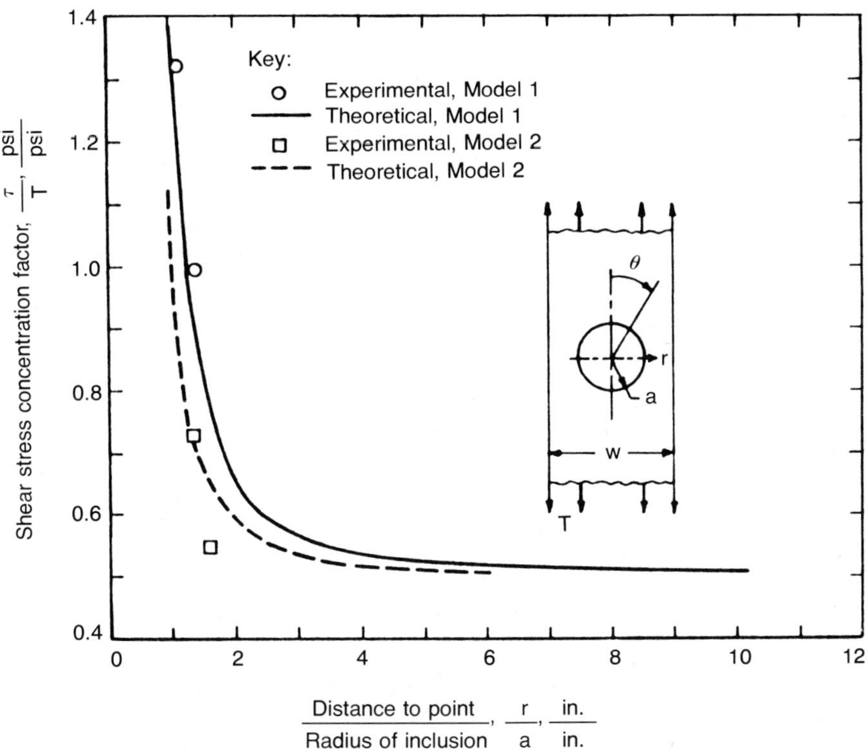

FIG. 6.20 Distribution of shear stress intensity across minimum section of plate, $\theta = 90°$, models 1 and 2

The normalized, theoretical shear stress across the minimum section was given for the particular conditions of these tests by the relation

$$\left(\frac{\tau}{T}\right)_{\theta=90°} = 0.5 - 0.044\,\frac{a^2}{r^2} + 0.336\,\frac{a^4}{r^4} + \frac{8.68a^2}{Tr^2} \qquad (6.40)$$

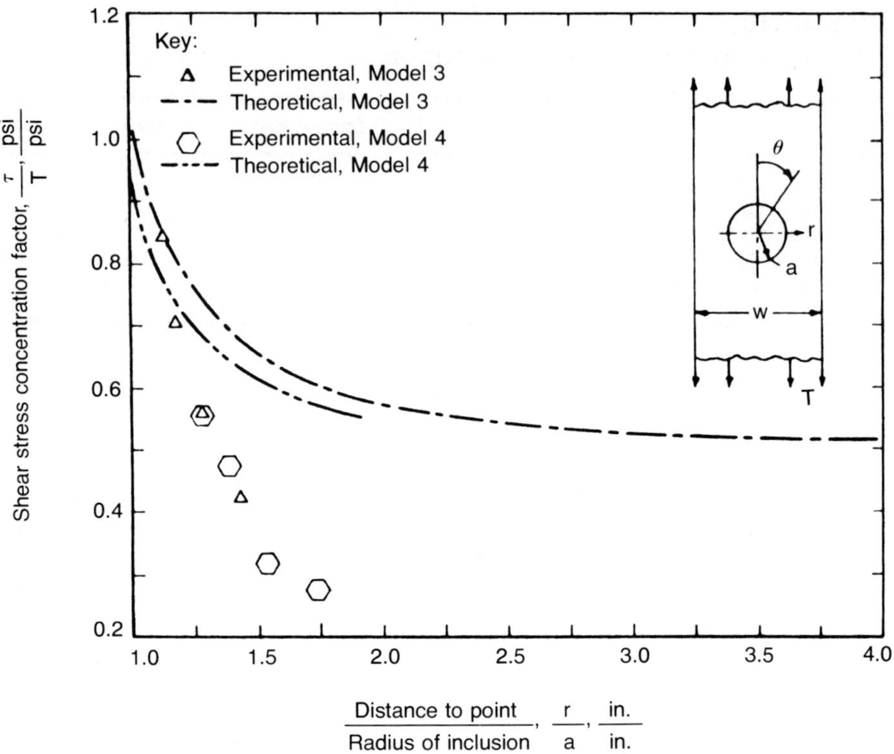

FIG. 6.21 Distribution of shear stress intensity across minimum section of plate θ = 90°, models 3 and 4

The normalized, theoretical shear stress around the boundary, r = a, was given by

$$\left(\frac{\tau}{T}\right)_{r=a} = -0.093 - 0.889 \cos 2\theta + \frac{8.68}{T} \qquad (6.41)$$

The results from the equations and the experimental data are plotted in figures 6.19 through 6.21. Considering the disparity between the experimental data and the theoretical curves drawn in these figures, it should be emphasized that the theoretical calculations were for the case of an infinite plate and zero lift at the inclusion, neither of which conformed to the conditions of the tests.

Since the birefringent coating was so thin, the usual correction factors for reinforcing effect, Poisson's contraction, parasitic birefringence, etc., were all of negligible order. The veridicality of data obtained from the birefringent coating technique, the sources of error, and means of reducing or avoiding these errors have been discussed extensively in other work.[71, 77, 90–96]

Conclusions

The primary objective of obtaining a specular specimen finish compatible with fringe multiplication for analysis with unmirrored birefringent coatings was achieved. In any further similar experiments, the optimum fringe multiplication mirror should be used for each particular multiplication and a more intense light source should be used if the resultant isochromatics are to be photographed. The specular finish required for fringe multiplication studies might be easily obtained with a hand-operated electroplating unit[(97)] available from Hipo-KG. The second objective of the particular problem application was achieved only in a cursory fashion, and further experiments and theoretical developments are necessary for determining the exact stress variation in finite tensile plates having lift and containing an inclusion.

REFERENCES

1. Williams, M. L., "Surface Singularities Resulting from Various Boundary Conditions in Angular Corners of Plates under Bending," U.S. Nat. Cong. of Appl. Mechs.; Ill. Inst. Tech.; 1951; p. 325.
2. Lardy, P., "Die Strenge Losung des Problems der schiefen Platte," Schweiz, Bauztg; 1949; p. 207.
3. Tungl, E., "Die Parallelogrammplatte mit Einzellast," Osterr. Ing. Archiv.; Vol. 13, 1959; p. 121.
4. Morley, L. S. D., "Bending of a Simply Supported Rhombic Plate under Uniform Normal Loading," *J. Mechs. and Appl. Math.;* Vol. XV, Part 4; Nov. 1962; pp. 413–427.
5. Favre, H., "Le calcul des plaques obliques par la methode des equations aux differences," *Publ. of Int'l. Assoc. Bridge and Struct. Engring.;* Zurich; 1943; p. 91.
6. Jensen, V. P., "Analyses of Skew Slabs," Engrng. Exp. Stat. Bull. No. 332, 1941; Univ. of Ill.; p. 39.
7. Timoshenko, S., and S. W. Krieger, *Theory of Plates and Shells,* 2nd edition, 1959; McGraw-Hill Book Co., Inc., New York, N.Y.; pp. 318–320.
8. Hoelscher, R. P., J. N. Arnold, and S. H. Pierce, *Graphic Aids in Engineering Computation,* 1st Edition; 1952; McGraw-Hill Book Co., Inc., New York, N.Y.
9. Seely, F. B., and J. O. Smith; *Advanced Mechanics of Materials,* 2nd Edition, 1932; John Wiley & Sons, Inc., New York, N.Y.; pp. 220–222.

10. Lambert, T. H., "Elastic Stresses due to an Interference Fit Pin in a Rectangular Plate of Finite Width," *J. Mech. Engr. Sci.;* Vol. 3, No. 3; 1961; pp. 236–241.
11. Muskhelishvili, N. I., *Some Basic Problems of the Mathematical Theory of Elasticity,* (Moscow, 1949, translation by J. R. M. Radok); P. Noordhoff Publishing Co., Ltd., Groningen, Holland; pp. 210–213.
12. Heywood, R. B., *Designing by Photoelasticity,* Chapman and Hall, Ltd., London; 1952; pp. 162 and 177.
13. Peterson, R. E., *Stress Concentration Design Factors,* John Wiley and Sons, Inc., N.Y.; 1953.
14. Savin, G. N., *Stress Concentration around Holes,* translated from Russian; Pergamon Press; 1961.
15. Frocht, M. M., *Photoelasticity,* Vol. 1, 1941; and Vol. 2; 1948; John Wiley and Sons, Inc., N.Y.
16. Bynum, D. J., "Fillet and Groove Stress Concentrations for Uniaxially and Biaxially Applied Loads," Southwest Research Institute Proj. To. T-1677-3, 27 Aug. 1962; Pressure Vessel Research Committee, Welding Research Council.
17. Bynum, D. J., and M. M. Lemcoe, "Stresses and Deflections in Laterally Loaded Perforated Plates," Bulletin No. 80, Aug. 1962; Welding Research Council, 345 E. 47th St., New York 17, N.Y.
18. Morgan, W. C., and P. T. Bizon, "Experimental Investigation of Stress Distributions Near Abrupt Change in Wall Thickness in Thin-Walled Pressurized Cylinders," Tech. Note D-1200; NASA, Washington, D.C.; June 1962.
19. Rastrelli, L. U., and R. C. DeHart, "Studies of Displacements in Solid Propellant Grains (Embedded Particle, X-Ray Detection System)," Interim Report, Proj. 1043-3, Contract No. NOnr-3363(FBM), Sept. 1, 1962, Structural Mechanics Branch, Office of Naval Research, Washington 25, D.C.
20. Rastrelli, L. U., and B. J. Pape, "Studies of Displacements in Solid Propellant Grains (Estimating Experimental Errors in X-ray Detection Facility)," Interim Report, Proj. 1043-3, Contract No. NOnr-3363(FBM), Sept. 5, 1962, Structural Mechanics Branch, Office of Naval Research, Washington 25, D.C.
21. Rastrelli, L. U., and R. C. DeHart, "Studies of Displacements in Solid Propellant Grains (Displacements in Cylindrical, Unfilled-Urethane Specimens under Uniaxial (Compression)," Interim Report, Proj. 1043-3, Contract No. NOnr-3363(FBM), Oct. 1, 1962, Structural Mechanics Branch, Office of Naval Research, Washington 25, D.C.

22. Bynum, D. J., and L. U. Rastrelli, "Studies of Displacements in Solid Propellant Grains (Surface Strains in Case-Bonded Models of Rocket Motors)," Interim Report, Proj. 1043-3, Contract No. NOnr-3363(FBM), Oct. 10, 1962, Structural Mechanics Branch, Office of Naval Research, Washington 25, D.C.
23. Rastrelli, L. U., and R. C. DeHart, "Studies of Displacements in Solid Propellant Grains (Longitudional Displacements in Free-Standing Solid Propellant Hollow Cylinders)," Interim Report, Proj. 1043-3, Contract No. NOnr-3363(FBM), March 1, 1963, Structural Mechanics Branch, Office of Naval Research, Washington 25, D.C.
24. Bynum, D. J., and L. U. Rastrelli, "Studies of Displacements in Solid Propellant Grains (Photoelastic Observations Using Tolylene Di-Isocyanate Polyurethane)," Interim Report, Proj. 1043-3, Contract No. NOnr-3363(00)(FBM), Aug. 20, 1962, Structural Mechanics Branch, Office of Naval Research, Washington 25, D.C.
25. Durelli, Lake, and Tsao, "Device for Applying Uniform Loading to Boundaries of Complicated Shape," SESA Proceed, XI, No. 1, 55 (1953).
26. Ordahl, D. D., and M. L. Williams, "Preliminary Photoelastic Design Data for Stresses in Rocket Grains," *Jet Propulsion*, 27, No. 6, 657–663 (June 1957).
27. Lee, G. H., *Experimental Stress Analysis*, John Wiley and Sons, Inc., New York, 172 (1950).
28. "A Research Study to Advance the State-of-the-Art of Solid Propellant Grain Design," Interim Summary Report, E 217-60, Contract No. AF33(616)-6530, Jan. 17, 1961, Thiokol Chemical Corp., Elkton Division, Elkton, Md., 57.
29. "A Research Study to Advance the State-of-the-Art of Solid Propellant Grain Design," Quarterly Progress Report No. 7, E 156-61, Contract No. AF33(616)-6530, Oct. 13, 1961, Thiokol Chemical Corp., Elkton Division, Elkton, Md., 30–33.
30. Bills, K. W., O. L. McCurdy, and R. B. Farnham, "The Development and Use of Birefringent Polymers for Determining Strain Distribution in Propellants," Bulletin of the Joint Meeting, JANAF Panels on Physical Properties and Surveillance of Solid Propellants, September 1960, Aerojet Corp., Azusa, Calif.
31. "A Research Study to Advance the State-of-the-Art of Solid Propellant Grain Design," Quarterly Progress Report No. 8, E 11-62, Contract AF33(616)-6530, Jan. 25, 1962, Thiokol

Chemical Corp., Elkton Division, Elkton, Md., 25–27.

32. Athey, R. J., J. G. Di Pinto, and J. S. Rugg, "Adiprene L, A Liquid Urethane Elastomer," Development Products Report No. 10, Mar. 15, 1958, Elastomer Chemicals Dept., E. I. DuPont De Nemours and Co., Inc., Wilmington 28, Del.
33. McCready, S. D., and J. G. Di Pinto, "Engineering Properties of Urethane Elastomers," Development Products Report No. 17, April 1960, Elastomer Chemicals Dept., E. I. DuPont De Nemours and Co., Inc., Wilmington 28, Del.
34. Miguel, A. S., and E. N. Duran, "The Formulation and Casting Techniques of Low-Modulus Photoelastic Coatings," Tech. Rep. No. 32-309, July 19, 1962, Jet Prop. Lab., Cal. Inst. Tech., Pasadena, Calif.
35. Reiner, Markus, "Deformation, Strain, and Flow," Interscience Publishers, Inc., New York, 138 (1960).
36. Daniel, I. M., and A. J. Durelli, "Photothermoelastic Analysis of Bonded Propellant Grains," *Experimental Mechanics*, 1 (3), 97–104 (1961).
37. Stein, R. S., S. Onogi, K. Sasaguri and D. A. Keedy, "Dynamic Birefringences of High Polymers II," *J. Appl. Phys.*, 34, No. 1, 80–90 (January 1963).
38. Mesnager, A. "Sur la détermination optique des tensions intérieures dans des solides a trois dimensions." *Competes Rendus*, Paris, V. 190, 1930, p. 1249.
39. Mabboux, Georges, "Applications de la photoélasticimetre a l'etude des ouvrages en béton," *Revue d'Optique*, V. II, 1932, pp. 501–507.
40. Oppel, G., "Das polarizationsoptische Schichtverfahren zur Messung der Oberflachenspannung am beanspruchten Bauteil ohne Modell," *Vereines Deutscher Ingenieure*, Zeitschrift, V. 81, 1937, pp. 803–804.
41. D'Agostino, J., D. C. Drucker, C. K. Liu, and Mylonas, "An Analysis of Plastic Behavior of Metals with Bonded Birefringent Plastic," *Proceed. SESA*, V. XII, No. 2, 1955, pp. 115–122.
42. Zandman, F., *Photostress, Principles and Applications*, Budd Co., Phoenixville, Pa.
43. Coker, E. G., L. N. G. Filon, and H. T. Jessop, *A Treatise on Photo-Elasticity*, 2nd Ed., 1957, Cambridge Univ. Press, N.Y.
44. Froche, M. M., *Photoelasticity*, V. 1, 1941, V. 2, 1948, John Wiley and Sons, Inc., New York.
45. Flugge, S., Editor, *Elasticity and Plasticity*, Encyclopedia of Physics, V. 6, 1958, Springer-Verlag, Berlin.

46. Heywood, R. B., *Designing by Photoelasticity*, 1952, Chapman and Hall, Ltd.
47. Bynum, D. J., and M. M. Lemcoe, "Birefringent Coating Analysis of Laterally Loaded Perforated Plates," *Exp. Mech.*, June 1963, pp. 140–148.
48. "Photostress Conversion Table," Bulletin IB 8023, 30 January 1959, Budd Company.
49. Hetenyi, M., Editor, *Handbook of Experimental Stress Analysis*, John Wiley and Sons, Inc., New York, p. 863.
50. Bynum, D. J., "Birefringent Coating Analysis of a Tensile Plate Containing a Heterogeneous Inclusion, Utilizing Fringe Multiplication," M.S. Thesis, Univ. Wash., 1961.
51. Day, E. E., A. S. Kobayashi, and C. N. Larson, "Fringe Multiplication and Thickness Effects in Birefringent Coating," *Exp. Mech.*, Vol. 2, No. 4, Apr. 1962, pp. 115–122.
52. Bynum, D. J., "On the Accuracy of Fringe Multiplication with Birefringent Coatings," Paper No. 815, SESA.
53. Post, D., and F. Zandman, "The Accuracy of the Birefringent Coating Method for Coatings of Arbitrary Thickness," Paper No. 557, presented at 1959 spring meeting, SESA.
54. Duffy, J., "Effects of the Thickness of Birefringent Coatings," Technical Report No. 34, Dept. of the Army Project 503-06-005, Apr. 1958, Brown University.
55. Lee, T. C., C. Mylonas, and J. Duffy, "Thickness Effects in Birefringent Coatings with Radial Symmetry," *Exp. Mech.*, Oct. 1961, pp. 134–142.
56. Zandman, F., S. S. Redner, and E. I. Riegner, "Reinforcing Effect of Birefringent Coatings," Paper No. 588, presented at 1959 spring meeting, SESA.
57. "Method of Correction for Parasitic Birefringence During Photostress Tests," Bul. PS-5051, 8 Mar. 1960, Budd Company.
58. Zandman, F., and H. N. Maier, "6 New Techniques for Photoelastic Coatings," *Product Engineering*, 24 July 1961, p. 45.
59. Miner, D. F., and J. B. Seastone, Editors, *Handbook of Engineering Materials*, 1st Ed., 1955, John Wiley & Sons, Inc., New York.
60. Bynum, D. J., and M. M. Lemcoe, "Stresses and Deflections in Laterally Loaded Perforated Plates," WRC Bul. 80, August 1962, Welding Research Council, New York.
61. "Instructions for the Use of Photostress Small Field Universal Meter," Bul. PS-3044, 21 Mar. 1960, Budd Company.
62. Bynum, D. J., "Fillet and Groove Stress Concentrations," *Exp. Mech.*, June 1964, pp. 160–167.

63. Bynum, D. J., L. U. Rastrelli, and R. C. DeHart, "Photoelastic Observations Using Tolylene Di-Isocyanate Polyurethane," *Exp. Mech.*, July 1964, pp. 191–200.
64. Pickett, A. G., D. J. Bynum, and W. R. Schmidt, "Experimental Stress Analysis of a Condenser Head to Determine Critical Orientation of External Piping Load," Interim Report, Project 1109-3, Southwest Research Institute, 28 June 1963.
65. "Instructions for Applying Photostress Sheet Plastic," Bul. IB-8004, 18 Feb. 1959, Budd Co.
66. Post, D., "Isochromatic Fringe Sharpening and Fringe Multiplication in Photoelasticity," SESA Proc., XII(2), 143–157 (1955).
67. Bynum, D. J., "On Fringe Multiplication with Unmirrored Birefringent Coatings," Technical Note submitted to SESA.
68. Day, E. E., A. S. Kobayashi, and C. N. Larson, "Fringe Multiplication and Thickness Effects in Birefringent Coatings," *Experimental Mechanics,* 2(4) 115–121 (1962).
69. Seely, F. B., and J. O. Smith, *Advanced Mechanics of Materials,* 2nd ed., John Wiley and Sons, Inc., New York, 139–167 (1952).
70. Larson, C. N., "On the Accuracy of the Birefringent Coating Method of Experimental Stress Analysis," MS Thesis, Dept. ME, Univ. Wash., Seattle, Wash., 1960.
71. Post, D., and F. Zandman, "The Accuracy of the Birefringent Coating Method for Coatings of Arbitrary Thickness," *Exp. Mech.*, Jan. 1961, p. 21–32.
72. Zandman, F., *Photostress, Principles, and Applications,* Budd Co., P.O. Box 245, Phoenixville, Pa.
73. Zandman, F., "Photoelastic-Coating Technique for Determining Stress Distribution in Welding Structures," *Welding Journal,* May 1960.
74. Post, D., "Isochromatic Fringe Sharpening and Fringe Multiplication in Photoelasticity," *SESA Proceed.*, V. 12, No. 2, 1955, p. 143–157.
75. Holen, J., "New Techniques in Photoelasticity," MS Thesis, Dept. ME, Univ. Wash., Seattle, Wash., 1957.
76. Holen, J., and E. E. Day, "New Techniques in Photoelasticity," *The Trend in Engineering,* V. 31, Apr. 1958, p. 10–16, Univ. Wash., Seattle, Wash.
77. Day, E. E., A. S. Kobayashi, and C. N. Larson, "Fringe Multiplication and Thickness Effects in Birefringent Coatings," *Exp. Mech.*, Apr. 1962, p. 115–122.
78. Jessop, H. T., C. Snell, and G. S. Holister, "Photoelastic Investigation on Plates with Single Interference-Fit Pins with

Load Applied to Plate Only," *Aero Qrtly*, London, V. 7, Part 4, Nov. 1956, p. 297–315.

79. Muskhelishvili, N. I., *Some Basic Problems of the Mathematical Theory of Elasticity*, 3rd Ed., 1949, Moscow, Russia, Translated by J. R. M. Radok, 1953, P. Noordhoff Ltd. Publ. Groningen-Holland, p. 210–218.

80. Coker, E. G. and L. N. G. Filon, *A Treatise on Photo-Elasticity*, 1957, Cambridge University Press.

81. Natsume, S., and B. Tanimoto, "A Note on the Stress Distribution in a Plate with Circular Plug or Concentric Ring under Edge Forces," *Proceed. 6th Jap. Nat. Cong. for Appl. Mech.*, 1956, p. 87–91.

82. Sezawa, K., and G. Nishimura, "Stresses under Tension in a Plate with a Heterogeneous Insertion," *Aero Res. Inst.*, V. 6, 1931–32, Tokyo Imperial Univ., Komaba, Tokyo, Japan, p. 25–44.

83. Sezawa, K., "The Effect of Local Heterogeneity on the Stress Distribution in Solids," *Engineering*, V. 135, 1933, p. 695–696.

84. Donnell, L. H., "Stress Concentrations due to Elliptical Discontinuities in Plates under Edge Forces," *Ann. Vol. T., Von Karman*, 1941, Calif. Inst. Tech.

85. Suyehiro, K., "The Distribution of Stress in a Tension Strip Having Circular Hole Filled with Plug," *Proceed. Jap. Soc. Mech. Engr.*, V. 17, No. 34, 1914, p. 231–232.

86. Goodier, J. N., "Concentration of Stress Around Spherical and Cylindrical Inclusions and Flaws," *ASME Trans.*, V. 55, 1933, p. A39–45.

87. Bynum, D. J., "Birefringent Coating Analysis of a Tensile Plate Containing a Heterogeneous Inclusion Utilizing Fringe Multiplication," MS Thesis, Dept. ME, Univ. Wash., Seattle, Wash., 1961.

88. *Alcoa Structural Handbook, A Design Manual for Aluminum*, Aluminum Company of America, Pittsburgh, Pa., 1958.

89. "Instructions for Applying Photostress Sheet Plastic," Bul. IB-8004, 18 Feb. 1959, Budd Co., Phoenixville, Pa.

90. Zandman, F., S. S. Redner, and E. S. Riegner, "Reinforcing Effect of Birefringent Coatings," *Exp. Mech.*, Feb. 1962, p. 55–64.

91. Zandman, F., S. S. Redner, and Post, D., "Photoelastic Coating Analysis in Thermal Fields," Instruments Div., Budd Co., Phoenixville, Pa.

92. Bynum, D. J., and M. M. Lemcoe, "Stresses and Deflections in Laterally Loaded Perforated Plates," *WRC Bul. No. 80*,

Aug. 1962, Welding Research Council.

93. Duffy, J., "Effects of the Thickness of Birefringent Coatings," *Exp. Mech.*, Mar. 1961, p. 74–82.
94. Lee, T. C., C. Mylonas, and J. Duffy, "Thickness Effects in Birefringent Coatings with Radial Symmetry," *Exp. Mech.*, Oct. 1961, p. 134–143.
95. Chen, W. T., and J. Duffy, "Effects of the Thickness of Birefringent Coatings for Two-Dimensional Variations in Surface Strain," DA Proj. No. 59901004, TR No. 2, Feb. 1962, U.S. Army Research Office (Durham).
96. "Method of Correction for Parasitic Birefringence during Photostress Tests," Bulletin PS-5051, 8 Mar. 1960, Budd Co., P.O. Box 245, Phoenixville, Pa.
97. "Developments to Watch," *Prod. Engr.*, 4 Feb. 1963, p. 57.

CHAPTER

7 DYNAMIC STRUCTURAL ANALYSIS

7.1 VIBRATION OF MULTISPAN BEAMS

The characteristic equation for free lateral vibrations of a multispan beam with intermediate simple supports was developed. The procedure for determining the characteristic equation and the mode participation factors for forced lateral vibrations of the given configuration was outlined. A separate coordinate system was adopted for each span since the use of a single generalized location coordinate would not give a harmonic response unless the boundary conditions were altered. Transformation of the boundary conditions to fit a single coordinate system would unduly complicate the analysis.

Free Vibrations of the Continuum

The differential equation of motion for the lateral vibration of a beam is, neglecting the effects of shear and rotary inertia:

$$EI \frac{\partial^4 y}{\partial x^4} = -\frac{\gamma A}{g} \frac{\partial^2 y}{\partial t^2} \tag{7.1}$$

where the coordinate system is as shown in figure 7.1. Letting

$$\alpha^2 \equiv \frac{EI\,g}{\gamma A} \tag{7.2}$$

equation (7.1) can then be expressed as

$$\frac{\partial^2 y}{\partial t^2} + \alpha^2 \frac{\partial^4 y}{\partial x^4} = 0 \tag{7.3}$$

In a normal mode, the general solution of equation (7.3) is

$$y = X\,(B_c \cos \omega t + B_s \sin \omega t) \tag{7.4}$$

The normal function, X, is determined by the boundary conditions, whereas the mode participation factors, B, are determined from the initial conditions. Substituting equation (7.4) into equation (7.3) yields

$$\frac{d^4 X}{dx^4} = \frac{\omega^2}{\alpha^2} X \tag{7.5}$$

Let

$$\kappa^4 \equiv \frac{\omega^2}{\alpha^2}$$

then equation (7.5) becomes

$$\frac{d^4 X}{dx^4} = K^4 X \tag{7.6}$$

Circular and hyperbolic sines and cosines are particular solutions of equation (7.6); therefore, the general solution can be written as

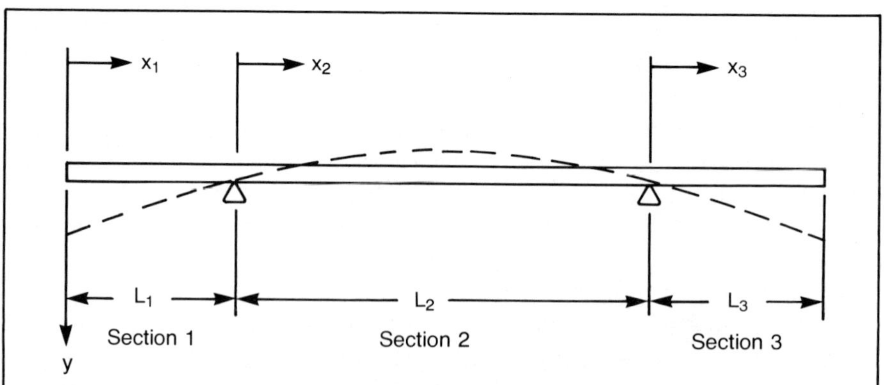

FIG. 7.1 Coordinate system

$$X = c_1^*(\sin Kx + \sin h\, Kx) + c_2^*(\sin Kx - \sin h\, Kx) + c_3^*(\cos Kx + \cos h\, Kx) + c_4^*(\cos Kx - \cos h\, Kx) \quad (7.7)$$

The alternate expression for equation (7.7) is

$$X = c_1 \sin Kx + c_2 \sin h\, Kx + c_3 \cos Kx + c_4 \cos h\, Kx \quad (7.8)$$

Considering all possible normal modes, the general expression for free lateral vibrations is

$$y = \sum_{i=1}^{\infty} X_i(B_{c_i} \cos \omega_i t + B_{s_i} \sin \omega_i t) \quad (7.9)$$

Note that since the expression for the moment, and hence the force, is different for each section depicted in figure 7.1, three normal functions will be required. A boundary condition is required for each order of a differential equation; therefore, 12 boundary conditions are necessary.

Using Newtonian notation and subscripts on the normal functions to indicate section, the 12 boundary conditions for the simply-supported, 3-span beam are as follows.

Section 1:

$$X_1''(@x_1 = 0) = 0 \quad (7.10)$$

$$X_1'''(@x_1 = 0) = 0 \quad (7.11)$$

$$X_1(@x_1 = L_1) = 0 \quad (7.12)$$

$$X_1'(@x_1 = L_1) = X_2'(@x_2 = 0) \quad (7.13)$$

Section 2:

$$X_2(@x_2 = 0) = 0 \quad (7.14)$$

$$X_2''(@x_2 = 0) = X_1''(@x_1 = L_1) \quad (7.15)$$

$$X_2(@x_2 = L_2) = 0 \quad (7.16)$$

$$X_2'(@x_2 = L_2) = X_3'(@x_3 = 0) \quad (7.17)$$

Section 3:

$$X_3(@x_3 = 0) = 0 \quad (7.18)$$

$$X_3''(@x_3 = 0) = X_2''(@x_2 = L_2) \quad (7.19)$$

$$X_3''(@x_3 = L_3) = 0 \quad (7.20)$$

$$X_3'''(@x_3 = L_3) = 0 \quad (7.21)$$

Using superscripts on the constants to denote section X from equation (7.8), and the first three derivatives become:

$$X_1 = c_1^1 \sin Kx + c_2^1 \sin h\, Kx + c_3^1 \cos Kx + c_4^1 \cos h\, Kx \tag{7.22}$$

$$X_2 = c_1^2 \sin Kx + c_2^2 \sin h\, Kx + c_3^2 \cos Kx + c_4^2 \cos h\, Kx \tag{7.23}$$

$$X_3 = c_1^3 \sin Kx + c_2^3 \sin h\, Kx + c_3^3 \cos Kx + c_4^3 \cos h\, Kx \tag{7.24}$$

$$X_1' = K(c_1^1 \cos Kx + c_2^1 \cos h\, Kx - c_3^1 \sin Kx + c_4^1 \sin h\, Kx) \tag{7.25}$$

$$X_2' = K(c_1^2 \cos Kx + c_2^2 \cos h\, Kx - c_3^2 \sin Kx + c_4^2 \sin h\, Kx) \tag{7.26}$$

$$X_3' = K(c_1^3 \cos Kx + c_2^3 \cos h\, Kx - c_3^3 \sin Kx + c_4^3 \sin h\, Kx) \tag{7.27}$$

$$X_1'' = K^2(-c_1^1 \sin Kx + c_2^1 \sin h\, Kx - c_3^1 \cos Kx + c_4^1 \cos h\, Kx) \tag{7.28}$$

$$X_2'' = K^2(-c_1^2 \sin Kx + c_2^2 \sin h\, Kx - c_3^2 \cos Kx + c_4^2 \cos h\, Kx) \tag{7.29}$$

$$X_3'' = K^2(-c_1^3 \sin Kx + c_2^3 \sin h\, Kx - c_3^3 \cos Kx + c_4^3 \cos h\, Kx) \tag{7.30}$$

$$X_1''' = K^3(-c_1^1 \cos Kx + c_2^1 \cos h\, Kx + c_3^1 \sin Kx + c_4^1 \cos h\, Kx) \tag{7.31}$$

$$X_2''' = K^3(-c_1^2 \cos Kx + c_2^2 \cos h\, Kx + c_3^2 \sin Kx + c_4^2 \cos h\, Kx) \tag{7.32}$$

$$X_3''' = K^3(-c_1^3 \cos Kx + c_2^3 \cos h\, Kx + c_3^3 \sin Kx + c_4^3 \cos h\, Kx) \tag{7.33}$$

Substituting equations (7.10) to (7.21) into equations (7.22) to (7.33) yields:

$$0 = 0 + 0 - c_3^1 + c_4^1 \tag{7.34}$$

$$0 = -c_1^1 + c_2^1 + 0 + 0 \tag{7.35}$$

$$0 = c_1^1 \sin KL_1 + c_2^1 \sin h\, KL_1 + c_3^1 \cos KL_1 + c_4^1 \cos h\, KL_1 \tag{7.36}$$

$$c_1^2 + c_2^2 = c_1^1 \cos KL_1 + c_2^1 \cos h\, KL_1 - c_3^1 \sin KL_1 + c_4^1 \sin h\, KL_1 \tag{7.37}$$

$$0 = 0 + 0 + c_3^2 + c_4^2 \tag{7.38}$$

$$-c_3^2 + c_4^2 = -c_1^1 \sin KL_1 + c_2^1 \sin h\, KL_1 - c_3^1 \cos KL_1 + c_4^1 \cos h\, KL_1 \tag{7.39}$$

$$0 = c_1^2 \sin KL_2 + c_3^2 \sin h\, KL_2 + c_3^2 \cos KL_2 + c_4^2 \cos h\, KL_2 \tag{7.40}$$

$$c_1^3 + c_2^3 = c_1^2 \cos KL_2 + c_2^2 \cos h\, KL_2 - c_3^2 \sin KL_2 + c_4^2 \sin h\, KL_2 \tag{7.41}$$

$$0 = 0 + 0 + c_3^3 + c_4^3 \tag{7.42}$$

$$-c_3^3 + c_4^3 = -c_1^2 \sin KL_2 + c_2^2 \sin h\, KL_2 - c_3^2 \cos KL_2 + c_4^2 \cos h\, KL_2 \qquad (7.43)$$

$$0 = -c_1^3 \sin KL_3 + c_2^3 \sin h\, KL_3 - c_3^3 \cos KL_3 + c_4^3 \cos h\, KL_3 \qquad (7.44)$$

$$0 = -c_1^3 \cos KL_3 + c_2^3 \cos h\, KL_3 + c_3^3 \sin KL_3 + c_4^3 \sin h\, KL_3 \qquad (7.45)$$

Equations (7.34) to (7.45) must be solved simultaneously to obtain the values of the constants. In matrix form, these equations can be expressed as

$$[F(L_i)]\{c_j^i\} = 0 \qquad (7.46)$$

Since $\{c_j^i\} = 0$ is trivial, (7.47)

then $[F(L_i)] = 0$ (7.48)

The solution of equation (7.43) yields the natural frequencies. The algebraic manipulations required to formulate equation (7.48) follow (from equations 34, 35, 38, and 42):

$$c_4^1 = c_3^1 \qquad (7.49)$$

$$c_2^1 = c_1^1 \qquad (7.50)$$

$$c_4^2 = -c_3^2 \qquad (7.51)$$

$$c_4^3 = -c_3^3 \qquad (7.52)$$

Substitute equations (7.49) to (7.52) into equations (7.36), (7.37), (7.39), (7.40), (7.41), (7.43), (7.44), and (7.45):

$$0 = c_1^1(\sin KL_1 + \sin h\, KL_1) + c_3^1(\cos KL_1 + \cos h\, KL_1) \qquad (7.53)$$

$$c_1^2 + c_2^2 = c_1^1(\cos KL_1 + \cos h\, KL_1) + c_3^1(-\sin KL_1 + \sin h\, KL_1) \qquad (7.54)$$

$$-2c_3^2 = c_1^1(-\sin KL_1 + \sin h\, KL_1) + c_3^1(-\cos KL_1 + \cos h\, KL_1) \qquad (7.55)$$

$$0 = c_1^2 \sin KL_2 + c_2^2 \sin h\, KL_2 + c_3^2(\cos KL_2 - \cos h\, KL_2) \qquad (7.56)$$

$$c_1^3 + c_2^3 = c_1^2 \cos KL_2 + c_2^2 \cos h\, KL_2 + c_3^2(-\sin KL_2 - \sin h\, KL_2) \qquad (7.57)$$

$$-2c_3^3 = -c_1^2 \sin KL_2 + c_2^2 \sin h\, KL_2 + c_3^2(-\cos KL_2 - \cos h\, KL_2) \qquad (7.58)$$

$$0 = -c_1^3 \sin KL_3 + c_2^3 \sin h\, KL_3 + c_3^3(-\cos KL_3 - \cos h\, KL_3) \qquad (7.59)$$

$$0 = -c_1^3 \cos KL_3 + c_2^3 \cos h\, KL_3 + c_3^3(\sin KL_3 - \sin h\, KL_3) \qquad (7.60)$$

Solve equations (7.53), (7.56), and (7.60) for c_3^1, c_3^2, c_3^3, respectively:

$$c_3^1 = -c_1^1 \frac{(\sin KL_1 + \sin h\, KL_1)}{(\cos KL_1 + \cos h\, KL_1)} \tag{7.61}$$

$$c_3^2 = \frac{c_1^2 \sin KL_2 + c_2^2 \sin h\, KL_2}{\cos h\, KL_2 - \cos KL_2} \tag{7.62}$$

$$c_3^3 = \frac{c_1^3 \cos KL_3 - c_2^3 \cos h\, KL_3}{\sin KL_3 - \sin h\, KL_3} \tag{7.63}$$

Substitute equations (7.61) to (7.63) into equations (7.54), (7.55), (7.57), (7.58), and (7.59):

$$c_1^2 + c_2^2 = c_1^1(\cos KL_1 + \cos h\, KL_1) + c_1^1\left(\frac{\sin KL_1 + \sinh KL_1}{\cos KL_1 + \cosh KL_1}\right)(\sin KL_1 - \sinh KL_1) \tag{7.64}$$

$$2\left(\frac{c_1^2 \sin KL_2 + c_2^2 \sinh KL_2}{\cos KL_2 - \cosh KL_2}\right) = c_1^1(\sinh KL_1 - \sin KL_1) + c_1^1\left(\frac{\sin KL_1 + \sin h\, KL_1}{\cos KL_1 + \cos h\, KL_1}\right)(\cos KL_1 - \cos h\, KL_1) \tag{7.65}$$

$$c_1^3 + c_2^3 = c_1^2 \cos KL_2 + c_2^2 \cos h\, KL_2 + \left(\frac{c_1^2 \sin KL_2 + c_2^2 \sin h\, KL_2}{\cos h\, KL_2 - \cos KL_2}\right)(-\sin KL_2 - \sin h\, KL_2) \tag{7.66}$$

$$2\left(\frac{c_1^3 \cos KL_3 - c_2^3 \cos h\, KL_3}{\sin h\, KL_3 - \sin KL_3}\right) = c_1^2 \sin KL_2 + c_2^2 \sin h\, KL_2 + \left(\frac{c_1^2 \sin KL_2 + c_2^2 \sin h\, KL_2}{\cos h\, KL_2 - \cos KL_2}\right)(-\cos KL_2 - \cos h\, KL_2) \tag{7.67}$$

$$0 = -c_1^3 \sin KL_3 + c_2^3 \sin h\, KL_3 + \left(\frac{c_1^3 \cos KL_3 - c_2^3 \cos h\, KL_3}{\sin KL_3 - \sin h\, KL_3}\right)(-\cos KL_3 - \cos h\, KL_3) \tag{7.68}$$

Solve equations (7.64) and (7.68) for c_1^1 and c_1^3, respectively:

$$c_1^1 = \frac{(c_1^2 + c_2^2)(\cos KL_1 + \cos h\, KL_1)}{(\sin KL_1 + \sin h\, KL_1)(\sin KL_1 - \sin h\, KL_1) + (\cos KL_1 + \cos h\, KL_1)^2} \tag{7.69}$$

$$c_1^3 = c_2^3 \frac{\left\{\sin KL_3 - (\cos h\, KL_3)\left(\dfrac{\cos KL_3 + \cos h\, KL_3}{\sin h\, KL_3 - \sin KL_3}\right)\right\}}{\left\{\sin KL_3 - (\cos KL_3)\left(\dfrac{\cos KL_3 + \cos h\, KL_3}{\sin h\, KL_3 - \sin KL_3}\right)\right\}} \tag{7.70}$$

Substitute equations (7.69) and (7.70) into equations (7.65) to (7.67):

$$c_1^2 \sin KL_2 + c_2^2 \sin h\, KL_2 = \left\{\frac{\cos KL_2 - \cos h\, KL_2}{2}\right\}$$

$$\times \left\{\frac{(c_1^2 + c_2^2)(\cos KL_1 + \cos h\, KL_1)}{(\sin KL_1 + \sin h\, KL_1)(\sin KL_1 - \sin h\, KL_1) + (\cos KL_1 + \cos h\, KL_1)^2}\right\}$$

$$\times \left\{\sin KL_1 - \sin KL_1 - \left(\frac{-\cos KL_1 + \cos h\, KL_1}{\cos KL_1 + \cos h\, KL_1}\right)(\sin KL_1 + \sin h\, KL_1)\right\} \tag{7.71}$$

$$c_2^3\left\{1 + \frac{\sin h\, KL_3 - (\cos h\, KL_3)\left(\dfrac{\cos KL_3 + \cos h\, KL_3}{\sin h\, KL_3 - \sin KL_3}\right)}{\sin KL_3 - (\cos KL_3)\left(\dfrac{\cos KL_3 + \cos h\, KL_3}{\sin h\, KL_3 - \sin KL_3}\right)}\right\} =$$

$$c_1^2\left\{\cos KL_2 - (\sin KL_2)\left(\frac{\sin KL_2 + \sin h\, KL_2}{-\cos KL_2 + \cos h\, KL_2}\right)\right\}$$

$$+ c_2^2\left\{\cos h\, KL_2 - (\sin h\, KL_2)\left(\frac{\sin KL_2 + \sin h\, KL_2}{-\cos KL_2 + \cos h\, KL_2}\right)\right\} \tag{7.72}$$

$$c_2^3\left\{-\cos h\, KL_3 + \cos KL_3\left[\frac{\sin h\, KL_3 - (\cos h\, KL_3)}{\sin KL_3 - (\cos KL_3)}\right.\right.$$

$$\left.\left.\frac{\left(\dfrac{\cos KL_3 + \cos h\, KL_3}{\sin h\, KL_3 - \sin KL_3}\right)}{\left(\dfrac{\cos KL_3 + \cos h\, KL_3}{\sin h\, KL_3 - \sin KL_3}\right)}\right]\right\} = \left\{\frac{\sin h\, KL_3 - \sin KL_3}{2}\right\}$$

$$\times \left\{c_1^2\left[-\sin KL_2 - (\sin KL_2)\left(\frac{\cos KL_2 + \cos h\, KL_2}{-\cos KL_2 + \cos h\, KL_2}\right)\right]\right.$$

$$\left.+ c_2^2\left[\sin h\, KL_2 - (\sin h\, KL_2)\left(\frac{\cos KL_2 + \cos h\, KL_2}{-\cos KL_2 + \cos h\, KL_2}\right)\right]\right\} \tag{7.73}$$

For convenience define the following functions.

$$D_1 \equiv \frac{\cos h\, KL_1 - \cos KL_1}{\cos h\, KL_1 + \cos KL_1} \tag{7.74}$$

$$D_2 \equiv \frac{\cos h\, KL_2 - \cos KL_2}{\cos h\, KL_2 + \cos KL_2} \tag{7.75}$$

$$D_3 \equiv \frac{\sin h\, KL_2 + \sin KL_2}{\cos h\, KL_2 - \cos KL_2} \tag{7.76}$$

$$D_4 \equiv \frac{\sin h\, KL_3 - \sin KL_3}{\cos h\, KL_3 + \cos KL_3} \tag{7.77}$$

Substituting equations (7.74) to (7.77) into equations (7.71) to (7.73):

$$c_1^2 \sin KL_2 + c_2^2 \sin h\, KL_2 = \left\{ \frac{\cos KL_2 - \cos h\, KL_2}{2} \right\} \times$$
$$\left\{ \frac{(c_1^2 + c_2^2)(\cos KL_1 + \cos h\, KL_1)}{(\cos KL_1 + \cos h\, KL_1)^2 + (\sin KL_1 + \sin h\, KL_1)(\sin KL_1 - \sin h\, KL_1)} \right\}$$
$$\times \{\sin h\, KL_1 - \sin KL_1 - D_1(\sin KL_1 + \sin h\, KL_1)\} \tag{7.78}$$

$$c_2^3 \left\{ 1 + \left(\frac{\sin h\, KL_3 - \dfrac{\cos h\, KL_3}{D_4}}{\sin KL_3 - \dfrac{\cos KL_3}{D_4}} \right) \right\} =$$
$$c_1^2\{\cos KL_2 - D_3 \sin KL_2\}$$
$$+ c_2^3\{\cos h\, KL_2 - D_3 \sin h\, KL_2\} \tag{7.79}$$

$$c_2^3 \left\{ -\cos h\, KL_3 + \cos KL_3 \left(\frac{\sin h\, KL_3 - \dfrac{\cos h\, KL_3}{D_4}}{\sin KL_3 - \dfrac{\cos KL_3}{D_4}} \right) \right\} =$$
$$\left\{ \frac{\sin h\, KL_3 - \sin KL_3}{2} \right\} \left\{ c_1^2(\sin KL_2)\left(-1 - \frac{1}{D_2}\right) \right.$$
$$\left. + c_2^2(\sin h\, KL_2)\left(1 - \frac{1}{D_2}\right) \right\} \tag{7.80}$$

For convenience again, define another function as follows.

$$D_5 \equiv \frac{\sin h\, KL_3 - \dfrac{\cos h\, KL_3}{D_4}}{\sin KL_3 - \dfrac{\cos KL_3}{D_4}} = \frac{D_4 \sin h\, KL_3 - \cos h\, KL_3}{D_4 \sin KL_3 - \cos KL_3} \tag{7.81}$$

Solve equation (7.79) for c_2^3 and then substitute this result into equation (7.80), using equation (7.81).

$$\left\{ \frac{c_1^2(\cos KL_2 - D_3 \sin KL_2) + c_2^2(\cos h\, KL_2 - D_3 \sin h\, KL_2)}{1 + D_5} \right\}$$
$$\times \{-\cos h\, KL_3 + D_5 \cos KL_3\}$$

$$= \left\{\frac{\sin h\, KL_3 - \sin KL_3}{2}\right\}\left\{c_1^2(\sin KL_2)\left(-1 - \frac{1}{D_2}\right)\right.$$

$$\left. + c_2^2(\sin h\, KL_2)\left(1 - \frac{1}{D_2}\right)\right\} \tag{7.82}$$

Solve equations (7.78) and (7.82) for c_1^2 and then equate the resulting relations:

$$c_2^2\left[\frac{(\sin h\, KL_2)\left(1 - \frac{1}{D_2}\right)\left(\frac{\sin h\, KL_3 - \sin KL_3}{2}\right)}{\frac{(\cos KL_2 - D_3 \sin KL_2)(-\cos h\, KL_3 + D_5 \cos KL_3)}{1 + D_5}}\right.$$

$$\left. - \frac{\frac{(\cos h\, KL_2 - D_3 \sin h\, KL_2)(-\cos h\, KL_3 + D_5 \cos KL_3)}{1 + D_5}}{+ (\sin KL_2)\left(1 + \frac{1}{D_2}\right)\left(\frac{\sin h\, KL_3 - \sin KL_3}{2}\right)}\right]$$

$$= c_2^2\left[\frac{\left\{\frac{\cos KL_2 - \cos h\, KL_2}{2}\right\}}{-\left\{\frac{\cos KL_2 - \cos h\, KL_2}{2}\right\}}\right.$$

$$\frac{\left\{\frac{\cos KL_1 + \cos h\, KL_1}{(\sin KL_1 + \sin h\, KL_1)(\sin KL_1 - \sin h\, KL_1) + (\cos KL_1 + \cos h\, KL_1)^2}\right\}}{\left\{\frac{\cos KL_1 + \cos h\, KL_1}{(\sin KL_1 + \sin h\, KL_1)(\sin KL_1 - \sin h\, KL_1) + (\cos KL_1 + \cos h\, KL_1)^2}\right\}}$$

$$\left. \times \frac{\{\sin h\, KL_1 - \sin KL_1 - D_1(\sin KL_1 + \sin h\, KL_1)\} - \sin h\, KL_2}{\{\sin h\, KL_1 - \sin KL_1 - D_1(\sin KL_1 + \sin h\, KL_1)\} + \sin KL_2}\right] \tag{7.83}$$

Note that this last step leads to the degenerate form of equation (7.48) since the column matrix in equation (7.46) now corresponds to the first order matrix in terms of c_2^2.

Rearranging equation (7.83):

$$\frac{(D_2 - 1)(\sin h\, KL_2)(\sin h\, KL_3 - \sin KL_3)(1 + D_5)}{(D_2 + 1)(\sin KL_2)(\sin h\, KL_3 - \sin KL_3)(1 + D_5)}$$

$$\frac{+ 2 D_2(-\cos h\, KL_2 + D_3 \sin h\, KL_2)(D_5 \cos KL_3 - \cos h\, KL_3)}{+ 2 D_2(-\cos KL_2 + D_3 \sin KL_2)(-D_5 \cos KL_3 + \cos h\, KL_3)}$$

$$= \{[(\cos KL_2 - \cos h\, KL_2)(\cos KL_1 + \cos h\, KL_1)$$

$$
\begin{aligned}
&\times (\sin h\, KL_1 - \sin KL_1 - D_1[\sin KL_1 + \sin h\, KL_1])] \\
&-[2(\sin h\, KL_2)([\sin KL_1 + \sin h\, KL_1][\sin KL_1 - \sin h\, KL_1] \\
&+ [\cos KL_1 + \cos h\, KL_1]^2)]\} \div \{-[(\cos KL_2 - \cos h\, KL_2) \\
&\times (\cos KL_1 + \cos h\, KL_1)(\sin h\, KL_1 - \sin KL_1 - \\
&D_1[\sin KL_1 + \sin h\, KL_1])] \\
&+ [2(\sin KL_2)([\sin KL_1 + \sin h\, KL_1] \\
&\times [\sin KL_1 - \sin h\, KL_1] + [\cos KL_1 + \cos h\, KL_1]^2)]\}
\end{aligned}
\tag{7.84}
$$

Substitute equation (7.81) into equation (7.84):

$$
\begin{aligned}
&\{[(0.5)(-1 + D_2)(\sin h\, KL_2)(\sin h\, KL_3 - \sin KL_3) \\
&\quad \times (D_4 \sin KL_3 - \cos KL_3 + D_4 \sin h\, KL_3 - \cos h\, KL_3)] \\
&\quad + [D_2(-\cos h\, KL_2 + D_3 \sin h\, KL_2) \\
&\quad \times (D_4 \sin h\, KL_3 \cos KL_3 - D_4 \sin KL_3 \cos h\, KL_3)]\} \\
&\quad \div \{[(0.5)(1 + D_2)(\sin KL_2)(\sin h\, KL_3 - \sin KL_3) \\
&\quad \times (D_4 \sin KL_3 - \cos KL_3 + D_4 \sin h\, KL_3 - \cos h\, KL_3)] \\
&\quad + [D_2(\cos KL_2 - D_3 \sin KL_2) \\
&\quad \times (D_4 \sin h\, KL_3 \cos KL_3 - D_4 \sin KL_3 \cos h\, KL_3)]\} \\
&= \{[(0.5)(\cos KL_2 - \cos h\, KL_2)(\cos KL_1 + \cos h\, KL_1) \\
&\quad \times (\sin h\, KL_1 - \sin KL_1 - D_1[\sin KL_1 + \sin h\, KL_1])] \\
&\quad - [(\sin h\, KL_2)([\sin KL_1 + \sin h\, KL_1][\sin KL_1 - \sin h\, KL_1] \\
&\quad + [\cos KL_1 + \cos h\, KL_1]^2)]\} \\
&\quad \div \{-[(0.5)(\cos KL_2 - \cos h\, KL_2) \\
&\quad \times (\cos KL_1 + \cos h\, KL_1)(\sin h\, KL_1 - \sin KL_1 - D_1 \\
&\quad \times [\sin KL_1 + \sin h\, KL_1])] \\
&\quad + [(\sin KL_2)([\sin KL_1 + \sin h\, KL_1] \\
&\quad \times [\sin KL_1 - \sin h\, KL_1] + [\cos KL_1 + \cos h\, KL_1]^2)]\}
\end{aligned}
\tag{7.85}
$$

Equation (7.86) (with the functionals, D, given by equations [7.74] to [7.77] is the final form of the characteristic equation for free lateral vibrations of the beam depicted in figure 7.1. The characteristic equation is transcendental, and each of the infinite number of roots, K_i, which satisfies the equation, corresponds to a principal frequency.

Consider

$$\omega = z\pi f \tag{7.86}$$

Commutating equations (7.86) and (7.82) with equation (7.6), the relation between the principal frequencies and the roots of equation (7.85) is

$$f_i = \frac{K_i^2}{2\pi}\left[\frac{EI\,g}{\gamma A}\right]^{1/2} \tag{7.87}$$

Forced Harmonic Vibration of the Continuum

Suppose the beam in figure 7.1 has equal circular functionals applied at each support. Let the functionals be simple harmonic accelerations, i.e.,

$$\ddot{y}(@x_1 = L_1) = -\ddot{Y}\sin pt \tag{7.88}$$

$$\ddot{y}(@x_2 = L_2) = \ddot{y}(@x_1 = L_1) \tag{7.89}$$

Integrating,

$$y_3 = \frac{Y_s \sin pt}{p^2} \tag{7.90}$$

The differential equation of motion is still valid, and its general solution is of the form

$$y = X \sin pt \tag{7.91}$$

Substituting equation (7.91) into the differential equation of motion yields

$$\frac{d^4X}{dx^4} = \frac{p^2}{a^2} X \tag{7.92}$$

Let

$$\alpha^4 \equiv \frac{p^2}{a^2} \tag{7.93}$$

The general solution of equation (7.93) is

$$X_1 = c_1^1 \sin \alpha x + c_2^1 \sin h\, \alpha x + c_3^1 \cos \alpha x + c_4^1 \cos h\, \alpha x \tag{7.94}$$

$$X_2 = c_1^2 \sin \alpha x + c_2^2 \sin h\, \alpha x + c_3^2 \cos \alpha x + c_4^2 \cos h\, \alpha x \tag{7.95}$$

$$X_3 = c_1^3 \sin \alpha x + c_2^3 \sin h\, \alpha x + c_3^3 \cos \alpha x + c_4^3 \cos h\, \alpha x \tag{7.96}$$

The boundary conditions are

$$X_1''(@x_1 = 0) = 0 \tag{7.97}$$

$$X_1'''(@x_1 = 0) = 0 \tag{7.98}$$

$$X_1(@x_1 = L_1) = \frac{Y_s}{p^2} \tag{7.99}$$

$$X_1'(@x_1 = L_1) = X_2'(@x_2 = 0) \tag{7.100}$$

$$X_2(@x_2 = 0) = \frac{Y_s}{p^2} \tag{7.101}$$

$$X_2''(@x_2 = 0) = X_1''(@x_1 = L_1) \tag{7.102}$$

TABLE 119
Nomenclature

Symbol	Variable	Typical units
A	Cross-sectional area	$inch^2$
B	Mode participation factor	inch
C	Constant	numeric
E	Young's modulus	psi
F( )	Function of	—
f	Frequency	cps
g	Conversion factor	$\frac{lb_m\text{-ft}}{lb_f\text{-sec}^2}$
I	Geometric moment of inertia	$inch^4$
L	Span length	inch
p	Forced angular velocity	rad/sec
t	Time	second
X	Normal function	numeric
x	Coordinate along the beam	inch
Y	Maximum excursion of y	inch
y	Deflection	inch
y_d	Displacement due to forcing functional	inch
y_f	Displacement due to free vibration	inch
y_3	Displacement at the support	inch
γ	Weight per unit volume	lb_f/in^3
ω	Principal angular velocity	rad/sec

$$X_2\,(@x_2 = L_2) = \frac{Y_s}{p^2} \tag{7.103}$$

$$X_2'\,(@x_2 = L_2) = X_3'\,(@x_3 = 0) \tag{7.104}$$

$$X_3\,(@x_3 = 0) = 0 \tag{7.105}$$

$$X_3''\,(@x_3 = 0) = X_2''\,(@x_2 = L_2) \tag{7.106}$$

$$X_3''\,(@x_3 = L_3) = 0 \tag{7.107}$$

$$X_3''\,(@x_3 = L_3) = 0 \tag{7.108}$$

The constants in the normal functions are found in the same manner as was done with free vibrations, i.e., substitute equations (7.97) to (7.108) into equations (7.94) to (7.96) and their deviations and solve the 12 equations simultaneously. After performing the indicated algebraic manipulations, the normal functions will be known explicitly. The general solution will be of the form

$$y_{d_1} = X_1 \sin pt \tag{7.109}$$

$$y_{d_2} = X_2 \sin pt \tag{7.110}$$

$$y_{d_3} = X_3 \sin pt \tag{7.111}$$

The complete solution then is

$$y = y_d + y_f \tag{7.112}$$

The component of displacement due to free vibration is determined by (a) compatibility of superposition and (b) initial conditions. To satisfy condition b, the normal function for free vibration is obtained by setting B = 0 in the boundary conditions, which has already been discussed.

Assume for the initial conditions (IC):

$$y(@t = 0) = 0 \tag{7.113}$$

$$\dot{y}(@t = 0) = 0 \tag{7.114}$$

Equations (7.109) to (7.111) satisfy IC 11. So the IC on y_f are

$$y_f(@t = 0) = 0 \tag{7.115}$$

$$\dot{y}_f(@t = 0) = -\dot{y}_d(@t = 0) \tag{7.116}$$

Since the free vibrations are given, then

$$B_{c_i} = 0 \tag{7.117}$$

Substituting equations (7.116) and (7.117)

$$-\dot{y}_d(@t = 0) = \sum_{i=1}^{\infty} X_{f_i} B_{s_i} \omega_i \tag{7.118}$$

Consider the orthogonality of the Eigen vectors associated with each Eigen value; to wit,

$$\int_o^L X_i X_j \, dx = 0, \; i \neq j \tag{7.119}$$

The mode participation factor, B_{s_i}, is then solved explicitly by multiplying both sides of equation (7.118) by X_{f_i} and integrating over L, using equation (7.119):

$$B_{s_i} = -\frac{\int_o^L \dot{y}_d(@t = 0) X_{f_i} \, dx}{\int_o^L X_{f_i}^2 \, w_i \, dx} \tag{7.120}$$

Equation (7.112) can now be written as

$$y = X_d \sin pt + \sum_{i=1}^{\infty} X_{f_i} (B_i \sin \omega_i t) \tag{7.121}$$

and the solutions for all of the constants have now been indicated.

7.2 FREQUENCY RESPONSE IN A NONLINEAR SYSTEM USING THE VARIATIONAL TECHNIQUE

The calculus of variations is a mathematical tool that can sometimes be used most efficaciously for a practical solution of problems in the discipline of engineering mechanics and more specifically in vibration analysis. The approach is to minimize a definite integral of a function which represents the action of a system. The differentials associated with these functions are perturbations of the dependent variables. To elucidate the approach of variational calculus, this technique is applied in a simple dynamic analysis. Intermediate steps are included for convenience.

Consider the singular system (figure 7.2) consisting of a weight and a nonlinear, elastic spring whose modulus character is

TABLE 120
Nomenclature

Symbol	Variable	Typical units
A	Maximum excursion in an oscillation	inch
B	Velocity at the initial condition	in/sec
E	Modulus of Elasticity	psi
f	Frequency	cps
I	Action integral	—
K	Spring constant	lb/in
L	Length of spring	inch
t	Time	sec
W	Weight	lb
x	Detrusion	inch
β	Action of the system	—
ϕ	Phase angle	radius
δ_x	Variation of X	inch
ω	Angular velocity	rad/sec
τ	Period of oscillation	sec
ψ	Functional parameter	radians
ε	Strain	in/in
σ	Stress	psi

portrayed in figure 7.3. Let the desideratum be to deduce the approximate natural frequency for any arbitrary bounds of linearity preceding the flat portion of the response.

To study the dynamic response of this simple system, consider the initial conditions:

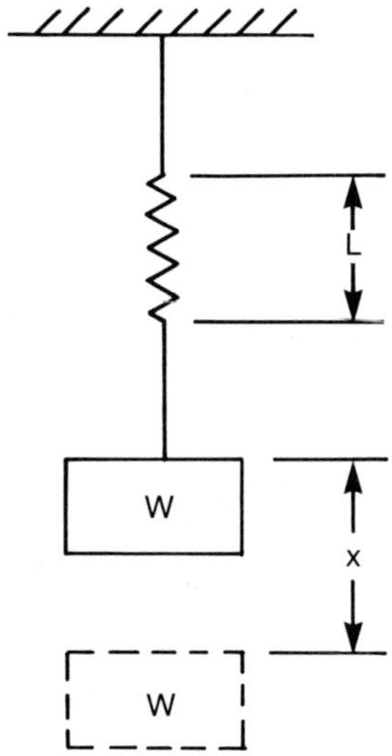

FIG. 7.2 Singular system

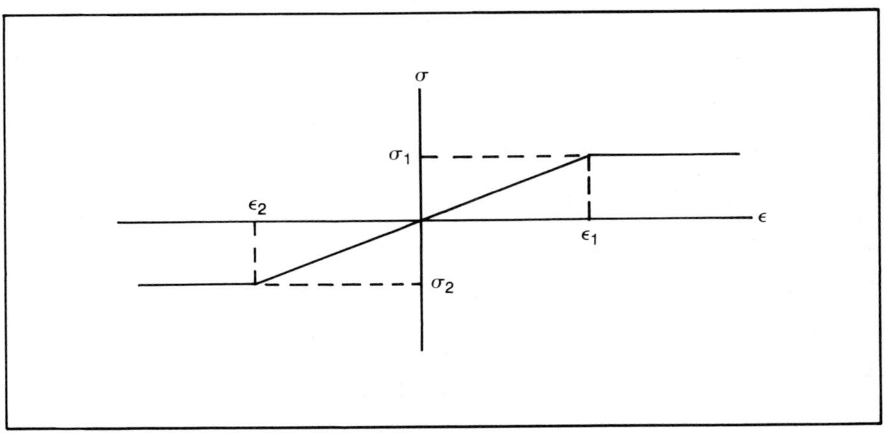

FIG. 7.3 Elastic modulus

$$x(0) = 0 \quad (7.122)$$

$$\dot{x}(0) = B \quad (7.123)$$

The resulting equation of motion is:

$$-\frac{W}{g}\ddot{x} = [K(x)]x \quad (7.124)$$

The spring constant is:

$$K(x) = \frac{C\,E(x)}{L} \quad (7.125)$$

With prescience for simplification, let

$$F(x) \equiv \frac{g\,C\,E(x)}{L\,W} \quad (7.126)$$

Substituting equations (7.125) and (7.126) into equation (7.124):

$$\ddot{x} + [F(x)]x = 0 \quad (7.127)$$

The Action Integral for the System

The action of the system is defined by:

$$\beta \equiv \ddot{x} + [F(x)]x \quad (7.128)$$

The solution of equation (7.127) will have the form:

$$x = A \sin(\omega t + \phi) \quad (7.129)$$

Because of initial condition 1, $\phi = 0$ (7.130)

Therefore, $x = A \sin \omega t$ (7.131)

$$\dot{x} = A\,w \cos \omega t \quad (7.132)$$

$$\ddot{x} = -A\,w^2 \sin \omega t \quad (7.133)$$

and the variation of x is

$$\delta x = \delta A \sin \omega t + Aw(\cos \omega t)\delta w \quad (7.134)$$

But the principal frequency is a constant so $\delta\omega = 0$ (7.135)

and equation (7.133) becomes $\delta x = \delta A \sin \omega t$ (7.136)

From initial condition 2 and equation (7.132), obtain

$$A = B/\omega \quad (7.137)$$

Considering the action over one period, τ

$$\int_0^{\tau} \beta\, \delta x\, dt = 0 \quad (7.138)$$

Substituting relations (7.128) and (7.136) into equation (7.138):

$$\int (\ddot{x} + [F(x)]x)(\sin \omega t)\delta A\, dt = 0 \tag{7.139}$$

Since the maximum excursion is a constant, δA is arbitrary.

Therefore, let $\delta A = 1$ (7.140)

and let $z = \omega t$ (7.141)

Hence $dz = \omega dt$ (7.142)

and $\sin \omega t = \sin z$ (7.143)

Substituting relations (7.123, 129, 140, 141, and 143) into equation (7.139) with the appropriate limit transformation yields:

$$I = \int_o^{2\pi} (-w^2 + F[x])(\sin^2 z)A\, dz \tag{7.144}$$

Taking the special case of $a = -c$ (7.145)

and $b = -d$ (7.146)

then equation (7.144) simplifies to

$$I = 4\int_o^{\pi/2} (-w^2 + F[x])(\sin^2 z)A\, dz \tag{7.147}$$

Solutions of the Action Integral

Case I, A > b. Introducing the functions (figure 7.4) into equation (7.147):

$$4\int_o^{\pi/2} - w^2 (\sin^2 z)\, Adz + 4\int_o^{\phi} \frac{b}{a}(\sin^3 z)\, A^2 dz$$
$$+ 4\int b(\sin^2 z)Adz = 0 \tag{7.148}$$

Consider the operators:

$$\int \sin^2 u\, du \equiv \frac{u}{2} - \frac{\sin 2u}{4} \tag{7.149}$$

$$\int \sin^3 u\, du \equiv -\left(\frac{\cos u}{3}\right)\left(\sin^2 u + 2\right) \tag{7.150}$$

Using these operators in equation (7.148):

$$-w^2 A\frac{\pi}{4} - \frac{bA^2}{\alpha}\left[\frac{\cos\phi}{3}(\sin^2\phi + 2) - \frac{2}{3}\right] + bA\left[\frac{\pi}{4} - \frac{\phi}{2} + \frac{\sin 2\phi}{4}\right] = 0 \quad (7.151)$$

Now observe from figure 7.4 that

$$A \sin\phi = b \quad (7.152)$$

and from equation (7.152) obtain

$$\sin\phi = \frac{b}{A} \quad (7.153)$$

and

$$\cos\phi = \frac{\sqrt{A^2 - b^2}}{A} \quad (7.154)$$

Then

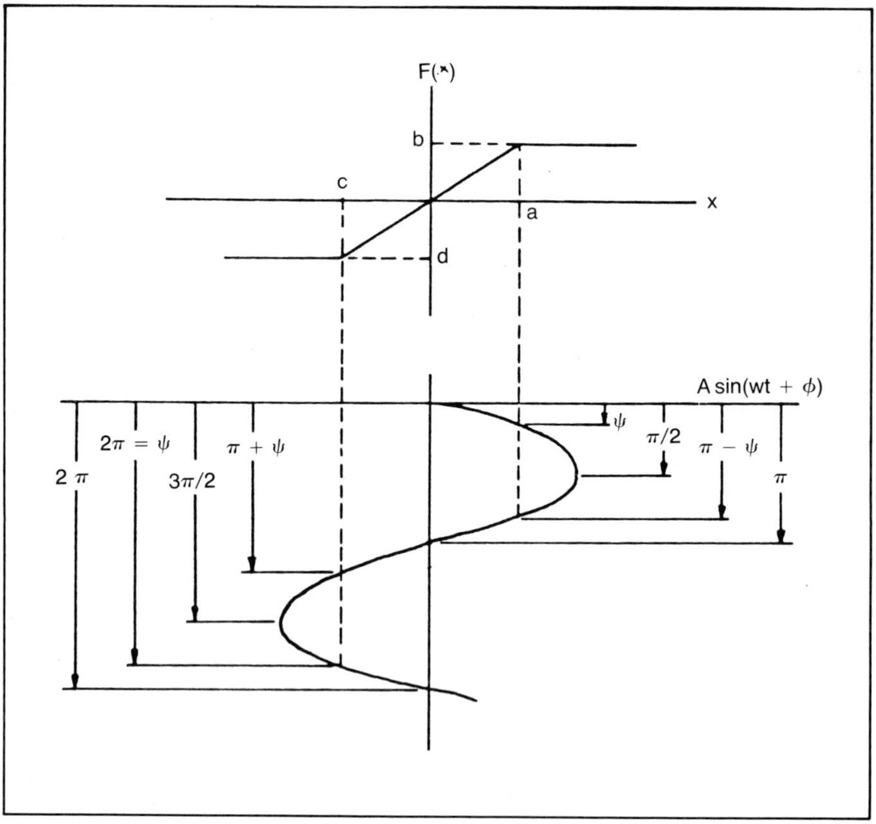

FIG. 7.4 Functionals

$$\phi = \text{arc sin}\left(\frac{b}{A}\right) \tag{7.155}$$

Also consider the identity

$$\sin 2\phi \equiv 2 \sin \phi \cos \phi \tag{7.156}$$

Inserting relation (7.137) into equations (7.153) to (7.156):

$$\sin \phi = \frac{bw}{B} \tag{7.157}$$

$$\cos \phi = \frac{\sqrt{B^2 - b^2w^2}}{B} \tag{7.158}$$

$$\phi = \text{arc sin}\left(\frac{bw}{B}\right) \tag{7.159}$$

$$\sin 2\phi = \frac{2bw}{B^2}\sqrt{B^2 - b^2w^2} \tag{7.160}$$

Substituting relations (7.137 and 157 to 160) into equation (7.151) now yields the characteristic equation:

$$(B^2 - b^2w^2)^{1/2}\left(\frac{b^2}{2B} - \frac{b^3}{3aB} - \frac{2bB}{3aw^2}\right) - \frac{w\pi B}{4} + \frac{2bB^2}{3aw^2} + \frac{6B\pi}{4w} - \frac{bB}{2w}\text{arc sin}\left(\frac{bw}{B}\right) = 0 \tag{7.161}$$

Given the numerical values for the constants a, b, and B, the single value of w which satisfies equation (7.161) can now be determined by trial and error. The natural frequency in cps is then obtained from

$$f = \frac{w}{2\pi} \tag{7.162}$$

Knowing the natural frequency, the maximum excursion is then determined from equation (7.137).

Case II, $A < b$

The second case is somewhat trivial since the physical premise is that the spring responds entirely linearly, i.e.,

$$f(x) = \frac{b}{a}x \tag{7.163}$$

Substituting relation (7.131) into equation (7.163)

$$[F(x)]\,x = \frac{b}{a} A \sin \omega t \tag{7.164}$$

Substituting relations (7.133 and 7.164) into equation (7.127):

$$-Aw^2 \sin \omega t + \frac{b}{a} A \sin \omega t = 0 \tag{7.165}$$

The natural frequency obtained from equation (7.165) is

$$f = \frac{1}{2\pi}\sqrt{\frac{b}{a}} \tag{7.166}$$

Conclusions

The dynamic analysis of a singular system having a nonlinear spring vividly demonstrates the utility of the variational calculus approach. This approach can be used in various studies of the dynamic response of systems wherein the nonlinear viscoelastic effects of the system must be considered.

7.3 FREQUENCY ENTRAINMENT ANALYSIS

The driven nonlinear system may have several unusual types of behavior which no purely linear system can experience. Some of these phenomena unique to the nonlinear system are discontinuous amplitude jumps, generation of subharmonics, and frequency entrainment. The latter was the subject of this problem study.

If the autoperiodic (natural) frequency of a system is near the heteroperiodic (driving) frequency, then the frequency of the system will be the same as the driver, i.e., the natural frequency is said to be entrained (figure 7.5). Frequency entrainment is, therefore, seen to be the limit of the beat frequency phenomenon in nonlinear systems. If the heteroperiodic frequency, w_o, is near an integral multiple of the autoperiodic frequency, w, the frequency of oscillation in the system can be w_o. This phenomena is referred to as frequency demultiplication and as frequency detuning.

In a conservative system whose behavior is mathematically represented by a DE (differential equation), only one solution of that DE exists where the constants in that solution are determined from the initial (or boundary) conditions. An infinite number of solutions exist for the nonlinear system having dissipative influences, however, since the solution is a complex function of the initial conditions. For that reason, the usual approach is to study the nature of these systems by observing their topology (i.e., phase planes, singularities, etc.).

The objective here was to study the system behavior and to determine the stability of the system represented by the DE

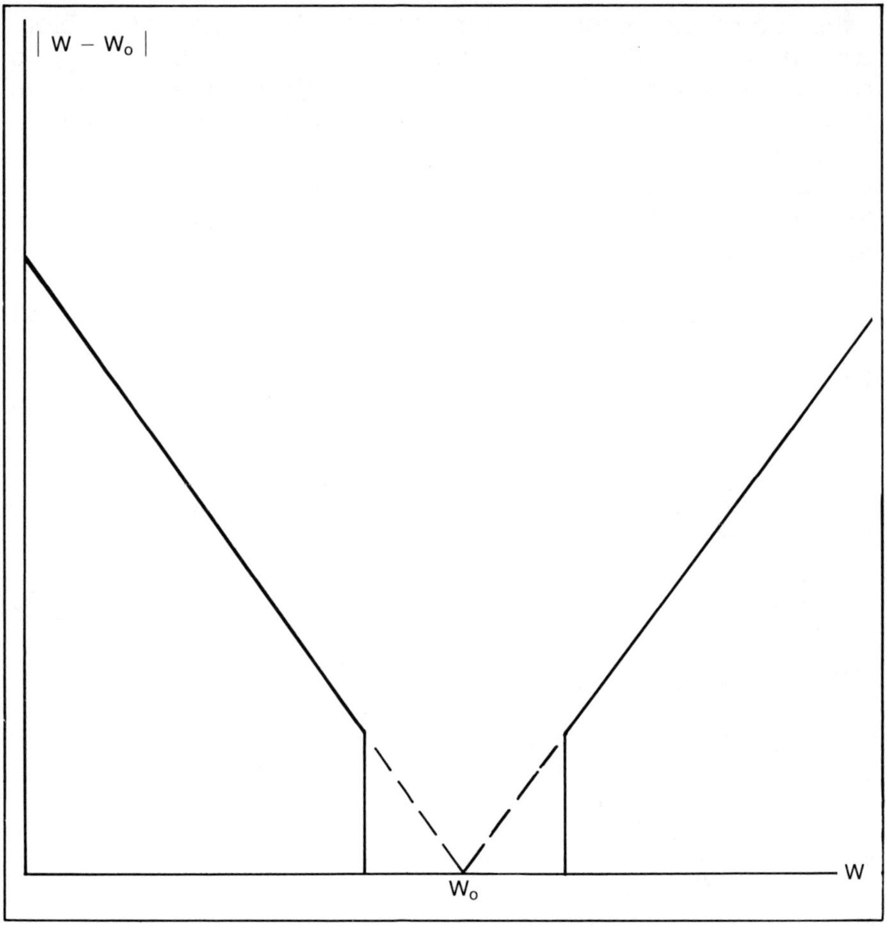

FIG. 7.5 Synchronization plane

$$\ddot{x} - \alpha x + \gamma x^3 + w_o^2 x = Bw_1^2 \sin w_1 t \tag{7.167}$$

for various values of the parameters. Stability here means periodic oscillations and unstable indicates near periodic oscillations. The procedure used was to utilize the Andronov-Witt topological approach[(2)] in conjunction with the experimental approach via the electronic analog computer.

Discussion

A Simple Ergchorad. As a prelude to the frequency entrainment study, a simple example of the ergodic (energy path) or ergchorad (energy area) problem was studied. The behavior exhibited by the parametric equations for these types of problems are the manifestation of a space filling curve, to wit, any system which has an almost periodic solution. Such a curve is now commonly known by mathematicians as Peano's curve Possibly the most simple demonstration of this behavior is that described by a nonautonomous conservative linear differential equation representing a system whose natural and forced frequencies are incommensurable, e.g.,

$$\frac{dx^2}{dt^2} + 2x = -2\cos 2t \tag{7.168}$$

Let the initial conditions be:

$$x(0) = 1,\ x(0) = \sqrt{2} \tag{7.169a, b}$$

To work the problem on an analog computer it was desirable to "slow" the time with a scale factor, i.e.,

$$t = \frac{\tau}{10} \tag{7.170}$$

which also gives

$$\frac{dx}{dt} = 10\frac{dx}{d\tau},\ \frac{d^2x}{dt^2} = 100\frac{d^2x}{d\tau^2} \tag{7.171a, b}$$

Substituting equation (7.171b) into equation (7.168) and rearranging yields a suitable machine equation:

$$\frac{d^2x}{d\tau^2} = -0.02x - 0.02\cos 0.2\tau \tag{7.172}$$

and equation (7.169) becomes

$$x(0) = 1,\ \frac{dx}{d\tau} = \frac{\sqrt{2}}{10} \doteq 0.1414 \tag{7.173a, b}$$

The analog computer circuit for equations (7.172) and (7.173) is given in figure 7.6, and the resultant phase plane plot (velocity vs. displacement) is given in figure 7.7. The arrowheads were added to show the direction of the action. The recording pen was stopped at point A since sufficient action was already shown. In an indefinitely long period of time, the path would completely cover the area of an ellipse whose semiaxis equals the period of the natural and forced frequencies (1), i.e.,

$$\text{semiaxis} \qquad = \frac{2\pi}{\omega} \tag{7.174a}$$

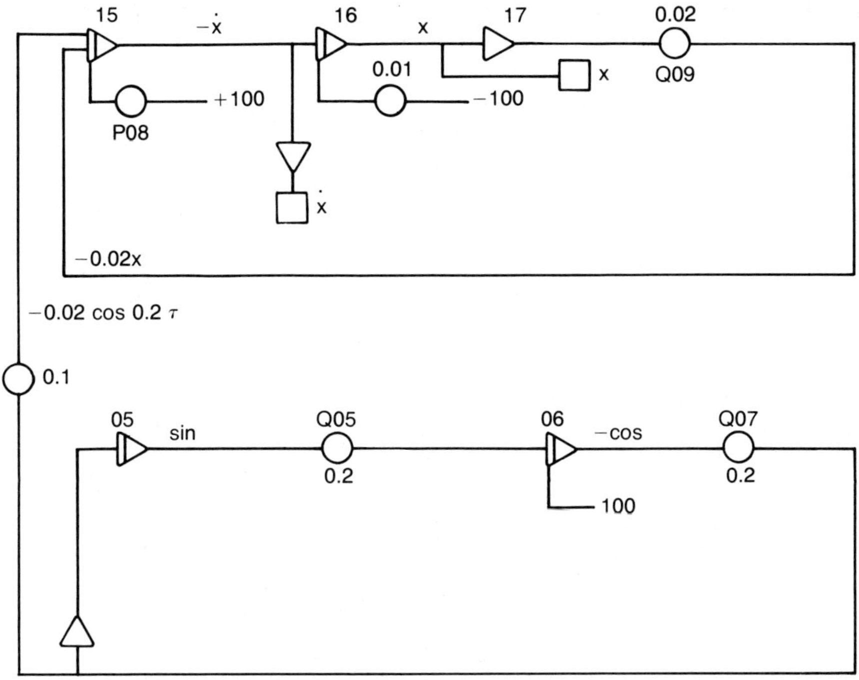

X : 50 volt/in., Y : 10 volt/in., τ = 10t

$\ddot{x} + 2x = -2 \cos 2t$

FIG. 7.6 Analog computer circuit for a simple Ergchorad problem

$$\text{major diameter} = 2\left(\frac{2\pi}{2}\right) \tag{7.174b}$$

$$\text{minor diameter} = 2\left(\frac{2\pi}{\sqrt{2}}\right) \tag{7.174c}$$

Andronov-Witt Topological Approach. Assume the Van der Pol type of solution for equation (7.167) in the form

$$x = b_1 \sin \omega_1 t + b_2 \cos \omega_1 t \tag{7.175}$$

where b_1 and b_2 are slowly varying functionals of time. Substituting equation (7.175) into equation (7.167), disposing of negligible second order derivatives of b_1 and b_2 (i.e., assuming α and γ are small), yields

$$2\dot{b}_1 + z\,b_2 - \alpha\, b_1\left(1 - \frac{b^2}{a_o^2}\right) = 0 \tag{7.176a}$$

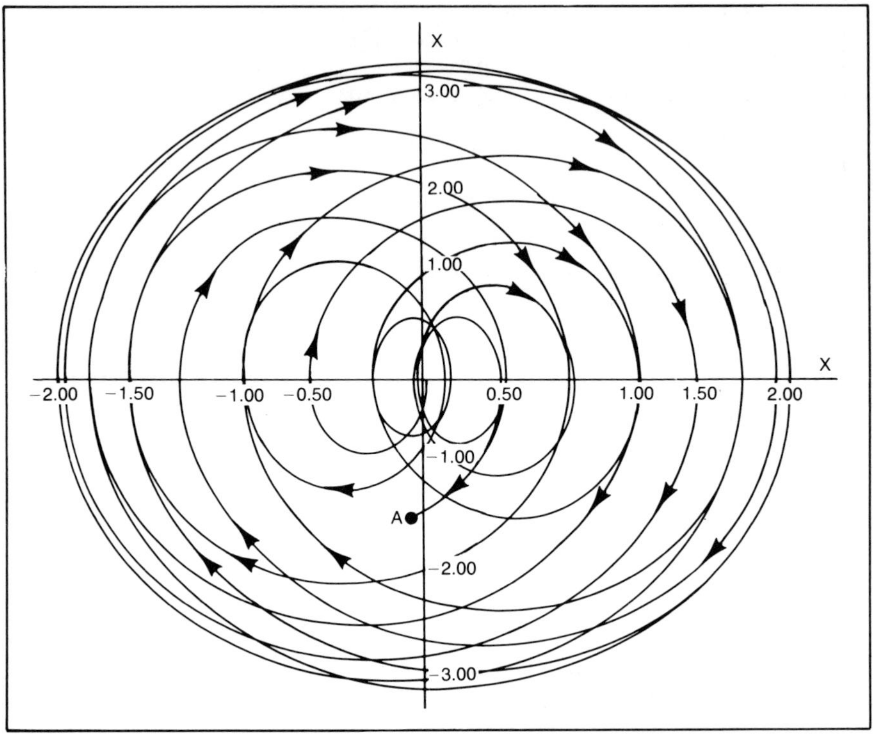

FIG. 7.7 Phase phane plot of simple Ergchorad problem

$$2\dot{b}_2 - z\,b_1 - \alpha b_2\left(1 - \frac{b^2}{a_o^2}\right) = 0 \qquad (7.176a)$$

where

$$z = 2(\omega_o - \omega_1),\ b^2 = b_1^2 + b_2^2,\ a_o^2 = \frac{4\alpha}{3\gamma} \qquad (7.177a, b, c)$$

Now let

$$\bar{x} = \frac{b_1}{a_o},\ \bar{y} = \frac{b_2}{a_o},\ a = \frac{z}{\alpha} \qquad (7.178a, b, c)$$

$$A = -\frac{B\omega_1}{a_o\alpha},\ r^2 = \bar{x}^2 + \bar{y}^2,\ \tau = \frac{\alpha t}{2} \qquad (7.178d, e, f)$$

Substituting equation (7.178) into equation (7.176) yields

$$\frac{d\bar{x}}{d\tau} = \bar{x}(1 - r^2) - a\bar{y} \qquad (7.179a)$$

$$\frac{d\bar{y}}{d\tau} = a\bar{x} + \bar{y}(1 - r^2) + A \qquad (7.179b)$$

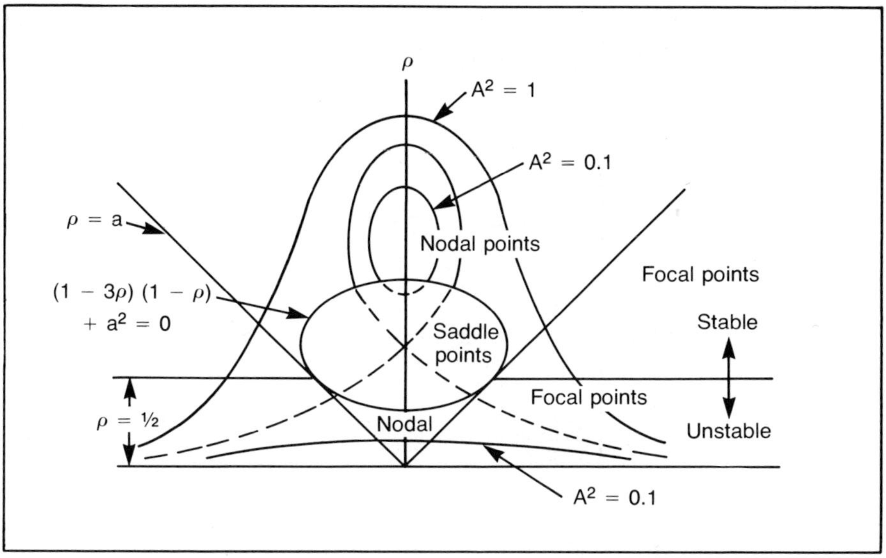

FIG. 7.8 Stability domain for periodicity

Equation (7.179) is then used to find the coordinates of the singular points, which are

$$\bar{x}_o = -\frac{\alpha\rho}{A}, \bar{y}_o = \frac{\rho(1-\rho)}{A} \tag{7.180a, b}$$

where

$$\rho = r_o^2 = x_o^2 + \bar{y}_o^2 \tag{7.181}$$

Substituting equation (7.180) into equation (7.181) yields

$$\rho[\alpha^2 + (1 - \rho^2)] = A^2 \tag{7.182}$$

which was found by Andronov-Witt (2) to be a suitable form of the variables from which the topology can be ascertained (figure 7.8).

The Andronov-Witt procedure was used to analyze the systems having the various values of the parameters as entered in table 121, and the results are given in the last column. There S and U mean stable and unstable, respectively, and N and F mean nodal and focal point, respectively. The natural frequency, w_o, was taken as unity in all of the systems. In general, the transient system response will be of shorter duration with nodal points as compared to focal points.

For nonlinear, dissipative systems which can be represented by equation (7.167), the Andronov-Witt topological approach conveniently yields approximate solutions when the parameters are small (compared with unity).

TABLE 121
Systems analyzed

Test	ρ	$\mathbf{a}$	$\mathbf{A}$	$-\mathbf{B}$	$\mathbf{w}_1$	$-\mathbf{B}_{w1}^2$	γ	α	$\mathbf{x}(0)$	$\mathbf{x}(0)$	Stability
1	3.00	2.00	4.90	5.65	0.50	1.41	0.50	0.50	0	0	SN
2	3.00	2.00	4.90	5.65	0.50	1.41	0.50	0.50	1.00	0	SN
3	3.00	2.00	4.90	5.65	0.50	1.41	0.50	0.50	0	1.00	SN
4	3.00	2.00	4.90	4.00	0.50	1.00	0.50	0.50	0	0	SN
5	3.00	2.00	4.90	4.20	0.50	1.05	0.50	0.50	0	0	SN
6	3.00	2.00	4.90	4.40	0.50	1.10	0.50	0.50	0	0	SN
7	3.00	2.00	4.90	4.60	0.50	1.15	0.50	0.50	0	0	SN
8	3.00	2.00	4.90	4.80	0.50	1.20	0.50	0.50	0	0	SN
9	3.00	2.00	4.90	4.80	0.50	1.20	0.50	0.50	3.00	3.00	SN
10	3.00	2.00	4.90	5.00	0.50	1.25	0.50	0.50	0	0	SN
11	1.00	2.00	2.00	2.30	0.50	0.58	0.50	0.50	0	0	SF
12	1.00	2.00	2.00	2.30	0.50	0.58	0.50	0.50	1.00	0	SF
13	1.00	2.00	2.00	2.30	0.50	0.58	0.50	0.50	0	1.00	SF
14	0.40	2.00	1.32	1.52	0.50	0.38	0.50	0.50	0	0	UF
15	1.50	1.00	1.37	4.46	0.50	1.12	0.50	1.00	0	0	SN

16	0.75	1.00	0.89	2.90	0.50	0.72	0.50	1.00	0	0	SF
17	0.25	1.00	0.625	2.04	0.50	0.51	0.50	1.00	0	0	UF
18	1.50	1.00	1.37	1.67	0.75	0.941	0.20	0.50	0	0	SN
19	0.75	1.00	0.89	1.08	0.75	0.608	0.20	0.50	0	0	SF
20	0.25	1.00	0.625	0.762	0.75	0.429	0.20	0.50	0	0	UF
21	0.75	1.00	0.89	2.06	0.80	1.32	0.20	0.80	0	0	SF
22	0.75	1.00	0.89	1.56	0.80	1.00	0.20	0.80	0	0	SF
23	0.75	1.00	0.89	1.17	0.80	0.75	0.20	0.80	0	0	SF
24	0.75	1.00	0.89	3.13	0.80	2.00	0.20	0.80	0	0	SF
25	1.50	0.25	0.584	0.756	0.90	0.612	1.00	0.80	0	0	SN
26	0.25	0.25	0.395	0	0.90	0	1.00	0.80	0	0	SN
27	0.25	0.25	0.395	0	0.90	0	1.00	0.80	2.00	2.00	SN
28	0.25	0.25	0.395	0.513	0.90	1.00	1.00	0.80	0	0	SN
29	0.25	0.25	0.395	0.513	0.90	2.00	1.00	0.80	0	0	SN
30	0.25	0.25	0.395	0.513	0.90	3.00	1.00	0.80	0	0	SN
31	0.25	0.25	0.395	0.513	0.90	5.00	1.00	0.80	0	0	SN

7.4 STABILITY ANALYSIS OF A NONLINEAR SYSTEM

Consider a single mass restrained by some combination of springs and dashpots so that the normalized effective spring constant is given by the relation

$$\frac{K}{m} = 3x - x^2 - 2 + \frac{\dot{x}}{x} \tag{7.183}$$

where Newtonian notation for ordinary differentiation, with respect to time, has been used. Whether this relation was developed from a discrete schematization of a solid propellant grain or some other physical possibility is unimportant for present purposes.

Using Hamilton's principle with equation (7.183), the resulting equation of dynamic equilibrium is

$$\ddot{x} + \dot{x} - 2x + 3x^2 - x^3 = 0 \tag{7.184}$$

This is seen to be a second order, first degree, nonlinear differential equation having constant coefficients. One of the peculiarities of this type of equation is that a different form of solution exists for each set of initial conditions. This characteristic is in contradistinction to any particular linear differential equation which has only one form of solution where the constants in the solution (but not the form) are a function of the initial (or boundary) conditions. For this reason, a topological configuration of a complex nonlinear equation, rather than the possibly large number of explicit solutions, is usually considered sufficient.

The problem then resolves to finding the type and location of singularities and describing the stability (or instability) in the propinquity of these singularities. The desideratum of this etude was to ascertain the topological configuration for the system whose mathematical representation is given by equation (7.184).

Discussion

The most commonly used procedure for analyzing a high-order nonlinear equation is to reduce the equation into several simultaneous first-order equations. Each of the first-order equations are generally limited to two dependent variables for convenience in graphical representations. The independent variable in the given formulation is time, t.

To reduce equation (7.184) to first-order equations, consider the identities

$$x_1 \equiv x \tag{7.185}$$

$$x_2 \equiv \dot{x}_1 \tag{7.186}$$

Substituting equation (7.186) into equation (7.184) yields

$$\dot{x}_2 + x_2 - 2x_1 + 3x_1^2 - x_1^3 = 0 \tag{7.187}$$

To eliminate the secular term, consider the manipulation

$$\dot{x}_2 = \frac{d}{dt} x_2 \frac{dt}{dx_1} \frac{dx_1}{dt} = \frac{dx_2}{dx_1} x_2 \tag{7.188}$$

Substituting equation (7.188) into equation (7.187) yields

$$\frac{dx_2}{dx_1} = \frac{x_2 - 2x_1 + 3x_1^2 - x_1^3}{-x_2} \tag{7.189}$$

A singularity can be defined as the point where the first derivatives of the dependent variables are simultaneously zero, i.e.,

$$\frac{dx_2}{dt} = 0 \tag{7.190}$$

$$\frac{dx_1}{dt} = 0 \tag{7.191}$$

A singular point is always a point of equilibrium, which may be stable or unstable. The stability concept is illustrated by the simple, classical example shown in figure 7.9. Now, dividing equation (7.190) by equation (7.191) and substituting in equation (7.189) yields

$$\frac{x_2 - 2x_1 + 3x_1^2 - x_1^3}{-x_2} = \frac{0}{0} \tag{7.192}$$

To satisfy this equation the only value that the dependent variable, x_2, can have is zero. Hence

$$x_1^3 - 3x_1^2 + 2x_1 = 0 \tag{7.193}$$

Factoring equation (7.193),

$$x_1(x_1^2 - 3x_1 + 2) = x_1(x_1 - 1)(x_1 - 2) = 0 \tag{7.194}$$

This equation shows that singularities occur when x_1 has values of 0, 1, and 2. The system therefore has 3 singular points whose phase-plane coordinates are (0, 0), (0, 1), and (0, 2).

The physical significance of the phase-plane plot can be explained quite simply. If the mass is given some initial condition which lies on any path that directs itself into the focus, the mass will first experience a few oscillations and then come to rest at the singularity. The stable domain will be indicated by parallel,

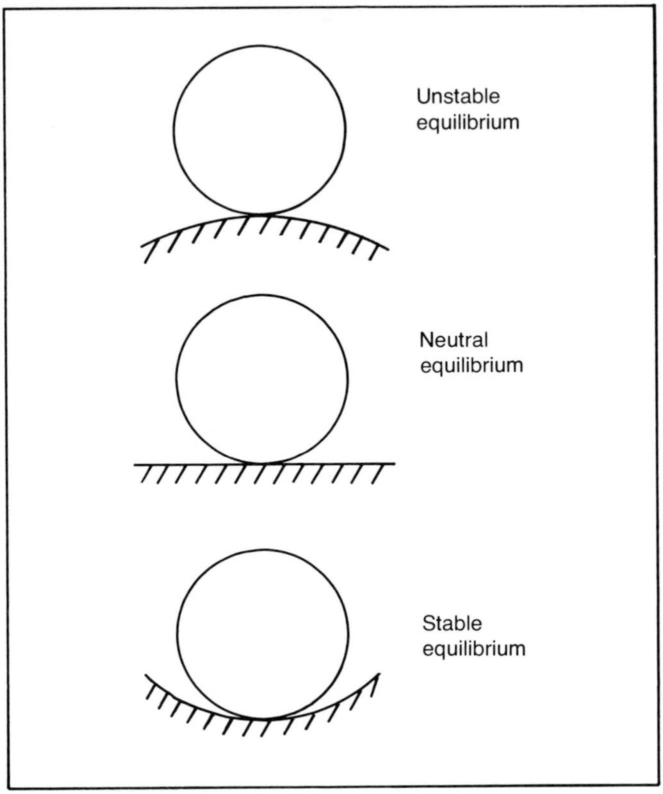

FIG. 7.9 Classes of equilibrium

diagonal lines. If the mass is given any initial condition which does not lie on one of the focus paths, the mass begins to oscillate and the detrusion becomes unbounded, i.e., the system experiences a failure.

Desinence

The system, $\ddot{x} + \dot{x} - 2x + 3x^2 - x^3 = 0$, has 3 singularities whose phase plane coordinates are (0, 0), (0, 2), and (0, 1). The first two singularities are unstable saddles whereas the third singularity is a stable focus.

An analog computer programmer could write and wire the circuit in less than two hours and obtain the phase plane plot in another two hours. To obtain the given dynamic respondence would require an unreasonable amount of time using digital computer routines, and the discipline of stability analysis of nonlinear systems is an application where the analog computer approach becomes most efficacious.

7.5 LIMIT CYCLE STABILITY IN NONLINEAR, NONCONSERVATIVE SYSTEMS

The limit cycle is a commonly used parameter which is useful in describing the stability of nonlinear systems. A limit cycle is an isolated closed trajectory on the phase plane diagram (x vs. $\dot{x}$) and represents a harmonic periodicity. Every trajectory in the neighborhood of a limit cycle either winds or unwinds upon the cycle. The limit cycle is stable if the nearby trajectories approach the limit cycle with increasing time, and the limit cycle is said to be unstable with the converse condition.

The difficulties ordinarily encountered in establishing the existence of limit cycles are such that first-order approximations are usually resorted to. In this way, some details of the problem, such as the existence of subharmonics, are not considered. However, the essential topological configuration can be described.

It should be noted that limit cycles can occur only in nonlinear, nonconservative systems. An interesting characteristic of these systems is that the stationary oscillatory state with self-sustained oscillations is determined by the differential equation itself, and the limit cycle is independent of the initial conditions. The stability of a nonlinear system does, however, depend upon the initial conditions. In the case of a conservative system, the stationary oscillatory state is always dependent upon the initial conditions.

The procedure that follows can be applied to any number of coupled differential equations. The manipulations with a single second-order, nonlinear homogeneous differential equation can be more easily followed, however, so a singular system was used for an example of the procedure.

Definition of the Problem

Consider the longitudinal vibrations of a weight, W, and represented by the Maxwell model shown in figure 7-10. With the detrusion, x, the spring constant of the case is

$$K = \frac{A_C E}{L} \tag{7.195}$$

Now equating the inertial force to the external spring force and the viscous resistance

$$-\frac{W}{g}\ddot{x} = Kx + [f(\dot{x})]\dot{x} \tag{7.196}$$

Consider the nonlinear viscoelastic material to be characterized by the relation:

TABLE 122
Nomenclature

Symbol	Variable	Typical units
A	Maximum excursion of an oscillation	inch
A_c	Cross sectional area of material 1	sq. in.
A_p	Cross sectional area of material 2	sq. in.
E	Young's modulus of the case	psi
E^*	Viscoelastic modulus	psi
exp	Naperian base	—
K	Spring constant	lbs./in.
L	Length	inch
t	Time	sec
W	Weight	pounds
x	Generalized coordinate	inch
$\dot{x}$	Newtonion notation for the derivative, $\frac{dx}{dt}$	in./sec.
c, α, β, γ	Constants	—
δx	Variation of x	inch
θ	Phase angle	radian
τ	Period of one oscillation	sec
ψ	Action of a system	—
φ	Deviation of A	inch

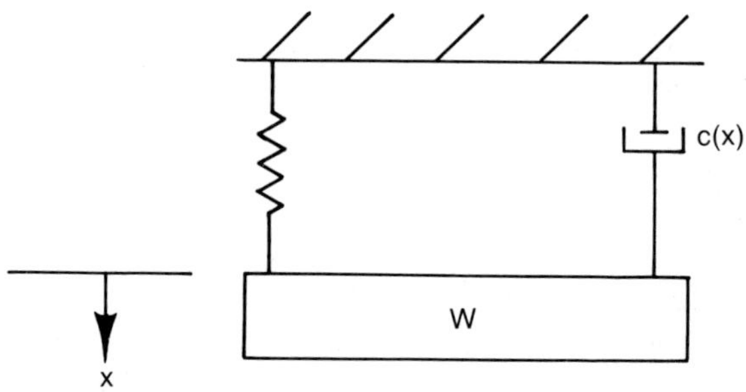

FIG. 7.10 Maxwell model

$$E^* = -c_1 + c_2\dot{x}^2 - c_3\dot{x}^4 \qquad (7.197)$$

Also consider the general relation for the viscoelastic modulus:

$$E^* = [f(\dot{x})]\frac{L}{A_p} \qquad (7.198)$$

Let

$$\alpha \equiv \frac{A_p g c_1}{LW},\ \beta \equiv \frac{A_p g c_2}{LW},\ \gamma \equiv \frac{A_p g c_3}{LW},\ \omega_0^2 \equiv \frac{K_g}{W} \qquad (7.199)$$

Substituting these identities, along with the result of the commutation of equations (7.197) and (7.198), into equation (7.196) yields:

$$\ddot{x} = \omega_0^2 x - \alpha\dot{x} + \beta\dot{x}^3 - \gamma\dot{x}^5 \tag{7.200}$$

Further, if the normalized spring constant of the case is taken as unity, the working differential equation is

$$x + \alpha\dot{x} - \beta\dot{x}^3 + \gamma\dot{x}^5 + x = 0 \tag{7.201}$$

The desideratum can now be explicitly stated:

1. Assuming γ, β, α are constants greater than zero (but very small), find the relations for the constants that will ensure the existence of limit cycles.
2. Determine the relations for the constants which indicate stability.
3. Provide the topological configuration.

Existence of Limit Cycles

Assume that equation (7.201) has a solution of the form

$$x = A(t)\cos[\omega t + \theta(t)] \tag{7.202}$$

where $A(t)$, $\theta(t)$ vary slowly. Then obtain

$$\dot{x} = \dot{A}\cos(\omega t + \theta) - A(\omega + \dot{\theta})\sin(\omega t + \theta) \tag{7.203}$$

$$\begin{aligned}\ddot{x} = {} & \ddot{A}\cos(\omega t + \theta) - \dot{A}(\omega + \dot{\theta})\sin(\omega t + \theta) - \dot{A}(\omega + \dot{\theta}) \\ & \times \sin(\omega t + \theta) - A(\ddot{\theta})\sin(\omega t + \theta) - A(\omega + \dot{\theta})^2 \\ & \times \cos(\omega t + \theta)\end{aligned} \tag{7.204}$$

Neglecting second-order terms, equation (7.204) becomes

$$\begin{aligned}\ddot{x} = {} & [\ddot{A} - A(\omega^2 + 2\,\omega\dot{\theta})][\cos(\omega t + \theta)] - [2\dot{A}\omega + A\ddot{\theta}] \\ & \times [\sin(\omega t + \theta)]\end{aligned} \tag{7.205}$$

The variation of x is

$$\delta x = \delta A\cos(\omega t + \theta) - A\delta\theta\sin(\omega t + \theta) \tag{7.206}$$

since $\delta\omega = 0$. Using the form (7.207)

$$(a + b)^3 = a^3 + 3a^2b + 3ab^2 + b^3 \tag{7.208}$$

and neglecting second-order terms:

$$\begin{aligned}\dot{x}^3 \doteq {} & [3\dot{A}A^2\omega^2][\cos(\omega t + \theta)\sin^2(\omega t + \dot{\theta})] \\ & - [A^3\omega^3 + 3A^3\omega^2\dot{\theta}][\sin^3(\omega t + \theta)]\end{aligned} \tag{7.209}$$

$$\begin{aligned}\dot{x}^5 = {} & [5\dot{A}A^4\omega^4][\cos(\omega t + \theta)\sin^4(\omega t + \theta)] \\ & - [A^5\omega^5 + 5A^5\omega^4\dot{\theta}] \\ & \times [\sin^5(\omega t + \theta)]\end{aligned} \tag{7.210}$$

The action of the system is given by the left-hand side of equation (7.201). Now considering the action, ψ, over one period, τ:

$$\int_0^\tau \psi \delta x \, dt = 0 \tag{7.211}$$

Letting

$$z = \omega t + \theta$$

then

$$\frac{dz}{dt} = \omega + \dot{\theta}$$

and

$$dt = \frac{dz}{\omega + \dot{\theta}} \tag{7.212}$$

The limits in equation (7.211) are 0 to 2π. Substituting equations (7.202), (7.203), (7.204), (7.209), (7.210) and (7.212), into equation (7.211) with the transformed limits:

$$\begin{aligned} 0 = \int_0^{2\pi} & \{(\cos z)(\ddot{A} - A\omega^2 - 2A\omega\dot{\theta} + \alpha\dot{A} + A) \\ & + (\cos z)(\sin^2 z)(-3\beta A \dot{A}^2 \omega^2) \\ & + (\cos z)(\sin^4 z)(5\gamma \dot{A} A^4 \omega^4) \\ & + (\sin z)(-2\dot{A}\omega - A\ddot{\theta} - \alpha A\omega - \alpha A\dot{\theta}) \\ & + (\sin^3 z)(\beta A^3 \omega^3 + 3\beta A^3 \omega^2 \dot{\theta}) \\ & + (\sin^5 z)(-\gamma A^5 \omega^5 - 5\gamma A^5 \omega^4 \dot{\theta})\} \\ & \begin{Bmatrix} (\cos z)\delta A \\ -A(\sin z)\delta\theta \end{Bmatrix} \left\{ \frac{dz}{\omega + \dot{\theta}} \right\} \end{aligned} \tag{7.213}$$

To integrate equation (7.213), consider the average values of θ, $\dot{\theta}$, $\ddot{\theta}$, A, $\dot{A}$, $\ddot{A}$ over one cycle and use the forms

$$\int_0^{2\pi} \sin^n u du = \frac{n-1}{n} \int_0^{2\pi} \sin^{n-2} u du \tag{7.214}$$

$$\int_0^{2\pi} \sin^2 u du = \int_0^{2\pi} \cos^2 u du = \pi \tag{7.215}$$

Then,

$$\int_0^{2\pi} \sin^4 u du = \frac{3}{4} \int_{-}^{2\pi} \sin^2 u du = \frac{3\pi}{4} \tag{7.216}$$

and

$$\int_0^{2\pi} \sin^6 u du = \frac{5}{6} \quad \int_0^{2\pi} \sin^4 u du = \frac{5\pi}{8} \tag{7.217}$$

$$\int_0^{2\pi} (\sin u)(\cos u)\, du = 0, \int_0^{2\pi} (\sin u)^2 (\cos u)^2 du = \frac{\pi}{4},$$

$$\int_0^{2\pi} (\sin u)^4 (\cos u)^2 du = \frac{\pi}{8} \tag{7.218}$$

The variations of A and θ, i.e., δA, $\delta\theta$, are arbitrary and are therefore taken as constants. Integrating equation (7.213) on $\delta\theta$:

$$\left(\frac{\pi}{\omega + \dot{\theta}}\right)\left(\ddot{A} - A\omega^2 - 2A\omega\dot{\theta} + \alpha A + \omega_o^2 \dot{A} - \frac{3}{4}\beta \dot{A} A^2 \omega^2 \right.$$
$$\left. + \frac{5}{8}\gamma \dot{A} A^4 \omega^4\right) = 0 \tag{7.219}$$

Integrating equation (7.213) on $\delta\theta$:

$$\left(\frac{-A\pi}{\omega + \dot{\theta}}\right)\left(-2\dot{A}\omega - A\ddot{\theta} - \alpha A\omega - \alpha A\dot{\theta} + \frac{3\beta A^3\omega^3}{4} + \frac{9}{4}\beta A^3 \omega^2 \dot{\theta}\right.$$
$$\left. + \frac{5}{8}[-\gamma A^5 \omega^5 - 5\gamma A^5 \omega^4 \dot{\theta}]\right) = 0 \tag{7.220}$$

For steady state, i.e., in the limit cycle,

$$\dot{A} = \ddot{A} = \dot{\theta} = \ddot{\theta} = 0 \tag{7.221}$$

Supplanting the derivatives in equation (7.219) by the relations in equation (7.221):

$$-A\omega^2 + \omega_o^2 A = 0 \tag{7.222}$$

which leads to

$$\omega^2 = \omega_o^2 = 1 \tag{7.223}$$

Supplanting the derivatives in equation (7.220) by the relations in equation (7.221):

$$-\alpha A\omega + \frac{3\beta A^3 \omega^3}{4} - \frac{5\gamma A^5 \omega^5}{8} = 0 \tag{7.224}$$

Multiplying equation (7.224) by $\frac{8}{5\gamma A}$ and using identity (7.223):

$$A^4 - \frac{6\beta A^2}{5\gamma} + \frac{8\alpha}{5\gamma} = 0 \tag{7.225}$$

Since equation (7.225) is a quadratic form,

$$A^2 = \frac{3\beta}{5\gamma} \pm \left(\frac{9\beta^2}{25\gamma^2} - \frac{8\alpha}{5\gamma}\right)^{1/2} \tag{7.226}$$

Observing equation (7.226), obviously one limit cycle exists when

$$\frac{9\beta^2}{25\gamma^2} = \frac{8\alpha}{5\gamma} \tag{7.227}$$

and two limit cycles exist when

$$\frac{9\beta^2}{25\gamma^2} > \frac{8\alpha}{5\gamma} \tag{7.228}$$

Stability of Limit Cycles

Consider a perturbation of the amplitude,

$$A = A_o + \varphi \text{ where } A_o \text{ is a constant.} \tag{7.229}$$

Substituting relation (7.229) into equations (7.219) and (7.220) using equation (7.223) and neglecting second-order terms:

$$-2A_o\dot{\theta} + \alpha\dot{\varphi} - \frac{3\beta\dot{\varphi}A_o^2}{4} + \frac{5\gamma\dot{\varphi}A_o^4}{8} = 0 \tag{7.230}$$

$$-2\varphi - \alpha A_o - \alpha\varphi - \alpha A_o\dot{\theta} + \frac{3\beta A_o^3}{4} + \frac{9\beta A_o^2\varphi}{4} + \frac{9\beta A_o^3\dot{\theta}}{4}$$

$$- \frac{5\gamma A_o^5}{8} - \frac{25\gamma A_o^4\varphi}{8} - \frac{25\gamma\theta A_o^5}{8} = 0 \tag{7.231}$$

Consider the identities:

$$\rho \equiv \alpha - \frac{3\beta A_o^2}{4} + \frac{5\gamma A_o^4}{8} \tag{7.232}$$

$$\Gamma \equiv + - \alpha + \frac{9\beta A_o^2}{4} - \frac{25\gamma A_o^4}{8} \tag{7.233}$$

Subrogating relations (7.232) and (7.233) into equations (7.230) and (7.231):

$$-2A_o\dot{\theta} + \dot{\varphi}\rho = 0 \tag{7.234}$$

$$-2\dot{\varphi} + \varphi\Gamma + \dot{\theta}A_o\Gamma + A_o\rho = 0 \tag{7.235}$$

From equation 40, $\dot{\theta} = \frac{\dot{\varphi}\rho}{2A_o}$ (7.236)

Inserting relation (7.236) into equation (7.235) and rearranging:

$$\dot{\varphi} + \varphi\left(\frac{2\Gamma}{\rho\Gamma - 4}\right) + \frac{2A_o\rho}{\rho\Gamma - 4} = 0 \tag{7.237}$$

The solution of equation (7.237) is

$$\varphi = N(\exp \lambda t) - \frac{2A_o\rho}{\rho\Gamma - 4} \tag{7.238}$$

Substituting equation (7.238) into equation (7.237) and rearranging, yields:

$$\lambda = \frac{2\Gamma}{4 - \rho\Gamma} \tag{7.239}$$

Observing equation (7.238) and considering equation (7.239), the limit cycle is seen to be stable if

$$\frac{2\Gamma}{4 - \rho\Gamma} < 0 \tag{7.240}$$

where Γ, ρ are evaluated from equations (7.232) and (7.233) and the constant A_o in Γ, ρ is evaluated using equation (7.226), i.e.,

$$A_o = \left[\frac{3\beta}{5\gamma} \pm \left(\frac{9\beta^2}{25\gamma^2} - \frac{8\alpha}{5\gamma}\right)^{1/2}\right]^{1/2} \tag{7.241}$$

Numerical Example

Case I, Single Limit Cycle. Consider equation (7.228) in the form:

$$\beta^2 = \frac{40\alpha\gamma}{9} \tag{7.242}$$

Let $\alpha = 0.10$, $\gamma = 0.25$ (7.243)

Then $\beta = \sqrt{\frac{40(0.10)(0.25)}{9}} = 0.333$ (7.244)

Using equation (7.226), the theoretical amplitude should be

$$A = \sqrt{\frac{3\beta}{5\gamma}} = \sqrt{\frac{3(0.333)}{5(0.25)}} = \sqrt{0.8} = 0.890 \tag{7.245}$$

Using equations (7.232) and (7.233):

$$\rho = 0.10 - 0.75(0.333)(0.8) + 0.625(0.25)(0.64) = 0 \tag{7.246}$$

$$\begin{aligned}\Gamma = &-0.10 + 2.25(0.333)(0.8)\\ &- 3.125(0.25)(0.64) = 0.0667\end{aligned} \tag{7.247}$$

From equation (7.239):

$$\lambda = \frac{2(0.0667)}{4 - 0(0.0667)} > 0 \tag{7.248}$$

Hence, by equation (7.240), the limit cycle is unstable.

Case II, Two Limit Cycles. Let $\alpha = 0.10, \gamma = 0.25, \beta = 0.50$ (7.249)

which satisfies equation (7.228) The theoretical amplitude is (using equation (7.226):

$$A^2 = \frac{3(0.50)}{5(0.25)} \pm \left[\frac{9(0.50^2)}{25(0.25^2)} - \frac{8(0.10)}{5(0.25)}\right]^{1/2}$$

$$= 1.20 \pm 0.890 = 2.09, 0.31 \quad (7.250)$$

Therefore,

$$A_1 = 1.46, A_2 = 0.555 \quad (7.251)$$

Again, using equations (7.232), (7.233), and (7.239) for A_1 and A_2:

$$\rho_1 = 0.10 - 0.75(0.50)(2.09) + 0.625(0.25)(2.09^2) = 0.003 \quad (7.252)$$

$$\rho_2 = 0.10 - 0.75(0.50)(0.31) + 0.625(0.25)(0.31^2) = 0.0071 \quad (7.253)$$

$$\Gamma_1 = -0.10 + 2.25(0.50)(2.09) - 3.125(0.25)(2.09^4) = -13.0 \quad (7.254)$$

$$\Gamma_2 = -0.10 + 2.25(0.50)(0.31) - 3.125(0.25)(0.31^4) = 0.247 \quad (7.255)$$

$$\lambda_1 = \frac{2(-13.0)}{4 - 0.003(-13.0)} = -6.55 < 0 \quad (7.256)$$

$$\lambda_2 = \frac{2(0.247)}{4 - 0.0071(0.247)} = 0.124 > 0 \quad (7.257)$$

Therefore, A_1 limit cycle is stable and A_2 limit cycle is unstable.

Discussion of Results

The wiring schematic of the analog computer is given by figure 7.11, and the phase-plane plots for Cases I and II are given by figures 7.12 and 7.13, respectively. A useful method for obtaining a complete trace of an unstable limit cycle is to reverse time, i.e., switch leads of odd-powered, time-dependent variables. This was done in the experiments, but the slightly unstable electronic components caused two distinct, but close, limit cycles as shown in figure 7.12, rather than a single cycle which should have been plotted. The average analog amplitude of 1.03 (1.01 and 1.05 from figure 7.12) was about 10% greater than the predicted theoretical value of 0.89.

The limit cycle of Case I is experimentally shown to be unstable since the trajectory unwinds from the limit cycle in the closed do-

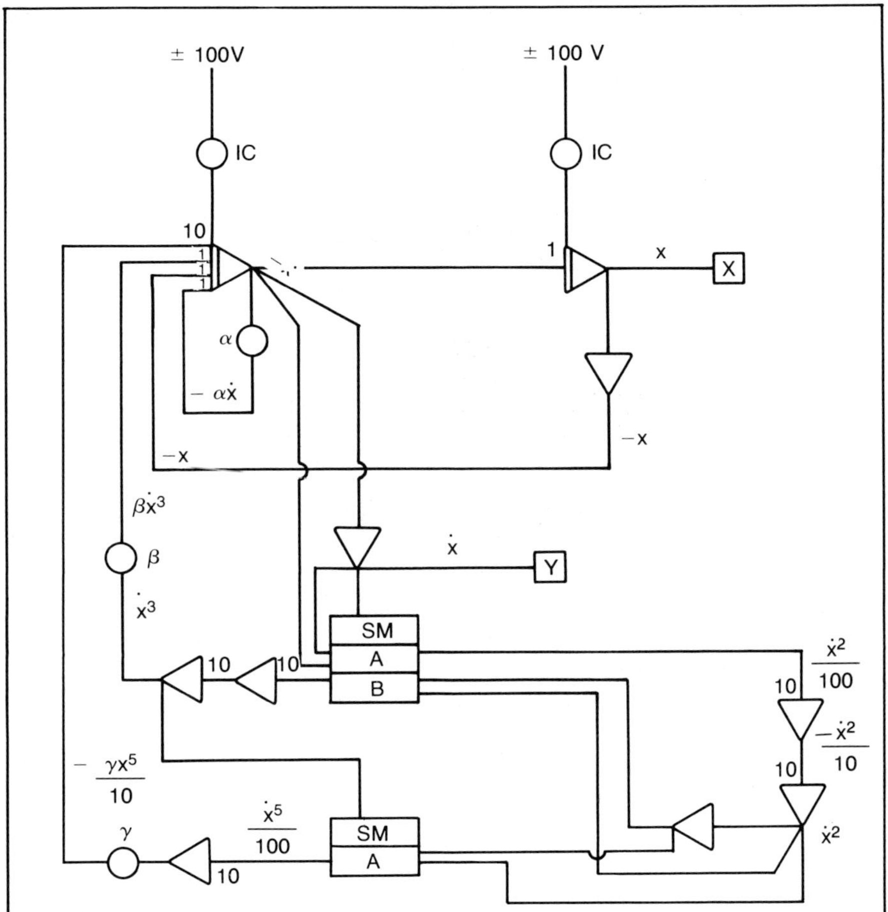

$\ddot{x} = -\alpha\dot{x} + \beta\dot{x}^3 - \gamma\dot{x}^5 - x$

XY plotter set at 0.5 volt/in.

FIG. 7.11 Analog wiring schematic

main. However, this type of a limit cycle is sometimes called semistable since the trajectory winds onto the cycle for initial conditions outside of the closed domain.

As shown by figure 7.13 (for Case II), the analog yielded two distinct and separated limit cycles, as predicted. The average experimental amplitudes were 0.65 and 1.40 which compare most favorably with the theoretical values of 0.56 and 1.46. Figure 7.13 shows that the trajectory winds onto the outer limit cycle for any initial conditions, which indicates stability as predicted. Since the tra-

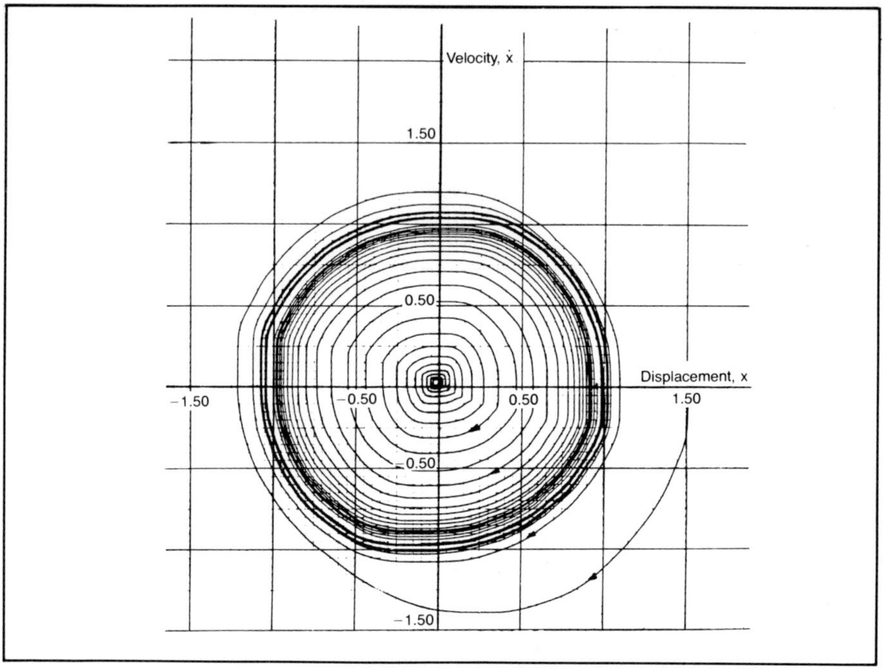

FIG. 7.12 Phase plane character, beta = 0.35

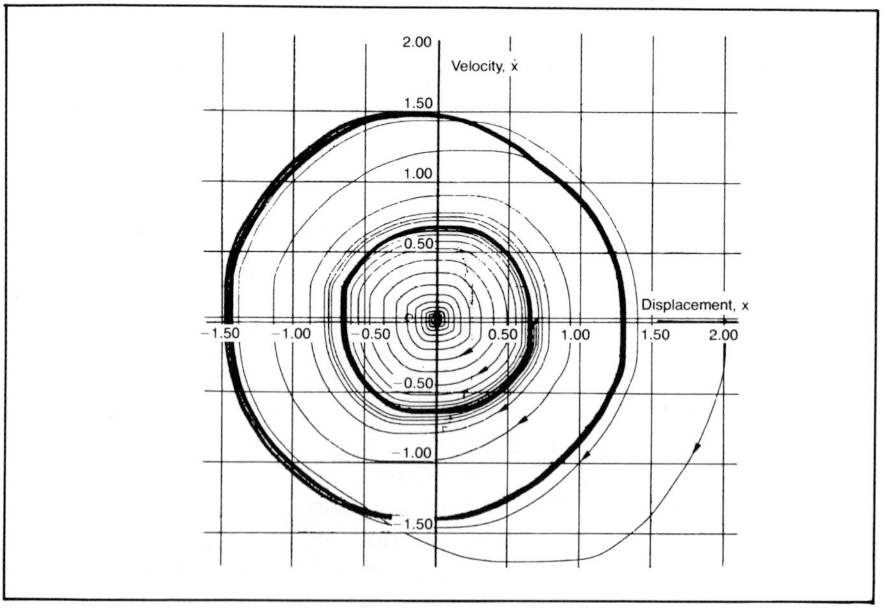

FIG. 7.13 Phase plane character, beta = 0.50

jectories unwind from the inner limit cycle within its domain, this cycle is said to be unstable, also as theoretically predicted.

7.6 EIGEN SUMMARIES USING MATRIX DEFLATION

In vibration problems, the equations of motion in terms of the stiffness matrix are given as

$$[S_{ij}]\{x_i\} = \lambda_n[m_i]\{V_i\} \tag{7.258}$$

and $\lambda_n\{v_i\}$ can be called the Eigen summary. In terms of the flexibility matrix, the usual notation is of the form

$$\{x_i\} = \lambda_n[m_i][F_{ij}]\{V_i\} \tag{7.259}$$

With the stiffness formulation, matrix iteration converges to the largest Eigen value, whereas convergence is to the smallest Eigen value with the flexibility matrix.

The idea of using matrix iteration to obtain Eigen values is merely an alternate procedure to the direct method of finding the roots of a characteristic equation. The advantage of the iterative process is that the complete Eigen summary is obtained at convergence. A disadvantage is that more computations are required than by the direct method. However, since only the Eigen values are obtained from the characteristic equation, further computations are required to obtain the Eigen vectors. The Eigen summaries must be known in order to conduct certain analyses, for instance, failure analysis.

The commonly used iterative procedure to obtain successive Eigen summaries is to repeatedly impose the orthogonality condition on the complete matrix.[(7)] For a system having n degrees of freedom (i.e., n generalized coordinates), the orthogonality condition is

$$\sum_{i=1}^{n} m_i[V_i^n][V_i^m] = 0 \tag{7.260}$$

where n and m indicate mode numbers and $n \neq m$.

For example, assume that the Eigen summary for the first mode has been obtained by iteration of a flexibility formulation. Then, equation (7.260) can be expressed as

$$\{x_i\} = [C_{ij}]\{V_i\} \tag{7.261}$$

Next, equation (7.261) is combined with the original formulation, i.e.,

$$\{x_i\} = \lambda_n[m_i][F_{ij}][C_{ij}]\{V_i\} \tag{7.262}$$

TABLE 123
Nomenclature

Symbol	Variable	Typical units
C	Orthogonal coefficient matrix	numeric
F	Flexibility matrix	numeric
i	Mass or coordinate subscript	numeric
k	Spring constant	lbs/inch
m	Mass	lb
n	Mode subscript	numeric
P	Force	lb
S	Stiffness matrix	numeric
T	Matrix trace	numeric
t	Time	sec
u	Excursion	inch
V	Eigen vector	inch
x	Detrusion	inch
λ	Eigen value	numeric
ω	Angular velocity	rad/arc

After the flexibility matrix and orthogonal coefficient matrix have been combined, iteration of the resultant formulation will converge to the second Eigen summary. The procedure is continued until all Eigen summaries have been calculated.

An alternative to the above procedure is a process called matrix deflation which is illustrated by two examples in the following sections. Comparing the two methods, the accumulative round-off error and the turn-around time on a digital computer should be less when using matrix deflation.

Eigen Summaries Using a Stiffness Matrix

Consider that all the springs in the system of figure 7.14 are initially in compression, i.e.,

$$x_1 > x_2 > x_3 > x_4 \qquad (7.263)$$

Let

$$K_1 = 2K, K_2 = 2K, K_3 = K \qquad (7.264)$$

Let

$$m_i = m, \text{ i.e., } m_1 = m_2 = m_3 = m_4 = m \qquad (7.265)$$

Using Newton's second law, $-m\ddot{x} = P$ (7.266)

the equations of motion, then, are

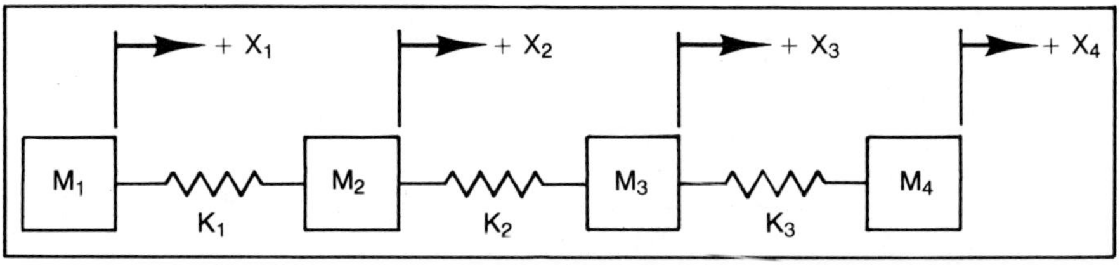

FIG. 7.14 Generalized coordinates for system A

$$m\ddot{x}_1 = -2K(x_1 - x_2) \tag{7.267}$$

$$m\ddot{x}_2 = 2K(x_1 - x_2) - 2K(x_2 - x_3) \tag{7.268}$$

$$m\ddot{x}_3 = 2K(x_2 - x_3) - K(x_3 - x_4) \tag{7.269}$$

$$m\ddot{x}_4 = K(x_3 - x_4) \tag{7.270}$$

Since there are four generalized coordinates, four Eigen summaries can be obtained. Let the mode subscripts be such that

$$w_4 > w_3 > w_2 > w_1 \tag{7.271}$$

Assuming harmonic motion, the solution of equations (7.267) to (7.270) will be of the form

$$u_i = x_i \cos w_n t \tag{7.272}$$

Therefore,

$$\ddot{u}_i = -\omega_n^2 x_i \cos w_n t \tag{7.273}$$

Substituting equation (7.273) into equations (7.267) to (7.270), the equations of motion in matrix notation are

$$-w_n^2 m \begin{bmatrix} 1 & 0 & 0 & 0 \\ 0 & 1 & 0 & 0 \\ 0 & 0 & 1 & 0 \\ 0 & 0 & 0 & 1 \end{bmatrix} \cos w_n t$$

$$= K \begin{bmatrix} -2 & 2 & 0 & 0 \\ 2 & -4 & 2 & 0 \\ 0 & 2 & -3 & 1 \\ 0 & 0 & 1 & -1 \end{bmatrix} \begin{Bmatrix} x_1 \\ x_2 \\ x_3 \\ x_4 \end{Bmatrix} \cos w_n t \tag{7.274}$$

Equation (7.274) simplifies to

$$\begin{bmatrix} 2 & -2 & 0 & 0 \\ -2 & 4 & -2 & 0 \\ 0 & -2 & 3 & -1 \\ 0 & 0 & -1 & 1 \end{bmatrix} \begin{Bmatrix} x_1 \\ x_2 \\ x_3 \\ x_4 \end{Bmatrix} = \lambda_n \begin{bmatrix} 1 & 0 & 0 & 0 \\ 0 & 1 & 0 & 0 \\ 0 & 0 & 1 & 0 \\ 0 & 0 & 0 & 1 \end{bmatrix} \begin{Bmatrix} x_1 \\ x_2 \\ x_3 \\ x_4 \end{Bmatrix} \tag{7.275}$$

where

$$\lambda_n \equiv \frac{w_n^2 m}{K} \tag{7.276}$$

If the Eigen vectors are known, the Eigen values can be calculated directly from equation (7.275), i.e.,

$$\begin{aligned} \lambda_n &= \frac{2x_1 - 2x_2}{x_1} = \frac{-2x_1 + 4x_2 - 2x_3}{x_2} \\ &= \frac{-2x_2 + 3x_3 - x_4}{x_3} = \frac{-x_3 + x_4}{x_4} \end{aligned} \tag{7.277a, b, c, d}$$

The smallest Eigen value in a system having no fixed point is zero, hence

$$\lambda_1 = 0 \tag{7.278}$$

This corresponds to equilibrium conditions. The resultant Eigen vectors (normalized with respect to x_1) for this mode and for the other modes are depicted in figure 7.15.

A node is a point in a system which does not move with dynamic conditions. The number of nodes in any particular mode equals one less than the mode order. The node locations for the given system are indicated by the circles in figure 7.15.

Assuming an initial Eigen vector of

$$\begin{Bmatrix} 1.0 \\ -1.5 \\ 2.0 \\ -0.5 \end{Bmatrix}$$

repeated iteration of equation (7.275) converges to the Eigen summary,

$$\lambda_4\{V_i\} = 6.25 \begin{Bmatrix} 1.00 \\ -2.12 \\ 1.39 \\ -0.264 \end{Bmatrix} \tag{7.279}$$

Now that the Eigen summary of the fourth mode has been determined, the stiffness formulation can be transformed by matrix deflation so that convergence of the iteration will be to the third mode. A consideration of the orthogonality relation[(8)] for the principal modes of oscillation yields

$$(\bar{P}_{n=4})(V_{n\neq 4}) = 0 \tag{7.280}$$

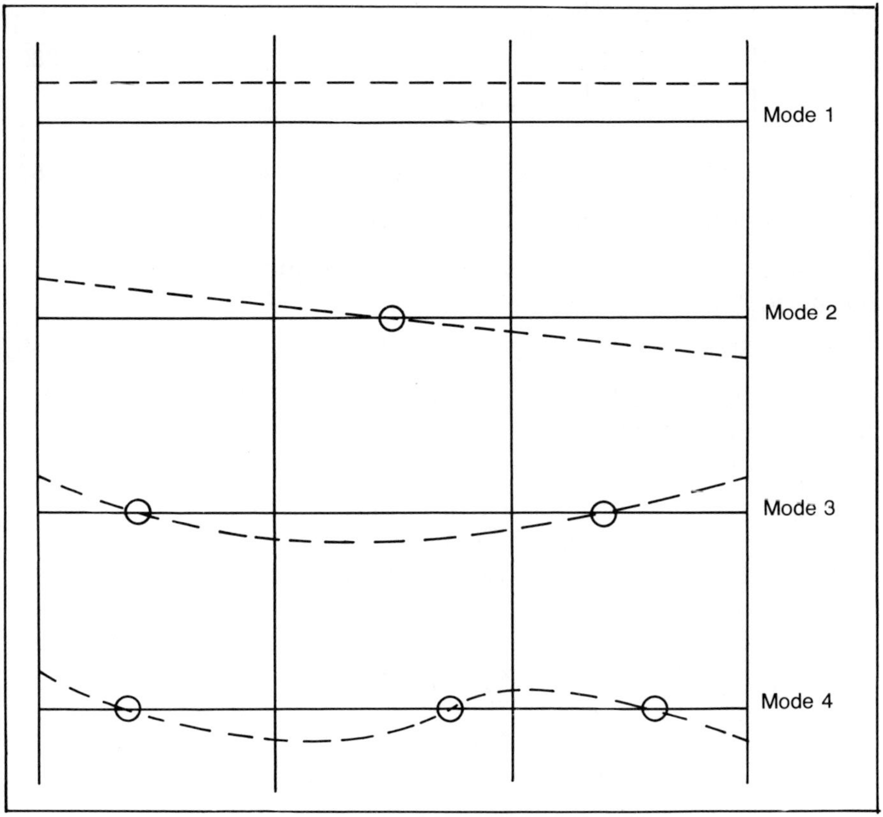

FIG. 7.15 Mode shapes of system A

$$\therefore w_4^2 m \begin{bmatrix} 1 & 0 & 0 & 0 \\ 0 & 1 & 0 & 0 \\ 0 & 0 & 1 & 0 \\ 0 & 0 & 0 & 1 \end{bmatrix} \begin{Bmatrix} x_1 \\ x_2 \\ x_3 \\ x_4 \end{Bmatrix}_{n=4} \cos w_4 t$$

$$\times \begin{bmatrix} 1 & 0 & 0 & 0 \\ 0 & 1 & 0 & 0 \\ 0 & 0 & 1 & 0 \\ 0 & 0 & 0 & 1 \end{bmatrix} \begin{Bmatrix} x_1 \\ x_2 \\ x_3 \\ x_4 \end{Bmatrix} \cos w_n t_{n=4} = 0 \tag{7.281}$$

$$\Rightarrow \begin{bmatrix} 1.00 & 0 & 0 & 0 \\ 0 & -2.12 & 0 & 0 \\ 0 & 0 & 1.39 & 0 \\ 0 & 0 & 0 & -0.264 \end{bmatrix} \begin{Bmatrix} 1 \\ 1 \\ 1 \\ 1 \end{Bmatrix}_{n=4}$$

$$\times \begin{bmatrix} 1 & 0 & 0 & 0 \\ 0 & 1 & 0 & 0 \\ 0 & 0 & 1 & 0 \\ 0 & 0 & 0 & 1 \end{bmatrix} \begin{Bmatrix} x_1 \\ x_2 \\ x_3 \\ x_4 \end{Bmatrix}_{n \neq 4} = 0 \tag{7.282}$$

$$\therefore 0 = 1.00\,x_1 = -2.12\,x_2 = 1.39\,x_3 = -0.264\,x_4 \tag{7.283}$$

$$\Rightarrow 1.00\,x_1 - 2.12\,x_2 + 1.39\,x_3 - 0.264\,x_4 = 0,\ n \neq 4 \tag{7.284}$$

Now normalize this equation so that the coefficient of x_4 is the negative of S_{34} in equation (7.275):

$$-3.78\,x_1 + 8.03\,x_2 - 5.25\,x_3 + 1.00\,x_4 = 0,\ n \neq 4 \tag{7.285}$$

Replacing the last equation of the original formulation, equation (7.275), with equation (7.285) yields

$$\begin{bmatrix} 2 & -2 & 0 & 0 \\ -2 & 4 & -2 & 0 \\ 0 & -2 & 3 & -1 \\ -3.78 & 8.03 & -5.25 & 1.00 \end{bmatrix} \begin{Bmatrix} x_1 \\ x_2 \\ x_3 \\ x_4 \end{Bmatrix}$$

$$= \lambda \begin{bmatrix} 1 & 0 & 0 & 0 \\ 0 & 1 & 0 & 0 \\ 0 & 0 & 1 & 0 \\ 0 & 0 & 0 & 0 \end{bmatrix} \begin{Bmatrix} x_1 \\ x_2 \\ x_3 \\ x_4 \end{Bmatrix},\ n \neq 4 \tag{7.286}$$

Adding the third and fourth rows,

$$\begin{bmatrix} 2 & -2 & 0 \\ -2 & 4 & -2 \\ -3.78 & 6.03 & -2.25 \end{bmatrix} \begin{Bmatrix} x_1 \\ x_2 \\ x_3 \end{Bmatrix} = \lambda \begin{Bmatrix} x_1 \\ x_2 \\ x_3 \end{Bmatrix},\ n = 4 \tag{7.287}$$

Assuming an initial Eigen vector set for the third mode of

$$\begin{Bmatrix} 1.00 \\ -0.50 \\ -1.25 \end{Bmatrix}$$

repeated iteration of equation (7.287) converges to the Eigen summary,

$$\lambda_3\{v_i\} = 2.85 \begin{Bmatrix} 1.00 \\ -0.427 \\ -1.24 \end{Bmatrix} \tag{7.288}$$

Using equation (7.277d),

$$2.85 = \frac{1.24 + x_4}{x_4} \tag{7.289}$$

or

$$x_4 = 0.671, \tag{7.290}$$

the complete Eigen summary of the third mode then is

$$\lambda_3\{v_i\} = 2.85 \begin{Bmatrix} 1.00 \\ -0.427 \\ -1.24 \\ 0.671 \end{Bmatrix} \tag{7.291}$$

To obtain λ_2, the matrix can be deflated again using the orthogonality condition,

$$(\bar{P}_{n=3})(V_{n\neq 3}) = 0 \tag{7.292}$$

$$\therefore \begin{bmatrix} 1.00 & 0 & 0 & 0 \\ 0 & -4.27 & 0 & 0 \\ 0 & 0 & -1.24 & 0 \\ 0 & 0 & 0 & 0.671 \end{bmatrix} \begin{Bmatrix} 1 \\ 1 \\ 1 \\ 1 \end{Bmatrix}_{n=3} \times \begin{bmatrix} 1 & 0 & 0 & 0 \\ 0 & 1 & 0 & 0 \\ 0 & 0 & 1 & 0 \\ 0 & 0 & 0 & 1 \end{bmatrix} \begin{Bmatrix} x_1 \\ x_2 \\ x_3 \\ x_4 \end{Bmatrix}_{n\neq 3} = 0 \tag{7.293}$$

Hence,

$$1.00\,x_1 - 0.427\,x_2 - 1.24\,x_3 + 0.671\,x_4 = 0,\ n \neq 3 \tag{7.294}$$

Normalizing this equation with respect to x_4,

$$1.49\,x_1 - 0.636\,x_2 - 1.85\,x_3 + 1.00\,x_4 = 0,\ n \neq 3 \tag{7.295}$$

Subtracting equations (7.285) and (7.295) to eliminate X_4:

$$-5.27\,x_1 + 8.66\,x_2 - 3.39\,x_3 = 0,\ n \neq 3, 4 \tag{7.296}$$

Normalize equation (7.296) so that the coefficient of X_3 is the negative of S_{13} in equation (7.287), i.e.,

$$3.10\,x_1 - 5.10\,x_2 + 2.00\,x_3 = 0,\ n \neq 3, 4 \tag{7.297}$$

Substituting equation (7.297) for the third relation in equation (7.287):

$$\begin{bmatrix} 2 & -2 & 0 \\ -2 & 4 & -2 \\ 3.10 & -5.10 & 2.00 \end{bmatrix} \begin{Bmatrix} x_1 \\ x_2 \\ x_3 \end{Bmatrix} = \lambda \begin{bmatrix} 1 & 0 & 0 \\ 0 & 1 & 0 \\ 0 & 0 & 0 \end{bmatrix} \begin{Bmatrix} x_1 \\ x_2 \\ x_3 \end{Bmatrix},\ n \neq 3, 4 \tag{7.298}$$

Adding the second and third relations of this equation yields

$$\begin{bmatrix} 2 & -2 \\ 1.10 & -1.10 \end{bmatrix} \begin{Bmatrix} x_1 \\ x_2 \end{Bmatrix} = \lambda_2 \begin{Bmatrix} x_1 \\ x_2 \end{Bmatrix},\ n \neq 3, 4 \tag{7.299}$$

Assuming an initial Eigen vector set for the second mode of

$$\begin{Bmatrix} 1.00 \\ 0.50 \end{Bmatrix}$$

repeated iteration of equation (7.299) converges to the Eigen summary,

$$\lambda_2\{v_i\} = 0.900 \begin{Bmatrix} 1.00 \\ 0.55 \end{Bmatrix} \tag{7.300}$$

Now find X_3 by equation (7.277b):

$$0.900 = \frac{-2(1.00) + 4(0.55) - 2x_3}{0.55} \tag{7.301}$$

$$\therefore x_3 = -0.295 \tag{7.302}$$

Find X_4 by equation (7.277d):

$$0.900 = \frac{0.295 + x_4}{x_4} \tag{7.303}$$

$$\therefore x_4 = -2.95 \tag{7.304}$$

The complete Eigen summary of the second mode, then, is

$$\lambda_2\{v_i\} = 0.900 \begin{Bmatrix} 1.00 \\ 0.55 \\ -0.295 \\ -2.95 \end{Bmatrix} \tag{7.305}$$

Consider,

$$T \equiv \Sigma\, s_{ii} \equiv \Sigma\, \lambda_n \tag{7.306}$$

In the given matrix,

$$\Sigma\, s_{ii} = 2 + 4 + 3 + 1 = 10 \tag{7.307}$$

To check the calculations, sum the computed Eigen vectors:

$$\Sigma\, \lambda_n = 0 + 0.900 + 2.85 + 6.25 = 10.00 \tag{7.308}$$

Trace Transforms

In very large order matrices, the roundoff error in the computations are ordinarily devastating. This error can be quite significant with matrix deflation. To obviate this difficulty, double precision arithmetic is commonly used and/or Eigen summaries for the lower frequencies are calculated from the flexibility formulation and the higher frequencies are calculated from the stiffness formulation.

The required two sets of influence coefficients can be obtained by three methods.

a. Initially formulate both the flexibility influence coefficients and the stiffness influence coefficients;

b. Initially formulate one set of the influence coefficients and then transform that formulation by matrix inversion to obtain the second set;

c. Initially formulate one set of the influence coefficients and then employ trace transforms on that set to obtain the second set.

Comparatively, formulation of the two influence matrices is inefficient for matrices of large order, and inverting a large matrix is not playing the game very well since the round-off error in the inversion can be most significant. The method of trace transforms is quite simple and, therefore, appears to be the most efficacious technique for obtaining all of the Eigen summaries in large order matrices. The method of trace transforms is as follows.

The stiffness formulation for the example above can be written as [see equations (7.310) to (7.325)].

$$\begin{bmatrix} 2 & -2 & 0 & 0 \\ -2 & 4 & -2 & 0 \\ 0 & -2 & 3 & -1 \\ 0 & 0 & -1 & 1 \end{bmatrix} \begin{Bmatrix} x_1 \\ x_2 \\ x_3 \\ x_4 \end{Bmatrix} = \lambda_n \begin{Bmatrix} x_1 \\ x_2 \\ x_3 \\ x_4 \end{Bmatrix} \tag{7.309}$$

(7.310)

If T = 10, define the transformed Eigen value as

$$\lambda_n^1 = \lambda_n - T = \lambda_n - 10 \tag{7.311}$$

Subtract the trace value from each element in the trace using equation (7.311) to obtain the transform of equation (7.309):

$$\begin{bmatrix} -8 & -2 & 0 & 0 \\ -2 & -6 & -2 & 0 \\ 0 & -2 & -7 & -1 \\ 0 & 0 & -1 & 1 \end{bmatrix} \begin{Bmatrix} x_1 \\ x_2 \\ x_3 \\ x_4 \end{Bmatrix} = \lambda_n^1 \begin{Bmatrix} x_1 \\ x_2 \\ x_3 \\ x_4 \end{Bmatrix} \tag{7.312}$$

Iteration of this formulation will now yield the largest transformed Eigen value, λ', but this corresponds to the smallest true Eigen values, λ. The n/2 Eigen values of larger magnitude can be calculated by matrix iteration and deflation using the formulation in equation (7.309), whereas the n/2 Eigen values of smaller magnitude can be conveniently calculated using the formulation of equation (7.312) with equation (7.311).

Note that the Eigen vectors in equations (7.309) and (7.312) are not commutative. The convergence of the first matrix iteration on equation (7.309) and (7.312) is to the Eigen vectors corresponding to the largest frequency and to the smallest frequency, respectively.

Eigen Summaries Using a Flexibility Matrix

Consider the system having four degrees of freedom with an end point fixed. For simplicity, let

$$K_i = K \quad (7.313)$$

$$\text{i.e., } K_1 = K_2 = K_3 = K_4 \quad (7.314)$$

$$\text{Also, let } m_i = m \quad (7.315)$$

The equations of motion for the system will be set up using flexibility coefficients. A flexibility coefficient, Fij, is defined as the deflection at i due to a unit load at j. The equations will be of the form

$$[Fij]\{P_i\} = \{\mathbf{x}_i\} \quad (7.316)$$

When m_4 deflects $F_{44} = 1/K$, then masses m_3, m_2, and m_1 simply detrude by the same amount, hence

$$F_{34} = F_{24} = F_{14} = F_{44} = \frac{1}{K} \quad (7.317)$$

By Maxwell's reciprocal theorem,

$$F_{43} = F_{42} = F_{41} = F_{34} \quad (7.318)$$

The equivalent spring constant for K_3 and K_4 is

$$\frac{1}{K_e} = \frac{1}{K_3} + \frac{1}{K_4} = \frac{2}{K} \quad (7.319)$$

Therefore, when m_3 deflects $F_{33} = 2/K$, then masses m_2 and m_1 detrude by the same amount, i.e.,

$$F_{23} = F_{13} = F_{33} = \frac{2}{K} \quad (7.320)$$

$$F_{32} = F_{31} = F_{23} \quad (7.321)$$

The remaining coefficients are obtained in a similar manner. The flexibility matrix is

$$[Fij] = \frac{1}{K}\begin{bmatrix} 4 & 3 & 2 & 1 \\ 3 & 3 & 2 & 1 \\ 2 & 2 & 2 & 1 \\ 1 & 1 & 1 & 1 \end{bmatrix} \quad (7.322)$$

Using equations (7.314) to (7.323), equation (7.316) becomes

$$\frac{w_n^2}{K}\begin{bmatrix} 4 & 3 & 2 & 1 \\ 3 & 3 & 2 & 1 \\ 2 & 2 & 2 & 1 \\ 1 & 1 & 1 & 1 \end{bmatrix}\begin{bmatrix} m_1 & 0 & 0 & 0 \\ 0 & m_2 & 0 & 0 \\ 0 & 0 & m_3 & 0 \\ 0 & 0 & 0 & m_4 \end{bmatrix}\begin{Bmatrix} x_1 \\ x_2 \\ x_3 \\ x_4 \end{Bmatrix} = \begin{Bmatrix} x_1 \\ x_2 \\ x_3 \\ x_4 \end{Bmatrix} \quad (7.323)$$

Let

$$\lambda_n = \frac{K}{mw_n^2} \tag{7.324}$$

then equation (7.323) becomes

$$\begin{bmatrix} 4 & 3 & 2 & 1 \\ 3 & 3 & 2 & 1 \\ 2 & 2 & 2 & 1 \\ 1 & 1 & 1 & 1 \end{bmatrix} \begin{Bmatrix} x_1 \\ x_2 \\ x_3 \\ x_4 \end{Bmatrix} = \lambda_n \begin{Bmatrix} x_1 \\ x_2 \\ x_3 \\ x_4 \end{Bmatrix} \tag{7.325}$$

The mode shapes and node locations are shown in figure 7.16. Iteration of equation (7.325) converges to the first Eigen summary,

$$\lambda_1 \{v_i\} = 8.30 \begin{Bmatrix} 1.00 \\ 0.880 \\ 0.653 \\ 0.346 \end{Bmatrix} \tag{7.326}$$

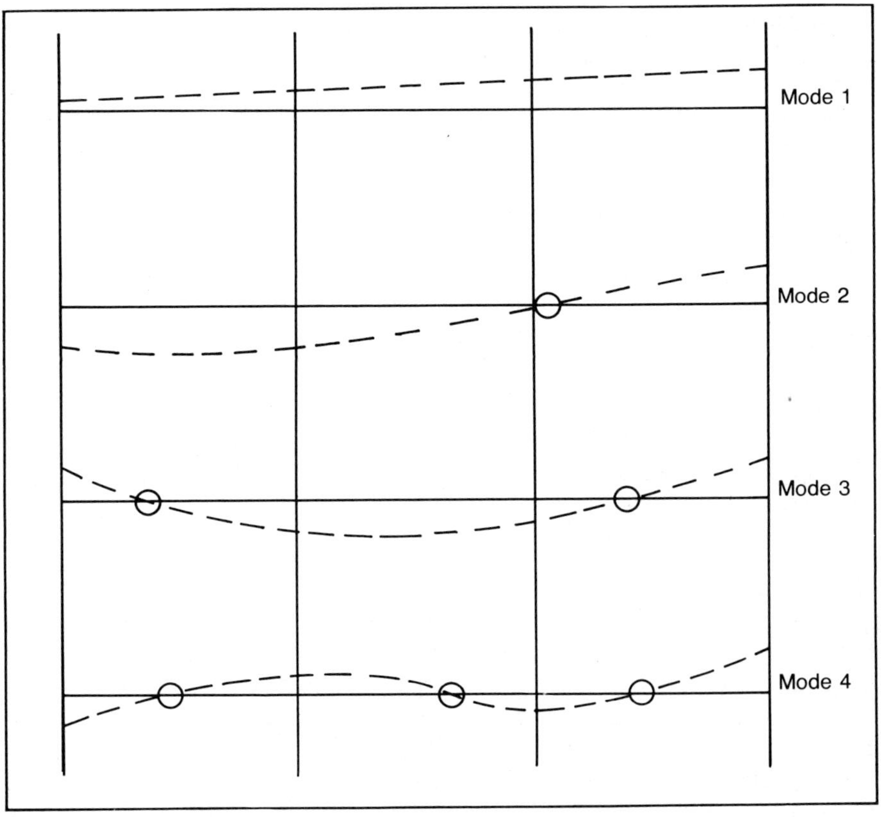

FIG. 7.16 Mode shapes of system B

Using the procedure above, the orthogonality condition is

$$x_1 + 0.880x_2 + 0.653x_3 + 0.346x_4 = 0,\ n \neq 1 \tag{7.327}$$

Using equation (7.327) to transform equation (7.325), the deflated matrix is

$$\begin{bmatrix} 0.360 & 0.040 & -0.030 \\ 0.240 & 0.690 & 0.308 \\ 0.120 & 0.347 & 0.654 \end{bmatrix} \begin{Bmatrix} x_2 \\ x_3 \\ x_4 \end{Bmatrix} = \lambda_2 \begin{Bmatrix} x_2 \\ x_3 \\ x_4 \end{Bmatrix},\ n \neq 1 \tag{7.328}$$

Iteration of equation (7.328) yields the Eigen summary

$$\lambda_2\{v_i\} = 1.00 \begin{Bmatrix} 0.015 \\ 1.00 \\ 1.00 \end{Bmatrix} \tag{7.329}$$

Substituting the iterant in equation (7.327), find that

$$x_1 = -1.012 \tag{7.330}$$

Normalizing the Eigen vectors with respect to X_1, the complete Eigen summary is

$$\lambda_2\{v_i\} = 1.00 \begin{Bmatrix} 1.00 \\ -0.015 \\ -0.99 \\ -0.99 \end{Bmatrix} \tag{7.331}$$

Continuing the process in the same manner, the remaining summaries are:

$$\lambda_3\{v_i\} = 0.442 \begin{Bmatrix} 1.00 \\ -1.42 \\ -0.353 \\ 1.38 \end{Bmatrix} \tag{7.332}$$

$$\lambda_4\{v_i\} = 0.228 \begin{Bmatrix} 1.00 \\ -2.28 \\ 2.10 \\ -1.06 \end{Bmatrix} \tag{7.333}$$

Conclusions

While only very simple examples were chosen to demonstrate the technique of matrix deflation, the process could be applied equally as well to large-order matrices. Computational effort with large matrices would probably be reduced by about 30% using matrix deflation, as opposed to continued operations on the complete matrix. The method of trace transforms is shown to be the most convenient alternative for calculating complete Eigen summaries of large-order matrices.

REFERENCES

1. Davis, H. T., *Introduction to Nonlinear Differential and Integral Equations*, 1962, Dover Publ. Inc.
2. Minorsky, N., *Nonlinear Oscillations*, 1962, Van Nostrand Publ. Inc.
3. Cunningham, W. J., *Introduction to Nonlinear Analysis*, 1958, McGraw-Hill.
4. Cunningham, W. J., *Introduction to Nonlinear Analysis*, 1958, McGraw-Hill Book Co., New York, N.Y.
5. Davis, H. T., *Introduction to Nonlinear Differential and Integral Equations*, 1962, Dover Publications, Inc., New York, N.Y.
6. Minorsky, N., *Nonlinear Oscillations*, 1962, D. Van Nostrand Co., Inc., Princeton, N.J.
7. Seto, W. W., *Theory and Problems of Mechanical Vibrations*, 1964, Schaum Publ. Co., 257 Park Avenue S., New York 10, N.Y., p. 92–98.
8. Karman, T. V., and M. A. Biot, *Mathematical Methods in Engineering*, 1940, McGraw-Hill Publ. Co., New York, p 172–178.

APPENDIX

EQUATIONS OF ELASTICITY

The strain-displacement relations for plane stress are

$$\varepsilon_z = u_z + (r - a)w_{zz} \tag{A1}$$

$$\varepsilon_\theta = \frac{V_\theta}{a} + \frac{(r - a)}{ar} w_{\theta\theta} - \frac{w}{r} \tag{A2}$$

$$\gamma_{z\theta} = \frac{u_\theta}{r} + \frac{r}{a} V_z + \frac{(r^2 - a^2)}{ar} w_{z\theta} \tag{A3}$$

$$\varepsilon_r = -\frac{\mu}{1 - \mu}(\varepsilon_z + \varepsilon_\theta) = 0 \tag{A4}$$

$$\gamma_{zr} = \gamma_{r\theta} = 0 \tag{A5}$$

The stress-strain relations are

$$\sigma_z = \frac{E}{1 - \mu^2}(\varepsilon_z + \mu\varepsilon_\theta) \tag{A6}$$

$$\sigma_\theta = \frac{E}{1 - \mu^2}(\varepsilon_\theta + \mu\varepsilon_z) \tag{A7}$$

$$\tau_{z\theta} = \frac{E}{2(1 + \mu)}(\varepsilon_z - \varepsilon_\theta) = \frac{\sigma_z - \sigma_\theta}{2} \tag{A8}$$

$$\sigma_r = \tau_{zr} = \tau_{r\theta} = 0 \tag{A9}$$